Chemical Vapour Deposition (CVD)

Series in Materials Science and Engineering

The series publishes cutting-edge monographs and foundational textbooks for interdisciplinary materials science and engineering. It is aimed at undergraduate- and graduate-level students, as well as practicing scientists and engineers. Its purpose is to address the connections between properties, structure, synthesis, processing, characterization, and performance of materials.

Aerospace Materials
Brian Cantor, H. Assender, P. Grant, Eds.

Fundamentals of Ceramics
Michel Barsoum, M.W. Barsoum

Fundamentals of Fibre-Reinforced Composite Materials
A.R. Bunsell, J. Renard

Automotive Engineering: Lightweight, Functional, and Novel Materials
Brian Cantor, P. Grant, C. Johnston

Multiferroic Materials: Properties, Techniques, and Applications
Junling Wang, Ed.

Skyrmions: Topological Structures, Properties, and Applications
J. Ping Liu, Zhidong Zhang, Guoping Zhao, Eds.

Conductive Polymers: Electrical Interactions in Cell Biology and Medicine
Ze Zhang, Mahmoud Rouabhia, Simon E. Moulton, Eds.

Computational Modelling of Inorganic Nanomaterials
Stefan T. Bromley, Martijn A. Zwijnenburg, Eds.

Physical Methods for Materials Characterisation, Third Edition
Peter E.J. Flewitt, Robert K. Wild

Advanced Thermoelectrics: Materials, Contacts, Devices, and Systems
Zhifeng Ren, Yucheng Lan, Qinyong Zhang

2D Materials for Nanoelectronics
Michel Houssa, Athanasios Dimoulas, Alessandro Molle, Eds.

Silicon Nanomaterials Sourcebook, Two-Volume Set
Klaus D. Sattler, Ed.

For more information about this series, please visit: https://www.crcpress.com/Series-in-Materials-Science-and-Engineering/book-series/TFMATSCIENG

Chemical Vapour Deposition (CVD)

Advances, Technology, and Applications

Edited by

Kwang Leong Choy

Director, UCL Institute for Materials Discovery
and
Professor of Materials Discovery
University College London, United Kingdom

CRC Press
Taylor & Francis Group
Boca Raton London New York

CRC Press is an imprint of the
Taylor & Francis Group, an **informa** business

CRC Press
Taylor & Francis Group
6000 Broken Sound Parkway NW, Suite 300
Boca Raton, FL 33487-2742

First issued in paperback 2021

ISBN 13: 978-0-367-78011-1 (pbk)
ISBN 13: 978-1-4665-9776-1 (hbk)

Library of Congress Cataloging-in-Publication Data

Names: Choy, Kwang Leong, editor.
Title: Chemical vapour deposition (CVD) : technology and applications /
Kwang Leong Choy, editor.
Description: First edition. | Boca Raton, FL : CRC Press/Taylor & Francis
Group, 2018. | Series: Series in materials science and engineering ; 22 |
"A CRC title, part of the Taylor & Francis imprint, a member of the Taylor
& Francis Group, the academic division of T&F Informa plc." | Includes
bibliographical references and index.
Identifiers: LCCN 2018013160 | ISBN 9781466597761 (hardback : acid-free paper)
Subjects: LCSH: Chemical vapor deposition. | Nanostructured materials.
Classification: LCC TS695 .C555 2018 | DDC 620.1/15--dc23
LC record available at https://lccn.loc.gov/2018013160

Visit the Taylor & Francis Web site at
http://www.taylorandfrancis.com

and the CRC Press Web site at
http://www.crcpress.com

Dedication

Thanks to Isaac, Emer and Albert for their great support and inspiration.

Contents

Preface

CVD and its variants have been used successfully as versatile tools to engineer surfaces that provide protective coatings (superthin, thin, and thick films; superlattices; and nanocomposite coatings) with advanced functionality, as well as for the processing of 0D (ultrafine particles), 1D (nanowires, nanotubes), 2D (graphene, layered materials), and 3D (hierarchical) structures, and bulk free-standing parts and composites. Engineers and scientists seeking to utilise CVD for this purpose need to understand the relationships between processing, structure, properties, and applications, while also taking into consideration cost and sustainability factors.

In recent years, we have seen a dramatic shift in the importance of CVD—it now plays a much more prominent role in the fabrication of nanomaterials, thin/thick films, nanocomposite coatings, and composites to address grand challenges in energy, engineering, and biomedical applications. Furthermore, the manufacturing industry has been investing heavily in research and development of high-performance, cost-effective nanostructured materials. These businesses recognize the benefits of thinking ahead and taking action to stay ahead in the global marketplace by applying cutting-edge, eco-friendly, sustainable processing technologies.

This book has been written as a way of providing a single source of information on the fundamentals of CVD, together with an outline of the most important recent technological advances and industrial applications. We have emphasized the broad range of practical uses for CVD and its cutting-edge role in the creation of novel, high-value products in ways that take into account environmental sustainability and cost-efficiency.

It is our hope that this book will provide an up-to-date introduction and overview of the state-of-the-art of advances in CVD and its applications. It should inspire readers to use CVD and its emerging variants to develop nanomaterials and thin/thick films with unique structures and optimised properties, as well as for the creation of new types of highly functional systems and efficient devices/products with greatly improved performance. It has been written for scientists, engineers, researchers, postgraduate students, and final-year undergraduate project students in materials science, materials chemistry, solid state physics, electrical engineering, chemical engineering, mechanical engineering, and biomedical engineering. It should also be useful for final-year undergraduates taking courses related to nanoscience and nanotechnology, materials processing, surface engineering, and surface processes. Finally, these contents should help to equip any student with a desire to enter an academic or industrial career in these technologically important fields.

I would like to express my sincere thanks to all contributing authors for their work on this book, and to the reviewers for helpful feedback on various chapters. The excellent assistance from the publisher, Taylor & Francis, led by Lou Chosen, and his team is gratefully acknowledged.

Kwang Leong Choy
University College London

Editor

Kwang Leong Choy [DPhil (oxon), DSc, FIMMM, FRSC, CSi] is a professor of Materials Discovery and Director of the UCL Institute for Materials Discovery. She has pioneered the novel electrostatic spray-assisted vapour deposition (a variant of CVD) technology, which is not only low cost, eco-friendly, and energy efficient, but also enhances coated product performance. Her contributions to the understanding of interactions between gaseous chemical species and solid substrates enables the control of the nucleation and growth of materials with a well-controlled structure and composition at the molecular level. This has led to the creation of high-performance ceramic films and new nanocomposite coatings. This technology has been successfully translated from the laboratory to industry, with the creation of two spin-off companies, and has led to Professor Choy being awarded the Grunfeld Memorial Award and Medal by the Institute of Materials for recognition of her professional contributions which have had significant impact on engineering applications.

Many of her advanced nanostructured ceramic coating materials and CVD-based coating technologies have led to significant impact and been exploited for industrial applications. These include ceramic-coated fibre-reinforced composites, ceramic thermal barrier coatings, superthin film inorganic transparent conducting films for displays and thin film solar cells, as well as multifunctional nanocomposite coatings for tribological applications. She has authored over 250 peer-reviewed publications, including 5 books and 20 patents in nanomaterials, superthin/thin/thick films, and nanocomposite coatings for structural, functional, and biomedical applications. She has also contributed to the *Handbook of Nanostructured* *Materials and Nanotechnology* by Academic Press (2000), which received an award of excellence from the Association of American Publishers. She has given over 150 plenary/keynote papers/invited lectures. She is a fellow of the Institute for Materials, Minerals and Mining (IOM3), a fellow of the Royal Society of Chemistry (FRSC), and is a Chartered Scientist, CSci. She has been awarded Guest Professorships at the University of Uppsala (2001/03), Ningbo Institute of Materials Technology and Engineering (NIMTE, 2010/2012), and the Chinese Academy of Sciences (CAS) Visiting Professorship for Senior International Scientist (2011/2013). She also participates in developing the European materials research roadmap and represents the UK in the Materials Experts Task Forces for High Level Group of EU Member States and H2020 Associated Countries on Nanosciences, Nanotechnologies (2016/17).

Professor Choy's work has led to her participation in numerous large research programmes. These include EU, EPSRC, and government flagship grants, as well as research contracts and collaboration to develop new engineering products with particular characteristics with industrial companies such as Rolls Royce, Aero Engine Control, EADS, PVi, QinetiQ, DSTL, BG/Advantica, and IMPT. She has been the international expert reviewer for Ontario Research Fund, Hong Kong Productivity Council, and Greek Ministry of Education/European Commission. She has been awarded Guest Professorships at the University of Uppsala (2001/03), Ningbo Institute of Materials Technology and Engineering (NIMTE, 2010/2012), and Chinese Academy of Sciences (CAS) Visiting Professorship for Senior International Scientist (2011/2013).

Contributors

Kwang Leong Choy
UCL Institute for Materials Discovery
University College London
London, United Kingdom

Maria-Beatrice Coltelli
Center for Materials Science and Engineering
University of Pisa
Pisa, Italy

María J. Hortigüela
Department of Physics
University of Aveiro
Aveiro, Portugal

Andrea Lazzeri
Center for Materials Science and Engineering
University of Pisa
Pisa, Italy

Gonzalo Otero-Irurueta
Department of Physics
University of Aveiro
Aveiro, Portugal

Guillaume Savelli
CEA-Liten
Département des Technologies des NanoMatériaux
Grenoble, France

Dhananjay K. Sharma
Department of Physics
University of Aveiro
Aveiro, Portugal

Manoj Kumar Singh
Department of Physics
University of Aveiro
Aveiro, Portugal

Cigang Xu
UCL Institute Materials Discovery
University College London
London, United Kingdom

current address

Institute of Materials
China Academy of Engineering Physics
Mianyang, China

Yuri Zhuk
Hardide plc
Oxfordshire, United Kingdom

Introduction

KWANG LEONG CHOY AND CIGANG XU

1.1 CHEMICAL VAPOUR DEPOSITION: FUNDAMENTALS AND PROCESS PRINCIPLES

1.1.1 Definition

Chemical vapour deposition (CVD) is a deposition process whereby vapour precursors undergo dissociation and/or chemical reactions to form a stable solid product in an activated (e.g., heat, plasma, light) environment. If the reaction occurs in the gas phase, i.e., homogeneous reaction, this will lead to the formation of fine particles. On the other hand, the reaction on or near a heated substrate would produce a dense film. A combination of homogeneous and heterogeneous reactions would tend to produce a porous film [1]. Figure 1.1 illustrates the CVD process [1].

Therefore, the CVD process differs from physical vapour deposition (PVD) process in that it only involves physical deposition such as evaporation, sputtering, and condensation without any chemical reactions.

In general, CVD equipment consists of three parts:

1. Chemical vapour precursor supply system: This is used to generate vapour precursors and then deliver to the reactor.
2. CVD reactor: The main function of the CVD reactor is to heat the substrate or the environment to the deposition temperature.
   The reactor consists of a reaction chamber, often equipped with a loadlock for the transport and placement of the substrate, a substrate

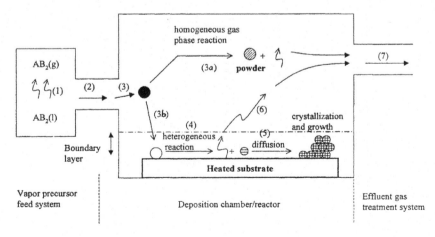

Figure 1.1 A schematic illustration of the key CVD steps during the deposition process. (From Choy, K., *Prog. Mater. Sci.*, 48, 57–170, 2003.)

holder, and a heating system with temperature control for the deposition of films and coatings. For the coating of powder, a fluidized bed reactor may be used.

3. Effluent gas handling system: This component consists of a neutralizing part for the exhaust gases and a vacuum system to provide the required reduced pressure for the CVD process that performs at low pressure or high vacuum during deposition.

Variants of CVD processes have been developed for a wider range of applications than any other deposition or coating technique. These various CVD processes offer versatility and the widest range of high-purity materials with well-controlled structure and composition at molecular levels. These include ultrafine powder (0-D), nanowires/nanotubes (1-D), 2D (nano)sheets, superthin/thin films and thick coatings, and 3D prototyping of near net shape structures. There are various types of CVD based on the followings, as shown in Figure 1.2. They include the following:

1. Activated methods used such as heat, plasma, light, ions which give rise to thermal CVD, plasma-assisted CVD, microwave (MW)-assisted CVD, photo-/laser-assisted CVD, and ion-assisted CVD, respectively. Atomic layer deposition (ALD) could operate via thermal or plasma mode.

2. Choice of precursors: normally, metal halides and hydrides are used in conventional CVD. The use of metalorganic precursors led to metalorganic CVD (MOCVD).

3. Deposition pressure: the deposition could be conducted in atmospheric/normal pressure, low pressure or ultrahigh vacuum (UHV), which led to normal pressure CVD and low-pressure CVD, respectively. UHV CVD processes include molecular-beam epitaxy (MBE) and chemical-beam epitaxy (CBE).

4. Deposition mechanism: The precursors in conventional CVD react at the same time on the substrate. On the other hand, in ALD, the precursors react separately.

These variants of CVD are summarised in Figure 1.2.

1.1.2 A brief history and development of CVD

Nowadays, CVD is an established industrial process to produce thin films and coatings for electronic, optoelectronic, engineering, and biomedical applications, surface modification, processing of ultrafine powder, ceramic fibres, and ceramic matrix composites (CMCs) as well as pyrometallurgy.

The CVD process can be dated back to prehistory, where carbon black from combustion processes was used for painting applications in China

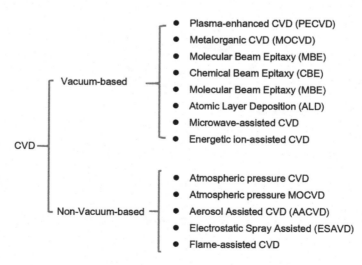

Figure 1.2 CVD and examples of its variants.

around 1500 BC [2]. Several reviews have described the historical development of CVD [3,4a]. This review article gives an overview of the evolution of CVD technology and processes, and it highlights the development of conventional CVD and the variants of CVD processes for the fabrication of technologically important materials, especially for practical applications.

CVD of tungsten (W) metal was reported in 1855 through the decomposition of WCl_6 with H_2 as a carrier gas [3]. To meet the demand for electrical lighting, a CVD process for carbon fibre filaments was developed and patented in 1880 [4a]. Further improvement was made by depositing metal (e.g., via CVD onto carbon fibre filaments) for more robust lamp filaments [4a]. The famous Mond Process to deposit nickel coating to allow the parts to stand a moderate heat from nickel tetracarbonyl, $Ni(CO)_4$, was developed in 1890 [5] and is still used nowadays as a route to produce high-purity Ni metal from nickel ore. The Mond process can be treated as a MOCVD process as it uses the metalorganic precursor.

The CVD deposition of Si by hydrogen reduction of $SiCl_4$ was reported in 1909 and 1927. These formed the foundation for the development of Si-based solar cells and rectifiers in the 1940s just after World War II [3]. In the 1950s the research continued on the preparation of other 1D materials such as Hg whisker [6], and further work on Si whisker from which the vapour-liquid-solid (VLS) process mechanism was proposed in 1964 [7], which was highly relevant to the subsequent considerable interests in 1D materials, in particularly the TEM work on carbon nanotubes (CNT) in 1991 [8]. In the 1940s, another variant of CVD process, flame synthesis, was used for the production of SiO_2, Al_2O_3, and TiO_2 nanopowders, which are even still used in various products nowadays from paints to sun-screen creams [9].

In the 1950s, the Ziegler–Natta process was developed to catalyse the polymerization of olefins with triisobutylaluminium [But3Al], which led to the coming of the plastic age [3]. It was found that triisobutylaluminium was a good precursor for the deposition of pure Al metal, and this process was used in the 1980s for the very-large-scale integration (VLSI) applications of integrated circuits in semiconductor industry [10].

In 1952, the growth of diamond by the CVD process was first performed and the subsequent work performed in the 1980s has led to industrialized CVD processes for synthetic diamond,

especially the MW-plasma-based CVD process, which was developed to produce synthetic diamonds for applications like jewellery [11].

Various important CVD processes were developed in the 1960s. The chemical vapour infiltration (CVI) process, a variant of the CVD process, was first developed to densify porous graphite preforms by infiltration within a carbon matrix in 1962; it has now been used commercially for the manufacture of carbon-carbon composite and ceramic fibre-reinforced CMCs like SiC_{fibre} (e.g., Nicalon)/SiC_{matrix} to provide high strength (about 400 MPa), fracture toughness (higher than 10 MPa m$^{1/2}$), corrosion and erosion resistance, and high-temperature structural materials. These can be used for reusable space vehicles, re-entry nose cones, heat exchangers, and aircraft brakes [1]. In the 1960s, Boron and SiC fibres with diameters of 100–140 μm were first produced by CVD on to a core (like W). Boron fibre was through a reaction of $H_2 + BCl_3$, and SiC fibre was using various chlorosilanes like CH_3SiCl_3 that gives SiC and 3HCl. These fibres can be used to reinforce metals [12]. In 1968, the term of MOCVD was used for the deposition of GaAs for III–V materials; almost at the same time, the MBE process was reported for the deposition of GaAs [13], followed by the development of CBE in the 1980s when both III and V solid sources were replaced by gas sources in the MBE process [14]. Such CVD of III–V films opened up optoelectronic device applications. In the 1960s, the ALD process was initiated in Russia with the name of 'molecular layering', whilst the current ALD process (previously named atomic layer epitaxy [ALE]) was developed in 1974 in Finland; the modern term of ALD was accepted in 1990. ALD process can be used for the deposition of high-K dielectric layers for the microelectronic industry [15]. In 1968, CVD coatings for cutting tools were first developed and marketed by Krupp in (West) Germany under the brand name Widia, followed by another big name in the cutting tool industry: Sweden's Sandvik [16].

In the 1970s, amorphous Si prepared through plasma-enhanced CVD (PECVD) was reported to behave as a semiconductor in 1975 [17], and the first amorphous Silicon solar cell with the efficiency of 2.4% in AM-1 sunlight was reported with the amorphous Si layers deposited through PECVD process in 1976 [18]. The desire for applying solar energy commercially has driven the development of large-area and large-scale PECVD processes for Si_3N_4 anti-reflective layers in the 1990s and 2000s, which is important to improve the silicon solar cell efficiency and widen the application of photovoltaics across the world [19]. Laser photo-deposition (laser-assisted CVD) was developed in 1979 to deposit metal features, with the potential application to etch or deposit materials [20].

In the 1980s, the laser ablation process of graphite indicated the presence of C_{60}, which was regarded as the discovery of C_{60} by the Nobel laureate Kroto and Smalley in 1985 [21], and the arc-discharge process was developed in 1990 as a process to prepare sufficient quantity of C_{60} and similar carbon material for further research on this type of carbon materials [22], in particular, from the remains of this type of arc-discharge process, the key work on carbon nanotubes (CNTs) was published in 1991 [8], followed by the growth of single-walled carbon nanotubes (SWNTs) in 1993 by arc discharge process [23a]. These works stimulated the field of 1D nanomaterials across the research community [23b], although multi-walled carbon nanotubes (MWNTs) were already prepared by CVD process in 1976 [24].

In 2004, the characterization of graphene obtained through the scotch tape method opened the door towards 2D materials [25]. This was followed by the CVD process reported in 2009 to grow high-quality single-layer graphene using Cu foil as the catalyst and CH_4 as the carbon source [26], and other 2D materials (BN, MoS_2) [27].

It is interesting to reflect that carbon materials including carbon black, diamond, CNTs, C_{60}, and graphene have all been prepared through CVD processes at different stages of human civilization.

Therefore, it is fair to say that CVD is an important process that evolves with human civilization up to now and which has created useful materials and high-value products for various structural, functional, and biomedical applications in transport, clean and renewable energy, electrical and optoelectronics, and engineering and healthcare sectors. CVD will continue to develop and evolve further for the benefit of the society in the future.

1.1.3 Fundamentals and process principles

As shown in Figure 1.2, the CVD process has evolved from thermal CVD to many variants as described in Section 1.1.1, including PECVD,

photo-assisted CVD (PACVD), MOCVD, MBE, CBE, ALD, MW-assisted CVD, energetic ion-assisted CVD, and laser-assisted CVD. Although each individual process may vary, the fundamental process principle remains similar, involving the following steps as shown in Figure 1.1 [1]:

1. Generation of gaseous reactants/precursors via evaporation or sublimation
2. Transport of the gaseous reactants into the reactor
3. Gas-phase reactions of precursors in the reaction zone to produce reactive intermediates:
   a. At a sufficiently high temperature in the gas phase, intermediates undergo reaction in the gas phase (homogeneous reaction) and produce stable solid fine powders and gaseous by-products.
   b. At a temperature below the dissociation of the intermediate phase, diffusion/convection of the intermediate species across the boundary layer occur and they subsequently undergo steps 4–7.
4. Adsorption of the reactants on the heated substrate surface and heterogeneous chemical reactions at the gas-solid interface, which leads to the formation of deposit and by-product
5. Surface diffusion of the deposits along the heated substrates to form the nucleation and crystallisation centres and growth of the film via further surface chemical reactions
6. Removal of the gaseous by-products from the boundary layer via diffusion and convection
7. Desorption and mass transport of the by-product and remaining fragments of the unreacted species away from the reaction zone

1.1.3.1 THERMODYNAMICS/KINETICS/MASS TRANSPORT OF CVD

As mentioned earlier, CVD processes involve many steps. In order to understand and have a good control on the process, it is important to understand (a) thermodynamics, (b) chemical kinetics, and (c) mass transport phenomena, as discussed in the following sections.

From the perspective of thermodynamics, in principle, for any reaction to occur, the Gibbs free energy of the reaction $\Delta G_r = \Delta G_f(products) - \Delta G_f(reactants) < 0$ should be true.

CVD process involves chemical reaction during the process; therefore, $\Delta G_r < 0$. Moreover, the equilibrium constant, K can be determined from the following equation

$$K = \exp (\Delta G_r/RT)$$

In which R is the gas constant and T is the process temperature.

As the CVD process needs thermal energy or another type of energy, the input external energy is essential for the reaction to occur. In principle, thermodynamic analysis can provide information about the feasibility of the reaction and process guidelines for choosing process conditions.

From the perspective of kinetics, the CVD process tends to be a non-equilibrium process, involving chemical reactions in the gas phase, or the substrate surface, absorption and desorption depending on the processes. The kinetics and possible reaction pathways of some important industrial chemical systems like Si and GaAs have been studied in detail, but there is no predictive model to analyse the general CVD processes. In some cases, an *in-situ* analysis tool, such as RHEED in MBE, can provide information and indication to control the process, but due to availability and capability of *in-situ* monitoring, generally CVD kinetic data is obtained experimentally through checking the reaction rate with the relationship of process parameters like process temperature and pressure, and identifying the possible rate-limiting steps. For example, for the deposition of TiB_2 from the reaction of $TiCl_4$, BCl_3 and H_2, the data shows that the relationship between deposition rate and reaction temperature complies with the Arrhenius law:

$$\text{Deposition rate} = A \exp (-Ea/RT)$$

Where A is a constant, Ea is the apparent activation energy, R is the gas constant, and T is the deposition temperature. Within the process temperature range 1050°C–1450°C, there are two different deposition mechanisms. From 1050°C to 1350°C, the deposition rate increases rapidly in an exponential manner with the increase of the process temperature, indicating the rate-limiting mechanism is surface chemical kinetics, like chemisorption, and/or chemical reaction, surface migration, lattice incorporation, and desorption, which depends on the deposition temperature. Above 1350°C, the surface kinetic processes become rapid so that the deposition is limited by the diffusion of the active gaseous species to the deposition surface, indicating the process is mass-transport limited, and the deposition rate depends weakly on temperature [1].

In the case of the growth of aligned CNTs, how to increase the height of aligned CNTs is still a challenge, so there is effort to understand why CNTs stop growing through *in-situ* monitoring of the growth process, as shown in Figure 1.3 [28]. During the atmospheric thermal CVD process of growing

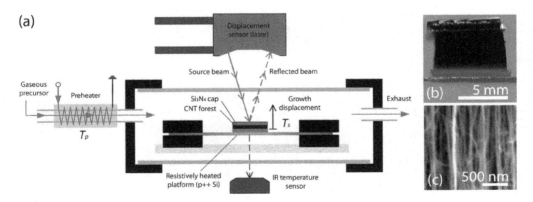

Figure 1.3 **(a)** Schematic representation of decoupled CVD apparatus where the reactant gases enter the reaction chamber after passing through a heated pipe maintained at Tp. The heated mixture deposits at the growth substrate, maintained at Ts. CNT forest height is measured in real time by a laser displacement sensor. **(b)** Digital photograph shows a CNT forest growing on the heated Si platform, while **(c)** shows the forest comprises well-aligned CNTs (Ts/Tp 825°C/1020°C). (From Meshot, E.R. et al., *ACS Nano*, 3, 2477–2486, 2009.)

aligned CNTs with a mixture of $C_2H_4/H_2/He$ at the temperature range of 675°C–875°C, the height of CNTs can be measured in real time by a laser displacement sensor, while the diameter distribution of CNTs can be measured nondestructively by transmission small-angle X-ray scattering (SAXS), as shown in Figure 1.4. Moreover, the comparison of SAXS result and TEM measurement shows that the resulting diameter is compatible between these two methods. The diameter of CNTs depends on the growth temperature, indicating that the treatment temperature of catalyst correlates with the diameter of CNTs [28].

In another case, it is found that graphene growth on Cu depends on the process conditions. High temperature and low methane flow rate and partial pressure were preferred to generate a low density of graphene nuclei, and high methane flow rate or partial pressure were preferred for continuous large-area graphene films, therefore, it is possible to develop an isothermal two-step growth process in which a low graphene nuclei density was set followed by achieving full graphene surface coverage by increasing the methane flow rate and partial pressure; moreover, the resulting graphene films with large domains

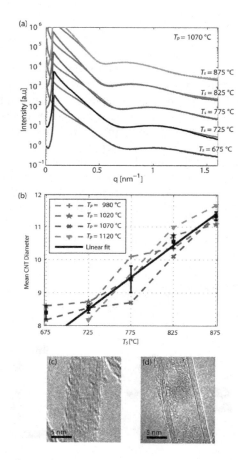

Figure 1.4 **(a)** Plots of scattered intensity *versus* scattering vector (grey) taken at the vertical midpoint of forests grown at different Ts values (constant Tp 1070°C), and corresponding fits (the upper curve of each set) that give the average CNT diameter within the X-ray beam path. **(b)** Correlation between Ts and diameter is observed over a wide range of Tp values, with a linear relationship persisting between Ts 725°C and 875°C. The black squares indicate the average diameters for all tested Tp values, and the error bars represent one standard deviation. The TEM images represent CNTs grown at **(c)** Ts 675°C and **(d)** Ts 875°C. (From (b) Meshot, E.R. et al., *ACS Nano*, 3, 2477–2486, 2009.)

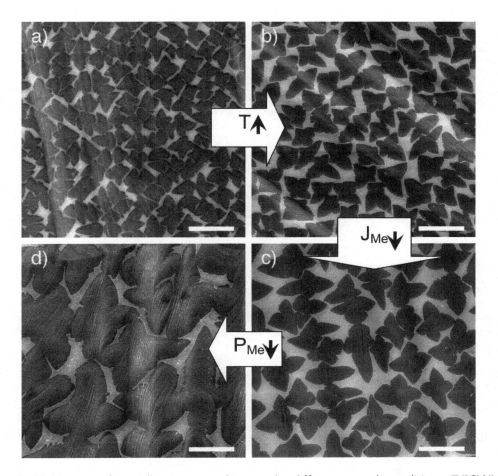

Figure 1.5 SEM images of partially grown graphene under different growth conditions: T (°C)/JMd (sccm)/PMe (mTorr): **(a)** 985/35/460, **(b)** 1035/35/460, **(c)** 1035/7/460, **(d)** 1035/7/160. Scale bars are 10 μm. (From Li, X. et al., *Nano Lett.*, 10, 4328–4334, 2010.)

can have higher mobility than those with small domains, predominantly due to a decrease in interdomain defects. Figure 1.5 shows SEM images of graphene on Cu under different growth conditions [26b].

Furthermore, the carbon isotope labelling (C^{13}) method implies that the growth kinetics of graphene is catalyst dependent. Using 700-nm-thick Ni film on SiO_2/Si wafer and 25-μm-thick Cu foil as the catalyst, with the controllable introduction of normal methane (CH_4) and $^{13}CH_4$ (99.95% pure), through the characterisation of the resulting graphene by Raman microscopy, optical microscopy and isotope labelling, it is proposed that growth of graphene differs with Ni and Cu catalyst, as shown in Figure 1.6. However, the binary phase diagrams of C-Ni and C-Cu are

similar as C has a limited solubility in the metal, and the solubility of C in Cu is much lower than that in Ni. The carbon source for growth of graphene on Cu is mainly from the CH_4, which is catalytically decomposed on the Cu surface with minimal carbon diffusion into the Cu. Once the surface is fully covered with graphene, the growth terminates with the absence of catalyst to decompose CH_4 further. As Ni can dissolve more carbon so that precipitation of extra C can occur during the cool-down stage, the result is nonuniform graphene films [26].

These examples indicate that kinetics of CVD processes are process-related.

From the perspective of mass transport phenomena, CVD process involves the delivery of vapour phase to the reactor, and sometimes to the

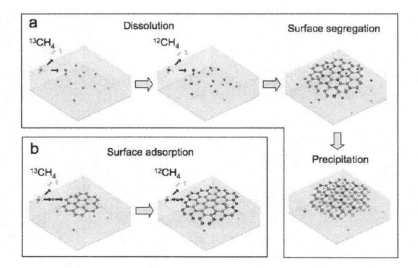

Figure 1.6 Schematic diagrams of the possible distribution of C isotopes in graphene films based on different growth mechanisms for sequential input of C isotopes. **(a)** Graphene with randomly mixed isotopes might occur from surface segregation and/or precipitation. **(b)** Graphene with separated isotopes might occur by surface adsorption. (From Li, X. et al., *Nano Lett.*, 9, 4268–4272, 2009.)

surface for the preparation of films, as indicated in Figure 1.1, the mass transport phenomena include:

1. Fluid dynamics, i.e., fluid flow, mass transfer, and heat transfer of the reactants from the vapour precursor supply unit into the reactor
2. Mass transport of reactants close to the substrate surface, diffusion through the boundary layer of the substrate, desorption of the by-products from the substrates, and transport of the by-products away from the substrate

From the perspective of fluid dynamics, apparently, the reactor geometry and thermal gradient characteristics between the inlet and the reactor influence the transport phenomena, which can be interpreted through the basic fluid mechanics. For example, Knudsen number (Kn),

$$Kn = \lambda/L$$

when λ is the mean free path and L is the characteristic length.

Kn defines the limits between laminar, intermediate, and molecular flow. This can explain the effect of the precursor flow rate.

For instance, regarding the growth of single-walled carbon nanotubes (SWNTs), the morphology of CNTs varies with the flow rate of the process gas, as shown in Figure 1.7; when the gas flow rate is low, SWNTs are in the format of serpentine shape (a–c), and with the increase of flow rate, the parallel segments become shorter and more spaced (c), and the formation of serpentine is inhibited (e) [29].

This means that the process is affected by the mass transport, like the flow rate of process gas, geometry of the reactor. In order to have a good control on the process, mass transport is an important factor.

1.1.3.2 CVD PRECURSORS DELIVERY, PROPERTIES, AND CHEMISTRY

1.1.3.2.1 Delivery

CVD process involves vapour phase, which can be from the solid, liquid, or gas precursor depending on the processes and the final products. When the precursor is gas, it can often be delivered directly to the reactor.

When the precursor is solid, depending on the process, it can be delivered to the reactor in different ways. For example, for the production of nanoparticles, solid precursor can be heated through thermal or arc discharge evaporation, as discussed in Chapter 2; for the growth of 1D nanostructures through conventional CVD process, as shown in Figure 1.8, the solid precursor

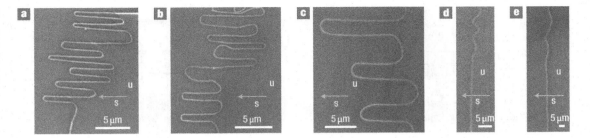

Figure 1.7 Flow rate and direction effects on serpentine shape. (a–e) SEM images of serpentines grown at different flow rates: u = 250 s.c.c.m. (a); u = 500 s.c.c.m. (b); u = 1,000 s.c.c.m. (c); u = 1,500 s.c.c.m. (d); u = 2,000 s.c.c.m. (e). As the flow rate increases, the parallel segments become shorter and more spaced (that is, larger U-turn diameters). Further increasing the flow rate (d, e) inhibits serpentine formation. (From Geblinger, N. et al., *Nat. Nanotech.*, 3, 195–200, 2008.)

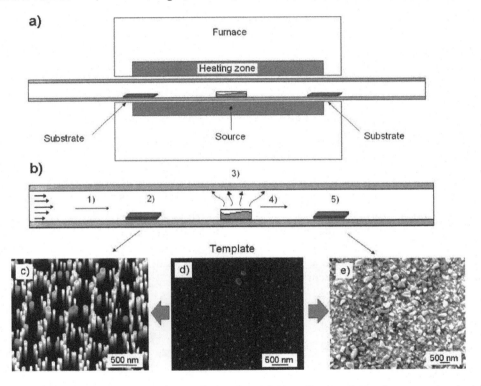

Figure 1.8 (a) Schematic of the experimental setup and (b) gas flow and reaction mapping inside the tube; numbers show the specific positions inside the tube as used in the paper. (c) NWs grown at position 2, (d) initial pattern before the growth experiment, (e) ZnOx film grown at position 5. (From Subannajui, K. et al., *Cryst. Growth Des.*, 10, 1585–1589, 2010.)

can be heated and then transferred to the substrate [30], or the solid precursor can be directly delivered to the reactor, as shown in Figure 1.9 [31]. Coal with the size distribution of 5–25 μm was injected by the carry gas Ar for the reaction to prepare CNTs through plasma-based CVD process; particularly, in the MBE process, solid precursor is heated in a specially designed

Knudsen cell for the delivery to the substrate. This will be discussed later on.

When the precursor is liquid, it can be delivered to the reactor directly through injection or spray, as shown in Figure 1.10. The liquid precursor Xylene is injected to the reactor directly with the catalyst ferrocene [32], or the liquid can be carried through the bubbler, as shown in Figure 1.11 [33];

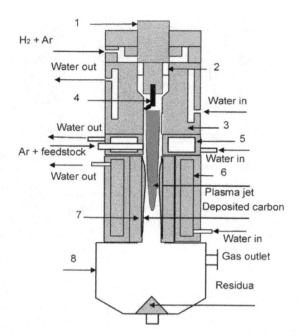

H₂ + Ar

Water out

Water out

Ar + feedstock

Water out

1

2

4

3

5

Water in

Water in

6

7

8

Water in

Gas outlet

Residua

Plasma jet
Deposited carbon

Figure 1.9   Schematic diagram of the experimental set-up: (1) cathode, (2) insulation, (3) anode, (4) arc, (5) injector, (6) reactor, (7) reactor wall, (8) accumulator. (From Tian, Y. et al., *Carbon*, 42, 2597–2601, 2004.)

this is often applied in the CVD, PECVD, ALD, and MOCVD for the delivery of the precursor. It can operate in two different modes: one mode is bubbling mode, in which the carrier gas will flow in from the inlet and carry the precursor vapour

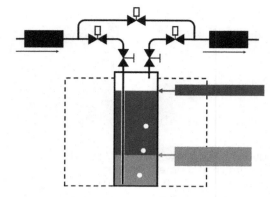

Figure 1.11   Typical arrangement used to transport vapours from a liquid or solid precursor chemicals into a chemical vapour deposition reaction chamber, the middle part is the scheme for the bubbler. (From Bour, D. et al., *J. Cryst. Growth*, 310, 2673–2677, 2008.)

to the reactor through the outlet; the other mode is vapour draw mode, in which the vapour pressure of precursor is high enough to allow the control through specially made mass flow controller (MFC) like conventional gas.

1.1.3.2.2 Precursor properties

As mentioned earlier, solid and liquid precursors can be delivered to the reactor for the reaction through various routes. There are several issues to be considered when identifying the proper precursors.

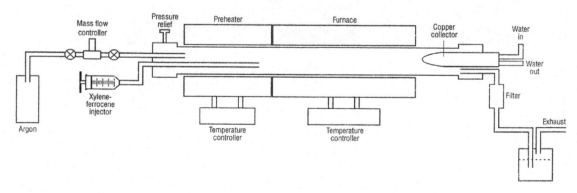

Figure 1.10   Schematic of the reactor used for the nanotube synthesis. The ferrocene–xylene liquid feed is injected into the preheater stage using a syringe pump. Using a mass flow controller, ArrH₂ gas is introduced at the entrance of the preheater to sweep the reactant vapors into the hot zone of the furnace. (From Andrews, R. et al., *Chem. Phys. Lett.*, 303, 467–474, 1999.)

1.1.3.2.3 Stability

For the solid and liquid precursor, it will involve storage, heating, evaporation, and decomposition steps. So, the ideal precursor should be stable for long-term storage at room temperature. Moreover, when it is used for the process, decomposition does not occur before it is needed. To analyse thermal stability of precursors, differential scanning calorimetry (DSC) is a common method which can provide information about the temperatures at which melting and decomposition occurs.

1.1.3.2.4 Volatility

As solid and liquid precursors need to be converted to vapour phase, the pretreatment step will depend on the individual precursor. For example, liquid precursor ethanol for the preparation of CNTs can be used in a vapour-drawing mode, i.e., it can be directly controlled through a MFC for its flow rate, or through Ar bubbler, and the vapour pressure of ethanol can be controlled through the heating or cooling of the bubbler, or varying the flow rate of carry gas, like Ar, to adjust the amount of precursor to the reactor.

The commercial supplier (like Sigma-Aldrich) of the liquid precursor has developed the hardware and protocol to measure the vapour pressure of liquid precursor at different temperatures, which will provide a good guide on the selection of precursor and process conditions.

1.1.3.2.5 Purity and purification

As CVD process is a chemical process, there is need for the precursor purity to avoid contamination. The contamination of precursor can result from the original source, storage, or delivery. This contamination will affect the properties of the CVD products or the process control. For example, in the commercial CO gas, there may be trace amount of ferrocene from the reaction between CO and metal container. Similarly, regarding metalorganic liquid precursor for MOCVD process, like the preparation of light-emission diodes (LEDs), the dopant level is essential for the device performance, therefore, the purity of precursor is important.

There are some characterisation tools developed to identify the impurities. For hydrocarbon impurities, mass spectrometry is a useful tool, for oxygen-containing organic impurities at the low ppm level, Fourier-transform (FT) NMR can be used.

Depending on the precursor, various purification processes can be utilized. For solid or liquid precursor, such as sublimation, recrystallization, fractional distillation (or rectification), preparative chromatography. For commercial gases, such as CO that will always contain a few ppm of $Fe(CO)_5$, the impurities can be filtered out using an in-line pretreatment setup prior to introduction, like the catalytic purifier (Oxyzorb, Matheson) for CO. However, for some types of precursor, conventional techniques may not be good enough, and new techniques may be required.

1.1.3.2.6 Chemistry of precursor

As a CVD process involves a broad range of materials, various types of precursor and chemistry have been demonstrated: thermal decomposition (pyrolysis), reduction, oxidation, hydrolysis, nitridation, disproportionation, and synthesis.

Here are some examples:

Thermal decomposition:	$SiH_4 \rightarrow Si + 2H_2$
Reduction:	$WF_6 + 3H_2 \rightarrow W + 6HF$
Oxidation:	$Zn(C_2H_5)_2 + 4O_2 \rightarrow ZnO + 5H_2O$
Hydrolysis:	$2AlCl_3 + 3CO_2 + 3H_2 \rightarrow Al_2O_3 + 6HCl + 3CO$
Nitridation:	$TiCl_4 + 1/2\,N_2 + 2H_2 \rightarrow TiN + 4HCl$
Disproportionation:	$2CO \rightarrow CO_2 + C$
Synthesis:	$TiCl_4 + 2\,BCl_3 + 5H_2 \rightarrow TiB_2 + 10HCl$

The more detailed description of various chemical reactions and general characteristics of these reactions can be found in the reference [1].

It is worth mentioning that for the same type of reaction, the different precursors can behave differently, indicating that the chemistry of precursor can have impact on the process. For example, for the preparation of $Pb(Zr, Ti)O_3$ (PZT) film through MOCVD process, Zr incorporation efficiency was significantly higher in PZT deposited using $Zr(OiPr)_2(thd)_2$ compared to $Zr(thd)_4$ [34].

From this description of the precursors, the characteristics of a desirable precursor can be summarised as follows:

- Good stability for storage and delivery
- Adequate volatility to allow proper control and delivery

- High chemical purity
- Availability with reasonable cost

1.1.3.3 CVD PROCESS PARAMETERS AND CONTROL/MONITORING TO ENSURE RELIABILITY AND REPRODUCIBILITY

As mentioned earlier, a CVD process involves many steps and is affected by many factors, such as the geometry of the reactor, temperature, and fluid dynamics. Therefore, there is need to ensure the process reliability and reproducibility. The typical technique is to *in-situ* monitor the process, and this depends on the process. For example, in MBE, RHEED is used in the UHV environment to characterise the film, from which the process can be controlled with good reliability and reproducibility [35].

In ALD, as shown in Figure 1.12, mass spectrometry is installed in the flow-type ALD reactor to detect the reaction product deuterated methane (CH_3D), in which $Al(CH_3)_3$ and deuterated water (D_2O) were used as the precursors. The reaction product from TMA and D_2O is expected to be CH_3D^+ ($m/e = 17$). Monitoring the mass 17 peak during the ALD process cycle can provide the information on the effect of process temperature and water dose, as shown in Figure 1.13 [36].

Other types of techniques, such as Raman spectroscopy, infrared spectroscopy, UV-visible absorption spectroscopy, laser-induced fluorescence spectroscopy, and optical emission spectroscopy have also been used to analyse the reaction species to monitor the process to ensure process reliability and reproducibility [1].

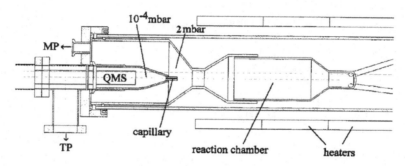

Figure 1.12 Schematics of the sampling in the MS-ALD equipment. The reactants and carrier gas are coming from the right and are pumped by a mechanical pump (MP). Only a small part of the flow goes through the capillary, and it is pumped by a turbo pump (TP). (From Juppo, M. et al., *Langmuir*, 16, 4034–4039, 2000.)

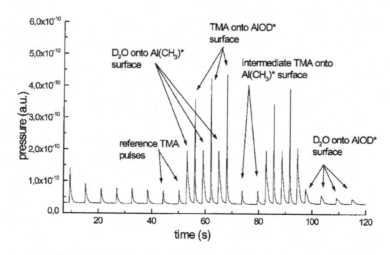

Figure 1.13 An example of the behaviour of mass 17 during the pulsing sequence. AlOD* and Al(CH₃)* represent all the possible adsorbed species after TMA pulse. (From Juppo, M. et al., *Langmuir*, 16, 4034–4039, 2000.)

On the other hand, the process parameters have obvious effect on the resulting product. From the perspective of the film deposition, the Movchan and Demichishin (MD) model was published to understand the structure zone of the film deposition, and then the zone model was extended to sputtered film in the Thorn model, which can be utilized to interpret experimental data with the observed structure in individual zones, as shown in Figure 1.14. The microstructure of vapour-deposited thin films is represented according to the zone model based on deposition through ion-beam-assisted deposition [37].

In the case of the preparation of 1D nanomaterial, such as aligned CNTs, the direction of gas flow can affect the height of CNTs, as shown in Figure 1.15; the precursor gas (a mixture of ethylene + water vapour) flows horizontally (lateral flow) and

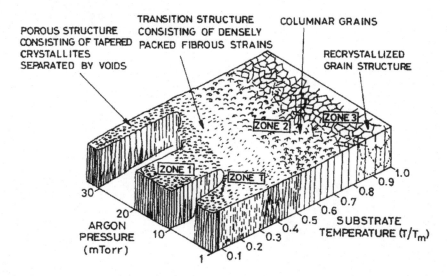

Figure 1.14 Representation of the microstructure of vapour-deposited thin films according to the zone model. (From Mohan, S. and Krishna, M.G., *Vacuum*, 46, 645–659, 1995.)

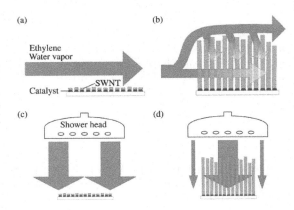

Figure 1.15 Schematics of gas flow direction to catalysts at initial stage of the growth and with the presence of a forest for **(a and b)** conventional lateral-flow and **(c and d)** top-flow growth, respectively. (From Yasuda, S. et al., *ACS Nano*, 3, 4164–4170, 2009.)

vertically (top flow). Top flow of the precursor can give highest CNTs, as shown in Figure 1.16, indicating that the process flow can affect the growth of 1D nanomaterials [38].

In the case of 2D materials, for the growth of graphene, in a general case, Cu catalyst gives single-layer graphene, Ni catalyst gives few-layer graphene, through the preparation of Cu-Ni bimetal catalyst by introduction of copper to nickel film to form Ni-Cu alloy as the segregation substrate, the amount of Ni will affect the layer of graphene obtained, as shown in Figure 1.17a. A 5.5 atom % ratio of nickel led to the coverage of monolayer graphene exceeding 95%. When the ratio was increased to 10.4%, coverage of bilayer graphene would be 89% (Figure 1.17b). This implies that graphene thickness can be controlled through the adjustment of the catalyst [39].

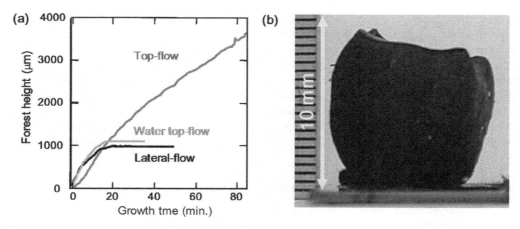

Figure 1.16 (a) Time evolution of the forest growth observed *in-situ* telecentric optical monitoring system for the three different growths. (b) Photograph of CNT forest with a height of 1 cm synthesised by top-flow growth. (From Yasuda, S. et al., *ACS Nano*, 3, 4164–4170, 2009.)

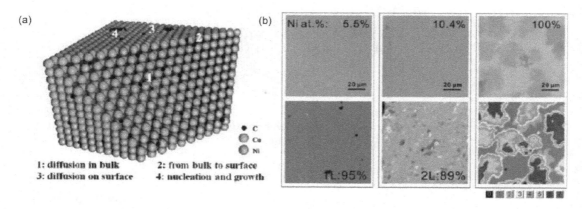

Figure 1.17 Layer-controlled segregation of graphene on synergistic Cu-Ni alloy substrate. (a) Schematic of the segregation process on the Cu-Ni alloy. (b) Optical microscope images of graphene segregated from Cu-Ni alloys with nickel atom concentration of 5.5%, 10.4%, and 100%, respectively. The bottom three panels indicated corresponding thickness analysis based on the optical contrast. (From Yan, K. et al., *Acc. Chem. Res.*, 46, 2263–2274, 2013.)

1.2 ADVANCES IN CVD DEPOSITION TECHNOLOGIES, GROWTH OF MATERIALS, AND APPLICATIONS

As mentioned earlier, CVD technologies evolve with the time and human civilization, and there are various CVD books [3a–c] and review articles [1,4] that describe the fundamental CVD process principles and mechanisms in detail. Therefore, this book will focus on the recent progress in the deposition technologies as well as the growth of materials and applications of CVD, especially towards practical industrial applications.

From the perspective of deposition technologies, CVD and its variant technologies have made significant advances from the following aspects towards new applications as highlighted below.

1.2.1 Large-area and high-volume production

The CVD technology is developed to fulfil the demand for large-area deposition or high-throughput production for the benefit of economies of scale. Several types of CVD technologies have been scaled-up accordingly for large-area and large-volume production.

With the development of high-brightness light-emitting diode (HBLED), the demand for III–V materials (e.g., GaN) increases. MOCVD technology has been developed to produce more III–V thin films on larger tools. PECVD processes are performed through in-line large-area PECVD tool for the production of silicon solar cells and liquid crystal displays (LCDs). Similarly, the larger MW-assisted CVD tool is used for the deposition of anti-reflective layers (e.g., Si_3N_4) for Si-solar cell production [19].

MBE technology is used to fabricate superthin/thin film materials required for optoelectronic devices, such as AlGaAs double heterostructure (DH) lasers. The wider usage of compact disc, video disc, and optical memory has produced a high demand for larger quantity of AlGaAs. Hence, larger MBE tools have also been developed [40].

ALD tools based on the spatial technique (steps are performed at different space of the reactor) for rigid substrates as well as roll-to-roll thin film deposition of Al_2O_3, for example, onto flexible substrates have also been developed recently [41].

1.2.2 Multiple functions

The functions of CVD technology have been expanded when more techniques are added/integrated, which can then be used for the preparation of materials with different configurations (e.g., bilayer, multilayer, superlattices, nanocomposite films). For example, ion beam technology is added to the MBE tool, such that the tool orientation relationship between GaN and the TiN/c-plane Al_2O_3 substrate, as shown in Figure 1.18 [42].

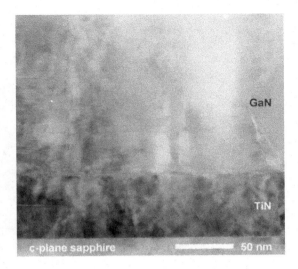

Figure 1.18 TEM cross-section micrograph showing the interface regions of the GaN/TiN bilayer (zone axis: Al_2O_3 [1100]). (From Gerlach, J. et al., *Thin Solid Films*, 459, 13–16, 2004.)

1.2.3 New materials and processes

In addition to the advances in CVD technologies, there are also advances in new processes to produce new materials. One example is CVD of graphene and other 2D materials (e.g., BN) [26,27]. Regarding graphene, a thermal CVD process has been developed for the deposition of high-quality single-layer graphene [26]. Moreover, in order to produce large-area and large-scale graphene, roll-to-roll CVD process has been developed to produce large-scale CVD of graphene, as shown in Figure 1.19 [43].

Instead of amorphous Si used in the conventional LCD, high-performance poly-Si transistor arrays have been developed for active-matrix organic light-emitting diodes (AMOLEDs) and ultra-high-resolution thin-film transistor LCD (TFT-LCD). The higher electron mobility of poly-silicon allows the transistors that control each pixel to be smaller and switch more quickly, hence enabling brighter, sharper displays that use less energy, which is desirable for mobile applications where image quality and battery life is critical [44].

Another example is the topological insulation materials (e.g., Bi_2Te_3) on GaAs(001) by MBE technology to avoid defect [45].

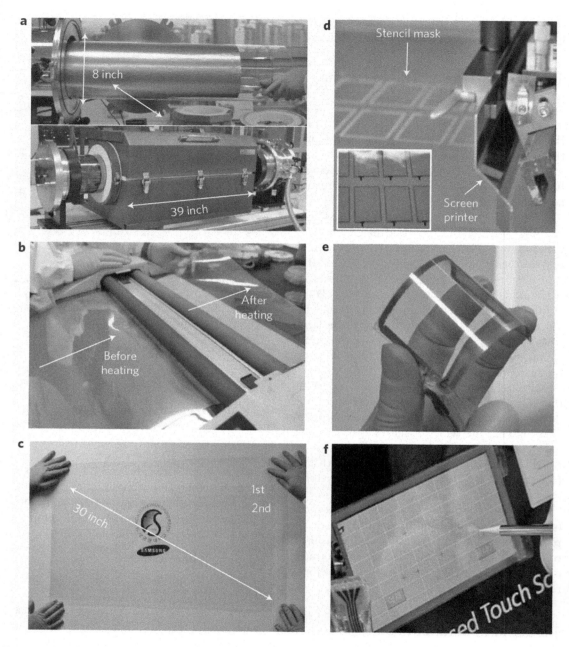

Figure 1.19 Photographs of the roll-based production of graphene films. (**a**) Copper foil wrapping around a 7.5-inch quartz tube to be inserted into an 8-inch quartz reactor. The lower image shows the stage in which the copper foil reacts with CH_4 and H_2 gases at high temperatures. (**b**) Roll-to-roll transfer of graphene films from a thermal release tape to a PET film at 120°C. (**c**) A transparent ultra large-area graphene film transferred on a 35-inch PET sheet. (**d**) Screen printing process of silver paste electrodes on graphene/PET film. The inset shows 3.1-inch graphene/PET panels patterned with silver electrodes before assembly. (**e**) An assembled graphene/PET touch panel showing outstanding flexibility. (**f**) A graphene-based touch-screen panel connected to a computer with control software. (From Bae, S. et al., *Nat. Nanotech.*, 5, 574–578, 2010.)

Other advances in CVD are detailed in individual chapters.

1.3 VARIANT VACUUM CVD METHODS

CVD methods can be classified as vacuum-based methods and non-vacuum-based methods according to the deposition pressure as follows:

1.3.1 Vacuum-based methods

Vacuum-based methods are categorized as the following methods based on individual characteristics:

1.3.1.1 PLASMA-ENHANCED CVD

The early PECVD process was dated back to 1869 and developed by a famous French synthetic chemist Marcellin Berthelot for the decomposition of gases like CH_4 in glow discharge, which is a typical reaction of modern PECVD. In 1876, J. Ogier reported the growth of Si:H and SiNx films from SiH_4 and SiH_4+N_2 precursors, respectively, through PECVD processes [4]. Nowadays, PECVD is widely used in many industries, for example, the production of SiNx film as passivation and antireflection layer for crystalline solar cells [46].

As compared with thermal CVD process, PECVD process involves the generation of plasma through electrical energy, unlike the use of thermal energy in thermal CVD. Plasma is a mixture of electrons, ions, and neutral species. Plasma can be classified into non-isothermal and isothermal plasma (or thermal plasma). In non-isothermal plasma, generated through electric field at reduced pressure, free electrons are accelerated to energies corresponding to several thousand degrees in the case of thermal activation; ions and neutral species either are not influenced by the fields or cannot follow changing fields fast enough, and their temperatures stays low. During the process, high-energy electrons collide with neutral gas molecules to result in dissociation processes of neutral molecules that would only occur at very high temperatures in the case of thermal equilibrium, so non-isothermal plasma allows the preparation of materials at the temperature, in many cases, lower than conventional thermal CVD processes. Comparatively, for the performance of SiO_2 dielectric layer, CVD process takes place at 600°C–800°C, whereas PECVD

deposition of SiO_2 below 400°C can function similarly [1]. So, for the film deposition, like SiNx layer, PECVD process is preferred [46]. The reactor for each type of plasma varies depending on the substrate size, plasma source, and processing conditions. DC glow discharge, 13.56 MHz radio frequency (RF), MW (900 MHz, 2.4 GHz) are commonly used frequencies for this type of PECVD process. In thermal plasma, all the species in the gas are heated to more than 5000 K and dissociated, and the electric energy is supplied to achieve high temperatures. Typical examples include arc discharge and low-RF-frequency thermal plasma at atmospheric pressure [47a]. This type of plasma process is good for the bulk material preparation, like plasma spray for nanoparticles [47b].

There are various types of non-thermal plasma, such as DC plasma, capacitively-coupled plasma (CCP), inductively-coupled plasma (ICP), and MW plasma. Figure 1.20 shows some typical plasma reactors and plasma sources [48]. The labelled RF plasma CVD is also named as CCP, the substrate (or sample) is also under the exposure of plasma, the density of CCP source is normally lower than that of ICP. As CCP-type PECVD process can produce larger-area film deposition with better uniformity than ICP-type PECVD processes, this type of PECVD process is widely used for the research and industrial applications for aSi and SiO_2 film deposition. A schematic PECVD reactor for production is shown in Figure 1.21 [44]; ICP-type PECVD process is applied when the effect of ion is less desirable for the film deposition, and the lower temperature PECVD process is needed, like remote plasma deposition of SiO_2 film.

Since the development of the PECVD process for dielectric films like SiO_2, and aSi film, in particular, the research shows dielectric film in the application of microelectronic industry and aSi for thin-film transistor (TFT), PECVD processes are gradually used in many industries over last several decade, including the deposition of dielectric film (like SiO_2, SiNx) in the microelectronic (as passivation layer), crystalline Si solar cell (as passivation and antireflection layer) [1], HBLEDs (as passivation layer) industries and their application as back lighting for LCD [49], and aSi (or polySi) film for display devices like LCD and AMOLED for consumer electronics such as TV, computer screens, and mobile phones [50]. The commercial plasma reactor was first developed in 1970s for the

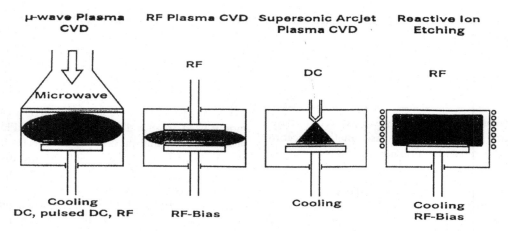

Figure 1.20 Principles of different plasma-assisted processes. (From Suchentrunk, R. et al., *Surf. Coat Tech.*, 97, 1–9, 1997.)

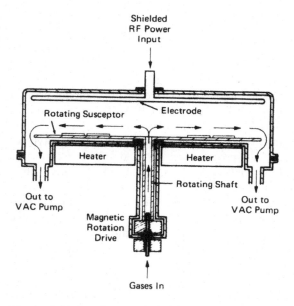

Figure 1.21 Production radial flow plasma-assisted CVD reactor. (From Sherman, A., *Thin Solid Films*, 113, 135–149, 1984.)

deposition of Si_3N_4 passivation layers [1], as each industry increase the scale, PECVD tools also increase the size. For example, the latest PECVD tool can process the dimension of 2200×2500 mm substrate, suitable for the deposition of dielectric layers SiO_2 for manufacturing high-quality metal oxide transistors for displays through minimizing hydrogen impurities to improve transistor stability

and provide optimized screen performance; for the deposition of aSi, PECVD tools can now cover the range of glass sizes used in the display industry from Gen 2 (0.2 m^2) all the way up to Gen 10 (9 m^2) by Applied Materials Inc [44].

In addition to the rigid substrate, PECVD tools and processes for flexible large-area substrate have also been developed. Figures 1.22 and 1.23 shows a roll-to-roll integrated tool for the deposition of thin films onto flexible stainless-steel foils for solar cells applications. The tooling system is 90 m long and 3 m tall with a web speed 60 cm/min capable of depositing nine layers of a-Si materials onto six 2.6 km, 36 cm wide on 127 μm thick stainless-steel foils [51].

The schematic diagram of a triple-junction cell structure is shown in Figure 1.24 [52], in which the top cell uses a-Si (80–100 nm thick) of an optical gap of 1.8 eV to capture the blue photos, the middle cell (150–200 nm thick) uses an a-SiGe alloy with a Ge-content of 10%–20%, and optical gap of 1.6 eV to capture the green photons, and the bottom cell (150–200 nm thick) uses a-SiGe alloy with a Ge-content of 40% to reduce the optical gap to 1.4 eV to capture the red photons. The film is deposited at the following conditions: 0.1–1 Torr, 150°C–300°C, power density 10–100 mW/cm^2, process gas of SiH_4-H_2-GiHe$_4$. The solar cell efficiency is about 13%. This type of solar cell was produced commercially.

It needs to be pointed out that the discussion focuses on the technological development of PECVD process; the related commercial

Figure 1.22 A 30-MW continuous roll-to-roll a-Si PECVD processor: side view. (From Izu, M. and Ellison, T., *Sol. Energy Mater. Sol. Cells*, 78, 613–626, 2003.)

Figure 1.23 A 30-MW continuous roll-to-roll a-Si PECVD processor: end view. (From Izu, M. and Ellison, T., *Sol. Energy Mater. Sol. Cells*, 78, 613–626, 2003.)

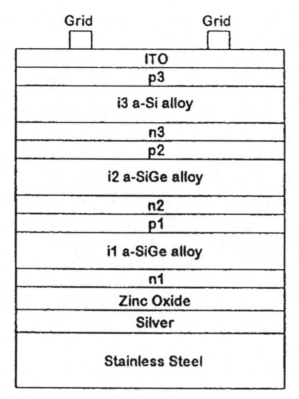

Figure 1.24 Schematic diagram of a triple-junction cell structure. (From Guha, S. et al., *Prog Photovoltaics*, 8, 141–150.)

side is outside of the coverage as the commercial success of flexible solar cell is related to the macroeconomic policy and overall commercial environment.

The detail MW-based plasma and its application will be discussed in the section of MW-based plasma.

Furthermore, the PECVD process has also been used for the deposition of polymer as non-stick layer [53]; this will be discussed in more detail in Section 1.4 (CVD of polymers).

Generally, PECVD process has now been widely used in many industries, including microelectronics, photovoltaics, optical film, and displays; the size of substrate has been increasing for the last several decades to obtain large-size devices and achieve large-scale economy. With the emerging application of AMOLED, it is expected that PECVD will develop further to meet future needs.

1.3.1.2 MICROWAVE-ASSISTED CVD

MW-assisted CVD is also a type of plasma-based CVD process; the principle of the process is similar to PECVD process. MW plasma is often generated with the frequency of 900 MHz or 2.45 GHz, and RF frequency is often 13.56 MHz. MW and RF discharges differ mostly in their electron energy distribution function (EEDF), and MW discharge often gives a higher ionization and dissociation rates. The electron densities can be in the order of 10^{13} cm^{-3} and the degree of ionization is higher (10% and above) in the MW plasma, which is also a rich source of intense emission in the vacuum and near-UV region, and the emission can stimulate plasmo-chemical and photo-chemical processes to affect the properties and growth rate of films. Since MW-based plasma is more efficient to produce active species than other types of plasma, it also gives more efficient film deposition [54].

MW plasma can be generated as isotropic plasma or magnetoactive plasma. Isotropic plasma can be generated through various types of systems, as shown in Figure 1.25 [54], such as waveguide discharge, cavity discharge, surfatron, and discharge excited by retardation structures. These systems vary in the way of introducing the MW power into discharge tube, but surface waves along the discharge tubes are excited in all the systems, and the energy is dissipated in the plasma in consequence of a collision and/or nonlinear absorption, and electron densities $Ne/Nc = \omega^2 pe / \omega^2 = 10{-}10^2$, where Ne is electron density and Nc is the critical density satisfying the condition $\omega = \omega pe$.

When there is the presence of an external magnetic field B, this type of MW plasma is regarded as magnetoactive plasma, which can be generated by (a) the absorption of UHR (ωUH), (b) the absorption at electron cyclotron resonance (ECR) (ωce), and (c) the collisionless absorption due to wave transformation at ωce/$\omega > 1$. Magnetoactive plasma can also generate dense plasma $Ne/Nc = 10{-}10^2$, and accelerate plasma in an inhomogeneous magnetic field [54].

There is a comparison for the preparation of SiNx film through N_2 and SiH_4+Ar between MW-plasma and RF-plasma process, it was shown that in the MW-plasma process, it was

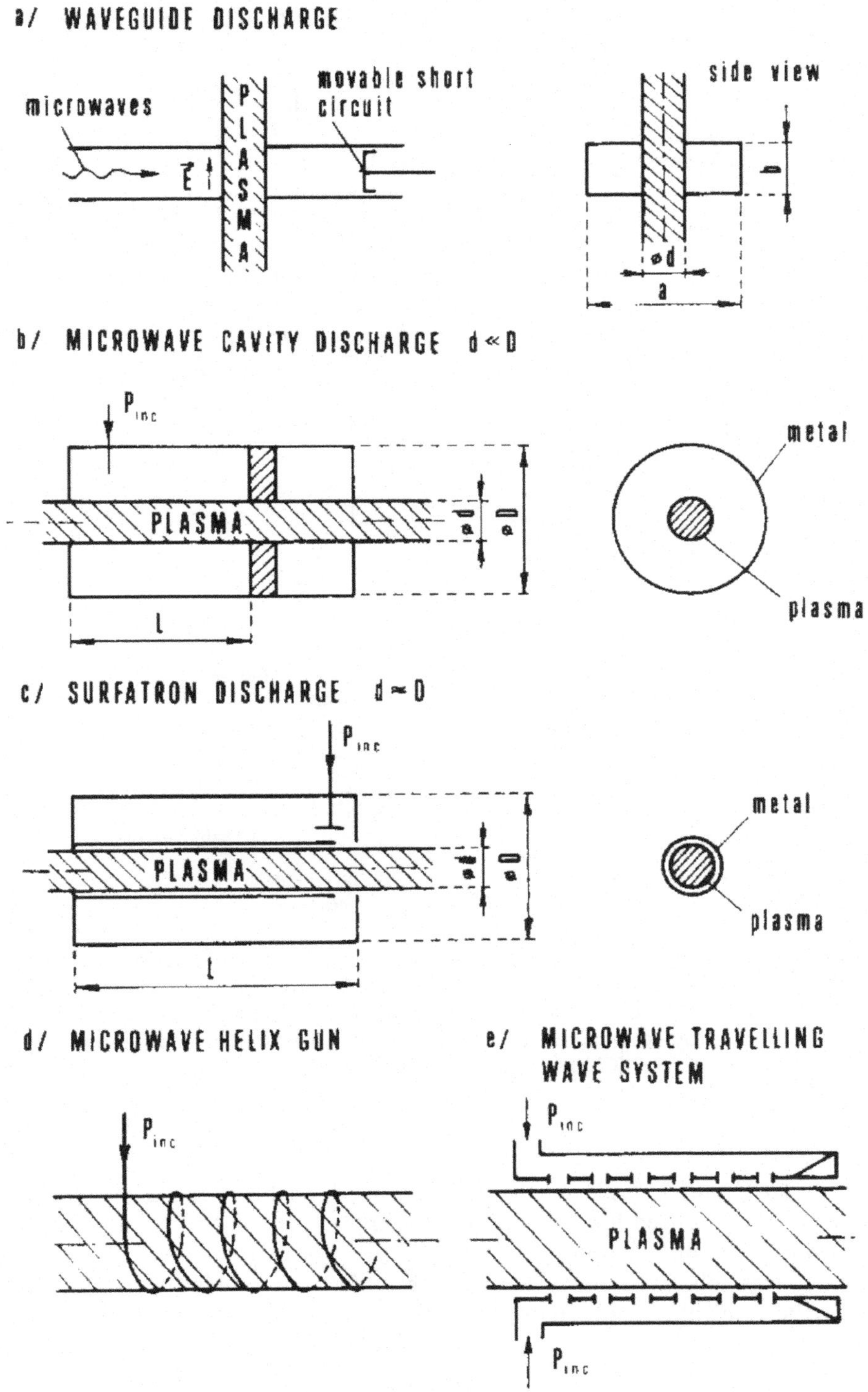

Figure 1.25 Basic systems used for the generation of an isotropic plasma: **(a)** waveguide discharge; **(b)** cavity discharge; **(c)** surfatron; **(d, e)** discharge excited by retardation structures. (From Musil, J., *Vacuum*, 36, 161–169, 1986.)

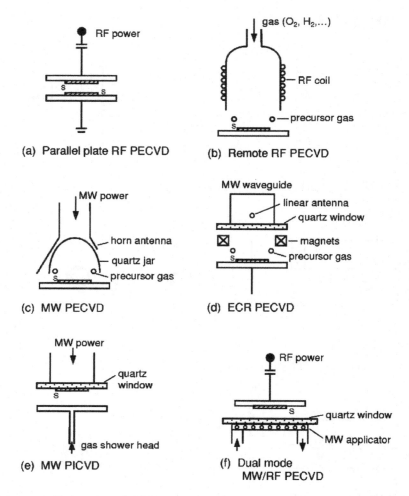

Figure 1.26 Schematic illustration of different types of microwave plasma systems: **(a)** MW plasma reactor with different excitation components (antenna, standing wave applicator, travelling wave applicator) used in a cw or pulsed modes; **(b)** ECR plasma reactor; **(c)** MW PICVD reactor for planar substrates; **(d)** dual-mode (pulsed) microwave/radiofrequency (MW/rf) plasma system; **(e)** remote MW/rf plasma system; **(f)** DECR PECVD system. (From Martinu, L. and Poitras, D., *J. Vac. Sci. Technol. A*, 18, 2619–2645, 2000.)

possible to generate remote N_2 plasma in the tube and then plasma was used for the deposition of SiNx film. In a similar mode, no SiNx film deposition was observed through RF plasma unless that RF electrode was shifted very close to the reactor [55].

Various types of MECVD reactors are shown in Figure 1.26 [56].

MW-assisted process not only can be used for the thin-film deposition but also for the growth of

diamond, as shown in Figure 1.27 [57]. The high electron density of MW plasma facilitates the dissociation of molecular hydrogen into atoms that react with the source hydrocarbon and create a complex mixture of hydrocarbon species including reactive carbon-containing radicals, and the H atoms created by the gaseous activation process may cause the abstraction of hydrogen from the surface CH bonds to create surface radical sites, some of which would react with gas-phase

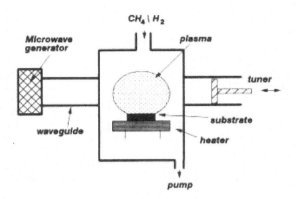

Figure 1.27 Microwave plasma enhanced CVD reactor. (From Ashfold, M. et al., *Chem. Soc. Rev.*, 23, 21–30, 1994.)

carbon-containing radicals to obtain an adsorbed carbon species. Figure 1.28 shows examples of diamond films deposited by MW-assisted CVD [58]. The detailed mechanism has been discussed in some reviews [57]. The MW-assisted CVD process is an established method for the production of diamond in the industry with its application as heat sink in the electronic devices [59].

A MW source can also be designed as a linear format for the large-area deposition of antireflection layer (Si_3N_4 from $NH_3 + SiH_4$) towards solar cell applications, as shown in the

Figure 1.29. This can allow the deposition to be operated continuously (see Figure 1.30) [19]. An industrial production system scheme is shown in Figure 1.31 [60].

In general, MW plasma can produce high concentrations of active species like electrons, ions, or excited molecules, and photons in vacuum UV, near UV, and the visible regions depend on the mechanism of plasma generation [54]. MW-assisted CVD process has already been applied in many industries, including MWCVD for the deposition of TiO_2 ($TiCl_4 + O_2$)/SiO_2($C_6H_{18}OSi_2 + O_2$) film as the cold-light mirror for projection lamps and IR reflectors for energy-efficient lamps on the scale of multi-millions [56] for the production of diamond film (H_2-CH_4) [59], solar cell panels applications [19]. In particular, a MW-plasma-based process has now been used to produce nanoscale graphene flake directly on the scale of tens of Kg per year using CH_4 as the carbon source [61].

1.3.1.3 METALORGANIC CVD

When metalorganic compounds are used during CVD, the process is generally called MOCVD. As mentioned previously, the Mond process uses $Ni(CO)_5$ for the preparation of Ni. This is possibly one of the early MOCVD processes [5]. In addition, metalorganic compounds have also been used

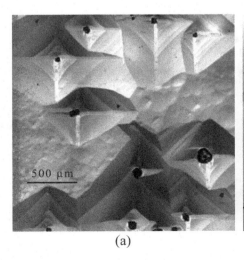
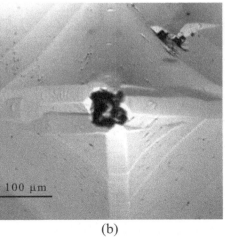

(a) (b)

Figure 1.28 Differential interference contrast microscopy (DICM) images of the epitaxial CVD diamond films grown at 1000°C. **(a)** Sample overview. **(b)** Magnification 20. (From Tallaire, A. et al., *Diam. Relat. Mater.*, 14, 249–254, 2005.)

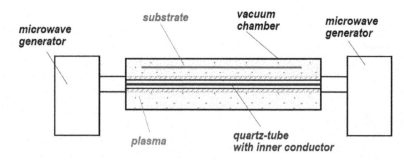

Figure 1.29 Principle of a linear microwave plasma source plasma line. (From Schlemm, H. et al., *Surf. Coat. Tech.*, 174, 208–211, 2003.)

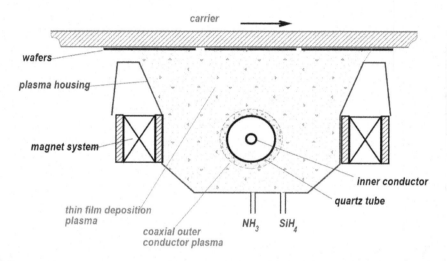

Figure 1.30 Typical cross-section of the magnetic field enhanced microwave plasma source. (From Schlemm, H. et al., *Surf. Coat. Tech.*, 174, 208–211, 2003.)

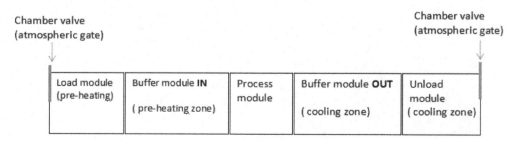

Figure 1.31 Schematic representation for MW PECVD coating system set-up for antireflection silicon nitride layer. (From SiNA®/MAiA® The modular inline system for PECVD anti-reflection coating and passivation layers, https://www.meyerburger.com/en/technologies/photovoltaics/upgrade-technologies/product-detail/product/maia/, accessed on January 3, 2019.)

in many other processes, such as CBE, and these processes can be regarded as MOCVD processes in a general term. Conventionally, thermal heating is generally used in the MOCVD process.

Others such as metalorganic vapour phase epitaxy (MOVPE), organo-metallic vapour phase epitaxy (OMVPE), and organo-metallic chemical vapour deposition (OMCVD) are also being used, and MOCVD is a more common term for these processes.

The MOCVD process was initially developed for the growth of III–V materials at lower temperatures with less contamination from the container as compared with the melt process such as the Czochralski method. Figure 1.32 shows the early example for the growth of GaAs via the MOVPE process [62a]. Figure 1.33 shows a schematic diagram of a simple MOCVD system [62b]. Although metalorganic-halide process was first performed in the early 1960s, however, Manasevit performed the later more detailed work and he was often regarded as the pioneer for MOCVD [13].

MOCVD process has now been widely used. But one of the main drivers is the development of LED towards solid state lighting, in particular, the development of high-brightness blue light LED in

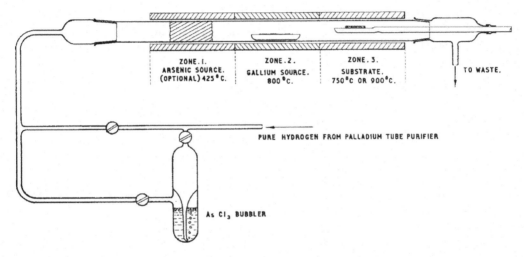

Figure 1.32 Epitaxial growth apparatus. (From Knight, J.R. et al., *Solid-State Electron.*, 8, 178–180, 1965.)

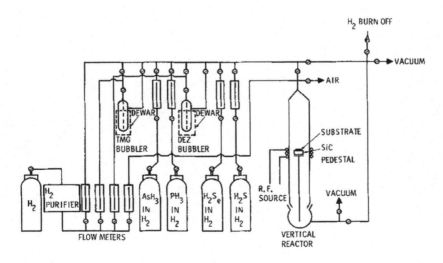

Figure 1.33 A schematic diagram of a simple MOCVD system. (From Manasevit, H.M. and Simpson, W.I., *J. Electrochem. Soc.*, 116, 1725–1732, 1969.)

the 1990s. The device is based on GaN material. In general, GaN can be deposited using trimethylgallium (TMG) and NH_3 as Ga and N source, respectively, at 1000°C–1050°C temperature range during the MOCVD process. The key requirements of the device are to obtain good quality of GaN material and p-type GaN material. In order to grow good high-quality GaN material, first it was found that a buffer layer on top of sapphire substrate can give better quality of GaN film, and further work by the Nobel laureate Nakamura showed that low temperature (450°C–600°C) growth of GaN layer can function as a good buffer layer, in particular, a two-flow MOCVD process, in which the main flow of reactant gas is parallel to the substrate, and the subflow of the inactive gas perpendicular to the substrate for the purpose of changing the direction of the main flow to bring the reactant gas into contact with the substrate. It is important to obtain uniform GaN films, without the subflow. Only a few island growths rather than a continuous film were obtained on the substrate, as shown in Figures 1.34 through 1.36. These show the micrograph of the surface of GaN film without and with GaN buffer layer [63]. This breakthrough work provides the good quality of GaN film (Hall Mobility values of 600 cm²/V·s at 300 K and 1500 cm²/V·s at 77 K for further blue LED device development led to the 2014 Nobel Prize in Physics); an example of device structure is shown in Figure 1.37 [64], and Figure 1.38 shows the electroluminescence (EL) spectra of the single quantum-well structure (SQW) LEDs at forward currents of 20 mA [64].

The chemical precursors for III–V materials are generally based on group III metals such as trimethyl Metal (Ga, Al, or In) or triethyl Metal (Ga). For group

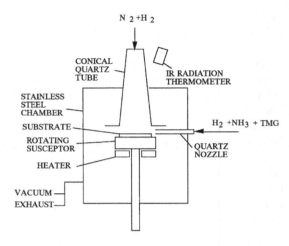

Figure 1.34 Schematic diagram of a MOCVD reactor with a two-flow process. (From Nakamura, S., *Jpn. J Appl. Phys.*, 30, L1705–L1707, 1991.)

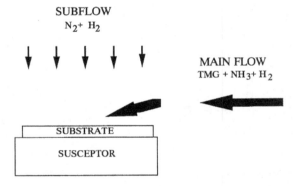

Figure 1.35 Schematic diagram of the process principle of the two-flow MOCVD. (From Nakamura, S., *Jpn. J Appl. Phys.*, 30, L1705–L1707, 1991.)

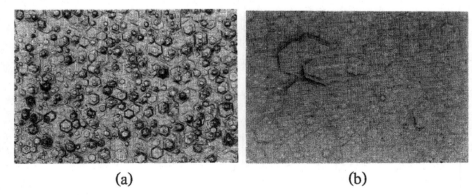

(a)　　　　　　(b)

Figure 1.36 Interference micrographs of the surface of the GaN films grown without **(a)** and with **(b)** the GaN buffer layer. (From Nakamura, S., *Jpn. J Appl. Phys.*, 30, L1705–L1707, 1991.)

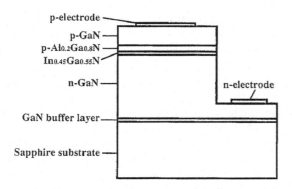

Figure 1.37 The structure of green single quantum-well structure (SQW) LED. (From Nakamura, S. et al., *Appl. Phys. Lett.*, 64, 1687–1689, 1994; Nakamura, S., *Mat. Sci. Eng B*, 43, 258–264, 1997.)

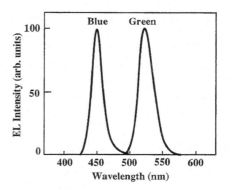

Figure 1.38 Electroluminescence of **(a)** blue and **(b)** green single quantum-well structure (SQW) LEDs at a forward current of 20 mA. (From Nakamura, S. et al., *Appl. Phys. Lett.*, 64, 1687–1689, 1994; Nakamura, S., *Mat. Sci. Eng B*, 43, 258–264, 1997.)

V-based gaseous precursors, NH_3, PH_3 or AsH_3 are often used as the precursor for N, P, As, respectively, Silane and bis(cyclopentadienyl) magnesium (Cp_2Mg) are used for n-type and p-type dopants for GaN. For II–VI materials, Diethyl Zinc (DeZn), Dimethyl Cadmium (DMCd), and Dimethyl Mercury (DMHg) are used as precursor for Zn, Cd, and Hg sources, respectively, and H_2S, Dimethyl Selenide (DMSe) is used as S and Se precursor [1]. There is also single-source precursor for the development of films, like

Sb_2Te_3 through the precursor of $(Et_2Sb)_2Te$ [65]. The list and example of metalorganic and organometallic precursors systems are used to grow III–V, II–Vi, and IV–Vi semiconducting materials as well as metallic films, and the development of new precursors have been reviewed [1,66].

Since Nakamura's work, great progress has been made to design a larger MOCVD tool towards production to handle multi-wafers. Figure 1.39 shows

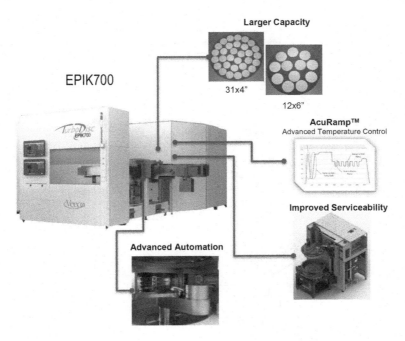

Figure 1.39 A modern MOCVD tool. (From TurboDisc EPIK 700 GaN MOCVD System for LED Production, Available at http://www.veeco.com/products/turbodisc-epik-700-gan-mocvd-system-for-led-production.)

a recent MOCVD process tool (accommodates $31 \times 4''$, $12 \times 6''$ and $6 \times 8''$ wafer carrier sizes) with automated loading/unloading and software control, which can provide homogeneous laminar flow and uniform temperature profile across the entire wafer carrier, and cost per wafer is lower than previous generation MOCVD systems [67].

1.3.1.4 MOLECULAR-BEAM EPITAXY

The MBE technique was developed for the growth of III–V thin films in the late 1960s at Bell Telephone Laboratories by J. R. Arthur and Alfred Y. Cho [68,69] and growth of III–V films (e.g., InAs) on single crystalline substrates for the study of Hall effect [70]. The term of 'molecular beam epitaxy' was first used in 1970 [70], with 'epitaxy' stemming from the Greek meaning 'stacked' or 'arranged in layers.' The crystallographic orientation is related to that of the base substrate [35]. The computer control was added to the MBE system to prepare GaAs-GaAlAs superlattice in IBM research centre in 1972. It is worth mentioning that to use the Al-doped GaAs sample from MBE in the Bell lab, two physicists at Bell Labs, Horst Störmer and Daniel Tsui, started to work on the sample in October 1981, and went on to discover the fractional quantum Hall effect, for which they shared the Nobel Prize in 1998 [71].

In CVD processes, the transport of gases is by the viscous flow when the gas pressure in the reactor is between 10^5 and 1 Pa; it becomes a molecular beam when the pressure is pumped down to 10^{-2} Pa. Generally, an MBE system includes a vacuum chamber that can reach a background high vacuum or UHV (at least 10^{-9} Torr); the process requires clean single-crystal substrates that can be difficult to obtain depending on the materials [72–74], effusion cells with individual heat shielding for the source, heat-able substrate holder, Auger spectroscopy (AES) for surface chemical analysis, low-energy electron diffraction (LEED), and/or reflection electron diffraction (RED) for in-situ structural analysis and mass spectrometers for analysis of residual gas and/or molecular beam. AES is mainly used for characterizing the initial substrate surface, which may have C or O impurities that can affect the film quality significantly, and RED is used to observe changes in the structure of the outermost layer of atoms in a crystal during the MBE growth process, qualitatively to detect changes in the surface symmetry. Figure 1.40 [35] shows the top view of an MBE tool. Figure 1.41 shows the cutaway view of effusion cell [75].

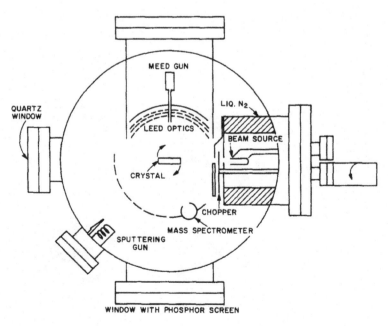

Figure 1.40 Top view of MBE system used for the measurement of beam reaction kinetics. (From Cho, A.Y. and Arthur, J.R., *Prog. Solid State Chem.*, 10, 157–191, 1975.)

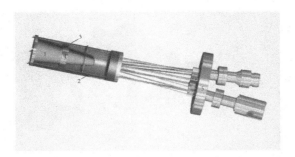

Figure 1.41 Cutaway view of an MBE thermal effusion furnace: (1) pyrolytic BN crucible, (2) resistive heater filament, (3) metal foil radiation shields, (4) wrap-around thermocouple and (5) mounting flange. (From Arthur, J., *Surf. Sci.*, 500, 189–217, 2002.)

Superthin film (like GaAs) on single-crystal substrates is built up atomic layer by atomic layer during the MBE growth process. The unique feature of MBE is the ability to prepare single-crystal layers with atomic dimensional precision.

The MBE growth process occurs on the substrate crystal surface, which plays a crucial role as it influences directly the arrangement of the atomic species of the growing film. Many studies have been performed on the interaction of substrates and growing films in order to understand the growth process. It is considered that the construction of the surface phase diagram of GaAs, by observing the As-stabilized and Ga-stabilized surface structures with high-energy electron diffraction (HEED), was the beginning of controlled epitaxial growth of GaAs thin film as shown in Figure 1.42, but the understanding of reconstruction of surface structures in MBE was not well accepted by the surface physics community in the early days, and is now well established [70]. Moreover, there is migration rate difference between adatoms (like Ga, In) on different facets to form 'kinks' in the channel and, as shown in Figure 1.43, the film in the ridge and groove is 25%–50% thicker than that on the side wall. This makes MBE on nonplanar substrate potentially useful for optoelectronic devices [75b].

The mechanisms of growth process have been studied through growth rate, surface morphology, and reaction kinetics of GaAs, as shown in Figure 1.44 [75]. The growth of GaAs may be described by the following:

$$Ga(g) + 1/2\ As_4\ (g) \rightarrow GaAs(s) + 1/2\ As_2(g)$$

Surface morphology has been studied by combined RED and electron microscopy, and the kinetics of the reaction of Ga and As_2 molecular beams incident have been investigated by modulated beam mass spectrometry. For the fabrication of devices, it needs to add controlled amounts of electrically active impurities; many elements have been added as dopant for GaAs film, including Ge [76], Sn [76], Mn [77], Te [78], Zn [79], C [77], Mg [80]. MBE can give precise control of doping, compositional profiles, and uniformity.

In its early days, MBE was used to produce low-dimensional structures; GaAs/AlGaAs superlattice was a classic example (see Figure 1.45 [70]), and (AlGa)As double-heterostructure (DH) laser, as shown in Figure 1.46 [81]. Other samples include two-dimensional electron gas (2DEG), and high mobility 2DEGs further lead to the discovery of fractional quantum hall effect in MBE-grown samples [82] and the observation of the quantisation of conductance in the units of $(2e^2/h)$ [83].

MBE method has evolved since its development in 1970. For example, the solid group III source can be replaced by metalorganic compounds; such technique is often termed as metalorganic MBE (MOMBE), which increases the source supply for MBE processes and simplifies the multiwafer scale-up. If the group V solid source is replaced by arsenic or phosphorous hydride, the techniques is often called gas-source MBE; when both III and V solid sources are replaced by gas sources, the technique is sometimes referred to as CBE [14,84].

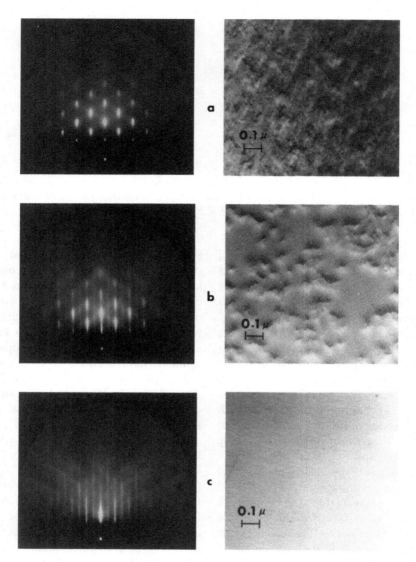

Figure 1.42 High-energy electron diffraction (HEED) patterns of (1 0 0) GaAs and the corresponding electron micrographs (38400X) of Pt-C replicas of the same surface. **(a)** Br2-methanol polish-etched (001) GaAs substrate heated in vacuum to 855K for 5min. **(b)** Deposition of an average thickness of 150Å of GaAs. **(c)** Deposition of 1μ GaAs. (From *Cho, A., J. Cryst. Growth*, 201–202, 1–7, 1999.)

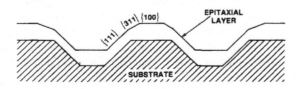

Figure 1.43 Schematic of cross-sectional view of non-planar substrate and epitaxial layer. (From Cho, A.Y., *J. Cryst. Growth*, 111, 1–13, 1991.)

Since MBE was developed in 1970, great progress has been made in other research fields, such as 1D nanostructures in the 1990s [23], and the applications of MBE method have also been extended into new areas such as topological insulator. The principle of MBE for the growth of III–V compound semiconductors can also be

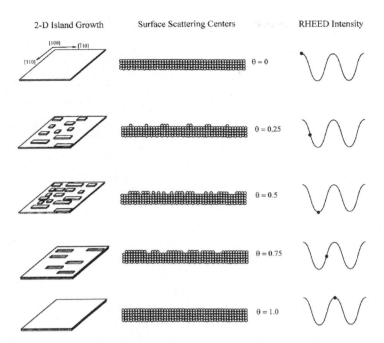

Figure 1.44 Schematic of the correlation of surface coverage of 2D clusters with idealised RHEED oscillations. (From Biasiol, G. and Sorba, L., *Crystal Growth of Materials for Energy Production and Energy-Saving Applications*, Fornari, R., Sorba, L., Eds., Edizioni ETS, Pisa, Italy, pp. 66–83, 2001.)

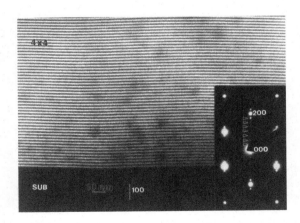

Figure 1.45 Scanning electron micrograph of the cross-sectional view of alternating layers of GaAs (dark lines) and AlxGa1-xAs (light lines) grown by MBE. (From Cho, A., *J. Cryst. Growth*, 201–202, 1–7, 1999.)

applied to the growth of group IV, II–VI, metal and insulation materials, oxide materials [85], as well as for the preparation of nanostructures such as GaAs/AlGaAs superlattice (SL) [86] and 2D materials (e.g., graphene) [87].

The research on topological insulators (e.g., Bi_2Se_3), a type of material that are insulators in the 'bulk' and have exotic metallic states present at their surfaces due to the topological order, has generated intense interest in this new state of quantum matter from both materials science and condensed matter physics. MBE is the major method used for such development due to the possibility to avoid defect formation by controlling the growth conditions [45], and vapour-solid synthesis method is utilized to grow few-layer nanoplates of Bi_2Se_3 on amorphous dielectric [88]. Further research on topological insulators may result in new spintronic, magnetoelectric devices and enhance the development of quantum computing [89]; this is an emerging field, and its development may drive forward the development of MBE processes.

Conventional MBE would use solid precursor sources. With the development of the MBE technique, various nomenclature was used to name the MBE process depending on the precursors used. The technique with only the metalorganic group III component is the gaseous source and a solid group V source was named as MOMBE. The method with solid group III component and gaseous group V hydride sources gas source was named as GSMBE [90].

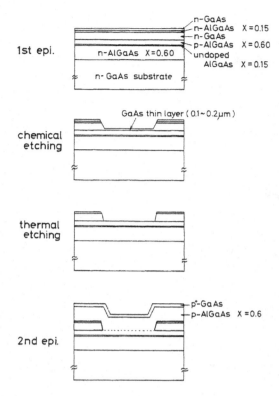

Figure 1.46 Schematic illustration of the process and structure of self-aligned laser. (From Tanaka, H. et al., *Jpn J. Appl. Phys.*, 24, L89–L90, 1985.)

Figure 1.47 A photo of CD Drive that uses laser made from the material grown by MBE. (From Dirkhb, Philips CDM210 CD Drive, 2006, https://commons.wikimedia.org/wiki/File:CDM210_cd_laufwerk.jpg, accessed on January 3, 2019.)

The MBE system can be clustered with other facilities, such as focused ion beam implantation and etching operation. The application of focused ion beam technology to maskless ion implantation in a MBE-grown GaAs or AlGaAs epitaxial layer for three-dimensional pattern doping crystal growth has been demonstrated [91] as an example of a multi-chamber system for the production of complicated device structures entirely in the UHV environment grown by MBE [92]. For the research of topological insulators, MBE is linked to an UHV system (e.g., Omicron), equipped with scanning tunnelling microscope (STM) and angle-resolved photo-emission spectroscopy (ARPES) [93]. In this type of equipment, the high-quality films grown by MBE would allow the investigation of topological surface states and the effect of those intrinsic defects directly by *in-situ* scanning tunnelling microscopy/spectroscopy (STM/STS) [94,95].

In addition to the fundamental research on the process, MBE method has been used for many device structures, such as a laser that is produced with MBE in commercial quantities for compact disc applications [40]. Figure 1.47 shows a photo of CD drive consisting of laser made from the material (AlGaAs) grown by MBE [96]. GaAs-based hetero-junction bipolar transistor (HBT) is used in mobile wireless devices as the RF power amplifier (PA), which is used to drive the transmitted signal through the antenna of the device, and GaAs-based p-HEMTs (high-electron-mobility transistors) is used as a low-noise amplifier (LNA) and an antenna switch for mobile wireless applications, as shown in Figure 1.48 [98], which play an important role in the development of production of MBE market [97].

Modern production MBE systems can provide high production rates using industry standard, large-area substrates, and by growing on multiple wafers simultaneously to maximize throughput for commercial fabrication of MBE devices. The latest features of the legendary MBE 7000 include the accommodation of up to 13 source ports, a load capacity of 70 of the 6-inch wafer or 140 of the 4-inch wafer, and a vast suite of *in-situ* monitoring tools. The high throughput of this industrial-leading tool is circa. 24,000 of 6-inch wafer per year [99a].

Since 1970, MBE research focused on the understanding of processes in the first decade; production of low-dimensional structures in the second decade; and the development of MBE for the mass market in the third decade. New developments of hardware, materials, and processes are continuing being developed in the twenty-first century; an example of a laser MBE system that is attached to a synchrotron beamline shown in Figure 1.49 [99b].

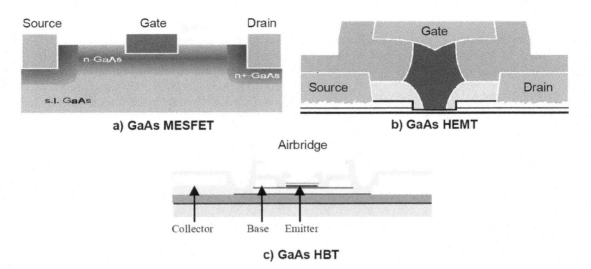

a) GaAs MESFET

b) GaAs HEMT

c) GaAs HBT

Figure 1.48   Scheme of GaAs-based device structures: **(a)** GaAs MESFET, **(b)** GaAs HEMT and **(c)** GaAs HBT. (From Berger, O., GaAs MESFET, HEMT and HBT competition with advanced Si RF technologies, *GaAs Mantech*, 4–7, 1999.)

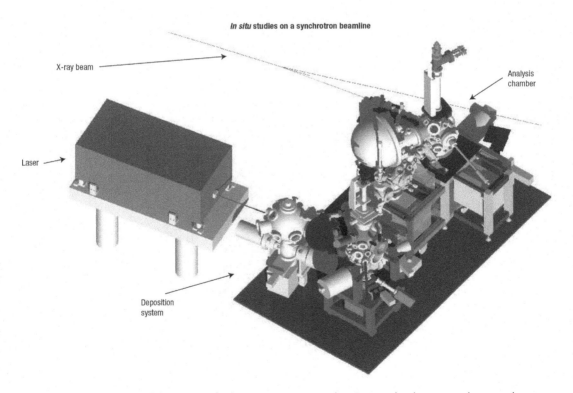

Figure 1.49 A schematic illustration of a laser MBE system that is attached to a synchrotron beamline. Such a system makes it easier to probe the electronic structure of surfaces and interfaces as the heterostructure is being grown. In the field of complex oxides, such *in-situ* facilities are just emerging. (From Ramesh, R. and Spaldin, N.A., *Nat. Mater*, 6, 21–29, 2007.)

There is growing need for the smart devices, like smart mobile phones, for wireless communications that need RF components, prepared from GaAs that is produced through the MBE process.

1.3.1.5 CHEMICAL-BEAM EPITAXY

CBE evolved from MBE for the preparation of III–V films was first demonstrated in the 1980s. Conventional MBE would use solid precursor sources for III and V element, and the technique with both components being gaseous sources was called CBE. In some literatures, it was also being called MOMBE by the early researchers, while the term of CBE has been gradually widely adopted [14,90].

To some extent, CBE was developed by combining the features of MOCVD and MBE; it is performed in an UHV system using molecular beams of reactive gases as sources. The original CBE has been used for the growth of III–V semiconductor layered systems. As compared with MOCVD processes that can be run under low-vacuum or even atmosphere and the precursor is viscous and the chemical reaches the substrate through diffusion, CBE requires UHV chamber ($<10^{-4}$ Torr), which is similar to MBE, and the mean-free paths between molecular collisions are longer than the

source inlet and substrate distance, so the gas transport is molecular beam. Although metalorganic precursors are used in both processes, in CBE process, the precursors for III and V sources are delivered separately, whereas precursors are delivered together in conventional atmospheric MOCVD.

Figure 1.50 shows the gas handling system and growth chamber with *in-situ* surface diagnostic capabilities incorporated into a CBE system [84]. MOCVD uses a large quantity of precursors, and the poor efficiency of source usage imposes the issue of handling toxic exhaust gases. CBE requires much lower level of AsH_3 and PH_3 precursors, and it is relatively easier to treat the exhaust; thus, CBE provides improvement over MOCVD, while MBE does not have the issue of toxic gases as solid sources are being used.

In CBE processes, the pressure is less than 10^{-4} Torr, the metalorganic precursor transport occurs as molecular beam due to the much longer mean-free paths, whereas a pressure within the range of 10^2 Torr–1 atm is being used for MOCVD, the transport of gas occurs by the flow and chemicals reach the surface by diffusion. CBE uses the source in vapour phase at room temperature; the gaseous sources react in complex reactions on the hot

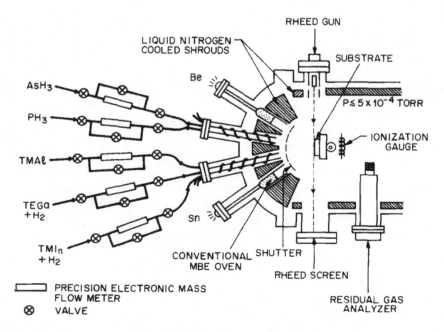

Figure 1.50 Gas handling system and growth chamber with *in-situ* surface diagnostic capabilities incorporated in a CBE system. (From Chiu, T. et al., *J. Appl. Phys.* 62, 2302–2307, 1987.)

growing surface to deliver the material needed for the layered growth, whereas MBE employs atomic beams (e.g., Ga) and molecular-beam (e.g., As$_4$) that are being evaporated from solid sources heated at high temperatures (e.g., 910°C for Ga) [100].

With the gaseous precursors, CBE process has many advantages as compared with MBE process:

1. CBE can provide a good control on the p-doping level of GaAs, using Triethyl Gallium (TEGa) instead of Trimethyl Gallium (TMGa) in order to deposit high purity GaAs material with the room temperature, hole concentrations between 10^{14} and 10^{16} cm^{-3} could be produced, whereas hole concentrations around 10^{20} cm^{-3} were obtained using TMGa. Consequently, TMGa could be used as a p-type carbon dopant to obtain acceptor (C) doping levels, by adding TMGA to the basic Ga source of TEGa, the desired p-type doing level can be achieved [101,102].

   This can also be used for the fabrication of high-quality AlGaAs/GaAs DH layers, which can give high threshold current densities comparable to MBE or MOCVD grown structures (<0.5 kA/cm²) [103a]. Moreover, this advantage can be used for the device fabrication to reach high and sharp p-type doping profiles using carbon. An example of such application is the fabrication of high-quality heterostructure bipolar transistors (HBTs) using heavily carbon-doped GaAs as the base [103b]. Figure 1.51 shows a schematic diagram of various processes for MBE, MOCVD, and CBE [84].

2. CBE process can be used for the selective growth of the epitaxy of III–V compounds (like InP), i.e., growth occurs only in the targeted area of the substrate and there is no deposition in other areas of the substrate. Such feature is of interest for the monolithic integration of electronic and optoelectronic devices. Since only surface reactions are important, the composition of the selectively deposited material does not depend on dimensions and directions of the mask openings, which is different from MOCVD for the selective growth of III–V films, CBE is the ideal technique for selective growth of III–V films (like In and GaInAs) on partially masked substrates, in particular when ternary

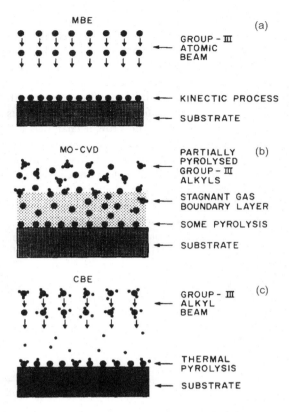

Figure 1.51 Growth kinetics involved in **(a)** conventional MBE, **(b)** OMCVD and **(c)** CBE. (From Chiu, T. et al., *J. Appl. Phys.*, 62, 2302–2307, 1987.)

compounds of a well-determined stoichiometry are to be deposited [103c].

3. CBE process enables a good process control, in which the use of room-temperature vapour sources for both group III's and V's with electronic MFCs would allow the precise, stable, and reproducible control of flows; this is particularly important in depositing lattice-matching materials such as Ga$_x$In$_{1-x}$A$_{1-y}$P$_y$ on InP, which is very difficult to be achieved by conventional MBE due to the requirements for high degree of accuracy, stability, and reproducibility in effusion cell temperature control for As and P. Moreover, in CBE process, the different III- or V- precursors and dopant gases can be pre-mixed to form a group-III and group-V beam instead of individual beam for each element. This can provide a high degree of material composition and thickness uniformity, which is very attractive for the scale-up of multi-wafer [104].

From the perspective of hardware, a CBE system has a dedicated precursor delivery component for different precursors, and a growth chamber equipped with a reflection high-energy electron diffraction (RHEED) for the *in-situ* diagnosis of surface reconstruction and smoothness, and with a residual gas analyser for measuring the gaseous species and amount in the growth chamber. For a CBE chamber, conventional MBE effusion cells containing charges of condensed phase elements of group III, group V, and dopants can also be incorporated into the same chamber. As a result, epitaxial growths with various combinations of gaseous and elemental sources can be made.

The measurement of RHEED intensity oscillations during CBE of IIIV and group-IV materials can provide the sub-monolayer precision thickness control of CBE, and RHEED oscillations in CBE also provided a very fast and accurate *in-situ* measurement of growth rate versus growth parameters, such as TEGa flow rate or substrate temperature,

which provides useful information regarding the growth mechanisms of CBE of GaAs [105].

In the 1990s, there was work to develop cluster and production system for CBE, as shown in Figure 1.52 [106].

There is a new type of chemical vapour beam deposition process, as shown in Figure 1.53 [107], it uses chemical beam approach to deposit films.

In general, CBE has demonstrated a wide range of advantages including easy deposition and doping of complex materials from various precursors, high composition, and thickness uniformity even on large substrates, high precursor conversion rate, high growth rates, high reproducibility, compatibility with UHV characterization techniques, compatibility with laser beam structuring, and surface selective growth.

As GaAs-related market is dominated by MBE, and CBE has some potential on this market, but the development of hardware (source) and control, and the scale-up for production with MBE tools. There

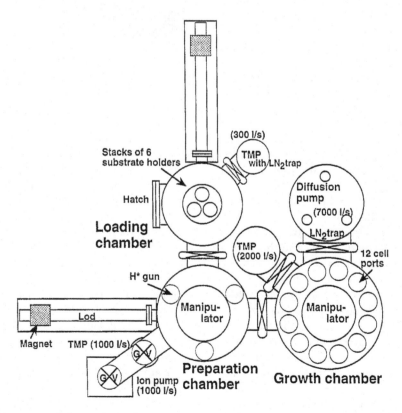

Figure 1.52 Schematic of the multi-wafer CBE system consisting of three chambers with a magnet-coupled semi-automatic transfer system. (From Ando, H. et al., *J. Cryst. Growth*, 164, 1–15, 1996.)

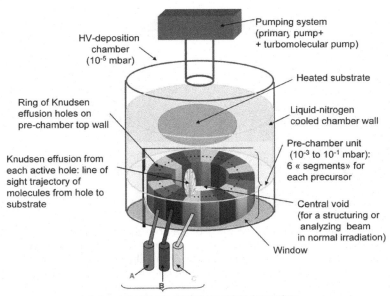

HV-deposition
chamber
(10^{-5} mbar)

Pumping system
(primary pump+
+ turbomolecular pump)

Heated substrate

Ring of Knudsen
effusion holes on
pre-chamber top wall

Liquid-nitrogen
cooled chamber wall

Knudsen effusion from
each active hole: line of
sight trajectory of
molecules from hole to
substrate

Pre-chamber unit
(10^{-3} to 10^{-1} mbar):
6 « segments» for
each precursor

Central void
(for a structuring or
analyzing beam
in normal irradiation)

Window

A   B   C

Independent thermostated precursor reservoirs

Figure 1.53 Schematic layout of the chemical beam vapour deposition system (Sybilla 150 from ABCD Technology). Three different precursors (labelled A, B and C) are evaporated from thermostatically controlled reservoirs into a compartmentalised prechamber, with 6 areas per precursor. Precursor molecules effuse from the prechamber into the deposition chamber through Knudsen holes with a cosine angular distribution and have ballistic trajectories to the heated substrate where they decompose. Liquid nitrogen–cooled walls trap all unreacted molecules and decomposition byproducts to maintain the deposition chamber under high-vacuum conditions. (From Wagner, E. et al., *ACS Comb. Sci.*, 18, 154–161, 2016.)

is almost no report for commercial CBE large-scale production tool, but CBE process is still used in the research and development for the new materials, like nanowires (GaAs) and quantum dots (InAs) [107].

1.3.1.6 ATOMIC LAYER DEPOSITION

ALD is also named 'atomic layer epitaxy' and 'molecular layering' in the literatures from different origins. In the 1960s, the Aleskovskiia group at the University of Leningrad (now St. Petersburg) independently developed the deposition process of oxide layers, which has the essence of ALD process, and the group referred to the process as 'molecular layering.' Their work was not known to the research community until 1990, and the related samples, equipment, and applications were not circulated. In 1974, the ALD process was developed by Tuomo Suntola at Instrumentarium Oy, Finland to prepare ZnS EL films for flat panel displays, and the term 'atomic layer epitaxy' was used, which was not termed as 'atomic layer deposition' until the late 1990s. Because 'epitaxy' originated

from Greek and means 'on-arrangement,' the semiconductor community devotes the term to single-crystal growth, and 'atomic layer epitaxy' was less well accepted by the semiconductor community. For a short period of time, the term of 'molecular layer epitaxy' was also used to refer to the ALD process. Generally, the current ALD method was attributed to Dr Tuomo Suntola, who is regarded as the inventor for the process [15]. The tool and schematic diagram for the ALD process used to deposit ZnS layer at 320°C for EL display is shown in Figure 1.54 [15].

The elemental Zinc and Sulphur were used as the precursor for Zn and S, respectively, and the substrate was rotated. The term of ALE was used to emphasize the surface control feature of the process in order to make a distinction with traditional source-controlled deposition methods such as evaporation and sputtering. Further optimisation was to replace the high-vacuum equipment with more compact flow-type reactors allowing the use of volatile reactants

(a)

(b)

(c)

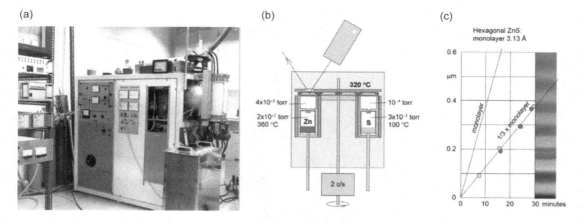

Figure 1.54 **(a)** The ALE reactor was built in the T piece on right of the pumping unit. **(b)** It consisted of a rotating carousel holding the substrates, and the sources for the reactants. Windows in the top flange allowed observation of the interference colour of the film building up in the course of the process. **(c)** By comparing the colour to an interference colour map the increase in thickness could be monitored during the process. (From Liu, X.W. et al., *J. Vac. Sci. Technol. A*, 32, 01A116, 2014.)

and a batch-type arrangement of the substrates, effectively utilizing the self-control feature of ALE, in which $ZnCl_2$ (heated at 380°C) was used as Zn source and H_2S was used as S source, $MnCl_2$ (heated at 510°C) as the doping source of Mn, and deposited at 450°C at 2 mBar, using Corning Glass 7059 as the substrate and undergoing the chemical reactions as follows [108]:

$$ZnCl_2 + H_2S \rightarrow ZnS + 2HCl$$

The resulting ZnS film was used to produce EL display in 1980, and the process was gradually spread out across the world. The first proof-of-concept EL display was installed in the Helsinki-Vantaa airport in 1983, as shown in Figure 1.55, and it has run continuously for 15 years, day and night, without a single character module being replaced. EL display is also the first product developed with ALD process [15].

ALD process can be performed in two modes: (a) temporal mode and (b) spatial mode. In temporal

Figure 1.55 Dr. Ralf Graeffe inspects the display modules in the test assembly in Helsinki-Vantaa airport underground cave in 1983. (From Liu, X.W. et al., *J. Vac. Sci. Technol. A*, 32, 01A116, 2014.)

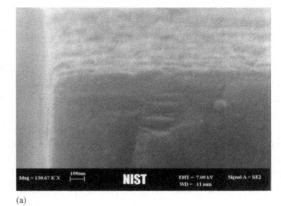

Figure 1.56 Schematic of vapour-phase, self-limiting atomic layer deposition (ALD) illustrating the cyclic process by which an Al_2O_3 surface coating is controllably constructed layer-by-layer from TMA and H_2O precursors. (From Peng, Q. et al., *Nano Lett.*, 7, 719–722, 2007.)

mode, ALD process can be treated as several steps; each step is followed by another step, for example, the deposition of Al_2O_3 film using Trimethylaluminium (TMA) and H_2O as the precursors, as shown in Figure 1.56, is proposed to have the following steps: (1) TMA reaction step, TMA vapour absorbs to the substrate surface that have hydroxylated functional group (–OH), and TMA reacts with these groups to form Al-O-Al$(CH_3)_2$ on the surface; (2) purge step, the excess or unreacted TMA and the resulting compound (like CH_4) will be purged, and the surface now contains –CH_3 group; (3) H_2O reaction step, H_2O reacts with Al-O-Al$(CH_3)_2$ to form Al-O-Al$(OH)_2$ on the surface; (4) Purge step, the excess or unreacted H_2O and the resulting compound (like CH_4) will be purged, and the surface now contains –OH group [109]. ALD is a self-limiting process; if the surface is covered in a step, no more surface reaction can occur during this step. Together these steps are treated as a cycle, and each cycle can deposit about 0.1 nm thick of film, so more cycles are needed, and the process can repeat after a cycle until the desired thickness is achieved. The overall process can take a long time. To address this time-consuming issue of temporal ALLD, spatial mode ALD was developed. The detail is discussed later on.

The surface reaction of Al_2O_3 can be summarised as [110]

(A) $AlOH + Al(CH_3)_3 \rightarrow Al\text{-}O\text{-}Al(CH_3)_2 + CH_4$

(B) $Al\text{-}O\text{-}Al(CH_3)_2 + 2H_2 \rightarrow Al\text{-}O\text{-}Al(OH)_2 + 2CH_4$

ALD can be used to deposit conformal coating, as shown in Figure 1.57, Al_2O_3 film can be deposited

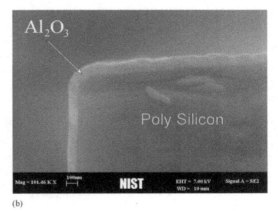

(a)

(b)

Figure 1.57 **(a)** SEM close-up of tip of uncoated cantilever beam. The beam was broken by a probe tip prior to investigation. **(b)** SEM close-up of the tip of a cantilever beam coated with 80 nm Al_2O_3. The picture clearly illustrates the conformability of the ALD film as it covers both top and sidewalls of the beam. (From Hoivik, N.D. et al., *Sens. Actuators A Phys.*, 103, 100–108, 2003.)

covering both top and sidewalls of devices [111]; this feature is important for the application of ALD process in the semiconductor industry.

Further development of the process was applied for the preparation of difference materials, such as III–V materials (e.g., GaAs) [15], the process principle is similar, but precursors would vary. In particular, with the requirement of a decrease in the size of Si-transistor, ALD process was used to deposit dielectric layers using the ALD production tool [112]. Figure 1.58 shows a variety of transistor structures, and ALD-deposited film ($SiON/HfO_2$) has already been used for the latest tri-gate device with 22 nm node [113].

This ALD process is a temporal process, in which the substrate does not move, a process step is followed by another step, and the film is deposited layer-by-layer, so the overall process is slow. In order to speed up the overall process, a spatial process has been developed. Figure 1.59 shows a scheme for atmospheric pressure spatial ALD (AP-SALD) [112b]. Figure 1.60 shows a detailed spatial ALD concept [41], the precursors are dosed simultaneously and continuously, but at different half-reaction zones, and the substrate moves between these zones where the half-reactions take place. When the substrate is

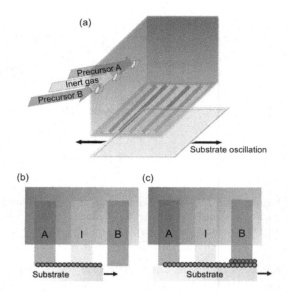

Figure 1.59 (a) AP-SALD manifold with alternating flows of precursor gases. (b) The substrate first moves underneath precursor A, absorbing a monolayer of the precursor, and then through the purging, inert (I) gas flow. (c) The substrate then moves underneath precursor B, which reacts with A to form a monolayer of the desired material. (From Musselman, K.P., *Chem. Mater.*, 28, 8443–8452, 2016.)

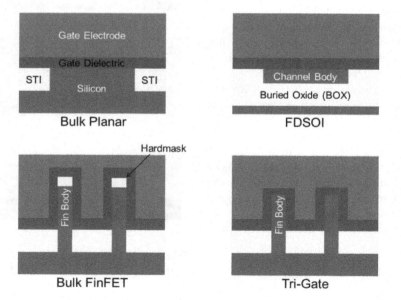

Figure 1.58 Schematic cross-sections across the channel, looking from source to drain, of the transistor comparing traditional bulk planar with fully depleted silicon on insulator (FDSOI), bulk FinFET and tri-gate device architectures, which have been or will be implemented at the 22 nm and below device nodes. (From Clark, R., *Materials*, 7, 2913–2944, 2014.)

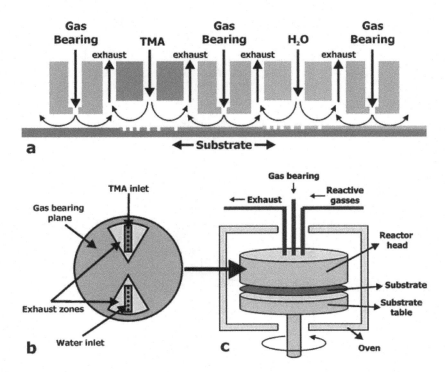

Figure 1.60 **(a)** Schematic drawing of the spatial ALD reactor concept, where the TMA and water half-reaction zones are separated by gas bearings. By moving the substrate underneath the reactor, the two half-reactions will take place sequentially to form an Al_2O_3 monolayer. **(b)** Schematic drawing of the bottom side of the spatial ALD reactor head, where the TMA and water half-reaction zones are integrated into inlets surrounded by exhaust zones and gas-bearing planes. **(c)** Schematic drawing of the reactor. The reactor head and rotating substrate table with the substrate in between are placed in an oven. The substrate table is rotated by a servo motor, connected by a drive shaft. The process- and waste gas lines are connected to the reactor head through an opening in the top of the oven. (From Poodt, P. et al., *J. Electrochem. Soc.*, 419–427, 2010.)

non-flexible, as shown in Figure 1.61 [41], a substrate would move along a linear row of repeating half-reaction zones (called 'cells') where the total number of cells determines the overall layer thickness. In a similar way, another type is to move the substrate back and forth in a reciprocating manner under a spatial ALD injector, where the overall thickness is determined by the number of passages underneath the injector. Similarly, when the substrate is flexible, roll-to-roll ALD process can be performed in a spatial mode, as shown in Figure 1.62, the substrate moves in a rotational way to pass the precursor zone, and the thickness of film can be controlled by adjusting the rotation frequency of the drum (typically rotating counter-clockwise) in combination with the translation speed of the foil. The process is explored to deposit Al_2O_3 film to act as

a barrier coating onto flexible polymer substrate for packaging applications [41d].

ALD process has already been used in the industry, for example, ASM's Pulsar ALD system has been used to deposit the high-k dielectric materials required for advanced CMOS transistor gates and other applications such as microelectromechanical systems (MEMS) applications [112]. It was the first ALD system to be used for high-volume production for high-k metal gate transistors [113].

The temporal process tool is often designed for batch process, and the spatial process tool can be designed for a continuous process, it is still at the stage for R&D and pilot production for the package applications of flexible electronics. One example is shown in Figure 1.63 [114].

The materials deposited through ALD processes include metal, oxide, nitride, and polymer, and the

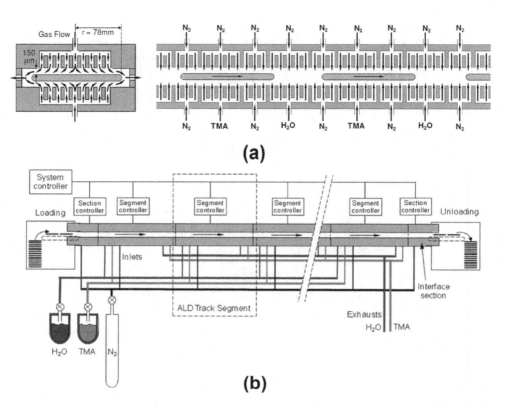

(a)

(b)

Figure 1.61 **(a)** Cross-section of the Levitrack spatial ALD reactor. The TMA (tri-methyl-aluminium) and H_2O vapours are mixed with N_2 to a concentration of a few percent. **(b)** Layout of the Levitrack system. (From Granneman, E.H.A. et al., *ECS Trans.*, 61, 3–16, 2014.)

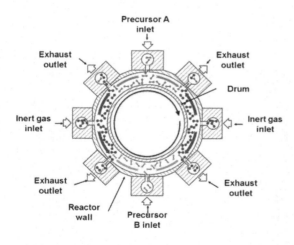

Figure 1.62 Schematic radial cross-section of the continuous ALD system. (From Maydannik, P.S. et al., *Chem. Eng. J.*, 171, 345–349, 2011.)

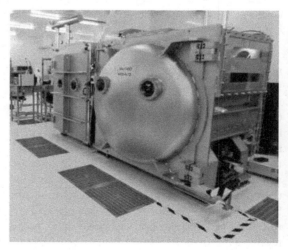

Figure 1.63 ALD thin film system for OLED encapsulation. (From Beneq Roll to Roll Atomic Layer Deposition Tool, Available at https://www.uk-cpi.com/technologies/printable-electronics/facilities-equipment/beneq-roll-to-roll-atomiclayer-deposition-tool.)

processes can provide deposition on surface, polymer, nanomaterials (like CNTs), and nanoparticles [115].

There are various types of ALD processes. The typical advantages for the ALD process are:

- Self-limiting process mechanism provides precise control on the process
- Low processing temperature (<150°C) for sensitive substrates
- Wide operating pressure range for easy line integration
- System deposition speed is highly scalable—could reach 10 m/min speeds
- Thickness is determined simply by number of deposition cycles. Precursors are saturatively chemisorbed => stoichiometric films with large-area uniformity and 3D conformality. Relatively insensitive to dust (films grow underneath dust particles)
- Intrinsic deposition uniformity and small source size => easy scaling
- Nanolaminates and mixed oxides possible
- Low-temperature deposition possible (RT-400°C)
- Gentle deposition process for sensitive substrates
- A broad range of substrates (or materials) can be used
- Conformal coating for broad features including high-aspect-ratio trench

1.3.1.6.1 The limitations of ALD

- Very low growth rate, it is not practical to grow thick films by ALD.
- Many important materials (Si, Ge, Si_3N_4, several multicomponent oxides, certain metals) cannot be deposited by ALD in a cost-effective way.
- Residues may be left from the precursors, which may contaminate the sample.
- Need to perform frequent maintenance to the equipment
- There is difficulty to obtain suitable precursors for some films

Compared with the conventional CVD processes, there are several features different between ALD and CVD, which may affect the selection of the proper process:

1. The way of precursor reaction with the surface
   As mentioned earlier, ALD precursor reacts separately on the surface, while CVD precursors react at the same time on the surface.

2. The stability of precursor
   As ALD reaction occurs on the surface, ALD precursor cannot decompose before reaching the surface. This is not essential in conventional CVD processes.

3. The process mechanism
   As ALD is self-limiting, excess precursor does not affect the growth rate, while CVD growth rate can depend on the supply of precursors.

These special features may have impact on the uniformity, thickness control, conformality, and other characteristics of processes.

Overall, the hardware for ALD processes evolved with the development of this processing method. Generally, the dimension of the substrate is one of the important factors for the design of the hardware. ALD has been used in the semiconductor industry for the production of films for microelectronics [116]. ALD process has continued to develop. In the early development stage, ALD was a thermal process; subsequently, plasma was introduced to the ALD process [117]. The temporal ALD process can be a thermal or plasma-based process, and the spatial ALD process is still the thermal process.

1.3.1.7 ENERGETIC ION-ASSISTED CVD

Ion-assisted process has evolved with the development of technology. Figure 1.64 shows a list of milestone developments of related processes, dating back to 1907. J. J. Thomson established the fundamental concept of the ion implantation equipment that is used nowadays in industry, which has a gas discharge type of ion source, an electrostatic deflection system, and a magnet. In the 1930s, the high voltage sources were developed to accelerate charged particles. In the 1950s, ion implantation process was developed with a BF_3 glow discharge source in order to implant ions into semiconductor wafers through mask apertures by Shockley, and ion implantation process was applied in the semiconductor industry in the 1960s, and became an accepted technology in the 1970s due to its special ability to produce devices that are unobtainable by other processes [118].

Ion beam technology has been used for implantation, etching, and ion-assisted film deposition. That is the focus of this part [119].

There are several forms of energetic ion-assisted CVD, of which ion beam-assisted deposition (IBAD) is a process consisting of the hybridisation of ion beams and PVD, and it can operate in PVD

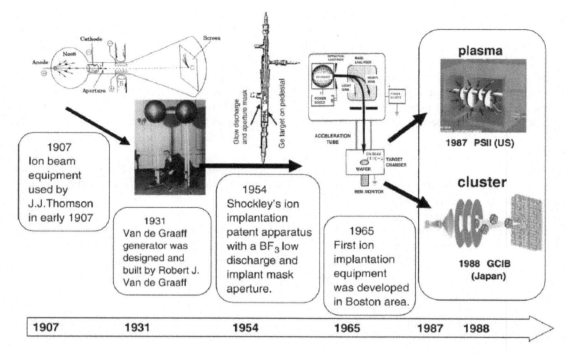

Figure 1.64 Industrial ion beam equipment development history. (From Yamada, I. and Toyoda, N., *Surf. Coat. Tech.*, 201, 8579–8587, 2007.)

mode without chemical reactions or CVD mode with chemical reactions. Besides IBAD, other terms include ion-assisted coating (IAC), ion-assisted deposition (IAD), ion vapour deposition (IVD), ion beam-enhanced deposition (IBED), and dynamic recoil mixing (DRM) at high energies [120]. The effect of ions on the properties of films was studied as early as 1963, and it was found that energetic ions within plasma have an important influence on coating properties in the early development of ion plating [119].

IBAD includes non-reactive processing and reactive processing. The former can be regarded as a PVD process, which is used for the deposition of thin film like NiFe, Co, Cu in the industry for the production of sensors and data storage devices [121], and the latter can be regarded as a CVD process. Figure 1.65 shows a schematic diagram of an ion-beam-assisted deposition system [122].

As shown in Figure 1.66 [127], the steps are proposed during the IBAD process for the

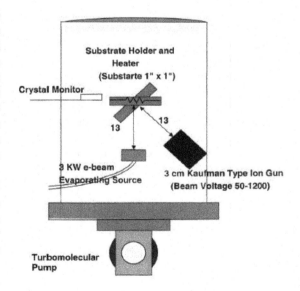

Figure 1.65 Schematic of an ion beam–assisted deposition IBAD system. (From Ma, C.-H. et al., *Surf. Coat. Technol.*, 133–134, 289–294, 2000.)

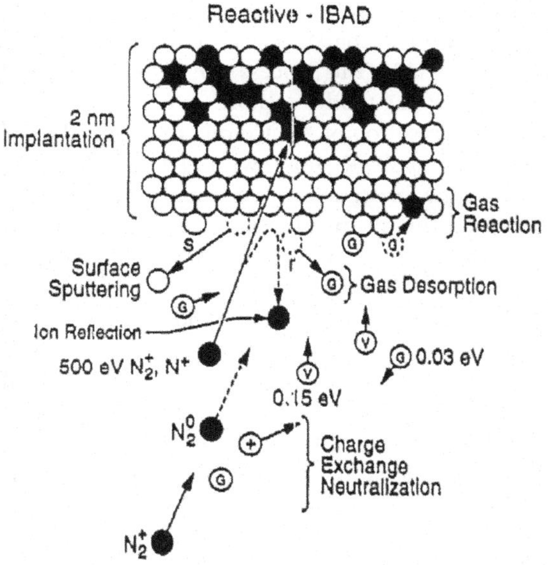

Reactive - IBAD

2 nm Implantation

Gas Reaction

Surface Sputtering

Ion Reflection

500 eV N_2^+, N^+

Gas Desorption

0.03 eV

0.15 eV

N_2^0

Charge Exchange Neutralization

N_2^+

Figure 1.66 Schematic representation of the IBAD processes. (From Cotell, C.M. and Hirvonen, J.K., *Surf. Coat. Technol.*, 81, 118–125, 1996.)

synthesis of Si_3N_4 film via a CVD process to predict the composition of silicon nitride using the evaporation of silicon and nitrogen bombardment from a Kaufman-type ion source. It involves six parameters: (1) the effective number of nitrogen atoms per ion, (2) the reflection coefficient of nitrogen ions from the substrate, (3) the sputtering yield of substrate atoms, (4) the ion beam neutralization factor, (5) the ion current, and (6) the Si deposition rate. Other films from reactive IBAD include TiN and AlN. IBAD processes have been used to deposit many types of coatings, in particular optical coatings (Al_2O_3 films) and magnetic thin films (Fe film) [120], B-C-N hard coating [123], and III–V materials like InN through localised ion source. For example, in the case of InN film, In-wire hot filament as localised ion source for In ions, and N_2 plasma, was generated through RF power (like 100W), so that InN film can be obtained [124].

There are different ways to extract ions from the source. One example is shown in Figure 1.67 [125]: ions are extracted from the plasma within the sources through apertures in aligned grids

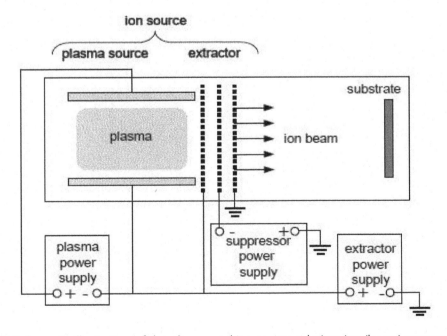

Figure 1.67 Schematic illustration of the plasma and ion source relation: ion (beam) sources are generally comprised of a plasma source (plasma generator) and an ion extraction system. (From Anders, A., *Surf. Coat. Technol.*, 200, 1893–1906, 2005.)

of the source. These sources are commercially available. Other types of sources include end-hall source, ECR ion source [120], and RF ion source [123]. Moreover, it is possible to obtain large-area ion source. For example, 600 mm wide ion source was produced using three independently controlled hot filaments and 5% beam uniformity can be achieved, as shown in Figure 1.68 [126].

Another type of ion-assisted CVD is ionized cluster beam (ICB) deposition, which has been developed since 1972. The process involves the deposition of loosely coupled clusters of atoms formed during passage through a nozzle from a heated crucible, as shown in Figure 1.69 [128]. The clusters pass through an ionizing electrode assembly located above the crucible, and a fraction of clusters are ionized and subsequently accelerated towards the substrate by a variable applied potential (1–10 KV). This process can allow low-energy deposition without the space charge problems often encountered with low-energy

ion beams, and un-ionized clusters are also deposited onto the substrate at thermal ejection velocities [119]. The process can also be operated through PVD mode without chemical reaction for the deposition of Au, Cu on glass, Ag, Al on Si substrate, or CVD mode with chemical reaction during the process for the deposition of FeOx films via the chemical reaction of using ionized O_2 with Fe clusters and atoms with the O_2 pressure of 0.12×10^{-4} Torr. Crystalline Fe_2O_3 was obtained without acceleration voltage and Fe_3O_4 with an accelerating voltage of 3 kV [129].

In the 1980s, the gas cluster ion beam (GCIB) process was developed; a schematic apparatus is shown in Figure 1.70 [118]. Gas clusters form through supersonic gas expansion from a high-pressure source into a vacuum via a small orifice or nozzle, which determines the cluster size and the beam flow pattern. The relative velocity of the gas atoms reduces due to the adiabatic expansion and condenses into clusters; the size of gas

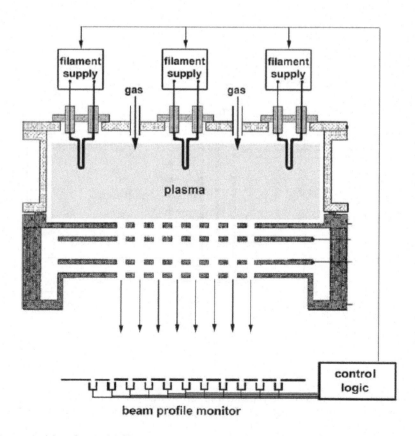

Figure 1.68 An example of a multi-filament ion source with a beam profile monitor. (From Anders, A., *Surf. Coat. Technol.*, 200, 1893–1906, 2005.)

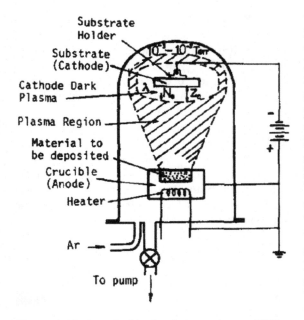

Substrate
Holder

Substrate
(Cathode)

Cathode Dark
Plasma

Plasma Region

Material to
be deposited

Crucible
(Anode)

Heater

Ar

To pump

Figure 1.69 A typical ionised cluster beam (ICB) system. The vacuum system and chamber as well as the power supplies are omitted for clarity. (From Takagi, T., *Thin Solid Films*, 92, 1–17, 1982.)

clusters ranges from 500 to 10,000 atoms held together by van der Waal's forces. A small aperture (or skimmer) allows the primary jet core of gas clusters to pass the neutral clusters, which are ionized by impact of electrons emitted and accelerated from a filament to form positive ion gas clusters (nominally one electron charge per cluster). The ionized clusters are extracted and accelerated by the potential of 2 KV–25 KV, and then the cluster ions are focused by electrostatic lenses, whiles monomers are filtered out through the use of transverse magnetic fields. The process has been developed to ultra-shallow B-doping of Si using Ar and B-containing molecular clusters from sources like B_2H_6 and BF_3 for ultra-shallow junction formation and doping towards semiconductor device applications, surface smoothing, low-k dielectric surface treatment for IC process applications, and reactive growth of thin films at low temperatures. One example is deposition of TiO_2 film; the chamber pressure was 2×10^{-5} Pa, a mixture of O_2 and He was introduced into vacuum through the nozzle, neutral O_2 cluster was ionized by electrons with ionization voltage of 300 V with the mean cluster size of about 2000 O_2 molecules. A metal Ti pellet was used as source material and evaporated through electron beam evaporator, with the acceleration voltage of 7 KV at 200°C, the well c-oriented rutile TiO_2 film was formed with average roughness of 0.4 nm, without assistance of GCIB rough amorphous film was formed [130]. Figure 1.71 shows AFM image of TiO_2 film from GCIB process [130].

From a general viewpoint, ion-assisted CVD processes also refer to processes that ions are involved, such as vacuum arc deposition and IBAD-MBE [131]. In addition, ion-assisted CVD process has been used to produce nanoparticles and this will be presented and discussed in Chapter 2. In addition, this variant of CVD has also been developed for the preparation of one-dimensional nanostructures [132].

From the perspective of industrial applications, IBAD process has been used in the industry to

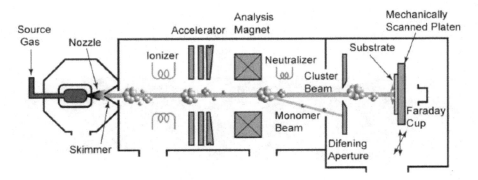

Figure 1.70 A schematic GCIB apparatus. (From Yamada, I. and Toyoda, N., *Surf. Coat.Technol.*, 201, 8579–8587, 2007.)

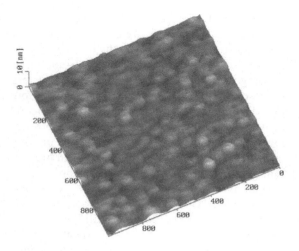

Figure 1.71 AFM image of TiO$_2$ film formed with 7 kV irradiation of gas cluster ion beam has ultra smooth surface (Ra = 0.4 nm). (From Nakatsu, O. et al., *Nucl. Instrum. Methods Phys. Res. B*, 206, 866–869, 2003.)

deposit different coatings [120]. For example, TiN coating was deposited on the martensitic stainless steel (SUS410) for the electric shaver blades for 'wet and dry' type men's shavers, as shown in Figure 1.72. The process can be operated continuously [133], and the resulting product is shown in Figure 1.73 [120]. The IBAD process involves an electron gun evaporation source for titanium, and a Kaufman type ion source (acceleration voltage 0.5–2 kV). Chamber pressure is approximately 1 × 10^{-3} Pa. Titanium was deposited under nitrogen ion beam irradiation, causing a reaction on the substrate that led to film formation with the reaction occurring as below:

$$2Ti + N_2^+ + e^- \rightarrow 2TiN$$

The thickness of TiN film is around 100 nm. In particular, comparison of TiN film from ion plating and IBAD process, the TiN film from IBAD

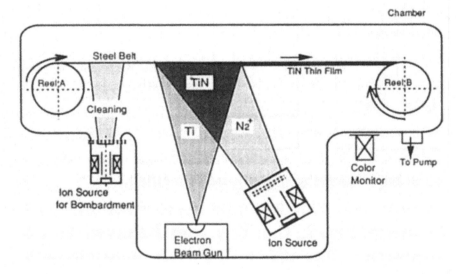

Figure 1.72 Schematic of the production equipment for the TiN coating of electric shavers by IBAD. (From Miyano, T. and Kitamura, H., *Surf. Coat. Technol.*, 65, 179–183, 1994.)

Figure 1.73 Photograph of electric shaver head sheet stock coated by ion beam–assisted deposition of TiN for providing corrosion protection and decorative gold coloration. (From Hirvonen, J.K., *Mater. Sci. Rep.*, 6, 215–274, 1991.)

shows better corrosion resistance due to ion irradiation. The following characteristics of IBAD process favours its mass production [133]:

1. It is easy to control the film composition, and it is superior in terms of operation safety
2. It is possible to improve the film integrity and adhesion
3. Bias voltage is not needed
4. Processing is possible under a high vacuum, and the efficiency of gas consumption is high

Moreover, there is industrial scale IBAD processing tool to provide the service for hard coatings like TiN, CrN, and SiNx applied for a variety of engineered components and manufacturing tooling [134].

In general, ion-based process has a long history of development, and energetic ion-assisted CVD covers a broad range of processes, including IBAD, GCIB process, and ionisation of source vapour deposition. The process can be used for the deposition of thin films and nanoparticles. Moreover, the impact of ions can be used to adjust the stress of the film. The ion-assisted CVD process has been used in the industry to deposit protective coatings like TiN for shavers; it is useful to mention that ion-beam-assisted deposition process is also used widely in industry like data storage, although it operates in PVD mode.

1.3.1.8 LASER/PHOTO-ASSISTED CVD

The photosensitised chemical reaction of hydrogen with ethylene was reported in 1910, and more work on photo-assisted CVD (PACVD) was done in the 1920s and 1930s [135]. In 1967, photo-assisted Si epitaxy through Hg lamp (100W) for the reduction of H_2 and Si_2Cl_6 at 700°C, and laser-assisted CVD, started in the 1970s. Since then, this type of process has attracted considerable interest [20].

In PACVD, photon energy is utilized as the energy source to induce reactions for the deposition, while gases are used as sources. Light, either traditional lamps or laser, can be exploited. The term photochemical vapour deposition (photo-CVD), photo-initiated or PACVD, or laser CVD is used in the literature depending on the light sources [136]. This part will start with laser-assisted CVD. The laser interaction and reactions can be classified as two types: (i) laser heats the substrate and induces the deposition on the surface. Such pyrolytic, thermochemical, or laser chemical vapour deposition reactions are analogues of thermal CVD

deposition of thin films, where precursors do not absorb the laser energy; (ii) laser energy is absorbed by the precursor, which is initially excited into non-dissociative states. After energy relaxation, the reactant gas can become too hot to decompose and undergo chemical reaction as in conventional CVD process; this is also classified as a pyrolytic process. In some cases, these two types of interactions can occur. If the substrate is transparent to the laser and the laser energy is absorbed directly by the reactant but without significant heating of the gas, the process is known as photolytic or photo-chemical deposition, or photo-deposition [137]. The apparatus scheme is shown in Figure 1.74 [137].

Since pyrolytic reaction is similar to the conventional CVD process, any type of precursors can be used. However, for photolytic deposition, it requires a good match between dissociative absorption transitions of the reactants and the available lasers. Thus, the choice of wavelength for photolysis is restrictive and often limited to the ultraviolet.

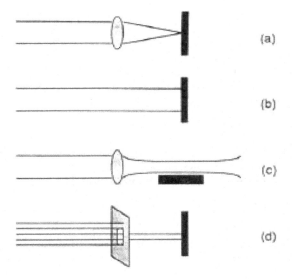

Figure 1.74 Schematic of various laser figurations used in laser-assisted deposition: **(a)** localised deposition with the laser focused onto a spot for direct writing; **(b)** large area processing with an unfocused or weakly focused beam impinging the surface; **(c)** large-area deposition with the laser focused to a line above and parallel to the surface with a cylindrical lens; **(d)** large-area patterned processing with the laser passing through a mask. (From Herman, I.P., *Chem. Rev.*, 89, 1323–1357, 1989.)

Laser-assisted CVD can be described through a six-step model of basic steps that occur simultaneously during thin film growth: (1) basic interaction of the laser with the medium, (2) transport of reactant gas to the laser interaction region, (3) primary decomposition step, (4) secondary decomposition of intermediates and transport to the film, (5) incorporation of deposit atoms into the film, and (6) transport of product gas from the film and the laser interaction region. The model is similar to the conventional CVD model as described in Figure 1.1, except that the step 1 involves the interaction of the medium with laser. These steps are not well characterised for many processes. There is more detailed discussion in the review in reference [137], for example.

The gas-phase reactants are usually alkyls, carbonyls, hydrides, halides, and AcAc's. Metal alkyls were among the first compounds used in laser deposition as molecules often photo-dissociate at 257 nm that is easily available from a frequency-doubled 514.5 nm argon laser line. The research on laser-assisted CVD concentrates on the materials currently used in the semiconductor industry, such as Al and Cu, Si, GaAs, SiO_2 and Si_3N_4 [137,138].

As compared with conventional CVD processes, laser-assisted CVD processes offer several distinct advantages. These include: (a) spatial resolution and control, (b) limited distortion of the substrate, (c) the possibility of cleaner films due to the small area being heated, (d) availability of rapid and non-equilibrium heating and cooling rates, and (e) the ability to interface easily with laser annealing and diffusion of semiconductor devices and laser processing of metals and alloys. These advantages may have potential for laser-assisted CVD to be used for the fabrication of localized reflecting coatings (e.g., Ni), and localized coatings (e.g., SiO_2) for waveguide optics [139].

Laser-assisted CVD process has been used to deposit coating on carbon fibres with industrial laser, as shown in Figure 1.75, it uses a 5KW continuous CO_2 laser operating at 2.5 KW or less at atmospheric pressure, splitting into two equal parts, CH_4 (0–12 slpm) was used as the carbon source for the deposition of pyrolytic carbon layer on carbon fibre (AS4C) at the rate of 600 nm/s, with the transport speed of 1 m/min. Pyrolytic carbon layer of 50 nm was deposited with the processing time about 100 ms. The resulting coated carbon fibre shows the increase of tensile strength with the increase of thickness of carbon coating, and the scatter of fracture strain is strongly reduced. These indicate

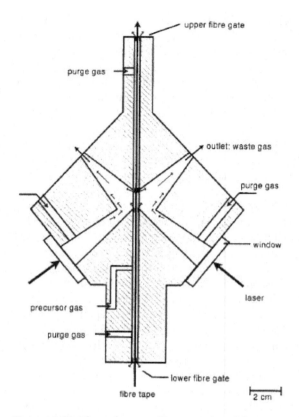

Figure 1.75 The scheme of laser-assisted CVD reactor. (From Hopfe, V. et al., *Appl. Surf. Sci.*, 106, 60–66, 1996.)

the improvement of mechanical properties of fibre. This work implies that laser-assisted CVD can be flexible for the coating system [140].

Besides thin films, laser-assisted CVD can also be used for the preparation of nanoparticles, and fibres of different materials, as shown in Figure 1.76. These include B, C, Si, W, Re, Ge, SiC,

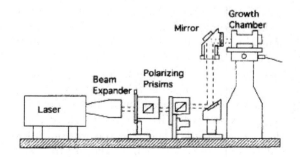

Figure 1.76 A schematic of the HP-LCVD apparatus. (From Wallenberger, F.T. et al., *Compos. Sci. Technol.*, 51, 193–212, 1994.)

Si$_3$N$_4$, ZnSe, GaAs et al. [141]. Figure 1.77 shows a spring boron fibre produced using laser CVD [141].

Except laser-style, photo-enhanced CVD tool, a schematic diagram is shown in Figure 1.78 [142], has been developed for the deposition of dielectric films, including SiO$_2$, Si$_3$N$_4$, and SiON. The process can also be used for the surface modification. For example, a recent example shows the PACVD process to tailor the wettability of singled-walled CNTs (SWNTs) using the mixture of H$_2$/CO (syngas) as the process gas. The experimental set-up is shown in Figure 1.79. The process started at 25°C, the total H$_2$/CO (ratio: 0.12) flow rate was 350 mL/min with the pressure at −15 KPa, the treatment time varies, 96 cm-long UVC germicidal lamps (Model T-97505-80, Cole-Parmer Inc.) were used for experimental treatments (main

emission peak at 253.7 nm, irradiance of 0.01 W/cm^2 at 3.5 cm). As shown in Figure 1.80, (A) shows TEM image of as-prepared purified SWNTs, and (B) shows TEM image of SWNTs with the reaction time of 120 min. It shows that there is deposition on the surface of SWNTs, as-prepared purified SWNTs have a contact angle (CA) of 79°, 5° for the treated SWNTs, which were measured through sessile drop CA measurements using an FDS OCA DataPhysics TBU 90E tensiometer [143].

There is already a commercial photo-enhanced CVD tool on the market, which uses ultraviolet light as an energy source for activating processing gases (e.g., SiH$_4$) for the deposition of dielectric films (e.g., SiO$_2$) at low temperatures (<150°C) and at 0.3–1 Torr [144].

As a whole, a pulse-laser deposition system has now been developed for large size with substrate diameter up to 560 mm, as shown in Figure 1.81 [144b]. PACVD process has been developed since 1910. Various wavelengths of light source, like Hg lamp and CO$_2$, have been used towards preparation of thin films and nanoparticles, and there is commercial equipment available for the PACVD, but few reports for the large-scale industrial applications for the process.

1.3.2 Non-vacuum methods

Non-vacuum CVD can be performed at atmosphere pressure for chemical reaction and/or deposition to occur in air for the synthesis of oxides and less oxygen sensitive materials such as oxides, II–VI semiconductor materials, or in an inert atmosphere for nitrides, carbides, and metals.

1.3.2.1 ATMOSPHERIC PRESSURE CVD

CVD performed at atmospheric pressure is called atmospheric pressure CVD. The CVD process was initially performed at atmospheric pressure in the 1880s for the preparation of carbon fibre [145], whereby a filament (e.g., pencil carbon) was heated in an organic solution (e.g., naphtha, turpentine, beeswax, balsam). Another example is the metal (e.g., Nb, Ta) deposited onto the carbon filament from the chemical reduction reaction of metal chloride (e.g., Niobium Chloride) gaseous precursor in hydrogen with a heating source such as a Bunsen burner, according to the equation

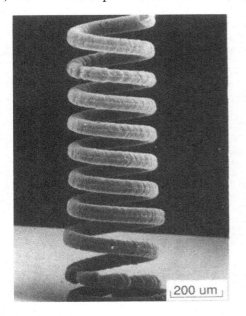

Figure 1.77 Photomicrograph of a boron microspring made with goniometer. (From Wallenberger, F.T. et al., *Compos. Sci. Technol.*, 51, 193–212, 1994.)

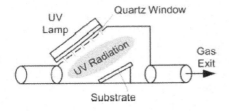

Figure 1.78 A schematic of a photo-assisted CVD set-up. (From Dion, C.A.D. and Tavares, J.A., *Powder Technol.*, 239, 484–491, 2013.)

$$NbCl_5 \text{ (s)} + 5/2 \ H_2 \text{ (g)} \rightarrow Nb \text{ (s)} + 5HCl \text{ (g)}$$

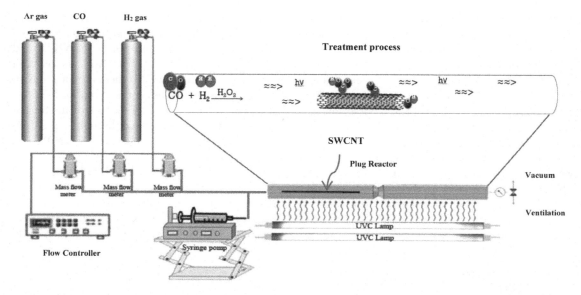

Figure 1.79 A schematic of the photo-assisted CVD set-up. (From Hosseininasab, S. et al., *Chem. Eng. J.*, 325, 101–113, 2017.)

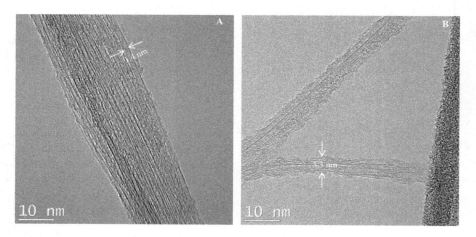

Figure 1.80 TEM micrographs of (**a**) purified SWNTs and (**b**) syngas-treated SWNTs. (From Hosseininasab, S. et al., *Chem. Eng. J.*, 325, 101–113, 2017.)

using CVD system as shown in Figure 1.82 [146]. Mond process for the deposition of Ni is also an atmospheric pressure CVD process [5]. Since metalorganic precursor is used in the Mond process, therefore, it will be discussed in the atmospheric pressure MOCVD section.

The widely used flame synthesis process that are performed at atmospheric pressure can also be considered as atmospheric pressure CVD process and further details will be presented in the Chapter 2 on the flame synthesis of nanoparticles on an industrial scale.

APCVD involves the chemical reaction and deposition to occur at atmospheric pressure. Thus, APCVD avoids the use of an expensive and sophisticated vacuum system and makes the coating manufacturing more economically viable. In addition, APCVD have added advantages such as large-area deposition, not restricted by vacuum deposition chamber, high throughput and good coating uniformity, and higher deposition rate than low-pressure CVD. However, high gas-flow rates tend to be required and the CVD tool would also need to be cleaned and maintained regularly.

Figure 1.81 A photo of hybrid pulse layer deposition tool. (From Lackner, J.M., *Surf. Coat. Technol.*, 200, 1439–1444, 2005.)

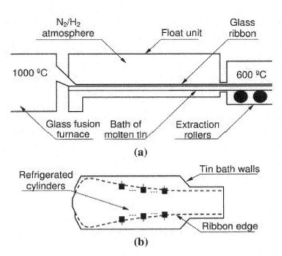

Figure 1.83 Sketch of the float process: **(a)** vertical section and **(b)** horizontal view. (From Prieto, M. et al., *Int. J. Therm. Sci.*, 41, 348–359, 2002.)

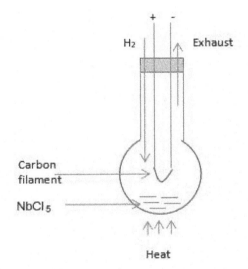

Figure 1.82 Schematic of the system used to deposit refractory (e.g. Nb, Ta) metal films onto carbon light bulb filaments by CVD from metal halide (e.g. Niobium chloride) precursors to make filament from refractory metal. (From Aylesworth, J.W., Art of manufacturing electrical incandescing conductors, US 553296 January 21, 1896.)

APCVD has been used in industry for the deposition of oxides, doped oxides, and non-oxide films.

1.3.2.1.1 APCVD of transparent conducting oxides on glass

It is worth mentioning that the floating process is important to the glass production, as shown in Figure 1.83, and APCVD process is widely used to coat glass. An example is the deposition of fluorine-doped tin oxide (FTO) (SnO$_2$:F) by Pilkington-NSG

[146b]. There are two different configurations such as turbulent flow CVD coater and laminar flow CVD coater that can be used in the APCVD process [146b]. Adherence and uniform SnO$_2$:F coating is being manufactured during on-line coating of glass by APCVD to reduce handling, and possible contamination or any subsequent washing step, thus enhancing coating adhesion onto glass substrates. Regarding the tin oxide deposition, the following tin precursors could be used including stannic chloride, monobutyl tin trichloride, and dimethyl tin dichloride. The fluorine dopant precursors could be hydrogen fluoride and trifluoro acetic acid. Equation (1.1) shows an example of the precursors used and chemical reaction for the deposition of SnO$_2$:F. SnO$_2$:F can be grown at rates (e.g., >100 nm/s) to match the line speed of the continuous glass translation speed to achieve high throughput and achieve the required thickness in less than 2s. Stable chemistry establish for continuous operation for many days high-efficiency, high-volume manufacturing for further cost reduction, in addition to the non-vacuum process. There is a lot of gas phase reaction and the turbulent flow reactor has a low precursor efficiency <10%. Hence, the laminar flow reactor has also been developed to increase the precursor efficiency [146b].

$$SnCl_4 + H_2O + HF \rightarrow SnO_2{:}F$$
$$+ HCl\ (\sim 1.5\ at\ \%F) \qquad (1.1)$$

Regarding SnO_2:F coating, the process can also be performed through batch process, as shown in Figure 1.84, and it is found that the morphology of coating depends on the process conditions and precursors (Figure 1.85) [146d].

In the industry, there is already proprietary process for the production of low emissivity coating (Low-E coating), such as Pilkington K Glass™. In the coating an underlayer of SiC_xO_y (circa. 70 nm) is deposited as a colour suppression and barrier

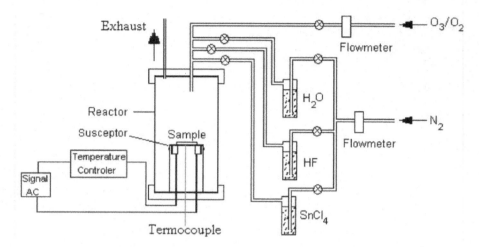

Figure 1.84 Schematic view of the APCVD reactor and flow system. (From Morales, C. et al., *Microelectronics J.*, 39, 586–588, 2008.)

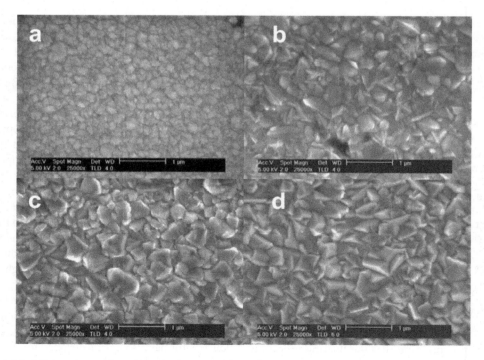

Figure 1.85 SEM images of SnO_2:F deposited with different precursors: **(a)** tetramethyltin (TMT), **(b)** monobutyltin trichloride (MBTC), **(c)** tin tetrachloride (TTC), **(d)** Asahi U-type. The RMS surface roughness measured on an area of 5 × 5 µm² is as follows: **(a)** 7 nm, **(b)** 22 nm, **(c)** 39 nm, **(d)** 40 nm. (From Prieto, M. et al., *Int. J. Therm. Sci.*, 41, 348–359, 2002.)

layer according to equation (1.2), with a top coat of SnO_2:F (350 nm thick) [146b].

$$SiH_4 + C_2H_4 + CO_2 \rightarrow SiC_xO_y + H_2O + \text{other by-product} \qquad (1.2)$$

For the industrial large-area on-line coating deposition, the following challenges need to be addressed and controlled. These include to: (a) deliver large amounts of chemical precursors at a constant temperature and flow continuously; (b) maintain coating uniformity across a 3-meter-wide glass sheet at a speed of 15 m/min as long as possible for cost-effective coating production. The longest coating run time is circa 60 hours; and (c) continuously inspect and monitor the coated product to ensure coating reliability, reproducibility, and quality control.

Since volatile precursors are required, this limits the range of materials that could be deposited by APCVD. For example, other transparent conducting oxides such as Al-doped ZnO (AZO) and indium-doped tin oxides (ITO) are more challenging to be deposited by APCVD.

AZO and ITO tend to be deposited by PVD methods such as sputtering. The CVD-deposited SnO_2:F is relatively cheaper and more durable and stable than the vacuum-sputtered Al-doped ZnO (AZO). However, AZO has better optical and electronic properties than SnO_2:F. AZO can be etched to enhance light scattering for silicon solar cells. In order to enhance light trapping, APCVD processing conditions can be tailored for surface texturing to increase the path length of incoming light, and greatly influence the solar cell device performance [147].

Figure 1.86a shows a SEM image of CVD-grown SnO_2:F coating with pyramidal texture on its surface, and (b) shows a solar cell grown on such a material. Such double-texture has shown to increase light trapping for amorphous/microcrystalline silicon tandem solar cells by increasing the path length of incoming light and the long wavelength sensitivity is improved (c) [148a]. Such direct surface texturing during the APCVD growth process could be cheaper than the sputtered transparent conducting oxide of Al-ZnO, which

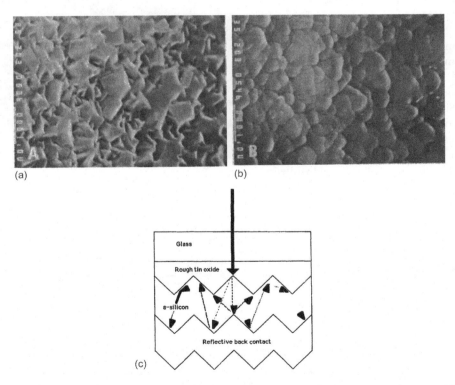

(a) (b) (c)

Figure 1.86 **(a)** Top view of a rough tin oxide film and **(b)** a solar cell grown on such a material. **(c)** Pictorial representation of the light-trapping effect. (From Gordon, R.G. et al., *Sol. Energy Mater.*, 18, 263–281, 1989.)

would require post processing of a wet-chemical etching step [148b].

One major application of atmospheric pressure CVD is to deposit film on glass in the production line of glass used in the building industry [149]; this is to prevent IR (up to 10 µm) radiation to the house. In Europe, most buildings use double-layer glazing glass windows, where the coating layer is inside to avoid damaging. As there is demand for environment-friendly building, this process is widely applied [150].

1.3.2.1.2 APCVD of films on wafers

APCVD is commonly used in the microelectronics industry, where they are used to apply thin layers of films onto wafers. With the development of improved tooling, the APCVD undergoes a renaissance which enables high wafer throughput, which is a big advantage of this process. Various companies and research groups have developed proprietary hardware to improve the CVD deposition and versatility and to reduce cost. For example, Figure 1.87 shows an example for the deposition of thin films onto continuous flow of substrates and process gas [151a].

The APCVD system consisting of modular multiple injector heads in series has been developed to maximize process throughput, uniformity, and flexibility while minimizing cost for the deposition of thin films (e.g., undoped, B-, P-doped SiO_2 and TiO_2). All process parameters can be monitored and controlled in a user-friendly touchscreen environment. The chemical injectors have been maintenance-consciously designed to allow exhaust ducting to be cleaned while in place on the system to minimize process interruption

[151b]. In the roller transport system, only the substrate is heated during the process. Such design greatly reduced the electricity and cooling requirements. An example of CVD system is shown in Figure 1.88 [115b].

1.3.2.1.3 APCVD of non-oxide films

Besides oxide films, APCVD can also be used to grow non-oxide films. Ultraclean APCVD has been used to grow non-oxide epitaxial films at low temperatures, 550°C–850°C, in several

Figure 1.88 A Cat-CVD apparatus for 12 inch Si process. (From Matsumura, H. and Ohdaira, K., *Thin Solid Films*, 516, 537–540.)

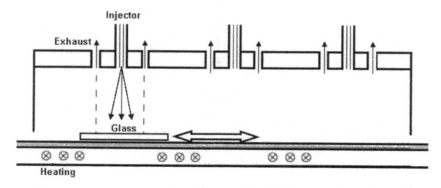

Figure 1.87 Schematic of an industrial atmospheric-pressure CVD system. (From Illiberi, A. et al., *Sol. Energy Mater. Sol. Cells*, 95, 1955–1959, 2011.)

reactors including a commercial production-type tool. For example, Sedgwick et al. reported the use of APCVD to grow Si and SiGe epitaxial films at low temperatures, 550°C–850°C [152]. The use of a load lock, gas purifiers, and careful gas-handling procedures would allow epitaxial films to be grown down to 550°C with undetectable levels of oxygen and carbon at the growth interface. This was achieved with the use of hydrogen to inhibit Si oxidation at the growth temperature. Doped layers with controlled high active levels of N-type doping is quite feasible at low temperatures.

A commercial APCVD system has been developed to grow epitaxial films of Si, compound semiconductors, where the system is applicable for high-volume, low-cost solar module production. Another type is roll-to-roll style system for flexible substrate; one example is shown in Figure 1.89 [151c].

Another important application for the atmospheric pressure CVD is to deposit hard coatings for a variety of applications, for example in tribological applications and the deposition of protective coatings onto cutting tools (e.g., cemented carbides) for protection against wear, corrosion, and high-temperature corrosion of functional surfaces in many industries. Various coatings such as TiC, TiN, Cr_7C_3, SiC, Si_3N_4, FeB, Fe_2B and Al_2O_3 have been prepared by CVD process [153].

For example, TiC coating can be deposited the reaction below [154]:

$$TiCl_4 (g) + 2 H_2 (g) + N_2 (g) = TiN (s) + 4 HCl (g)$$

The CVD process can generate film of a few μm. It is realized that the use of pressures of less than 1 atm improves the layer quality and coating homogeneity [153].

In 1968, CVD coatings for cutting tools were first developed and marketed by Krupp in (West) Germany under the brand name Widia, followed by another big name in the cutting tool industry-Sweden's Sandvik. The further work shows that multi-layer structure performs better than single- or double-layer coatings in metal-cutting operations at higher speeds. Figure 1.90 shows a multi-layer coating from USA Kennametal tool [16].

However, the CVD process requires high temperatures and it is good for honed edges, although not for sharp edges. Other processes, such as plasma-based CVD or PVD process, can be employed. Atmospheric pressure plasma-based CVD process is also being widely applied. For example, the Fraunhofer team in Germany has developed a CVD deposition system (see Figure 1.91) whereby vapour can be activated by plasma or reactions in the vapour phase (water, ozone) to deposit coatings with thicknesses ranging from 10 to 1000 nm [155]. Oxide coatings including conductive oxides of

Figure 1.89 A roll-to-roll Cat-CVD system. (From Matsumura, H. and Ohdaira, K., *Thin Solid Films*, 516, 537–540.)

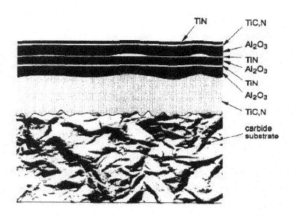

Figure 1.90 A schematic of a multilayer coating architecture of Kennametal's KC990 grade cutting tool. (From Subramanian, C., *Wear*, 165, 85–95, 1993.)

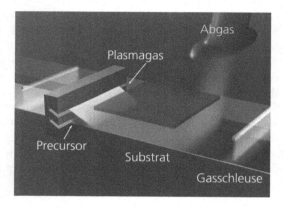

Figure 1.91 A schematic of the APCVD system by Fraunhofer. (From Althues, H., Chemical vapor deposition at atmospheric pressure AP-CVD–Fraunhofer IWS, https://www.iws.fraunhofer.de/en/business_fields/chemical_surface_reaction_technology/chemical_surface_technology/technologies/ap-cvd.html, accessed on July 18, 2017.)

fluorine-doped tin dioxide (SnO_2:F) can be deposited. In addition, photocatalytically active coatings based on TiO_2 have been fabricated using such APCVD system at comparatively low temperatures on large-area deposition onto many types of substrate materials (polymers, metals, glass, composite materials) [151d].

Besides the use of APCVD at an industrial scale, this process is also being explored for the synthesis of novel materials, such as graphene [156a] and SiC films using SiH_4-C_3H_8-H_2 reaction gas system [156b].

1.3.2.2 ATMOSPHERIC PRESSURE MOCVD

As mentioned previously, Mond Process is an atmospheric pressure MOCVD process, in which $Ni(CO)_4$ decomposed to form high-purity Ni. This becomes a good process to extract and purify Ni, or deposit Ni thin films [5].

Similarly, thermal decomposition of organo-metals is used for the metal deposition, in particular in the semiconductor industry. For example, W layer is deposited by the decomposition of $W(CO)_6$ [3].

One of the industrial applications of the atmospheric pressure MOCVD process is to deposit TiO_2 coatings onto the glass containers from the chemical reaction of $(RO)_yTi(OCH_3CCHCOR')_{4-y}$ (R: 2-4 C atoms, y:1-2; R':group-containing methyl, ethyl, methoxy, ethoxy) at 510°C–670°C. This is one of the largest applications for thin-film deposition on glass. With the development of technology, glass container becomes thinner, to avoid surface scratch and maintain outlook, the surface coating is needed. It can be applied by spray coating of metalorganic chemicals and followed by heating [157]. Modern process may vary, and the precursor may be dimethyltin, tin tetrachloride, monobutyltin trichloride, and titanium tetrachloride. If metal organics, like dimethyltin or monobutyltin trichloride, are used, the process is atmospheric pressure MOCVD when applied through spray [158].

1.3.2.3 AEROSOL-ASSISTED CVD

Aerosol-assisted CVD (AACVD) is a variant of the CVD process whereby chemical precursor in aerosol form can be generated via:

1. Ultrasonics (e.g., Pyrosol [1,159])
2. Pneumatics (1,159,160,161)
3. Electrostatic or electrospraying methods (e.g., ESAVD [1,159,162,163], EAAJD [1,164])

The aerosol precursor is subsequently delivered to a CVD reactor and undergoes decomposition and chemical reaction leading to the desired deposition as shown in Figure 1.92. The various aerosol generator methods and relevant equations to describe the aerosol generation have been reviewed in references [1,159].

The chemical precursor can be prepared by dissolving solids (e.g., chlorides, nitrates, acetates, alkoxides) or liquid starting chemicals into a solvent (e.g., organic solvent with a high boiling point)

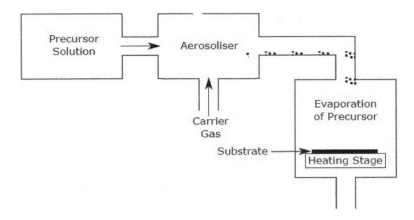

Figure 1.92 A schematic illustration of the AACVD process.

or a mixture of solvents to assist the vaporisation of the chemicals and provide additional thermal energy for the dissociation or decomposition of the chemicals at lower temperatures. Organometallic precursors have also been used in this process and it can also be known as aerosol-assisted metalorganic CVD [165].

The benefits of AACVD include:

1. Simplifying chemical precursor delivering system and vapour precursor generation as compared to the conventional CVD method which tends to use a bubbler or vaporiser and dedicated liquid precursor delivery lines
2. Enabling a lower deposition temperature with the inclusion of solvent in the precursor mixtures
3. Allowing different chemical precursors to be formulated into a single precursor mixture, thus enabling the deposition of multicomponent materials with well-controlled stoichiometry and rapid deposition with close intermixing and small diffusion distances between reactant and intermediates
4. Allowing the ease of the inclusion of ultrafine materials/fillers, thus facilitating the deposition of (nano)composite coatings

1.3.2.3.1 Deposition mechanism

The aerosol generated could undergo one of the possible deposition mechanisms depending on the substrate or deposition temperature as shown in Figure 1.93 [159]:

Process I: The aerosol precursor droplets are sprayed or directly applied onto a heated substrate, followed by the removal of the solvent through evaporation and decomposition of the precursor to the finished product. Occasionally, the deposited film will be subjected to a further sintering step to achieve a dense crystalline film. Dense thin films (<1 μm) with an ultrafine crystalline structure can be obtained using this deposition mechanism. In order to obtain a thick film, Process I has to be repeated several times in order to obtain the required thickness. Obviously, this is a time-consuming process for the deposition of thick films. Moreover, as the film thickness increases, the deposited film tends to be porous, and cracking or spalling of the film can occur because of repeated drying, decomposition, and/or sintering procedures.

Process II: The solvent is evaporated prior to arriving onto the substrate surface, and the precursor precipitate is subsequently deposited onto the heated substrate and decomposed and/or undergoes chemical reactions to yield the desired materials.

Process III: The solvent is evaporated while approaching the substrate surface, and the precursor precipitate formed subsequently undergoes volatilisation near the vicinity of the substrate surface and adsorption of the vapour onto the heated substrate surface, followed by the decomposition and/or chemical reactions to yield the desired materials. This mechanism is similar to the heterogeneous

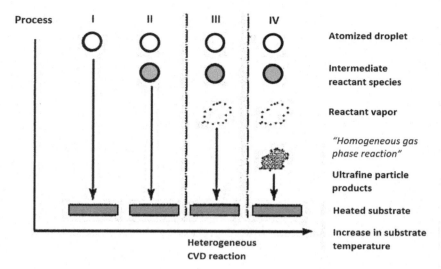

Figure 1.93 Influence of the temperature of the aerosol assisted deposition process. (From Choy, K.L., Vapour processing of nanostructured materials, in *Handbook of Nanostructured Materials and Nanotechnology*, Academic Press, San Diego, CA, pp. 533, 2000; Poodt, P. et al., *J. Electrochem. Soc.*, 419–427, 2010.)

CVD deposition process, which tends to produce dense films with excellent coating adhesion.

Process IV: As the deposition/substrate temperature is very high, the decomposition and/or chemical reaction occur in the vapour phase, leading to homogeneous nucleation (similar to the homogeneous CVD reaction) and, hence, formation of stable fine particles in the gas phase, which are then deposited onto the heated substrate. The particles are then sintered on the heated substrate, leading to the formation of porous films, with poor adhesion. Alternatively, the powders can be collected in the gas phase for the production of ultrafine powders.

Among all these processes, only Processes III and IV can be considered as aerosol-assisted CVD because in the true CVD process, chemical precursor must be in the form of chemical vapour that undergo chemical reactions (i.e., homogeneous and/or heterogeneous chemical reactions). Pyrosol [159,166,167], AACVD [161,168,169], AAMOCVD [165,170,171], ESAVD [1]; [162,163], and EAAJD [1,164] processes are examples of aerosol-assisted CVD processes that would meet such classification.

1.3.2.3.2 AACVD from ultrasonically or pneumatically generated aerosol

AACVD, AAMOCVD, and Pyrosol have been used for the synthesis of ultrafine particles, thin films, nanocomposite coatings as shown in Figure 1.94. An example of Au nanoparticles (NPs) deposited from the AACVD reaction of auric acid (0.02 mol) in methanol (20 mL) with air (2 L min^{-1}) at 450°C [172] is shown in Figure 1.94a. Thin film such as Cr_2O_3 as shown in Figure 1.94b was deposited from the aerosol precursor containing chromium nitrate nonahydrate in ethanol at 280°C–300°C followed by annealing at 500°C. The nanocomposite of WS2 in Cr_2O_3 matrix in Figure 1.94c could be produced by AACVD by adding inorganic fullerene-like WS$_2$ particles (80–220 nm) into a 0.05 M precursor containing a mixture of chromium nitrate nonahydrate in ethanol to obtain uniform suspension in an ultrasonic bath. The precursor was then atomized to generate fine aerosol droplets using an ultrasonic generator, at a frequency of 1.7 MHz, with nitrogen as carrier gas [173,174]. The heterogeneous reaction would lead to the formation of a Cr_2O_3 shell on the nanoparticle surface, which would greatly affect the microstructure and surface roughness of the nanocomposite coating as shown in Figure 1.91c.

Figure 1.94 AACVD of **(a)** nanoparticles (e.g. Au NPs) (From Chew, C.K.T., Chemical vapour deposition of gold nanoparticle and metal oxide composites, PhD Thesis, University College London, London, UK, 2016.); **(b)** thin films (i.e. Cr_2O_3) and **(c)** nanocomposite film (e.g. WS_2 fullerene in Cr_2O_3 matrix). (From Hou, X. and Choy, K.L., *Thin Solid Films*, 516, 8620–8624, 2008.)

From the tribological test using a 1.6-mm-diameter alumina ball with a constant loading 0.2 N and a reciprocating moving speed 2 mm/min, the friction coefficient of Cr_2O_3 was significantly reduced by 25% with the incorporation of IF-WS_2 nanoparticles [174].

In the last 10 years, AACVD has been increasingly exploited as a tool for the development of nanocomposite coatings. Unlike the vacuum processes, the ease of incorporating nanomaterials (e.g., nanoparticles, fullerene, 2D materials, nanotubes, nanowires) into precursor solution mixture to form aerosol has facilitate the deposition of nanocomposite films cost-effectively. Besides co-deposition of nanomaterials and matrix simultaneously, AACVD also has the flexibility for the *in-situ* synthesis of nanoparticles in matrix (e.g., [175]). AACVD of nanocomposite films has been reviewed in more detail with further examples and applications in Chapter 7.

AACVD has also been used to grow a seed layer to direct and promote full film growth via an atmospheric pressure CVD (APCVD) overlay of transparent conducting oxide film. The AACVD of SnO_2 seed layer was deposited at 500°C from pneumatically generated aerosol consisting of a mixture of MBTC (0.2 mol dm^{-3}), MeOH, and TFA using compressed air (constant 2 bar pressure) and cooled under air to room temperature. The SnO_2 base film was subsequently formed using APCVD from the reaction of MBTC and EtOAc alongside the TFA fluorine-dopant precursor. EtOAc was used to provide an oxidant source in order to help formation of the oxide. Such two-step CVD method would allow for reproducible control over morphology

and denser, rougher, higher-performing TCO at a relatively low growth temperature (500°C) [161].

In addition, AACVD has been explored for the growth of nanotubes and nanowires. The growth mechanisms are presented in Chapter 3. Such nanostructures have a variety of potential functional applications in sensors, displays, electrodes, nanoscale electronic devices, catalyst supports, etc. Figure 1.95a shows the TEM image of CNTs grown by AACVD of ethanol with ferrocene at 850°C. The CNTs also grown onto the low cost, flexibility, and conducting steel substrates by AACVD and were tested for potential use in a high-energy supercapacitor which exhibited a high geometric capacitance of 3.4 m Fcm^{-2}; however, a low specific capacitance of 0.34 F g^{-1} due to excessive CNT loading on the substrate could be optimized further [176].

AACVD also has a huge potential for producing large volumes of CNTs with excellent areal selectivity, vertical alignment selectivity, and selective doping with well-controlled stoichiometry [177]. The undesirable carbonaceous materials and impurity due to contamination of solvent and chemical reagents, or those formed from precursors in the gas phase during AACVD, can be hindered via mild oxidizing agents (e.g., water vapour) [177,178,179,180]. Aligned multi-walled carbon nanotubes (MWNTs) with high purity and bulk yield were achieved on a silicon substrate (Figure 1.95b) by AACVD via the decomposition of an aerosol generated ultrasonically which consisted of a liquid hydrocarbon source and an evaporable catalyst and transported by wet argon gas with a flow rate of 1500 sccm for the CNTs growth

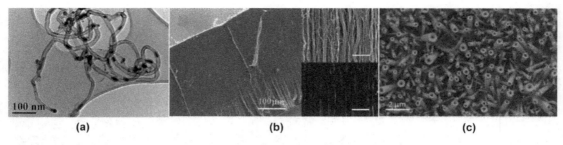

(a) **(b)** **(c)**

Figure 1.95 AACVD of nanostructures: **(a)** TEM image of CNTs grown by AACVD (From Sagu, J.S. et al., *Adv. Eng. Mater.*, 18, 1059–1065, 2016.); **(b)** SEM image of vertically aligned multi-walled carbon nanotubes with bulk yield on a silicon substrate (From Liu, P., *Abstr. Pap. Am. Chem. Soc.*, 240,2010.) and **(c)** SEM of WO3 nanostructures. (From Vallejos, S. et al., *Chemosensors*, 4, 4, 2016; Annanouch, F.E. et al, *Sens. Actuators B Chem.*, 216, 374–383, 2015.)

at 850°C for 30mins under hydrogen at a flow rate of 1800 sccm. A 30 nm-thick aluminium buffer layer was first sputtered onto a silicon wafer with a 600-nm-thick SiO_2 layer. The lengths of the aligned CNTs were circa. 400 μm. It seems that the introduction of specific amounts of water vapour played a key role in *in-situ* controlling the length, purity, and surface defects of the nanotubes [177]. Non-carbon-based such as WO3 nanostructures have also been grown by AACVD onto Si-based micro-hotplates (Figure 1.95c) to explore for gas-sensing applications [181,182].

Although the non-vacuum AACVD is a promising low-cost and scalable option for larger-scale applications, however, the ultrasonic and pneumatic methods for the aerosol generation still need to be refined in order to produce a reliable aerosol generation and delivery to ensure the reproducible and uniform nanomaterials, films, or coatings so that the lab results can be adopted by industry for production.

1.3.2.4 ELECTROSTATIC SPRAY-ASSISTED VAPOUR DEPOSITION

Electrostatic spray-assisted vapour deposition (ESAVD) has been developed and reviewed by Choy and it is a form of AACVD [1]. ESAVD is a true CVD process which involves spraying atomised aerosol precursor across an electric field into a heated environment where the charged aerosol will undergo decomposition and chemical reaction in the vapour phase as shown in Figure 1.96. The aerosol can be sprayed upward or downward during deposition. The main process parameters for the ESAVD of films include deposition temperature, field strength, and stand-off distance. For

the synthesis of nanostructured films, for example, the deposition temperature is preferably below 550°C, the field strength was within the range of 4–25 kV, and the precursor flow rate varied from 10 to 30 mL/h [1].

ESAVD is an atomic deposition method, which can produce highly pure materials with structural control at the molecular scale level at relatively low processing temperatures. The structure, stoichiometry, crystallinity, and texture of films can be controlled by optimising the process parameters. If the chemical reaction occurs in the gas phase (Process IV), this would lead to the synthesis of nanosize powder (Figure 1.97a) [183]. Nanostructured porous films can be deposited by tailoring a combination of homogeneous and heterogeneous reactions to occur as shown in Figure 1.97b [183]. Whereas a dense and adherent nanofilm (Figure 1.97c) will be deposited onto a substrate [183], if the aerosol would undergo decomposition and chemical reaction near the vicinity of the heated substrate (i.e., similar to Process III of the AACVD in Figure 1.93). The ESAVD process can also lead to the deposition of thin/thick films with various dense, porous, or multilayer porous/dense/ porous microstructure by tailoring the appropriate chemical reactions to occur as shown in Figure 1.98a, b, and c, respectively [1].

In addition, ESAVD has also been explored as the low cost, eco-friendly, and non-vacuum method for the fabrication of high-performance Cu(In, Ga)(S, Se)₂ (CIGSSe)-based absorbers for thin film chalcogenide solar cells. Currently, the vacuum PVD method is used to manufacture CIGS absorber which accounted for circa 40% of the total photovoltaic device cost. In addition

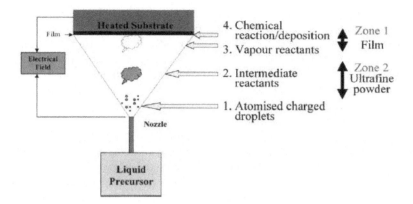

Figure 1.96 A schematic illustration of the ESAVD process. (From Choy, K., *Prog. Mater. Sci.*, 48, 57–170, 2003.)

Figure 1.97 Transmission electron micrographs of various structures of Eu-Y$_2$O$_3$ synthesised by tailoring chemical reactions: **(a)** gas phase reaction to form nanosized powder; **(b)** a combination of homogeneous and heterogeneous reactions to form porous structure and **(c)** heterogeneous reaction to form dense film. (From Choy, K.L., Process principles and applications of novel and cost-effective ESAVD based methods, in *Innovative Processing of Films and Nanocrystalline Powders*, Imperial College Press, London, UK, pp. 15–69, 2002.)

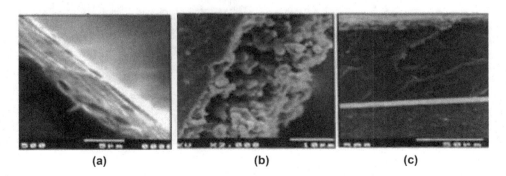

Figure 1.98 Scanning electron micrographs of various microstructures of films deposited using the ESAVD process: (a) dense films, (b) porous films and (c) multilayer coatings. (From Choy, K., *Prog. Mater. Sci.*, 48, 57–170, 2003.)

to the non-vacuum and low-cost benefits of the ESAVD process, the composition and stoichiometry of the CIGS absorber can be tailored and treated accordingly. For example, Copper (Cu)-deficient (Cu/In + Ga = 0.76) CIGS films were designed to avoid the rather dangerous KCN treatment step for the removal of conductive minor phases of Cu_2S/Cu_2Se [184]. During the ESAVD process, rather than using organic solvents which form strong complex with metal solvents, an environmentally friendly and abundant mix solvent of water/ethanol is used as solvent for the mixture of metal salts (i.e., Cu, In, Ga) and thiourea precursor and the deposition performed at substrate temperatures ranging from 250°C to 450°C. The CIGS films were subsequently selenised in an argon-filled tube furnace at 550°C for 30 minutes. Figure 1.99a shows the SEM top-view and cross-sectional images of the ESAVD-deposited CIGS film followed by selenisation to produce CIGSSe film with efficiency above 8% [184,185].

Furthermore, the morphology and electronic properties of the ESAVD-deposited absorber could be modified via a simple sodium (Na) treatment method to enhance the solar cell performance and achieve efficiency above 10% [184]. The Na incorporation via 0.2 M dippling into NaCl prior to selenisation seems to increase the grain size as shown in Figure 1.99b and lead to a strong (112) orientation of the CIGS film increase in the hole concentration and p-type conductivity due to defect passivation increased open circuit-voltage (Voc) due to the increased p-type conductivity. Thus, this would lead to overall improvement in the crystal structure, electronic characteristics, and photovoltaic properties of CIGSSe solar cells. A cross-section SEM of the ESAVD-deposited CIGS solar cells with 10.83% efficiency (with NaCl treatment) is shown in Figure 1.100 [184]. The effect of sodium treatment on the ESAVD-fabricated CIGSSe solar cells and their j-V characteristics as measured under simulated AM 1.5 100 mW cm^{-2} illuminations is shown in Figure 1.101 [184].

Figure 1.99 SEM top-view and cross-sectional images of the ESAVD-deposited CIGS film followed by selenisation to produce CIGSSe film: **(a)** without treatment (From Wang, M. et al., *Sci. Rep.*, 7, 2017.) and **(b)** with 0.2 M NaCl treatment. (From Wang, M. et al., *Sci. Rep.*, 7, 2017; Mungan, E.S. et al., *IEEE J. Photovolt.*, 3, 451–456, 2013.)

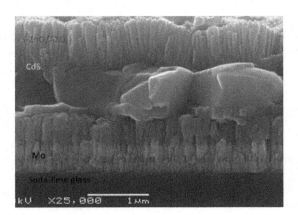

Figure 1.100 Cross-sectional SEM image of the ESAVD-deposited CIGS solar cells with 10.83% efficiency (with NaCl treatment). (From Wang, M. et al., *Sci. Rep.*, 7, 2017.)

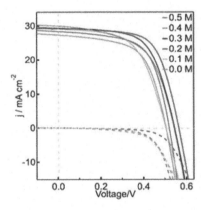

Figure 1.101 j-V characteristics of CIGSSe solar cells measured under simulated AM 1.5 100 mW cm^{-2} illuminations, with and without Na doping. (From Wang, M. et al., *Sci. Rep.*, 7, 2017.)

1.3.2.4.1 Benefits of ESAVD

Besides the advantages of the AACVD, the ESAVD method also has the following added benefits as compared to the conventional CVD methods [1]:

1. ESAVD-based method does not involve the use of scrubber and effluent gas handling system to remove any unreacted precursors or toxic by-products. This is because the starting materials are relatively more environmentally friendly. A low consumption of chemical precursors and relatively low concentration of precursor (e.g., 0.05 M) is used. Furthermore, the process has a higher precursor conversion.

2. High deposition efficiency (>90%) because the precursor is directed to the substrate under the electric field. Thus, this minimises the loss of precursor to the surrounding.

3. It is a versatile technique which can produce thin or thick films in the form of single layer, multilayer, and compositionally graded coatings, and composite coatings. Adherent thick films (e.g., >250 μm) can be easily deposited using the ESAVD-based technique which tends to be difficult to be achieved using the conventional CVD and those variants that have been described earlier.

4. Highly pure materials can be produced at relatively low processing temperatures (e.g., a fraction of their melting point). For example, cubic YSZ can be deposited at 550°C.

5. It is a one-step process for optimized deposition, without the need for further heat treatment.

6. It is a simple and flexible technique which allows the incorporation of the deposition technique into on-line processing.

The comparison of ESAVD-based methods with other electrospraying techniques has been discussed and reviewed in detail in the reference [1].

A wide range of materials has been produced by ESAVD method, both oxides and non-oxide materials for structural, functional, and biomedical applications.

These materials are summarized below:

Binary oxides: TiO_2, ZnO, Al_2O_3, SiO_2 [1,145,186,187]

Doped oxides: In_2O_3-SnO_2, Y_2O_3-ZrO_2, Eu-Y_2O_3 [188,189,190,191].
Multicomponent oxides: $PbTiO_3$, $BaZrO_3$, La(Sr)MnO_3, CaO-P_2O_5-SiO_2 [1,192,193,194].
Sulphides and selenides: CdS, ZnS, CdSe [1,195,196].
Chalcogenides and kesterite: CIS, CIGS, CZTS [1,185].

ESAVD is a versatile process and effort has been focused on the exploitation of ESAVD for the deposition of both thin and thick films (e.g., 250–400 μm). These films and coatings can be deposited onto both conducting and non-conducting substrates (e.g., metal ceramic and plastics) for a wide range of potential applications. These include transparent conducting oxide thin films [188,189], optical films [45], absorbers for thin film solar cells [197], solid oxide fuel cells [190], ferroelectric films for sensors and memory devices [192], reforming catalysts [187], thermal barrier coatings, and bioactive coatings [1,194]. In addition to the examples of ESAVD of thin and thick films given in the review [1], advances and further case studies of ESAVD of thin/thick films, especially for the fabrication of multicomponent absorbers for thin films chalcogenide and kesterite solar cells, are presented in Chapter 6.

This chapter also highlights recent advances in AACVD for the deposition of superthin or nanothin (<100 nm) films as illustrated in the case study of transparent conducting ITO film [198]. Figure 1.102 shows the ESAVD of ITO (circa 80 nm thick) onto a large substrate (50 mm) Si <100> wafer from a mixture of metal chlorides of indium and tin chloride in alcoholic solvents at a temperature within the range of 400°C–550°C in air [199]. The as-deposited film with a thickness of circa. 80 nm is typically dark blue. Figure 1.102a shows an example of the crystalline ITO film deposited by ESAVD with high-optical transparency (>88%) and it is highly uniform (Figure 1.102b). The variations in thickness and sheet resistance are 3% and 6.4%, respectively (Figure 1.103) [199]. The carrier concentration of the ESAVD of ITO film is 5.4×10^{20}/cm^3, mobility 27.3 cm^2/Vs and a refractive index of 1.8–2.0.

After heat treatment and plasma hydrogenation (plasma power: 100 W, 13.56 MHz frequency; temperature: 200°C; Flow: 200 sccm H_2 and treatment time: 5–10 minutes), the sheet resistance of ESAVD of ITO can reach <20 ohm/square. These results of the non-vacuum ESAVD of ITO are comparable with those sputtered ITO samples, which is suitable as the alternative cost-effective electrode for thin film silicon solar cells and transparent conducting electrode for display applications.

Furthermore, the ESAVD-deposited ITO exhibited good thermal stability (Figure 1.104) during fast-firing process at high temperature (circa. 850°C) as compared to the sputtered samples [199]. The firing study of ITO using various conditions confirms that the ESAVD ITO films are capable of maintaining the desired sheet-resistant properties during the fire process. The sheet resistances at fast firing at 850°C for 15 s and 150 s are 55 and 51 Ω/·, respectively. On the other hand, the sputtered samples would have deteriorated and

(a)

(b)

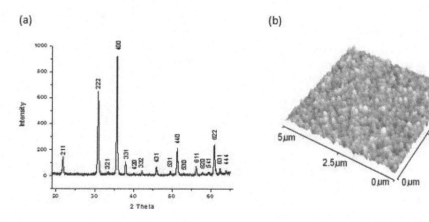

Figure 1.102 **(a)** XRD and **(b)** AFM of ESAVD of ITO film onto Si wafer. (From Choy, K.L. et al., ESAVD of transparent conducting oxide films, EU-FP7ThinSi project (241281) final report, 2013.)

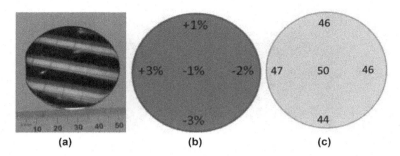

(a)   (b)   (c)

Figure 1.103 **(a)** ESAVD of highly optical transparent (>90%) ITO film with circa. 80 nm thick onto 50 mm Si wafer; and the schematic illustration of the variation in thickness **(b)** and sheet resistance (in Ω/square) across silicon wafer. (From Choy, K.L. et al., ESAVD of transparent conducting oxide films, EU-FP7ThinSi project (241281) final report, 2013.)

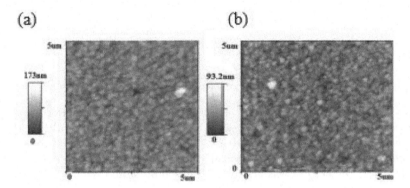

Figure 1.104 AFM of the as-deposited ITO by ESAVD before **(a)** and after **(b)** thermal annealing at 850°C for 60 s. (From Choy, K.L. et al., ESAVD of transparent conducting oxide films, EU-FP7ThinSi project (241281) final report, 2013.)

some of the surface features decomposed to become island-like features as shown in Figure 1.105 [199]. The thermal degradation of ITO by sputtering has also been reported by other researchers [200]. The highly stable ITO by ESAVD could be due to the deposition temperature of the conducting films as compared to the lower deposition temperature by the vacuum sputtering process. The higher deposition temperature would lead to a better crystallinity. Therefore, the thermally stable ITO by ESAVD could potentially be used as the replacement of the normal anti-reflection layer Si_3N_4 in silicon solar cells. In addition, non-indium-based nanothin (circa. 80 nm) TCOs such as Sb-doped SnO_2 (ATO) and FTO have also been deposited using ESAVD. The electrical properties of the ESAVD of these TCO films are summarised in Table 1.1 [198].

1.3.2.4.2 Electrostatic-assisted aerosol jet deposition

In order to widen the choice of chemical precursor and include aqueous precursor, a variant called electrostatic-assisted aerosol jet deposition (EAAJD) has also been developed [1]. EAAJD is also a non-vacuum AACVD and can be scalable. It has a similar deposition mechanism as ESAVD; however, the key difference between the ESAVD and the EAAJD method is that the aerosol atomisation and electrostatic discharge functions in EAAJD are separated unlike the ESAVD process. In addition, EAAJD not only preserves the benefits of ESAVD but also enables higher deposition efficiency (>95%) and deposition rate (>0.2 µm/min) [1]. Such high deposition efficiency and high growth rate of the EAAJD have been used for the synthesis of dense films [164] as well as porous films [201].

(a)   **(b)**

Figure 1.105 AFM of the as-deposited ITO by sputtering before (a) and after (b) thermal annealing at 850°C for 60 s. (From Choy, K.L. et al., ESAVD of transparent conducting oxide films, EU-FP7ThinSi project (241281) final report, 2013.)

Table 1.1 Electrical properties of TCO films

As-deposited TCOs	Sheet resistance (Ω/)	Carrier concentration ($\times 10^{20}/cm^3$)	Mobility (cm^2/Vs)
ITO	14.0	2.74	54
ATO	38.5	7.08	8.77
FTO	47.7	4.35	14.3

Source: Choy, K.L. et al., EU FP-7 ThinSi project report (Project n. 241281), 2012.

EAAJD can produce highly textured and uniform films onto amorphous substrate (e.g., glass) as demonstrated in the deposition of II–VI semiconductor films from a mixture of metal chloride and thiourea. The aerosol was generated ultrasonically at a frequency of 1.7 MHz with nitrogen as a carrier gas (flow rate 500 to 2000 mL/min) and the field strength was 2–5 kV/cm with a deposition time of 2 mins.

XRD traces in Figure 1.106 of CdS and ZnS films deposited using the EAAJD process showing a strong preferred orientation [164]. Fine crystalline CdS (grain size <200 nm) film was formed at low temperature 300°C as shown in Figure 1.107a with a typical bandgap energy value of 1.65 eV. Whereas ZnS films would require a higher deposition temperature (450°C) which resulted in large grain size (grain size ca. 500 nm) with a band gap energy value of 3.31eV (Figure 1.107b) [164]. The good optical properties of these II–VI semiconducting films make EAAJD attractive for large-scale fabrication of thin films for photovoltaic applications.

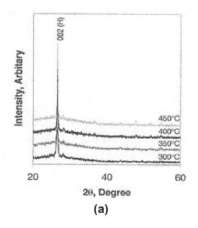

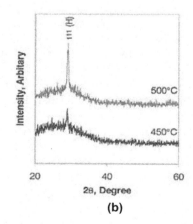

(a)   **(b)**

Figure 1.106 XRD traces of CdS (a) and ZnS (b) films deposited using the EAAJD process at different temperatures showing a strong preferred orientation (EAAJD2). (From Su, B. and Choy, K.L., *Thin Solid Films*, 361, 102–106, 2000; Su, B. and Choy, K.L., *J. Mater. Chem.*, 10, 949–952, 2010.)

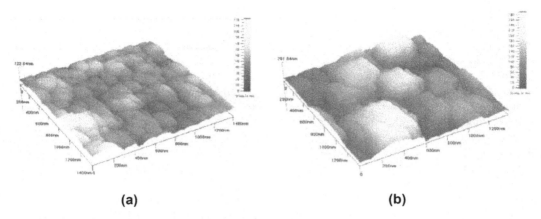

(a)   **(b)**

Figure 1.107 AFM images of **(a)** CdS and **(b)** ZnS films (EAAJD2). (From Su, B. and Choy, K.L., *Thin Solid Films*, 361, 102–106, 2000; Su, B. and Choy, K.L., *J. Mater. Chem.*, 10, 949–952, 2010.)

1.3.2.5 FLAME-ASSISTED CVD

Flame-assisted CVD is a variant of CVD process, in which flame is used as heat source for the reactions; precursors can be gas or liquid precursors [202].

Flame synthesis process has been used to prepare ultrafine particles, such as carbon black, fumed silica, and TiO_2 at industrial scales. In addition, the flame spray process has also been developed for the production of special chemicals. This will be discussed in more detail in Chapter 2.

Flame-assisted CVD has been used to prepare optical fibres, via vapour-phase axial deposition (VAD) process to realize the continuous fabrication of high-silica fibre preforms in the axial direction. The schematic diagram is shown in Figure 1.108.

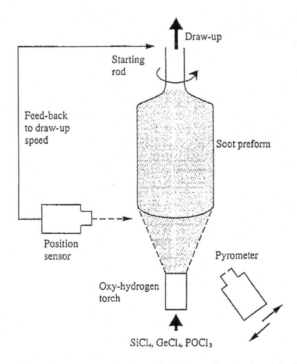

Figure 1.108 Schematic of soot formation process of VAD method using H_2–O_2 torch. (From Sanada, K. et al., *J. Non-Cryst. Solids*, 188, 275–284, 1995.)

During the process, precursor materials such as $SiCl_4$, $GeCl_4$, $POCl_3$, and BBr_3 are fed into oxy-hydrogen torches. The resulting fine glass particles through the flame hydrolysis reaction are deposited onto the end surface of a starting seed silica glass rod to obtain a porous preform in the axial direction, which is consolidated into a transparent preform by zone melting with a carbon ring heater. The reaction and deposition are enclosed in a clean environment obtained by connecting the glass chamber and the water-cooled stainless-steel vessel. The starting rod is pulled upward and rotated, and maintains cylindrical symmetry in the chemical deposition to produce a porous preform with a constant diameter [203].

Another flame-assisted CVD process for optical fibre is the outside vapour deposition (OVD) process, as shown schematically in Figure 1.109.

A source material mixture such as $SiCl_4$, $GeCl_4$, (BCl_3), $POCl_3$ and O_2 is reacted in a flame of CH_4/O_2 or H_2/O_2 mixtures to produce small particles of very high-purity glass of the desired composition. These particles, called soot, are typically 10–100 nm diameter with surface areas of >20 m²/gm [204]. For the OVD process, the morphology of the soot depends on the target temperature, as shown in Figure 1.110 [205].

Further details about the flame-assisted CVD of optical fibres are in Chapter 6.

Flame-assisted CVD process has also been used to prepare novel materials such as CNTs and graphene. It is worth mentioning that almost all types of carbon, such as amorphous carbon [206], diamond [207], MWNTs [208], SWNTs [209], and graphene can be produced by this process [206].

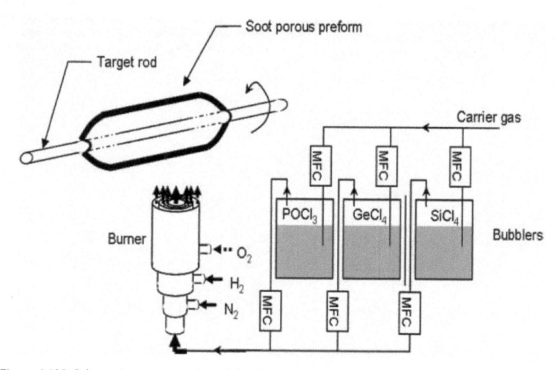

Figure 1.109 Schematic representation of the OVD method. (From Petit, V. et al., *Aerosol Sci. Technol.*, 44, 388–394, 2010.)

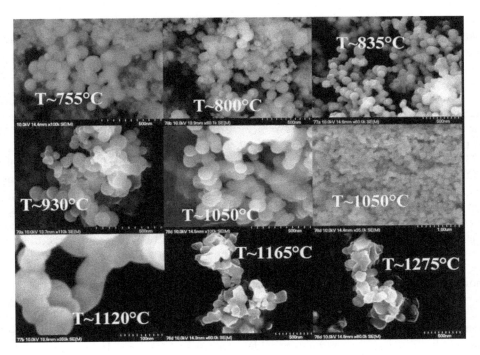

Figure 1.110 SEM pictures of the soot layers deposited at various target temperatures. (From Petit, V. et al., *Aerosol Sci. Technol.*, 44, 388–394, 2010.)

1.4 CVD OF POLYMERS

Besides the use of CVD for the deposition of metal, ceramic, and semiconductor materials, CVD is also an established method used in industry for the deposition of polymers. The CVD of polymer involves vapour phase monomers to react and/or polymerise to deposit organic thin films directly onto mildly heated substrates. There are various types of processes for CVD of polymers. In the 1980s, the need to address the processing difficulties encountered in the solution process of polyimides for its applications in the semiconductor industry has driven the development of vapour deposition polymerisation (VDP) as an alternative process for the fabrication of high-temperature polymeric films. VDP is a dry process; the films produced do not exhibit the rheological effects and solvent retention of solution of the conventional cast films. The process was developed independently by Iijima and Salem [210,211].

As shown in Figure 1.111 [211], the process involves the evaporation of different monomers,

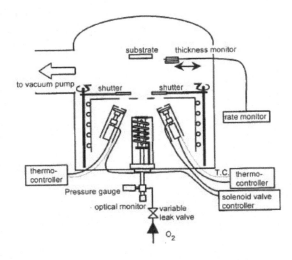

Figure 1.111 Schematic of the experimental system for vapour deposition polymerisation. (From Miyamae, T. et al., *Langmuir*, 17, 8125–8130, 2001.)

for example, pyromellitic dianhydride (PMDA) and oxydianiline (ODA) separately in an evacuated container at 200°C under the system pressure $1–5 \times 10^{-6}$ Torr. The process and hardware set-up are to ensure that there is no condensation or gas-phase reaction, and the reaction occurs when the monomers condensed in the substrate to form polyamic acid, which can be cured *in-situ* or after removal from the system in a dry N_2 atmosphere above 175°C in order to produce the polyimide of PMDA/ODA. Further characterisation indicated that the resulting films possessed properties similar to spin-coated films, and the VDP process can be utilized for patterning of polyimide. It is proposed that the polycondensation reaction between PMDA and ODA takes place on the surface of the substrate even at 25°C, and the functional groups of the monomers had a fairly high frequency of collision in vapour deposition polymerisation [210].

In the 1990s, hot-wire CVD (HWCVD) of polymer started with a negative result; however, the deposition of film from hexafluoropropylene oxide (HFPO) by HWCVD resulted in initiated CVD (iCVD) and oxidative CVD (oCVD) process for the polymer deposition [212].

For the chain growth polymers, a volatile initiator can be introduced in the vapour phase together with the monomer(s), and the reaction can be named accordingly depending on the process. For iCVD where the initiator (e.g., tert-butyl peroxide) is activated thermally, whereas photo-iCVD

is activated by UV exposure, or initiated PECVD (iPECVD) if activated by plasma excitation [213].

In a typical iCVD reactor as shown in Figure 1.112a, the filament wires (e.g., 28 AWG Nichrome wire, 80%Ni/20% Cr) were heated to about 250°C, which can produce free radicals from the initiator, such as tert-butyl peroxide, without cracking of the monomer (e.g., HFPO). With the presence of the initiator, monomer absorption and polymerisation proceeds onto a cooled substrate. If there is no initiator, the film growth is negligible. This iCVD process is versatile for chain growth polymers, such as PTFE, acrylates, methacylates, styrenes, vinylpyrrolidone, and maleic anhydride (Figure 1.112b). The process requires precursor molecules with a high vapour pressure to allow the delivery of precursors into the vacuum chamber and on the substrate surface, this restricts the use of monomers and initiators with high molecular weight or polarity [214].

iCVD process can be scaled-up towards a difference scale as shown in Figure 1.113 [215]. In particular, the process can also be used as a roll-to-roll (R2R) deposition process.

Another type of CVD of polymer is through step-growth polymerisation, and the process includes oCVD, the previously mentioned vapour phase polymerisation, molecular layer deposition, and oxidative molecular layer deposition. An example of oCVD process is shown in Figure 1.114 [214]. During the oCVD process, the oxidant and

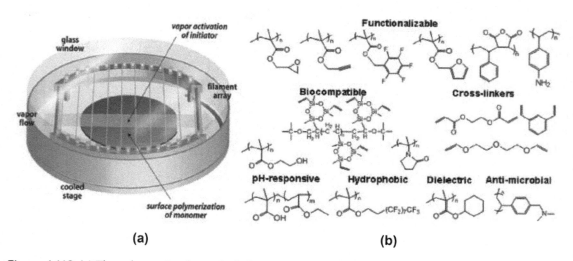

(a)   (b)

Figure 1.112 (a) The schematic of a typical iCVD reactor. (b) Examples of some functional monomers being polymerised by iCVD and piCVD. (From Asatekin, A. et al., *Mater. Today*, 13, 26–33, 2010.)

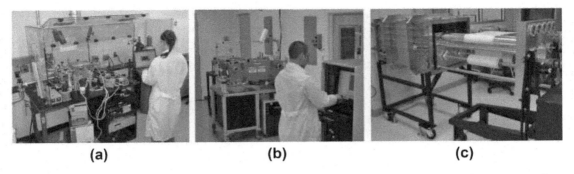

Figure 1.113 GVD has developed a number of standard iCVD coating systems, including **(a)** a small R&D coating system (with 6 in. coating area), **(b)** a mid-sized production batch coater (20 in. × 30 in. coating area) and **(c)** a roll-to-roll web coating system (30 in. in width). (From Pryce Lewis, H.G. et al., *Thin Solid Films*, 517, 3551–3554, 2009.)

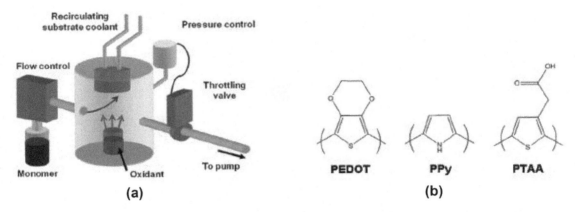

Figure 1.114 **(a)** A schematic of a typical oCVD reactor. **(b)** Chemical structures of the conducting polymers deposited by oCVD. PEDOT: poly(ethylenedioxythiophene), PPy: polypyrrole, PTAA: poly (3-thiopheneacetic acid). (From Asatekin, A. et al., *Mater. Today*, 13, 26–33, 2010.)

monomer are delivered to the substrate through the vapour phase; absorption and spontaneous reaction proceed directly on the substrate. Polymers that can be prepared through this process include PEDOT, polypyrrole (PPy), and poly(3-thiopheneacetic acid) (PTAA) [216]. oCVD process also requires precursors with sufficient volatility. There is issue on commercial availability of functional precursors, and these factors have effects on the range of chemistries achievable through oCVD processes.

oCVD process can also be scaled up, as shown in Figure 1.115 [217], and also can be performed in a roll-to-roll reactor.

The applications for iCVD process are being considered for surface engineering, optoelectronic devices and sensors and patterning of surfaces [214]. As shown in Figure 1.116, a conformal layer of poly (4-vinyl pyridine) (P4VP) is deposited in the trench to be a part of circuit. Upon exposure to nitroaromatic explosives, P4VP would swell to bridge the trench and complete the circuit [214].

Besides CVD-based polymerisation, plasma-based process has also been utilized for the preparation of polymers. In 1874, the phenomenon that organic compounds formed polymers in plasma (ionized gas) was observed, and until 1955, such polymers were recognized as by-products of the phenomena associated with electrical discharge. Therefore, the progress was not treated as a process to form polymers and little attention has been paid to the properties of these polymers. From 1960, plasma (glow discharge) polymerization approach was employed to make special coatings, for example, the deposition of perfluorocarbon on

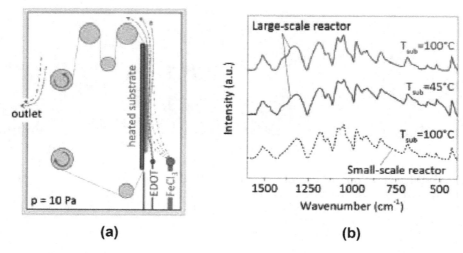

(a)   **(b)**

Figure 1.115 **(a)** Illustrated side profile of the R2R oCVD reactor. Monomer and oxidant flows (EDOT and FeCl₃, respectively) are directed upwards along the heated substrate. Rotation of the rolls indicates direction of the substrate movement in the R2R mode. **(b)** Comparison of FTIR spectra of the PEDOT films deposited in the R2R large-scale and small-scale reactors. (From Kovacik, P. et al., *Materials Horiz.*, 2, 221–227, 2015.)

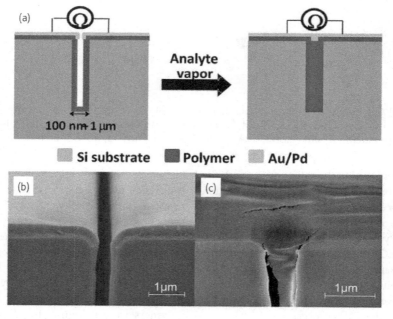

Figure 1.116 **(a)** Operating principle of sensor based on CVD polymer-coated microtrenches. Microtrenches are conformally coated by a functional polymer, followed by nonconformal deposition of Au/Pd. When exposed to analyte vapour, the polymer swells, completing the circuit. **(b)** SEM of microtrench coated with P4VP and Au/Pd. **(c)** SEM of microtrench after exposure to nitroaromatic vapour. Upon swelling of P4VP, the two sides of the trench meet, and the circuit is completed, turning the circuit "on". (From Asatekin, A. et al., *Mater. Today*, 13, 26–33, 2010.)

metals (e.g., 5 μm Al) for potential applications in capacitors and transmission lines from a mixture of monomeric fluorethylenes at 0.7–0.9 Torr at room temperature [218]. This demonstrated advantageous features, including good adhesion to the substrate and low dielectric constant. Hence, this approach was investigated towards applied research. It is often being called 'glow discharge polymerisation' or 'plasma polymerisation.' Polymers from glow discharge polymerisation are in most cases highly branched and highly cross-linked and adhere to solid surfaces, and solvent-soluble polymers can also form by tuning the conditions of the glow discharge. The mechanism of the process involves free radicals, and low-temperature plasma is often used because organic compounds decompose at high temperatures and 'hot plasma' cannot be used. The type of electric power source and mode of coupling are two major factors involved in the design of a reaction vessel [219]. Figure 1.117 shows a schematic representation of competitive ablation and polymerization (CAP) scheme of polymer formation in plasma [220].

Detailed investigation on the various types of monomers, reactor vessel, and process conditions have been performed. These focus on continuous-wave-style plasma polymerisation [219]. When the process was conducted through pulsed-plasma polymerization [221], in the example of maleic anhydride, there is greater retention of the anhydride functionality originating from the parent monomer during pulsing of the electrical discharge, as lower average power values result in less fragmentation of the precursor molecule and in reduced damage of the growing plasma polymer layer during the duty cycle on-time, combined with radical-initiated polymerisation of maleic anhydride during the off-period. Figure 1.118 shows the schematic diagram for the following steps:

On-time: Plasma polymerization R (incorporated into plasma polymer)
Off-time: radical initiated chain growth

As pulsed-plasma polymerisation can provide greater retention of the functional group, the method has been utilized to generate hydrophobic or hydrophilic surface, which can be applied in different sectors. For example, hydrophobic surface has been prepared through pulsed-plasma polymerisation process of 1H,1H,2H,2H-heptadecafluorodecylacylate under a pressure of 2×10^{-3} bar and the chamber temperature of 50°C through a duty cyle of plasma (13.56 MHz, 200 W), for example, on-time of 20 μs

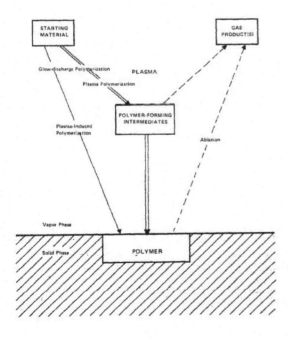

Figure 1.117 Competitive ablation and polymerisation (CAP) scheme of polymer formation in plasma. (From Yasuda, H. and Hsu, T., *Surf. Sci.*, 76, 232–241, 1978.)

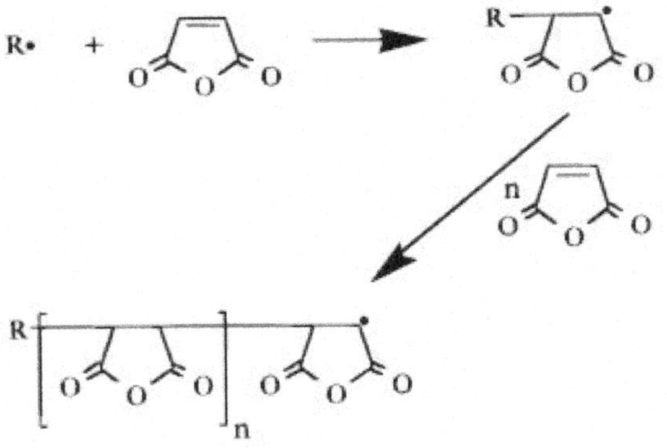

Figure 1.118 A schematic for the steps of on-time and off-time of duty cycle for pulsed plasma polymerisation of maleic anhydride in an electrodeless cylindrical glass reactor (4.5 cm diameter, 460 cm³ volume, base pressure of 5.2×10^{-3} mbar, 13.56 MHz RF generator for the peak power level in the range of 5–90W, pulse on-time is around 20–800 μs and off-time is around 20–1200 μs). (From Ryan, M.E. et al., *Chem. Mater.*, 8, 37–42, 1996.)

and off-time of 20000 μs [222]. This type of process and film can be applied to mobile phones to improve water-proof properties of the phones, and it is also applicable in electronics, filtration, and military. The process and technology have now been applied on an industrial scale [223]. Further examples of the use of CVD in industry for the deposition of polymer such as Parylene is presented in Chapter 6.

1.5 CVD MODELLING AND SIMULATION

CVD is a process whereby chemical vapours react and undergo heterogeneous reaction to deposit a stable solid film onto a heated substrate. CVD can also undergo homogeneous reaction in the gas phase to form ultrafine particles. CVD is important for many industries, such as semiconductor industry, and there is a continuous need for new CVD processes and equipment. Trial-and-error approaches have been used for the process and equipment development. However, a more systematic approach has been recognized, which led to the development and application of predictive models, such as CVD reactor modelling for the description of hydrodynamics and transport phenomena, and the first detailed models for CVD gas-phase chemistry-based on elementary reactions rather than lumped chemistry, and the modelling of the surface chemistry in CVD [224]. Figure 1.119 shows the history of CVD modelling development [224].

In general, the modelling for CVD processes can be categorized into three types: (1) the modelling for CVD reactor, which simulates multi-dimensional flow and transport phenomena in relation to process operation and equipment design issues, combining with simple descriptions of the CVD chemistry. The studies for different types of reactors have been reported; (2) the modelling of CVD chemistry, which focuses on unravelling the detailed reaction mechanisms in the gas phase and at the surface, combining with simple 0-D or 1-D hydrodynamic models. The gas phase chemistry is described as many finite-rate, elementary reaction steps, assuming each reaction is reversible, and the parameters for rate constants, activation energies, and thermo-chemical data are obtained from experimental data, or classical theoretical models using statistical thermodynamics, bond dissociation enthalpies, transition state theory and RRKM theory, or density functional theory and ab initio computational chemistry calculations. The surface chemistry involves gaseous molecules impinging on the surface, adsorbed molecules, free sites, and dangling bonds on the surface. Various approaches can be utilized to model these steps; (3) Microscopic models, which describes the interaction between (molecular) transport phenomena and deposition within sub-micron trenches and contact holes on wafers for semiconductor manufacturing [224].

Further development on the theoretical progress and computer performance makes it possible to combine detailed descriptions of transport phenomena and reaction chemistry into a single

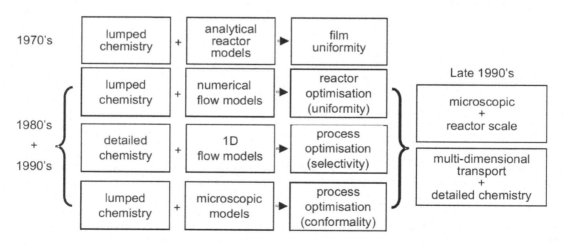

Figure 1.119 History of CVD models development and integration. (From Kleijn, C.R. et al., *J. Cryst. Growth*, 303, 362–380, 2007.)

computational model. The commercial computational fluid dynamics (CFD) codes can be used to design and optimize reactor configurations and process conditions, simulate typical CVD phenomena, such as multi-component gas diffusion, wall-to-wall thermal radiation and detailed multi-reaction gas and surface chemistry. There are various codes that may produce results with differences of an order of magnitude. This is the issue requiring the attention for the modelling.

CVD involves complex and strongly coupled phenomena occurring at multiple time and length scales. Numerous numerical models have been developed to simulate the CVD process, but most of them focused only on particular scales including molecular dynamics simulation [225], Monte Carlo simulation [226], finite element and CFD, etc. A simulation model accurately describing all the physical phenomena involved in CVD should implement the equations governing electronic interactions and molecular motions over various time and length scales. The development of such a model remains an elusive goal, even with the ever-increasing power of computers. In contrast, multiscale modelling strategies are now becoming available to perform simulations of the entire CVD process. Predicting the microstructure and properties of materials is a tremendous scientific and economic challenge which might become a reality with the advent of multiscale simulation, which has recently emerged as one of the most promising interdisciplinary fields of investigation in Computational Materials Science. The literature on the subject is becoming abundant, but only a few reviews have focused on multiscale modelling of thin-film growth by vapour deposition techniques [227–229].

Multiscale modelling strategies are now becoming available to perform simulations of the entire CVD processes [224]. There are two types of multiscale modelling: one is referred to as hierarchical or sequential analysis, in which first simulations at the higher resolution are performed and the resulted properties are extracted and used as input in the next level method, and the other is referred to as concurrent multiscale analysis allowing for the consideration of multiple scales simultaneously. Different methods are used to describe the given regions of the material with the appropriate time and length scale resolution [230]. The methods suitable for particular time and length scales are used to treat aspects of materials phenomena that operate only over those scales, as shown in Figure 1.120 [230]. There are good reviews (e.g., [228a,228b]) on the main characteristics of

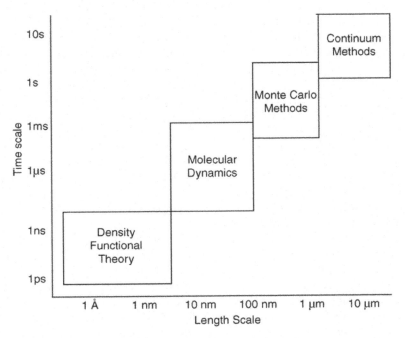

Figure 1.120 Characteristic time and length scales for simulations methods of various materials. (From Karakasidis, T.E. and Charitidis, C.A., *Mat. Sci. Eng. C Biomim. Supramol. Syst.*, 27, 1082–1089, 2007.)

individual methods, including Ab-initio methods, molecular dynamics, kinetics Monte Carlo methods and reactor model, and how they are combined within multiscale modelling strategies.

Multiscale models have been developed to investigate the CVD growth of nanotubes [231a,231b], nanocrystals [232a,232b], nanoparticles [233] and 2D materials (e.g., graphene) [234].

Recently, most of the modelling regarding CVD focuses on the growth mechanism of silicon [235,236,237], CNT [238,239], graphene [239], ceramic materials SiC, TiC, etc. [240], or the heat loss [236], influencing process parameters such as gas pressure, temperature distribution using CFD [235], COMSOL [241], mathematic model, or atomistic simulations [239]. For example, Ni et al., performed modelling and simulation of silicon epitaxial growth in a Siemens reactor using commercial CFD software ANSYS FLUENT. The distributions of gas velocity, temperature and species concentrations in the reactor were predicted numerically [235].

Ramos et al. studied heat losses in a CVD reactor for polysilicon production after taking into account radiative heat loss, conductive heat loss, and convective loss and a comprehensive model for heat loss is presented and validated by comparison with experimental results obtained using a laboratory-scale CVD reactor [236]. Liu et al. investigated the evolution of silicon particle growth by chemical vapour deposition of silane pyrolysis in a three-dimensional slugging fluidized bed reactor using CFD software ANSYS FLUENT [237].

Raji et al. investigated the reacting flow dynamics and temperature distributions inside the CVD reactor during the formation of CNTs. The theoretical approach solves the momentum and energy equations, in conjunction with the reaction kinetics involved. The mathematical model is numerically solved using a two-dimensional CFD formulation, utilizing the COMSOL Software [241].

Elliott et al. discussed the synthesis of CNTs and graphene by catalytic CVD and PECVD, summarising the state-of-the-art understanding of mechanisms controlling their growth rate, chiral angle, number of layers (walls), diameter, length, and quality (defects), and presented a new model for 2D nucleation of a graphene sheet from amorphous carbon on a nickel surface by comparing the results of tight-binding, semi-empirical molecular orbital theory and reactive bond order force field calculations (atomistic modelling) [239].

Geiser et al. proposed a multiscale model based on two different software packages to simulate thin films of metallic or ceramic materials, such as SiC or a mixture of SiC and TiC. The large scales are simulated with CFD software based on the transport reaction model (or macroscopic model), and the small scales are simulated with ordinary differential equations [240].

It was reported from the literature that sometimes remarkable agreement is achieved between Kinetic Monte Carlo (KMC) simulation and experiments [242], and sometime the observed trend is similar to the experiment [228]. Whereas current multi-scale CVD models largely succeed in relating the set macroscopic process conditions of macroscopic and feature scale physical and chemical properties of the deposited films, little progress has been in making a connection to properties of CVD products at the atomic and crystal grain scales, and their influence on the functional properties of the CVD products.

Modelling becomes an important approach for the development of new CVD reactor and processes. For example, since the progress made in the nitride blue laser diode in the 1990s, the demand for new MOCVD reactors and process optimisation arise, the number of wafers that can be processed has increased with the development of new production tools. This has driven the work on the modelling of the reactor and process. The earlier work is to model some aspects related to the reactor or process, such as heat transfer behaviour in a certain type of the reactor, and gradually some types of the reactors (e.g., horizontal reactor or rotating disk reactor), became the main types of the reactors, and further modelling work is to optimise the process window based on these reactors, or the design of larger scale reactors.

The team from Veeco Instruments, one of two major MOCVD production tool providers, for example, used the modelling results to modify the alkyl injection system, which improves growth uniformity and alkyl efficiency while optimizing uniformity tuning for a variety of reactor conditions during LED process development. In their work, they developed calculation models that represent the effects of process parameters and reactor features on the growth rate and uniformity of GaN/InGaN in the EMCORE's D180 and E300 reactors. Figure 1.121 presents the detailed numerical grid of the short jar D180 reactor with a 'bow-tie' flow

(a) (b)

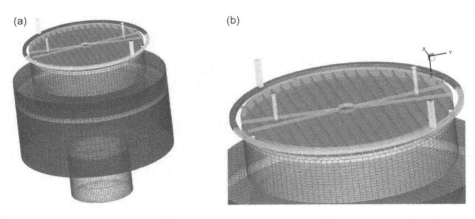

Figure 1.121 Detailed numerical grids of D180 EMCORE reactor (**a**) and flow flange (**b**). (From Kadinski, L. et al., *J. Cryst. Growth*, 261, 175–181, 2004.)

flange (flow injector with a sectorial alkyl injector schematics) [243]. Series of experiments has been performed in this reactor to verify the GaN computational models. It shows that the use of very detailed geometrical models is crucial for accurate predictions, so it is the detailed model approach. From the perspective of flow optimization ad reactor design, as shown in Figure 1.122a, computer modelling shows that if very different inflow velocities are established between the alkyl and hydrides zones, recirculation patterns are introduced. (The length of the velocities vectors is proportional to the velocity magnitude). When the alkyl zone and hydride injector velocities are properly adjusted, the modelling result is almost perfect vortex-free flows, as shown in Figure 1.122b [243].

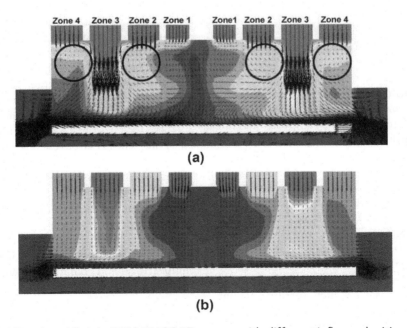

Figure 1.122 (**a**) Simulated flow in E300 EMCORE reactor with different inflow velocities through alkyls zones (circles represent the recirculation areas); (**b**) simulated flow in E300 EMCORE reactor with the adjusted inflow velocities through alkyls zones. (From Kadinski, L. et al., *J. Cryst. Growth*, 261, 175–181, 2004.)

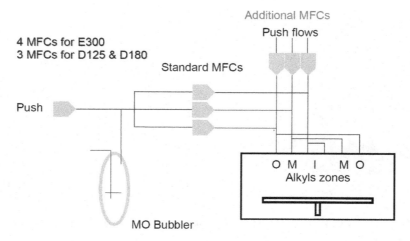

4 MFCs for E300
3 MFCs for D125 & D180

Figure 1.123 An example of a new hardware modification. (From Kadinski, L. et al., *J. Cryst. Growth*, 261, 175–181, 2004.)

The modelling work has indicated to add an additional push-flow MFC to each alkyl injection zone (see Figure 1.123) in order to achieve alkyl and hydride zone flow velocity matching within the reactor. The additional MFC per injection zone provides the ability to control the velocity of the gas injected into each zone, while being able to independently control the alkyl concentration injected into each zone. This flexibility allows one to take advantage of superposition approach to optimize the growth rate deposition uniformity across the susceptor for a given growth condition. Moreover, the experimental results using the modification showed marked improvements in thickness, compositional, and brightness uniformity of InGaN MQW LED structures, and the alkyl utilization efficiency was improved [243].

The processing conditions can be optimized for the maximum deposition rate and uniformity in vertical rotating disc MOCVD reactors based on CFD modelling [244].

As MOCVD process involves reactors and many process parameters, the modelling approach can be combined with the experimental results for further optimisation, such as transport phenomena and the effect of geometry of the reactor on the growth rate and uniformity of III–V films [245]. This modelling approach is also valuable for the optimal design of horizontal reactor [246].

1.6 CVD AND ITS VARIANTS AS TOOLS FOR PROCESSING ADVANCED MATERIALS TO ADDRESS ENERGY, ENGINEERING, AND BIOMEDICAL GRAND CHALLENGES

CVD and its variants provide the capability for the preparation of various types of materials, including 0D nanoparticles, 1D nanostructures, 2D materials and thin films, and 3D structures.

These materials can be prepared as standalone materials, such as nanoparticles or 1D nanostructures, or as a component of functional structures.

These tools and materials can provide solutions for future challenges from the following perspectives:

1. Improvement on the current technological development
2. Delivering the technology for future development

In this part, the applications of CVD for energy, engineering, and biomedical applications and challenges will be highlighted.

1.6.1 Energy

The grand challenges in energy involve energy saving through higher efficiency or new products, energy generation/harvesting, and energy storage. From the perspective of energy saving, one example is the development of LEDs, in particular HBLED, which can be used as back light for display to replace the old format display technology, and used as lighting. The development of blue LED is an important step to obtain HBLED, as shown in Figure 1.124, for which the Nobel Prize was awarded for Nakamura and other scientists for their work on blue LED [247]. The application of LED can save more energy than previous products, such as incandescent light bulbs.

MOCVD process is the key process to grow the III–V-based functional thin films, and PECVD process involves the deposition of passivation layers. There is ongoing development on these

processes to support further development of LED products.

From the perspective of energy generation/harvesting, solar cells have been used widely to produce clean energy from sun to reduce the use of fossil energy. Currently, crystalline Si-based solar cells are widely used, in which PECVD process can be used to deposit amorphous Si, crystalline Si layer and anti-reflective layers [19,52]. As the current solar cell efficiency (e.g., 26.3% ± 0.5, [248]) is still lower than theoretical efficiency (33.16% for AM 1.5 G illumination) [249], considerable effort is made to achieve higher efficiency solar cells.

From the perspective of energy storage, lithium ion battery and supercapacitor are the main portable energy storage devices. CVD/PECVD can be used to produce CNTs and graphene nanoplatelets that can be used in the lithium ion battery electrode to enhance lithium ion battery performance [250]. Figure 1.125 shows graphene-based transparent conductive film and graphene touch panel product [251].

Apparently, there is always a need for more efficient and greener energy. CVD and its variants will keep developing for the future need.

1.6.2 Engineering

From the engineering perspective, most functions can be derived from the modified surfaces or coatings. There are several key areas involving CVD and its variants.

In the tool industry, hard-coatings, such as multilayer coatings $TiC/Al_2O_3/TiN$ (see Chapter 6),

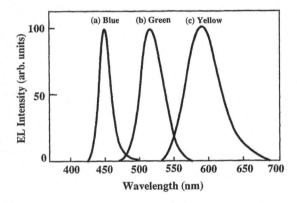

Figure 1.124 Electroluminescence of **(a)**, **(b)** and **(c)** SQW LEDs at a forward current of 20 mA. (From Nakamura, S. et al., *Jpn. JAppl. Phys. Part 2*, 34, L797–L799, 1995.)

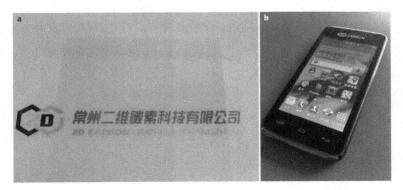

Figure 1.125 Mass production and application of large-area graphene films. **(a)** A large transparent conductive graphene film (about 20 × 20 cm²) manufactured by 2D Carbon Tech. **(b)** Prototype of a mobile phone (2D Carbon Tech) using a graphene touch panel. (From Ren, W. and Cheng, H.-M., *Nat. Nanotechnol.*, 9, 726–730, 2014.)

or the nanostructured Tungsten Carbide-based nanocomposite coatings (see Chapter 7), and DLC deposited by CVD, ion-assisted CVD, these coating tools and processes have been used widely, and further development is still ongoing [16].

Diamond coating has also been used in many industries due to its special properties. In particular, MW-plasma or hot filament plasma processes have been used to produce the diamond films, further optimisation of processes and tool development is still required [11]. Further description of the industrial applications of CVD of diamond films is presented in Chapter 6.

Nanoparticle-based TiO_2 coatings used in marine coating or paint for built environment, for example, are now widely applied. Flame aerosol-based CVD processes have been used for the large-scale production of ultrafine particles of TiO_2 (see Chapter 2). There is a need for more functional coatings, and further process improvement will be required.

Optical engineering involves high-demand coatings, such as infrared detectors. Ion-assisted process are the key process to achieve such high-demand coating layers. These process and tool are being developed further [119].

The key component of electrical engineering is HBT. The device is made based on MBE processes. This is now widely used in mobile for telecommunication. The requirement for better performance device and lower price would drive the process and tool development [97].

Another important area is the thin-film magnetic head for data storage; this would require conformal coating for the magnetic layers and DLC layers. In particular, the thickness of the DLC layer is very thin, about 2 nm. CVD process and ion-assisted CVD/PVD process is used, as shown in Figure 1.126, as the demand for the capacity increase of storage medium is always there; therefore, the development of process and tool is needed accordingly [252].

1.6.3 Biomedical

Biomedical areas are very broad, and various materials and related processes have been used or under development. DLC coating has been widely applied in medical implant parts (e.g., hips, heart valves), to protect the part and

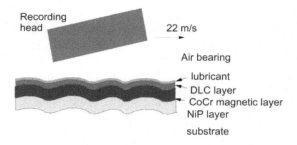

Figure 1.126 A schematic representation of a disk and head. (From Robertson, J., *Thin Solid Films*, 383, 81–88, 2001.)

improve the compatibility of the part with the body. Various CVD-based processes are used to produce DLC coatings, as shown in Figure 1.127. Further development is required to improve the performance of the coatings [253]. Figure 1.128 shows an example of the CVD-coated surface of component [254].

Nanoparticle-based cream has been widely used as anti-UV protection. As mentioned previously, flame aerosol-based CVD process are used to produce the nanoparticle (e.g., TiO_2) on a large scale. Further development is to improve the functions of the coating.

The emerging area is the development of magnetic particles for the imaging and drug delivery towards diagnostics and therapy. CVD-related

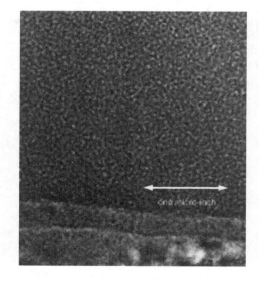

Figure 1.127 TEM micrograph of DLC coating showing graphitic nanostructure. (From Dearnaley, G. and Arps, J.H., *Surf. Coat. Technol.*, 200, 2518–2524, 2005.)

Figure 1.128 DLC-coated knee condylar component and femoral head. (From Oñate, J.I. et al., *Surf. Coat. Technol.*, 142–144, 1056–1062, 2001.)

processes are used to produce magnetic particles, as this field is still under the development. Further investigation is still needed. This part will be covered in Chapter 2.

1.7 ADVANTAGES AND LIMITATIONS OF CVD

Variants of CVD technologies and their special applications have been highlighted. The review on the comparison of different CVD technologies has been covered in reference [1], for example. In general, these CVD technologies vary from the process parameters (see Figure 1.129) and scalability, which in turn would affect the thin film and coating properties.

There are various CVD technologies, including atmospheric pressure CVD process, low vacuum process like PECVD, and high vacuum process like MBE. The selection of specific CVD technology will depend on the requirement of the resulting CVD products, either coating/films or nanoparticles. The following is the list of some criteria [1]:

1. Be capable of depositing the required type of coating and thickness

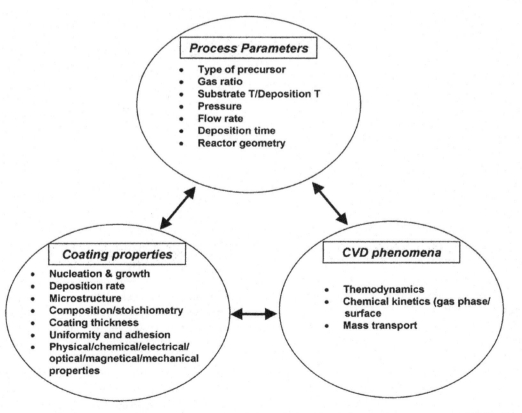

Figure 1.129 shows a schematic representation of the relationship of process parameters, CVD phenomena and coating properties. (From Choy, K., *Prog. Mater. Sci.*, 48, 57–170, 2003.)

2. Be capable of coating the engineering components uniformly with respect to both size and shape
3. Be cost-effective in terms of costs of the substrate, coating material, coating technique, and minimised equipment down time and improved the quality of the coated end-product

Although CVD is a complex chemical system, it has the following distinctive advantages [1]:

1. CVD and its variants can be used for the preparation of a broad range of materials, including nanoparticles, 1D nanomaterials, film and 2D materials and 3D structures.
2. Producing uniform films with good reproducibility and adhesion at reasonably high deposition rates.
3. CVD, particularly thermal CVD, is a non-line-of-sight process with good throwing power. Therefore, it can be used to uniformly coat complex-shaped components and deposit films with good conformal coverage. Such distinctive feature outweighs the PVD process.
4. It has the ability to control crystal structure, surface morphology, and orientation of the CVD products by controlling the CVD process parameters.
5. Deposition rate can be adjusted readily through CVD and its variant, from low-deposition-rate process favoured for the growth of epitaxial thin films for microelectronic applications, to the process for the deposition of thick protective coatings.
6. Reasonable processing cost for the conventional CVD technique, and some CVD variants are being scaled-up to reduce the cost.
7. The flexibility of using a wide range of chemical precursors such as halides, hydrides, organometallics which enable the deposition of a large spectrum of materials including metal, carbides, nitrides, oxides, sulphides, and III–V and II–VI materials.
8. Relatively low-deposition temperatures, and the desired phases can be deposited *in-situ* at low energies through vapour phase reactions, and nucleation and growth on the substrate surface. This enables the deposition of refractory materials at a small fraction of their melting temperatures. For example, refractory materials such as SiC (melting point: 2700°C)

can be deposited at 1000°C using the following chemical reaction:

$$CH_3SiCl_3 \rightarrow SiC + 3HCl$$

However, the drawbacks of CVD include:

1. Chemical and safety hazards caused by the use of toxic, corrosive, flammable and/or explosive precursor gases. However, these drawbacks have been minimized using variants of CVD such as electrostatic spray-assisted vapour deposition (ESAVD) and CCVD methods that use more environmentally friendly precursors.
2. Difficult to deposit multicomponent materials with well-controlled stoichiometry using multi-source precursors because different precursors have different vaporisation rates. However, this limitation can be minimised using single-source chemical precursors.
3. The use of more sophisticated reactor and/or vacuum system by CVD variants such as low pressure or UHV CVD, plasma-assisted CVD, and PACVD tends to increase the cost of fabrication. However, there are also other variants of CVD such as AACVD and flame-assisted CVD (FACVD) that do not use sophisticated reactor and/or vacuum systems. Such variants may provide alternatives for applications where production cost is an issue.

1.8 COMPARISON OF CVD WITH OTHER PROCESSING TECHNIQUES: STRUCTURE, PROPERTIES, COST, SUSTAINABILITY, AND VERSATILITY

Apart from CVD processes, there are other methods that may be able to produce similar structures (e.g., columnar, polycrystalline, equiaxed, and texture films), such as PVD process, electro-deposition, and sol-gel process. The following part gives a brief of description and comparison of these processes.

The PVD process is referred to the process that the species to be deposited is transformed into the gaseous state by a physical process through thermal evaporation or another impact process. The deposition usually occurs at a gas pressure below about 10^{-1} mbar in a vacuum chamber. No chemical reactions take place in PVD processes. PVD process covers a broad a range of processes, including thermal evaporation, arc evaporation, and sputtering [255]. PVD can be dated back to 1852, a Welsh physicist, William Robert Grove, described the sputter deposition and ion-etching experiments. Vacuum was achieved with a mechanical piston pump, the power was voltaic pile, the electrode consisted of a copper plate, with an electroplated and polished silver surface, and a rod that passes through a leather stopper in the top of the glass vacuum chamber, with a steel needle attached to its end. The pressure ranged 100–500 mTorr, the distance between the steel needle and the silver plate was around 0.25 cm, using a mixture of hydrogen and air as the process gas, the silver plate as positive electrode, and the steel needle as the cathode. There was a thin-film deposition (primarily iron oxide) on the silver substrate. In 1877, A. W. Wright at Yale University reported the growth of adherent noble-metal films sputter-deposited from wire targets onto glass slide. Further work on Pt film deposition lay the foundation for the manufacturing of commercial mirrors by the late 1800s. In 1891, William Crooks reported deposition rate and sputtering yields for 23 different metals. In the 1930s, the commercial equipment was available for sputter-depositing gold. The earliest report of RF discharge was by J. J. Thomson in 1891. The first roll-to-roll web coating was installed in 1934 to sputter deposit gold on glossy paper to create foil for hot stamping in specialty printing processes to produce shiny decorative designs on textiles, wood, plastics, leather, and other materials [4].

Another type of PVD process, arc deposition, was demonstrated for the growth of metal oxide films by cathodic-arc deposition in air powered by high voltage, relatively low current with Leiden-jar type batteries in the mid-1760s. In 1857, Faraday reported arc deposition of films on glass substrate to investigate the optical properties of metals [4]. The PVD process (sputtering) is now widely used in the industry to deposit ITO film for flat panel display devices [256].

Further work on PVD includes the evaporation of material, like Si to deposition film [257], and preparation of nanoparticles [258]. Generally, PVD is a line-of-sight process; it can only coat one side per run and cannot coat 3-dimensional structures uniformly. The commercial applications include deposit ITO layer for the solar cell panel.

Electrodeposition is also called electroplating; the variant processes are electrotyping and electroforming. Electrodeposition is a process that uses electric current to reduce metal cations (like Cu^{2+}) in a solution to form a deposition layer on a substrate (usually electrode).

The electrodeposition process dates back to the work of Italian chemist Luigi Valentino Brugnatelli in 1803. He used Alessandro Volta's invention (1800), the voltaic pile, to facilitate the first electrodeposition gold films onto silver metal parts, which was not used in general industry for the following 30 years. By 1839, scientists in Britain and Russia had independently devised the copper electroplating processes similar to Brugnatelli's for printing press plates. In 1838, a Russian physicist invented electrotyping (also named galvanoplasty); one of the notable applications was to deposit coating on sculptures and domes for the churches in Russia. In Birmingham of Britain, the electroplating industry was founded by George Elkington and Henry Elkington, who were awarded the first patents for electroplating in 1840, and spread around the world. The further development of electroplating was facilitated by the development of electric generators in the late nineteenth century [4]. One of the important applications of electroplating is for the preparation of printing circuit boards (PCB) that are a key development of modern electronics [259]. The electroplating process has also been widely used to deposit coating to improve corrosion protection and enhanced wear properties, along with better appearance for metal or hardware parts in many industries, like transportation, automotive, and aviation sectors. The processes include copper, hard chromium plating, bronze alloy, sulfamate nickel, gold, and silver-plating processes. Plating equipment can be with different scales, operated manually or automatically. Electrodeposition can be used for the deposition of powder, thin film, and 3D structures [260]. One latest industrial application is electrodeposited nanocrystalline Ni (100 nm) for *in-situ* repairing nuclear steam generator tube to address localised corrosion and intergranular degradation [261].

The solution-gelation ('sol-gel') process refers to the process involving steps of preparation of 'sol' and 'gel' form, consisting of the following steps: (i) preparing a homogeneous solution of precursor(s), usually in an organic solvent miscible with water; (ii) converting the solution to the 'sol' form by treating with a suitable reagent, like H_2O with acid (e.g., HF, HCl) for oxide; (iii) inducing the sol to become 'gel' through polycondensation; (iv) shaping the gel to the desired forms or shape like thin films and converting the gel to final materials usually through sintering [262]. The sol-gel work can be dated back to 1846, Ebelman observed the formation of a transparent glass after leaving a mixture of silica orthoester and water untouched for a month [263]. In 1864, the term of 'sol-gel' was used by Graham during his work on silica sols [264]. During his doctoral study in 1912–1915, Patrick developed an economically viable and rapid sol-gel method to make silica gel from sodium silicate (Na_2SiO_3) in large quantities, and in 1931, Kistler used the term of 'aerogel' to describe the first synthesis of a highly porous silica (SiO_2) form by supercritically drying the gel obtained through hydrolytic polycondensation of silicic acid [$Si(OH)_4$]. During 1930–1950, more work was done to convert hydrolysable metal compounds into well-defined transparent layers of metal oxides, and dip-coating method was developed to form thin films of various oxide, which suited to fabrication on a large industrial scale. There are good reviews on the historical development of the sol-gel technology.

Sol-gel process involves reaction of alkoxides in solution under the formation of element (Si or others)-oxygen-element (Si or others) bonds. The synthesis through sol-gel involves the following steps [265]:

1. Reaction of metal or non-metal alkoxides (depending on the material)
2. Bonding the components to one another through hydrolysis
3. Transfer of OR substituents to OH substituents (partial), this induces the condensation reaction into chain or three-dimensionally cross-linked polymers

Polycondensation

1. Removal of OR and OH substituents
2. Transition into oxidic state

The process depends on the quantity of water, temperature, time, type of substituents, type of solvent, partial hydrolysis-decelerating chelatisation, and other factors. To form coatings from sol-gel solution, centrifugal, solution lowering or dip-coating procedures can be performed.

Sol-gel process can be used to prepare oxides, nitrides, oxinitrides, sulphides, and carbides, and the process can have the advantages of broad chemistry, and the product in the format of monolithics, fibres and films, coatings can be obtained through sol-gel processes [266].

Sol-gel process has been used to develop coating for glass for sun-shielding windows, vehicle rear-view mirrors and antireflection layers [263], abrasion resistant coatings for plastic lenses [267], coatings for the decoration of crystal glass [268] on industrial scale, and many new applications like wrap biomolecules [264].

As described, both CVD and PVD processes are vapour-based processes; there is a good comparison done between these two processes, as shown in Table 1.2.

Both PVD and CVD are vapour-phase processes. Thus, they can produce highly pure materials as compared to the electrodeposition and the wet-chemical sol-gel methods.

One example of direct comparison of CVD and PVD process is the process and function of poly silicon. Polycrystalline Si layer can be produced in many ways, such as electron-beam evaporation, sputtering and CVD. For its application as the gate, the work shows that the electron-beam evaporation can provide a clean layer with excellent thickness control (± 10 percent from wafer to wafer of the same run). The resulting silicon with the substrate at 300°C is amorphous, no grain is detectable in scanning electron microscope (SEM), but this process had a major problem. Thermal stresses caused by nonuniform heating led to frequent explosions of the silicon evaporation source to result in droplets of molten silicon hitting the wafer to affect the following procedure: the sputtering process can give uniform layer, but the layer has heavy ionic contamination giving rise to leakage current failure; CVD process to deposit silicon through decomposition of silane (SiH_4) at temperature ranging 650°C–750°C gives a good silicon layer with the best overall compromise between uniformity and cleanliness of the silicon layer [257]. This

Table 1.2 Comparison of CVD and PVD process

CVD	PVD
Sophisticated reactor and/or vacuum system	Sophisticated reactor and vacuum system
Simpler deposition rigs with no vacuum system has been adopted in variants of CVD such as pyrosol, ESAVD, FAVD and CCVD	Vacuum system is definitely required to facilitate the generation of vapour species and increase the mean free path of the vapour species to arrive at the substrate
Expensive techniques for LPCVD, plasma assisted CVD, photo assisted CVD, MOCVD, ALD, EVD	Expensive techniques
Relatively low-cost techniques for AACVD and FACVD	
Non-line-of-sight process. Therefore, it can • coat complex shaped components • deposit coating with good conformal coverage	Line-of-sight process. Therefore, it has • difficulty in coating complex shaped components • conformal coverage problem
Tend to use volatile/toxic chemical precursors	Tend to use expensive sintered solid targets/sources which have difficulties in:
Less volatile/more environmentally friendly precursors have been adopted in variants of CVD such as ESAVD and CCVD	• large area deposition • varying the composition or stoichiometry of the deposits
Multi-source precursors tent to produce	Multi targets or single targets have difficulty in controlling the stoichiometry of the films because different elements will evaporate or sputter at different rates, except the laser ablation method
Non-stoichiometric films	
Single source precursors have overcome such problems	
High deposition temperatures in conventional CVD	Low to medium deposition temperature
Low to medium deposition temperatures can be achieved using variants of CVD such as plasma assisted CVD, photo assisted CVD, MOCVD, ESAVD	
Indispensable for certain material (like single-layer grapheme, blue LED structures)	Impossible for certain materials, like graphene, there is limitation for the material range

Source: Choy, K., Prog. Mater. Sci., 48, 57–170, 2003.

indicates that CVD process is more suitable for this application.

A brief comparison of CVD with PVD, electro-deposition and sol-gel processes from the perspective of deposition of films that are commercially made, is shown in Table 1.3.

The selection of the process method will depend on the film properties, process conditions, process control, scale, continuity, and environmental impact. For instance, large-scale production of antireflection SiNx coating can only be produced through PECVD process, due to the film

Table 1.3 Comparison of different processes on the deposition of films

	Materials and structure	Properties	Cost	Sustainability	Versatility and examples of industrial application
CVD	Amorphous, crystalline Example of materials (metal (like W), ceramic (like Al_2O_3), II–VI (like ZnSe NWs), III–V (like GaN), polymer (like PTFE), composites like doped Si nanowire), 2D (like grapheme), Bi_2Se_3 topological insulator)	Optical (like LED), mechanical (like cutting tool coating), Electrical (like polySi), dielectrical (like Al_2O_3)	Main cost is on the equipment, high installation cost	Good, as the scale increase, the cost for each piece reduces	Very Good, as it has many variant CVD processes, MOCVD for production of GaN for LED [247], PECVD for aSi for TFT application in display [50]
PVD	Amorphous, Crystalline TiN, ITO film	Optical (like antireflection film on glass) Electrical (like ITO)	Main cost is on the equipment, high installation cost	Good, as the scale increase, the cost for each piece reduces	Fair, there is limited material range ITO for flat panel display [256]
Electrode- position	Amorphous, Nanocrystalline (like nanocrystalline Ni)	Electrical (like Cu plating for PCB) Chemical (anti-corrosion)	Low cost on the equipment	Good, the regulation may affect some processes	Fair, there is limited material range Electrodeposition of nanocrystalline Ni for in-situ repairing nuclear steam generator tube [261]
Sol-gel	Amorphous (like SiO_2) Process needs to be repeated to achieve the desired thickness. Materials, mostly oxides	Optical (like antireflection) Mechanical (like abrasion resistant)	Low cost on the equipment, main cost is on the precursor	Good, the film for plastic lenses will increase	Fair, there is limited material range Hybrid sol-gel coating for plastic lenses [267]

thickness, scale, and environment impact. In particular, III–V film obtained from MBE cannot be obtained through the PVD, electrodeposition, or sol-gel processes. Similarly, diamond films can only be deposited using the CVD hot-filament or plasma-assisted CVD processes and they cannot be achieved using PVD, electrodeposition, or sol-gel processes.

Overall, the CVD process has very good versatility. However, other processes may have process advantages for some applications, such as low-temperature abrasion-resistant coating for plastic lenses (CR 39, polycarbonate, poly(methyl methacrylate) etc.) through sol-gel processing of (3-glycidoxypropyl) trimethoxysilane (GPTMS), tetramethoxysilane (TMOS), and titanium tetraethylate ($Ti(OEt)_4$, or the nanocomposite coating with the incorporation of boehmite nanoparticles in the film [267]. This type of low-temperature nanocomposite coating is difficult through CVD processes, or deposition of nanocrystalline Ni through electrodeposition to *in-situ* repair nuclear steam generator tube to address localised corrosion and intergranular degradation, where localised deposition of nanocrystalline material is difficult through CVD processes [261].

1.9 SUMMARY

This chapter has given a background on the development of CVD processes for various materials and CVD technologies. The variants of CVD methods are discussed individually. For each CVD method, the principle, process, and tool are described in order to provide a brief overview and highlight its latest progress. CVD modelling and simulation is also highlighted to describe various approaches and their development, especially related to the time and length scales. The advantages and limitations of CVD are also being discussed and compared with other deposition technologies. The applications of CVD and its variants in environment, energy, engineering, and biomedical sectors are being highlighted.

In general, CVD is a versatile technology. Its tool and process have been developed with human civilisation, there is always new demand for new technologies and processes, and effort has always been made to improve current technologies and processes, or use CVD to synthesise new materials. It is expected that such trend will be continued in the future.

ACKNOWLEDGEMENT

The authors would like to thank Dr. Matthew Jackson for proofreading the chapter.

REFERENCES

1. K. Choy, Chemical vapour deposition of coatings, *Progress in Materials Science* 48 (2003), pp. 57–170.
2. S. Pratsinis, History of manufacture of fine particles in high-temperature aerosol reactor, in *Aerosol Science and Technology: History and Reviews*, Ensor, D.S. (Ed.), RTI International Research Triangle Park, NC, 2011.
3. (a) A.C. Jones and M.L. Hitchman, Overview of chemical vapour deposition, in *Chemical Vapour Deposition: Precursors, Processes and Applications*, Royal Soc Chemistry, Cambridge, UK, 2009, pp. 1–37; (b) H.O. Pierson, *Handbook of Chemical Vapour Deposition, Principles, Technology and Applications*, Noyes Publications, 1999, 506; (c) H. Yasuda, *Luminous Chemical Vapor Deposition and Interface Engineering*, Marcel Dekker, 2005, 805.
4. (a) J.E. Greene, Tracing the 5000-year recorded history of inorganic thin films from ~3000 BC to the early 1900s AD, *Applied Physics Reviews* 1 (2014), p. 041302; (b) Y. Kajikawa, S. Noda, Growth mode during initial stage of chemical vapor deposition, *Applied Surface Science*, 2005, 245(1–4), 281–289.
5. L. Mond, C. Langer and F. Quincke, Action of carbon monoxide on nickel, *Journal Chemical Society Transactions* 57 (1890), pp. 749.
6. G.W. Sears, A growth mechanism for mercury whiskers, *Acta Metallurgica* 3 (1955), pp. 361–366.
7. W. Wagner, KCL-Whiskers aus der dampfphase, *Zeitschrift fur Naturforschung Part A-Astrophysik Physik und Physikalische Chemie A* 19 (1964), p. 1490.
8. S. Iijima, Helical microtubules of graphitic carbon, *Nature* 354 (1991), pp. 56–58.

9. W. Stark and S. Pratsinis, Aerosol flame reactors for manufacture of nanoparticles, *Powder Technology* 126 (2002), pp. 103–108.

10. M.L. Green, R.A. Levy, R.G. Nuzzo and E. Coleman, Aluminum films prepared by metal-organic low pressure chemical vapor deposition, *Thin Solid Films* 114 (1984), pp. 367–377.

11. P.M. Martineau, S.C. Lawson, A. Taylor and M.J. Crowder, Identification of synthetic diamond grown using chemical vapor deposition (CVD), *Gems and Gemology* 40 (2004), pp. 2–25.

12. A.R. Bunsell and M.H. Berger, Inorganic fibers for composite materials, *Composites and Science and Technology* 51 (1994), pp. 127–133.

13. H.M. Manasevit, Recollections and reflections of MO-CVD, *Journal of Crystal Growth* 55 (1981), pp. 1–9.

14. W. Tsang, Chemical beam epitaxy of InP and GaAs, *Applied Physics Letters* 45 (1984), pp. 1234–1236.

15. X.W. Liu, E. Haimi, S.P. Hannula, O.M.E. Ylivaara and R.L. Puurunen, On the reliability of nanoindentation hardness of Al_2O_3 films grown on Si-wafer by atomic layer deposition, *Journal of Vacuum Science & Technology A* 32 (1) (2014), p. 01A116.

16. C. Subramanian, Strafford and KN, Review of multicomponent and multilayer coatings for tribological applications, *Wear* 165 (1993), pp. 85–95.

17. W.E. Spear and P.G. Le Comber, Substitutional doping of amorphous silicon, *Solid State Communications* 17 (1975), pp. 1193–1196.

18. D.E. Carlson and C.R. Wronski, Amorphous silicon solar cell, *Applied Physics Letters* 28 (1976), pp. 671–673.

19. H. Schlemm, A. Mai, S. Roth, D. Roth, K. Baumgartner and H. Muegge, Industrial large scale silicon nitride deposition on photovoltaic cells with linear microwave plasma sources, *Surface & Coatings Technology* 174 (2003), pp. 208–211.

20. T.F. Deutsch, D.J. Ehrlich and R.M. Osgood, Laser photodeposition of metal films with microscopic features, *Applied Physics Letters* 35 (1979), pp. 175–177.

21. H.W. Kroto, J.R. Heath, S.C. O'Brien, R.F. Curl and R.E. Smalley, C60: Buckminsterfullerene, *Nature* 318 (1985), pp. 162–163.

22. W. Krätschmer, L.D. Lamb, K. Fostiropoulos and D.R. Huffman, Solid C60: A new form of carbon, *Nature* 347 (1990), pp. 354–358.

23. (a) S. Iijima, T. Ichihashi, Single-shell carbon nanotubes of 1nm diameter, *Nature* 363(1993), 603–605; (b) Y.N. Xia, P.D. Yang, Y.G. Sun, Y.Y. Wu, B. Mayers, B. Gates et al., One-dimensional nanostructures: Synthesis, characterization, and applications, *Advanced Materials* 15 (2003), pp. 353–389.

24. A. Oberlin, M. Endo and T. Koyama, Filamentous growth of carbon through benzene decomposition, *Journal of Crystal Growth* 32 (1976), pp. 335–349.

25. K.S. Novoselov, Electric field effect in atomically thin carbon films, *Science* 306 (2004), pp. 666–669.

26. (a) X. Li, C.W. Magnuson, A. Venugopal, J. An, J.W. Suk, B. Han et al., Graphene films with large domain size by a two-step chemical vapor deposition process, *Nano Letters* 10 (2010), pp. 4328–4334; (b) X. Li, W. Cai, L. Colombo and R.S. Ruoff, Evolution of graphene growth on Ni and Cu by carbon isotope labeling, *Nano Letters* 9 (2009), pp. 4268–4272.

27. F. Bonaccorso, A. Lombardo, T. Hasan, Z.P. Sun, L. Colombo and A.C. Ferrari, Production and processing of graphene and 2d crystals, *Materials Today* 15 (2012), pp. 564–589.

28. (a) X. Li, X. Zhang, L. Ci, R. Shah, C. Wolfe, S. Kar et al., Air-assisted growth of ultra-long carbon nanotube bundles, *Nanotechnology* 19 (2008), p. 455609; (b) E.R. Meshot, D.L. Plata, S. Tawfick, Y. Zhang, E.A. Verploegen and A.J. Hart, Engineering vertically aligned carbon nanotube growth by decoupled thermal treatment of precursor and catalyst, *ACS Nano* 3 (2009), pp. 2477–2486.

29. N. Geblinger, A. Ismach and E. Joselevich, Self-organized nanotube serpentines, *Nature Nanotechnology* 3 (2008), pp. 195–200.

30. K. Subannajui, N. Ramgir, R. Grimm, R. Michiels, Y. Yang, S. Müller et al., ZnO nanowire growth: A deeper understanding based on simulations and controlled oxygen experiments, *Crystal Growth Design* 10 (2010), pp. 1585–1589.

31. Y. Tian, Y. Zhang, B. Wang, W. Ji, Y. Zhang and K. Xie, Coal-derived carbon nanotubes by thermal plasma jet, *Carbon* 42 (2004), pp. 2597–2601.

32. R. Andrews, D. Jacques, A.M. Rao, F. Derbyshire, D. Qian, X. Fan et al., Continuous production of aligned carbon nanotubes: A step closer to commercial realization, *Chemical Physics Letters* 303 (1999), pp. 467–474.

33. D. Bour, Z. Yang and C. Chua, Simple technique for measuring the filled volume of liquid or solid CVD precursor chemicals in bubblers, *Journal of Crystal Growth* 310 (2008), pp. 2673–2677.

34. J.F. Roeder, T.H. Baum, S.M. Bilodeau, G.T. Stauf, C. Ragaglia, M.W. Russell et al., Liquid delivery MOCVD: Chemical and process perspectives on ferroelectric thin film growth, *Advanced Materials for Optics and Electronics* 10 (2000), pp. 145–154.

35. A.Y. Cho and J.R. Arthur, Molecular beam epitaxy, *Progress in Solid State Chemistry* 10 (1975), pp. 157–191.

36. M. Juppo, A. Rahtu, M. Ritala and M. Leskela, In situ mass spectrometry study on, surface reactions in atomic layer deposition of Al_2O_3 thin films from trimethylaluminum and water, *Langmuir* 16 (8) (2000), pp. 4034–4039.

37. (a) J.A. Thornton, *Structure-Zone Models of Thin Films*, (1988), pp. 95–105; (b) S. Mohan and M.G. Krishna, A review of ion beam assisted deposition of optical thin films, *Vacuum* 46 (1995), pp. 645–659.

38. S. Yasuda, D.N. Futaba, T. Yamada, J. Satou, A. Shibuya, H. Takai et al., Improved and large area single-walled carbon nanotube forest growth by controlling the gas flow direction, *ACS Nano* 3 (2009), pp. 4164–4170.

39. K. Yan, L. Fu, H. Peng and Z. Liu, Designed CVD growth of graphene via process engineering, *Accounts of Chemical Research* 46 (2013), pp. 2263–2274.

40. H. Tanaka and M. Mushiage, MBE as a production technology for AlGaAs lasers, *Journal of Crystal Growth* 111 (1991), pp. 1043–1046.

41. (a) P. Poodt, D.C. Cameron, E. Dickey, S.M. George, V. Kuznetsov, G.N. Parsons et al., Spatial atomic layer deposition: A route towards further industrialization of atomic layer deposition, *Journal of Vacuum Science & Technology A* 30 (2012), p. 010802; (b) P. Poodt, A. Lankhorst, F. Roozeboom, V. Tiba, K. Spee, D. Maas et al., Ultrafast atomic layer deposition of alumina layers for solar cell passivation, 2010, pp. 419–427; (c) E.H.A. Granneman, V.I. Kuznetsov and P. Vermont, Spatial ALD, deposition of Al_2O_3 films at throughputs exceeding 3000 Wafers per hour, *ECS Transactions* 61 (2014), pp. 3–16; (d) P.S. Maydannik, T.O. Kääriäinen and D.C. Cameron, An atomic layer deposition process for moving flexible substrates, *Chemical Engineering Journal* 171 (2011), pp. 345–349.

42. J. Gerlach, T. Hoche, F. Frost and B. Rauschenbach, Ion beam assisted MBE of GaN on epitaxial TiN films, *Thin Solid Films* 459 (2004), pp. 13–16.

43. S. Bae, H. Kim, Y. Lee, X.F. Xu, J.S. Park, Y. Zheng et al., Roll-to-roll production of 30-inch graphene films for transparent electrodes, *Nature Nanotechnology* 5 (8) (2010), pp. 574–578.

44. A. Sherman, Plasma-assisted chemical vapor deposition processes and their semiconductor applications, *Thin Solid Films* 113 (1984), pp. 135–149.

45. J.P. Liu, K.L. Choy and X.H. Hou, Charge transport in flexible solar cells based on conjugated polymer and ZnO nanoparticulate thin films, *Journal of Materials Chemistry* 21 (2011), pp. 1966–1969.

46. A.G. Aberle, Surface passivation of crystalline silicon solar cells: A review, *Prog. in Photovoltaics: Research and Applications* 8 (2000), pp. 473–487.

47. (a) P.K. Bachmann, G. Gartner and H. Lydtin, Plasma-assisted chemical vapor deposition processes, *MRS Bulletin* (1988), pp. 52–59; (b) J. Karthikeyan, C.C. Berndt et al, Plasma spray synthesis of nanomaterial powders and deposits, *Materials Science and Engineering: A*, 238 (2) (1997), pp. 275–286.

48. (a) A.R. Reinberg, Plasma deposition of inorganic thin films, *Annual Review of Materials Science* 9 (1979), pp. 341–372; (b) R. Suchentrunk, G. Staudigl, D. Jonke and H.J. Fuesser, Industrial applications for plasma processes—examples and trends, *Surface and Coatings Technology* 97 (1997), pp. 1–9.

49. C.A. Yuan, C.N. Han, H.M. Liu and W.D. van Driel, Solid-state lighting technology in a nutshell, in *Solid State Lighting Reliability*, W.D. van Driel and X.J. Fan, eds., Springer New York, 2013, pp. 13–41.

50. H. Lee, I. Park, J. Kwak, D.Y. Yoon and C. Lee, Improvement of electron injection in inverted bottom-emission blue phosphorescent organic light emitting diodes using zinc oxide nanoparticles, *Applied Physics Letters* 96 (2010), p. 153306.

51. M. Izu and T. Ellison, Roll-to-roll manufacturing of amorphous silicon alloy solar cells with in situ cell performance diagnostics, *Solar Energy Materials and Solar Cells* 78 (2003), pp. 613–626.

52. S. Guha, J. Yang and A. Banerjee, Amorphous silicon alloy photovoltaic research-present and future, *Progress in Photovoltaics: Research and Applications* 8 (2000), pp. 141–150.

53. S.R. Coulson, I.S. Woodward, J.P.S. Badyal, S.A. Brewer and C. Willis, Ultralow surface energy plasma polymer films, *Chemistry of Materials* 12 (2000), pp. 2031–2038.

54. J. Musil, Microwave plasma: Its characteristics and applications in thin film technology, *Vacuum* 36 (1986), pp. 161–169.

55. L. Bardos, J. Musil and P. Taras, Differences between microwave and RF activation of nitrogen for the PECVD process, *Journal of Physics D: Applied Physics* 15 (1982), pp. L79–L82.

56. (a) H.W. Etzkorn et al., Plasma CVD process for coating a dome-shaped substrate, 1991, US5154943; (b) L. Martinu and D. Poitras, Plasma deposition of optical films and coatings: A review, *Journal of Vacuum Science & Technology A*, 18 (6) (2000), pp. 2619–2645.

57. M. Ashfold, P. May, C. Rego and N. Everitt, Thin-film diamond by chemical-vapor-deposition methods, *Chemical Society Reviews* 23 (1994), pp. 21–30.

58. A. Tallaire, J. Achard, F. Silva, R.S. Sussmann and A. Gicquel, Homoepitaxial deposition of high-quality thick diamond films: Effect of growth parameters, *Diamond and Related Materials* 14 (2005), pp. 249–254.

59. E. Sevillano, Microwave-plasma deposition of diamond, in *Low-Pressure Synthetic Diamond: Manufacturing and Applications*, Springer Science & Business Media, Berlin, Germany, 2013, pp. 11–40.

60. SiNA®/MAiA®. The modular inline system for PECVD anti-reflection coating and passivation layers, https://www.meyerburger.com/en/technologies/photovoltaics/upgrade-technologies/product-detail/product/maia/, accessed on January 3, 2019.

61. A.T.H. Chuang, J. Robertson, B.O. Boskovic and K.K.K. Koziol, Three-dimensional carbon nanowall structures, *Applied Physics Letters* 90 (2007), pp. 123107.

62. (a) J.R. Knight, D. Effer and P.R. Evans, The preparation of high purity gallium arsenide by vapour phase epitaxial growth, *Solid-State Electronics* 8 (1965), pp. 178–180; (b) H.M. Manasevit, W.I. Simpson, The use of metalorganics in the preparation of semiconductor materials, *Journal of The Electrochemical Society*, 116 (12) (1969), 1725–1732.

63. S. Nakamura, GaN growth using GaN buffer layer, *Japanese Journal of Applied Physics* 30 (1991), pp. L1705–L1707.

64. (a) S. Nakamura, T. Mukai and M. Senoh, Candela-class high-brightness InGaN/AlGaN double-heterostructure blue-light-emitting diodes, *Applied Physics Letters* 64 (1994), pp. 1687–1689; (b) S. Nakamura, First laser diodes fabricated from III–V nitride based materials, *Materials Science and Engineering B* 43 (1997), pp. 258–264.

65. G. Bendt, S. Schulz, S. Zastrow and K. Nielsch, Single-source precursor-based deposition of Sb_2Te_3 films by MOCVD **, *Chemical Vapor Deposition* 19 (2013), pp. 235–241.

66. (a) C. Vriamont, M. Devillers, O. Riant and S. Hermans, Catalysis with gold complexes immobilised on carbon nanotubes by pi-pi stacking interactions: Heterogeneous catalysis versus the Boomerang effect, *Chemistry-A European Journal* 19 (36) (2013), pp. 12009–12017; (b) S. Koponen, P. Gordon, S. Barry, Principles of precursor design for vapour deposition methods, *Polyhedron* 108 (2016), pp. 59–66.

67. TurboDisc EPIK 700 GaN MOCVD System for LED Production. Available at http://www.veeco.com/products/turbodisc-epik-700-gan-mocvd-system-for-led-production.

68. J. Arthur, Interaction of Ga and As_2 molecular beams with GaAs Surfaces, *Journal of Applied Physics* 39 (1968), p. 4032.

69. A.Y. Cho, Morphology of epitaxial growth of GaAs by a molecular beam method: The observation of surface structures, *Journal of Applied Physics* 41 (1970), pp. 2780–2786.

70. A. Cho, How molecular beam epitaxy (MBE) began and its projection into the future, *Journal of Crystal Growth* 201–202 (1999), pp. 1–7.

71. W.P. McCray, MBE deserves a place in the history books, *Nature Nanotechnology* 2 (2007), pp. 259–261.

72. D. Haneman, Structure and adsorption characteristics of (111) and (111) surfaces of InSb cleaned by ion bombardment and annealing, *Journal of Physics and Chemistry of Solids* 14 (1960), pp. 162–168.

73. D. Haneman, Surface structures and properties of diamond-structure semiconductors, *Physical Review* 121 (1961), pp. 1093–1100.

74. G. Biasiol and L. Sorba, Molecular beam epitaxy: Principles and applications, *Crystal Growth of Materials for Energy Production and Energy-Saving Applications,* R. Fornari, L. Sorba, Eds., Edizioni ETS, Pisa, Italy (2001), pp. 66–83.

75. (a) J. Arthur, Molecular beam epitaxy, *Surface Science* 500 (2002), pp. 189–217; (b) A.Y. Cho, Advances in molecular beam epitaxy (MBE), *Journal of Crystal Growth* 111 (1991), pp. 1–13.

76. A. CHO, Film deposition by molecular-beam techniques, *Journal of Vacuum Science and Technology,* 8(5) (1971), pp. S31–S38.

77. M. Llegems, R. Dingle and L.W. Rupp, Optical and electrical properties of Mn-doped GaAs grown by molecular-beam epitaxy, *Journal of Applied Physics* 46 (1975), pp. 3059–3065.

78. J.R. Arthur, Surface stoichiometry and structure of GaAs, *Surface Science* 43 (1974), pp. 449–461.

79. J.R. Arthur, Adsorption of Zn on GaAs, *Surface Science* 38 (1973), pp. 394–412.

80. A.Y., Cho and M.B. Panish, Magnesium-doped GaAs and AlxGa1−xAs by molecular beam epitaxy, *Journal of Applied Physics,* 43(12) (1972), pp.5118–5123.

81. H. Tanaka, M. Mushiage, Y. Ishida and H. Fukada, Single-longitudinal-mode self-aligned (AlGa) as double-heterostructure lasers fabricated by molecular beam epitaxy, *Japanese Journal of Applied Physics* 24 (1985), pp. L89–L90.

82. D.C. Tsui, H.L. Stormer and A.C. Gossard, Two-dimensional magnetotransport in the extreme quantum limit, *Physical Review Letters* 48 (1982), pp. 1559.

83. (a) B.J. van Wees, L.P. Kouwenhoven, H. van Houten, C.W.J. Beenakker, J.E. Mooij, C.T. Foxon et al., Quantized conductance of magnetoelectric subbands in ballistic point contacts, *Physical Review B* 38 (1988), p. 3625; (b) D.A. Wharam et al., One-dimensional transport and the quantisation of the ballistic resistance, *Journal of Physics C: Solid State Physics,* 21 (8) (1988), pp. L209–L214.

84. T. Chiu, W. Tsang and J. Cunningham, Gallium-induced and arsenic-induced oscillations of intensity of reflection high-energy electron-diffraction in the growth of (001) GaAs by chemical beam epitaxy, *Journal of Applied Physics* 62 (1987), pp. 2302–2307.

85. J. Cheng, V.K. Lazarov, G.E. Sterbinsky and B.W. Wessels, Synthesis, structural and magnetic properties of epitaxial MgFe$_2$O$_4$ thin films by molecular beam epitaxy, *J. Vac. Sci. Tech. B: Microelectronics and Nanometer Structures* 27 (2009), pp. 148–151.

86. L.L. Chang, L. Esaki, W.E. Howard and R. Ludeke, The growth of a GaAs–GaAlAs superlattice, *Journal of Vacuum Science and Technology* 10 (1973), pp. 11–16.

87. F. Maeda and H. Hibino, Growth of few-layer graphene by gas-source molecular beam epitaxy using cracked ethanol, *Physica Status Solidi (b)* 247 (2010), pp. 916–920.

88. D. Kong, W. Dang, J.J. Cha, H. Li, S. Meister, H. Peng et al., Few-layer nanoplates of Bi$_2$Se$_3$ and Bi$_2$Te$_3$ with highly tunable chemical potential, *Nano Letters* 10 (2010), pp. 2245–2250.

89. J.E. Moore, The birth of topological insulators, *Nature* 464 (2010), pp. 194–198.

90. H. Luth, Chemical beam epitaxy—A child of surface science, *Surface Science* 299 (1994), pp. 867–877.

91. E. Miyauchi and H. Hashimoto, Application of focused ion beam technology to maskless ion implantation in a molecular beam

epitaxy grown GaAs or AlGaAs epitaxial layer for three-dimensional pattern doping crystal growth, *Journal of Vacuum Science & Technology A: Vacuum, Surfaces, and Films* 4(1986), pp. 933–938.

92. J.P. Harbison, Summary Abstract: III–V molecular beam epitaxy: Toward a concept of all ultrahigh vacuum processing, *Journal of Vacuum Science & Technology A: Vacuum, Surfaces, and Films* 4 (1986), pp. 1033–1034.

93. Y. Zhang, K. He, C.-Z. Chang, C.-L. Song, L.-L. Wang, X. Chen et al., Crossover of the three-dimensional topological insulator Bi_2Se_3 to the two-dimensional limit, *Nature Physics* 6 (2010), pp. 712–712.

94. P. Cheng, C. Song, T. Zhang, Y. Zhang, Y. Wang, J.-F. Jia et al., Landau quantization of topological surface states in $Bi_2 Se_3$, *Physical Review Letters* 105 (2010).

95. Y. Jiang, Y. Wang, M. Chen, Z. Li, C. Song, K. He et al., Landau quantization and the thickness limit of topological insulator thin films of $Sb_2 Te_3$, *Physical Review Letters* 108 (2012).

96. Dirkhb, Philips CDM210 CD Drive (2006), https://commons.wikimedia.org/wiki/File:CDM210_cd_laufwerk.jpg, accessed on January 3, 2019.

97. (a) P. Frigeri, L. Seravalli, G. Trevisi and S. Franchi, Molecular beam epitaxy: An overview, in *Comprehensive Semiconductor Science and Technology*, Elsevier, Amsterdam, the Netherlands, 2011, pp. 480–522; (b) M. O'Steen et al., Chapter 29 Systems and technology for production-scale molecular beam epitaxy, *Molecular Beam Epitaxy, From Research to Mass Production*, Elsevier, 2013, 657–678.

98. O. Berger, GaAs MESFET, HEMT and HBT competition with advanced Si RF technologies, *Technology* (1999), pp. 4–7.

99. R. Ramesh and N.A. Spaldin, Multiferroics: Progress and prospects in thin films, *Nature Materials* 6 (2007), pp. 21–29.

100. J.E. Davey and T. Pankey, Epitaxial GaAs films deposited by vacuum evaporation, *Journal of Applied Physics* 39 (1968), pp. 1941–1948.

101. N. Pütz, E. Veuhoff, H. Heinecke, M. Heyen, H. Luth and P. Balk, GaAs growth in metal–organic MBE, *J. Vac. Sci. Tech. B: Microelectronics and Nanometer Structures* 3 (1985), pp. 671–671.

102. N. Pütz, H. Heinecke, M. Heyen, P. Balk, M. Weyers and H. Luth, A comparative study of $Ga(CH_3)_3$ and $Ga(C_2H_5)_3$ in the MOMBE of GaAs, *Journal of Crystal Growth* 74 (1986), pp. 292–300.

103. (a) W. Tsang and J. Campbell, Very low dark current and high quantum efficient INGaAs/InP p-i-n photodiodes grown by chemical-beam epitaxy, *IEEE Transactions on Electron Devices* 33 (1986), p. 1862; (b) C.R. Abernathy et al., Carbon doping of III–V compounds grown by MOMBE, *Journal of Crystal Growth*, 105 (1–4) (1990), pp. 375–382; (c) O. Kayser, Selective growth of InP/GaInAs in LP-MOVPE and MOMBE/CBE, *Journal of Crystal Growth*, 107 (1991), pp. 989–998.

104. W. Tsang, Chemical beam epitaxy, *IEEE Circuits and Devices Magazine* 4 (1988), pp. 18–24.

105. A. Robertson, T.H. Chiu, W.T. Tsang and J.E. Cunningham, A model for the surface chemical kinetics of GaAs deposition by chemical-beam epitaxy, *Journal of Applied Physics* 64 (1988), pp. 877–887.

106. H. Ando, S. Yamaura and T. Fujii, Recent progress in the multi-wafer CBE system, *Journal of Crystal Growth* 164 (1996), pp. 1–15.

107. E. Wagner, C.S. Sandu, S. Harada, C. Pellodi, M. Jobin, P. Muralt et al., Geometry of chemical beam vapor deposition system for efficient combinatorial investigations of thin oxide films: Deposited film properties versus precursor flow simulations, *ACS Combinatorial Science* 18 (2016), pp. 154–161.

108. T. Suntola, A.J. Pakkala and S.G. Lindfors, Method for performing growth of compound thin films, U.S. Patent No. 4,413,022. November 1, 1983.

109. Q. Peng, X.-Y. Sun, J.C. Spagnola, G.K. Hyde, R.J. Spontak and G.N. Parsons, Atomic layer deposition on electrospun polymer fibers as a direct route to Al_2O_3 microtubes with precise wall thickness control, *Nano Letters* 7 (2007), pp. 719–722.

110. S.M. George, A.W. Ott and J.W. Klaus, Surface chemistry for atomic layer growth, *The Journal of Physical Chemistry* 100 (1996), pp. 13121–13131.

111. (a) M. Ritala, M. Leskela, J.P. Dekker, C. Mutsaers, P.J. Soininen and J. Skarp, Perfectly conformal TiN and Al_2O_3 films

deposited by atomic layer deposition, *Chemical Vapor Deposition* 5 (1) (1999), pp. 7–9; (b) N.D. Hoivik, J.W. Elam, R.J. Linderman, V.M. Bright, S.M. George and Y.C. Lee, Atomic layer deposited protective coatings for micro-electromechanical systems, *Sensors and Actuators A: Physical* 103 (2003), pp. 100–108.

112. (a) Pulsar XP ALD. Available at http://www.asm.com/solutions/products/atomic-layer-deposition-products/pulsar-xp-ald; (b) K.P. Musselman, C.F. Uzoma and M.S. Miller, Nanomanufacturing: High-throughput, cost-effective deposition of atomic scale thin films via atmospheric pressure spatial atomic layer deposition, *Chemistry of Materials* 28 (2016), pp. 8443–8452.

113. (a) C. Auth, C. Allen, A. Blattner, D. Bergstrom, M. Brazier, M. Bost et al., A 22 nm high performance and low-power CMOS technology featuring fully-depleted tri-gate transistors, self-aligned contacts and high density MIM capacitors, in *VLSI Technology (VLSIT), 2012 Symposium on* (2012), pp. 131–132; (b) R. Clark, Emerging applications for high K materials in VLSI technology, *Materials* 7 (2014), pp. 2913–2944.

114. Beneq Roll to Roll Atomic Layer Deposition Tool. Available at https://www.uk-cpi.com/technologies/printable-electronics/facilities-equipment/beneq-roll-to-roll-atomic-layer-deposition-tool.

115. H.C. George, A.O. Orlov and G.L. Snider, Platinum single-electron transistors with tunnel barriers made by atomic layer deposition, *Journal of Vacuum Science & Technology B* 28 (2010), pp. C6L6–C6L8.

116. R.W. Johnson, A. Hultqvist and S.F. Bent, A brief review of atomic layer deposition: From fundamentals to applications, *Materials Today* 17 (2014), pp. 236–246.

117. J. Choi, Y. Kim, K. Chang and D. Tomanek, Itinerant ferromagnetism in heterostructured C/BN nanotubes, *Physical Review B* 67 (2003), p. 125421.

118. I. Yamada and N. Toyoda, Nano-scale surface modification using gas cluster ion beams—A development history and review of the Japanese nano-technology program, *Surface and Coatings Technology* 201 (2007), pp. 8579–8587.

119. J.K. Hirvonen, Ion beam assisted thin film deposition, in *Materials and Processes for Surface and Interface Engineering*, Springer, Dordrecht, the Netherlands, 1995, pp. 307–346.

120. J.K. Hirvonen, Ion beam assisted thin film deposition, *Materials Science Reports* 6 (1991), pp. 215–274.

121. J.F. Londoño et al., Advanced sensor fabrication using integrated ion beam etch and ion beam deposition processes, Veeco Instruments, Plain View, NY, 2000.

122. C.-H. Ma, J.-H. Huang and H. Chen, A study of preferred orientation of vanadium nitride and zirconium nitride coatings on silicon prepared by ion beam assisted deposition, *Surface and Coatings Technology* 133–134 (2000), pp. 289–294.

123. I. Jimenez, R. Torres, I. Caretti, R. Gago and J.M. Albella, A review of monolithic and multilayer coatings within the boron–carbon–nitrogen system by ion-beam-assisted deposition, *Journal of Materials Research* 27 (2012), pp. 743–764.

124. M. Alizadeh, B.T. Goh, A.K. Pandey, C.F. Dee and S.A. Rahman, Low-RF-power growth of InN thin films by plasma-assisted reactive evaporation with a localized ion source, *Materials Chemistry and Physics* 199 (2017), pp. 408–415.

125. H.R. Kaufman, Broad-beam ion sources: Present status and future directions, *Journal of Vacuum Science & Technology A: Vacuum, Surfaces, and Films* 4 (1986), pp. 764–771.

126. A. Anders, Plasma and ion sources in large area coating: A review, *Surface and Coatings Technology* 200 (2005), pp. 1893–1906.

127. (a) D. Van Vechten, G.K. Hubler, E.P. Donovan and F.D. Correll, Fundamentals of ion-beam-assisted deposition. I. Model of process and reproducibility of film composition, *Journal of Vacuum Science & Technology A: Vacuum, Surfaces, and Films* 8 (1990), pp. 821–830; (b) C.M. Cotell and J.K. Hirvonen, Effect of ion energy on the mechanical properties of ion beam assisted deposition (IBAD) wear resistant coatings, *Surface and Coating Technology* 81 (1996), pp. 118–125.

128. T. Takagi, Role of ions in ion-based film formation, *Thin Solid Films* 92 (1982), pp. 1–17.

129. T. Ina, Y. Minowa, N. Koshirakawa and K. Yamanishi, Development of an ionized cluster beam system for large-area deposition, *Nuclear Instruments and Methods in Physics Research Section B: Beam Interactions with Materials and Atoms* 37–38 (1989), pp. 779–782.

130. O. Nakatsu, J. Matsuo, K. Omoto, T. Seki, G. Takaoka and I. Yamada, Titanium-dioxide film formation using gas cluster ion beam assisted deposition technique, *Nuclear Instruments and Methods in Physics Research Section B: Beam Interactions with Materials and Atoms* 206 (2003), pp. 866–869.

131. (a) J. Kirschner, H. Engelhard and D. Hartung, An evaporation source for ion beam assisted deposition in ultrahigh vacuum, *Review of Scientific Instruments* 73 (2002), pp. 3853–3860; (b) D. Manova et al., Thin film deposition using energetic ions, *Materials* 3 (8) (2010), pp. 4109–4141.

132. M. Adachi, S. Tsukui and K. Okuyama, Nanoparticle formation mechanism in CVD reactor with ionization of source vapor, *Journal of Nanoparticle Research* 5 (2003), pp. 31–37.

133. T. Miyano and H. Kitamura, Coating on the cutting edge of an electric shaver by ion beam assisted deposition, *Surface and Coatings Technology* 65 (1994), pp. 179–183.

134. A.H. Deutchman and R.J. Martyka, Industrial Scale Ion Beam Enhanced Deposition (IBED) processing system, in *ASM International Surface Engineering Congress*, (2002).

135. H.S. Taylor and D.G. Hill, The reactions of ethylene, hydrogen and the saturated hydrocarbons under the influence of excited mercury[1], *Journal of the American Chemical Society* 51 (1929), pp. 2922–2936.

136. M. Hanabusa, Photoinduced deposition of thin films, *Materials Science Reports* 2 (1987), pp. 51–97.

137. I.P. Herman, Laser-assisted deposition of thin films from gas-phase and surface-adsorbed molecules, *Chemical Reviews* 89 (1989), pp. 1323–1357.

138. R.M. Osgood and T.F. Deutsch, Laser-induced chemistry for microelectronics, *Science* 227 (1985), pp. 709–714.

139. S. Allen, Laser chemical vapor-deposition–a technique for selective area deposition, *Journal of Applied Physics* 52 (1981), pp. 6501–6505.

140. V. Hopfe, R. Jäckel and K. Schönfeld, Laser based coating and modification of carbon fibres: Application of industrial lasers to manufacturing of composite materials, *Applied Surface Science* 106 (1996), pp. 60–66.

141. F.T. Wallenberger, P.C. Nordine and M. Boman, Inorganic fibers and microstructures directly from the vapor phase, *Composites Science and Technology* 51 (1994), pp. 193–212.

142. C.A.D. Dion and J.R. Tavares, Photo-initiated chemical vapor deposition as a scalable particle functionalization technology (a practical review), *Powder Technology* 239 (2013), pp. 484–491.

143. S. Hosseininasab, N. Faucheux, G. Soucy and J.R. Tavares, Full range of wettability through surface modification of single-wall carbon nanotubes by photo-initiated chemical vapour deposition, *Chemical Engineering Journal* 325 (2017), pp. 101–113.

144. J.M. Lackner, Industrially-scaled large-area and high-rate tribological coating by pulsed laser deposition, *Surface and Coatings Technology* 200 (2005), pp. 1439–1444.

145. W.E. Sawyer and A. Man, Carbon for electric lights, US0229335A, June 29, 1880, p. 2.

146. (a) J.W. Aylesworth, Art of manufacturing electrical incandescing conductors, US 553296 January 21, 1896; (b) M. Prieto, J. Díaz and E. Egusquiza, Analysis of the fluid-dynamic and thermal behaviour of a tin bath in float glass manufacturing, *International Journal of Thermal Sciences* 41 (2002), pp. 348–359; (c) C. Morales, H. Juarez, T. Diaz, Y. Matsumoto, E. Rosendo, G. Garcia et al., Low temperature SnO_2 films deposited by APCVD, *Microelectronics Journal* 39 (2008), pp. 586–588; (d) A. de Graaf, J. van Deelen, P. Poodt, T. van Mol, K. Spee, F. Grob et al., Development of atmospheric pressure CVD

processes for high-quality transparent conductive oxides, *Energy Procedia* 2 (2010), pp. 41–48.

147. T. Oyama, Past, present and future of dry coating technology, Res. Reports Asahi Glass Co., 57 (2007), pp. 83–90.

148. (a) R.G. Gordon, J. Proscia, F.B. Ellis and A.E. Delahoy, Textured tin oxide films produced by atmospheric pressure chemical vapor deposition from tetramethyltin and their usefulness in producing light trapping in thin film amorphous silicon solar cells, *Solar Energy Materials* 18 (1989), pp. 263–281; (b) J. Müller et al., TCO and light trapping in silicon thin film solar cells, *Solar Energy*, 77 (6) (2004), pp. 917–930.

149. P. Chesworth, Coating glass, 1984, US493657.

150. A. Arnaud, Industrial production of coated glass: Future trends for expanding needs, *Journal of Non-Crystalline Solids* 218 (1997), pp. 12–18.

151. (a) A. Illiberi, B. Kniknie, J. van Deelen, H.L.A.H. Steijvers, D. Habets, P.J.P.M. Simons et al., Industrial high-rate (similar to 14 nm/s) deposition of low resistive and transparent ZnOx: Al films on glass, *Solar Energy Materials and Solar Cells* 95 (2011), pp. 1955–1959; (b) H. Matsumura and K. Ohdaira, Recent situation of industrial implementation of Cat-CVD technology in Japan, *Thin Solid Films* 516 (2008), pp. 537–540; (c) H. Althues, Chemical vapor deposition at atmospheric pressure AP-CVD - Fraunhofer IWS, https://www.iws.fraunhofer.de/en /business_fields/chemical_surface_reaction_technology/chemical_surface_technology /technologies/ap-cvd.html, accessed on July 18, 2017.

152. T.O. Sedgwick and P.D. Agnello, Atmospheric pressure chemical vapor deposition of Si and SiGe at low temperatures, *Journal of Vacuum Science Technology A: Vacuum, Surfaces, and Films* 10 (1992), pp. 1913–1919.

153. H.E. Hintermann, Tribological and protective coatings by chemical vapour deposition, *Thin Solid Films* 84 (1981), pp. 215–243.

154. A. Kato and N. Tamari, Crystal growth of titanium nitride by chemical vapor deposition, *Journal of Crystal Growth* 29 (1975), pp. 55–60.

155. Chemical vapor deposition at atmospheric pressure AP-CVD-Fraunhofer IWS. Available at https://www.iws.fraunhofer.de/en/business_fields/chemical_surface_reaction_technology/chemical_surface_technology/technologies/ap-cvd.html.

156. (a) B. Hu, H. Ago, Y. Ito, K. Kawahara, M. Tsuji, E. Magome et al., Epitaxial growth of large-area single-layer graphene over Cu(111)/sapphire by atmospheric pressure CVD, *Carbon* 50 (2012), pp. 57–65; (b) Y.Y. Yang et al., Growth and characterization of Si/SiN/SiC structures by APCVD process, *ICSICT 9th International Conference on Solid-State and Integrated-Circuit Technology*, 2008, pp. 730–733.

157. C.L. Gray and R.H. Dettre, Process for increasing the scratch resistance of glass, October 17, 1961.

158. I.P. Parkin and R.G. Palgrave, CVD of functional coatings on glass, in *Chemical Vapour Deposition: Precursors, Processes and Applications*, Royal Society of Chemistry, London, UK, 2009, pp. 451–476.

159. K.L. Choy, Vapour processing of nanostructured materials, in *Handbook of Nanostructured Materials and Nanotechnology*, Academic Press, San Diego, CA, 2000, pp. 533.

160. X. Hou and K.-L. Choy, Processing and applications of aerosol-assisted chemical vapor deposition, *Chemical Vapor Deposition* 12 (2006), pp. 583–596.

161. N. Noor, C.K.T. Chew, D.S. Bhachu, M.R. Waugh, C.J. Carmalt and I.P. Parkin, Influencing FTO thin film growth with thin seeding layers: A route to microstructural modification, *Journal of Materials Chemistry C* 3 (2015), pp. 9359–9368.

162. G. Altamura, M.Q. Wang and K.L. Choy, Influence of alkali metals (Na, Li, Rb) on the performance of electrostatic spray-assisted vapor deposited $Cu_2ZnSn(S,Se)_4$ solar cells, *Scientific Reports* 6 (2016), p. 22109.

163. R. Chandrasekhar and K.L. Choy, Electrostatic spray assisted vapour deposition of fluorine doped tin oxide, *Journal of Crystal Growth* 231 (2001), p. 215.

164. (a) B. Su and K.L. Choy, Electrostatic assisted aerosol jet deposition of CdS, CdSe and ZnS thin films, *Thin Solid Films* 361 (2000), pp. 102–106; (b) B. Su and K.L. Choy, Synthesis, Microstructure and optical properties of ZnS films by electrostatic assisted aerosol jet deposition technique, *Journal of Materials Chemistry* 10 (2010), pp. 949–952.

165. (a) P.B. Tavares, V.S. Amaral, J.P. Araújo, A.A.C.S. Lourenço, J.M. Vieira and J.B. Sousa, Deposition of LaCaMnO$_3$ thin films using aerosol-assisted metalorganic chemical vapor deposition (MOCVD) substrate and annealing effects, *Journal of Magnetism and Magnetic Materials* 196–197 (1999), pp. 490–492; (b) O. Monteiro et al., Aerosol-assisted metallo-organic chemical vapour deposition of Bi$_2$Se$_3$ films using single-molecule precursors, *Journal of Materials Chemistry* 13 (2003), pp. 3006–3010; (c) S. Mlowe et al., Heterocyclic dithiocarbamato-iron(III) complexes: Single-source precursors for aerosol-assisted chemical vapour deposition (AACVD) of iron sulfide thin films, *Dalton Transactions* 45 (6) (2016), pp. 2647–2655.

166. (a) K.L. Choy, Special issue on aerosol-assisted chemical vapor deposition, *Chemical Vapor Deposition* 12 (10) (2006), pp. 581–582; (b) X.H. Hou and K.L. Choy, Processing and applications of aerosol-assisted chemical vapor deposition, *Chemical Vapour Deposition* 12 (2006), pp. 583–596.

167. G. Blandenet, M. Court and Y. Lagrade, Thin layers deposited by the pyrosol process, *Thin Solid Films* 77 (1981), pp. 81–90.

168. C. Chabrol, E. Rouvière and F. Schuster, The Pyrosol® Process, in *Sol-Gel Technologies for Glass Producers and Users*, M.A. Aegerter and M. Mennig, Eds., Springer, Boston, MA, 2004, pp. 69–75.

169. J.L. Deschanvres, M. Langlet, M. Labeau and J.C. Joubert, Growth of garnet thin films for magneto-optic memories by pyrosol CVD process, *IEEE Transactions on Magnetics* 26 (2002), pp. 187–189.

170. O.C. Monteiro, T. Trindade, F.A.A. Paz, J. Klinowski, J. Watersc and P. O'Brien, Aerosol-assisted metallo-organic chemical vapour deposition of Bi$_2$Se$_3$ films using single-molecule precursors, *Journal of Materials Chemistry* 13 (2003), pp. 3006–3010.

171. S. Mlowe, D.J. Lewis, M.A. Malik, J. Raftery, E.B. Mubofu, P. O'Brien et al., Heterocyclic dithiocarbamato-iron(III) complexes: Single-source precursors for aerosol-assisted chemical vapour deposition (AACVD) of iron sulfide thin films, *Dalton Transactions* 45 (2016), pp. 2647–2655.

172. C.K.T. Chew, Chemical vapour deposition of gold nanoparticle and metal oxide composites, PhD Thesis, University College London, London, UK, 2016.

173. X. Hou, K.-L. Choy, N. Brun and V. Serín, Nanocomposite coatings codeposited with nanoparticles using aerosol-assisted chemical vapour deposition, *Journal of Nanomaterials* 2013 (2013), p. 1.

174. X. Hou and K.L. Choy, Synthesis of Cr$_2$O$_3$-based nanocomposite coatings with incorporation of inorganic fullerene-like nanoparticles, *Thin Solid Films* 516 (2008), pp. 8620–8624.

175. M. Labeau, B. Gautheron, F. Cellier, M. Vallet-Regi and E.G. and J.M. Gonzalez, Calbert Pt Nanoparticles dispersed on SnO$_2$ thin films: A microstructural study, *Journal of Solid State Chemistry* 102 (1993), pp. 434–439.

176. J.S. Sagu, K.G.U. Wijayantha, M. Bohm, S. Bohm and T.K. Rout, Aerosol-assisted chemical vapor deposition of multi-walled carbon nanotubes on steel substrates for application in supercapacitors, *Advanced Engineering Materials* 18 (2016), pp. 1059–1065.

177. P. Liu, Statistics in multiresolution modeling: An application of Bayesian interference in multiscale coarse-graining, *Abstracts of Papers of the American Chemical Society* 240 (2010).

178. Y.J. Lee, Low-impurity, highly conformal atomic layer deposition of titanium nitride using NH$_3$-Ar-H$_2$ plasma treatment for capacitor electrodes, *Materials Letters* 59 (2005), pp. 615–617.

179. J.L. Hu, C.C. Yang and J.H. Huang, Vertically-aligned carbon nanotubes prepared by water-assisted chemical vapor deposition, *Diamond and Related Materials* 17 (2008), pp. 2084–2088.

180. C.L. Pint, S.T. Pheasant, A.N.G. Parra-Vasquez, C. Horton, Y. Xu and R.H. Hauge, Investigation of optimal parameters for oxide-assisted growth of vertically aligned single-walled carbon nanotubes, *Journal of Physical Chemistry C* 113 (2009), pp. 4125–4133.

181. S. Vallejos, F. Di Maggio, T. Shujah and C. Blackman, Chemical vapour deposition of gas sensitive metal oxides, *Chemosensors* 4 (2016), p. 4.

182. F.E. Annanouch, I. Gràcia, E. Figueras, E. Llobet, C. Cané and S. Vallejos, Localized aerosol-assisted CVD of nanomaterials for the fabrication of monolithic gas sensor microarrays, *Sensors and Actuators B: Chemical* 216 (2015), pp. 374–383.

183. K.L. Choy, Process principles and applications of novel and cost-effective ESAVD based methods, in *Innovative Processing of Films and Nanocrystalline Powders*, Imperial College Press, London, UK, 2002, pp. 15–69.

184. M. Wang, M.A. Hossain and K.-L. Choy, Effect of sodium treatment on the performance of electrostatic spray assisted vapour deposited copper-poor Cu(In, Ga)(S, Se)$_2$ solar cells, *Scientific Reports* 7 (2017).

185. (a) M.A. Hossain, M. Wang and K.L. Choy, Ecofriendly and non vacuum electrostatic spray-assisted vapor deposition of Cu(In,Ga)(S,Se)$_2$ thin film solar cells, *ACS Applied Materials and Interfaces* 7 (2015), pp. 22497–22503; (b) G. Altamura, M.Q. Wang, K.L. Choy, Improving efficiency of electrostatic spray-assisted vapor deposited Cu$_2$ZnSn(S,Se)$_4$ solar cells by modification of Mo/absorber interface, *Thin Solid Films* 597 (2015), pp. 19–24; (c) E.S. Mungan et al., Modeling the effects of Na incorporation on CIGS solar cells, *IEEE Journal of Photovoltaics*, 2013, 3(1), pp. 451–456.

186. J. Du and K.L. Choy, Electrostatic spray assisted vapour deposition of TiO$_2$-based films, *Solid State Ionics* 173 (2004), p. 119.

187. X.H. Hou and K.L. Choy, Photocatalytic activity of nanocrystalline TiO$_2$-based films produced by ESAVD method, *Materials Science & Engineering: C* 25 (2005), pp. 669–674.

188. E.S. Raj and K.L. Choy, Microstructure and properties of indium tin oxide films produced by electrostatic spray assisted vapour deposition process, *Materials Chemistry and Physics* 82 (2003), pp. 489–492.

189. X. Hou, K.L. Choy and J.P. Liu, Electrical and optical performance of transparent conducting oxide films deposited by electrostatic spray assisted vapour deposition, *Journal of Nanoscience and Nanotechnology* 11 (2011), pp. 8114–8119.

190. K. Choy, W. Bai, S. Clarojrochkul and B. Steele, The development of intermediate-temperature solid oxide fuel cells for the next millennium, *Journal of Power Sources* 71 (1998), pp. 361–369.

191. K.L. Choy, J.P. Feist, A.L. Heyes and B. Su, Eu-doped Y$_2$O$_3$ phosphor films produced by electrostatic-assisted chemical vapor deposition, *Journal of Materials Research* 14 (1999), pp. 3111–3114.

192. J. Du, Y. Wu, K.L. Choy and P.H. Shipway, Structure evolution and stoichiometry control of Pb(Zr, Ti)O$_3$ thick films fabricated by electrospray assisted vapour deposition, *Applied Surface Science* 256 (2010), pp. 4606–4611.

193. J.W. Yan, X.H. Hou and K.L. Choy, The electrochemical properties of LSM-based cathodes fabricated by electrostatic spray assisted vapour deposition, *Journal of Power Sources* 180 (2008), pp. 373–379.

194. X.H. Hou, K.L. Choy and S.E. Leach, Processing and in vitro behavior of hydroxyapatite coatings prepared by electrostatic spray assisted vapor deposition method, *Journal of Biomedical Materials Research Part A* 83A (2007), pp. 683–691.

195. K.L. Choy and B. Su., Growth behaviour and microstructure of CdS thin films deposited by electrostatic assisted chemical vapour deposition process, *Thin Solid Films* 388 (2001), p. 9.

196. M. Wei and K.L. Choy, Deposition of highly oriented ZnS thin films on Si(100) substrate using electrostatic spray assisted vapour deposition, *Chemical Vapour Deposition* 8 (2002), p. 15.

197. G. Altamura, M.Q. Wang and K.L. Choy, Improving efficiency of electrostatic spray-assisted vapor deposited $Cu_2ZnSn(S,Se)_4$ solar cells by modification of Mo/absorber interface, *Thin Solid Films* 597 (2015), pp. 19–24.

198. K.L. Choy, X. Hou and J. Peng, EU FP-7 ThinSi project report (Project n. 241281), 2012.

199. K.L. Choy, X. Hou, J. Liu, ESAVD of transparent conducting oxide films, EU-FP7ThinSi project (241281) final report (2013).

200. Y.N. Kim, H.G. Shin, J.K. Song, D.H. Cho and H.S. Lee, Thermal degradation behavior of indium tin oxide thin films deposited by radio frequency magnetron sputtering, *Journal of Materials Research* 20 (2005), pp. 1574–1579.

201. B. Su and K.L. Choy, Electrostatic assisted aerosol-gel deposition of porous silica films, *Journal of Materials Science Letters* 8 (1999), p. 1705.

202. S. Li, Y. Ren, P. Biswas and S.D. Tse, Flame aerosol synthesis of nanostructured materials and functional devices: Processing, modeling, and diagnostics, *Progress in Energy and Combustion Science* 55 (2016), pp. 1–59.

203. K. Sanada, T. Shiota and K. Inada, Studies of the chemical kinetics on the vapor phase axial deposition method, *Journal of Non-Crystalline Solids* 188 (1995), pp. 275–284.

204. P.C. Schultz, Fabrication of optical waveguides by the outside vapor deposition process, *Proceedings of the IEEE* 68 (1980), pp. 1187–1190.

205. V. Petit, A. Le Rouge, F. Béclin, H. El Hamzaoui and L. Bigot, Experimental study of SiO_2 soot deposition using the outside vapor deposition method, *Aerosol Science and Technology* 44 (2010), pp. 388–394.

206. J.-S. Lee, A. Chandrashekar, B.M. Park, L.J. Overzet and G.S. Lee, Effects of oxygen plasma on optical and electrical characteristics of multiwall carbon nanotubes grown on a four-probe patterned Fe layer, *Journal of Vacuum Science & Technology B* 23 (2005), pp. 1013–1017.

207. S.H. Baek, D.F. Mihec and J.B. Metson, The deposition of diamond films by combustion assisted CVD on Ti and Ti-6Al-4V, *Chemical Vapor Deposition* 8 (2002), p. 29.

208. L. Yuan, K. Saito, W. Hu and Z. Chen, Ethylene flame synthesis of well-aligned multi-walled carbon nanotubes, *Chemical Physics Letters* 346 (2001), pp. 23–28.

209. R.L. Vander Wal, T.M. Ticich and V.E. Curtis, Diffusion flame synthesis of single-walled carbon nanotubes, *Chemical Physics Letters* 323 (2000), pp. 217–223.

210. M. Iijima and Y. Takahashi, Vapor deposition polymerization: A study on film formation in reaction of pyromellitic anhydride and bis(4-aminophenyl) Ether, *Macromolecules* 22 (1989), pp. 2944–2946.

211. (a) J.R. Salem, F.O. Sequeda, J. Duran, W.Y. Lee and R.M. Yang, Solventless polyimide films by vapor deposition, *Journal of Vacuum Science & Technology A: Vacuum, Surfaces, and Films* 4 (1986), pp. 369–374; (b) T. Miyamae, K. Tsukagoshi, O. Matsuoka, S. Yamamoto and H. Nozoye, Surface characterization of polyamic acid and polyimide films prepared by vapor deposition polymerization by using sum-frequency generation, *Langmuir* 17 (2001), pp. 8125–8130.

212. A.M. Coclite, Smart surfaces by initiated chemical vapor deposition, *Surface Innovations* 1 (2013), pp. 6–14.

213. C. Chen, Multiscale imaging, modeling, and principal component analysis of gas transport in shale reservoirs, *Fuel* 182 (2016), pp. 761–770.

214. A. Asatekin, M. Barr, S. Baxamusa, K. Lau, W. Tenhaeff, J. Xu et al., Designing polymer surfaces via vapor deposition, *Materials Today* 13 (2010), pp. 26–33.

215. H.G. Pryce Lewis, N.P. Bansal, A.J. White and E.S. Handy, HWCVD of polymers: Commercialization and scale-up, *Thin Solid Films* 517 (2009), pp. 3551–3554.

216. S. Vaddiraju, K. Senecal and K. Gleason, Novel strategies for the deposition of -COOH functionalized conducting copolymer films and the assembly of inorganic nanoparticles on conducting polymer platforms, *Advanced Functional Materials* 18 (2008), pp. 1929–1938.

217. P. Kovacik, G. del Hierro, W. Livernois and K.K. Gleason, Scale-up of oCVD: Large-area conductive polymer thin films for next-generation electronics, *Materials Horizons* 2 (2015), pp. 221–227.

218. J.H. Coleman, Polymerizing method and apparatus, U.S. Patent 3,068,510, issued December 18, 1962.

219. H. Yasuda, Glow discharge polymerization, *Journal of Polymer Science: Macromolecular Reviews* 16 (1981), pp. 199–293.

220. H. Yasuda and T. Hsu, Plasma polymerization investigated by the comparison of hydrocarbons and perfluorocarbons, *Surface Science* 76 (1978), pp. 232–241.

221. M.E. Ryan, A.M. Hynes and J.P.S. Badyal, Pulsed plasma polymerization of maleic anhydride, *Chemistry of Materials* 8 (1996), pp. 37–42.

222. S.R. Coulson, I. Burnett and J.H. Sambell, Coating of a polymer layer using low power pulsed plasma in a plasma chamber of a large volume. U.S. Patent Application 13/752,474, filed May 30, 2013.

223. P2i, P2i Splash-proof, P2i, https://www.p2i.com/solutions/splash-proof/, accessed on August 11, 2017.

224. C.R. Kleijn, R. Dorsman, K.J. Kuijlaars, M. Okkerse and H. van Santen, Multi-scale modeling of chemical vapor deposition processes for thin film technology, *Journal of Crystal Growth* 303 (2007), pp. 362–380.

225. X.W. Zhou, R.A. Johnson and H.N.G. Wadley, A molecular dynamics study of nickel vapor deposition: Temperature, incident angle, and adatom energy effects, *Acta Materialia* 45 (1997), pp. 1513–1524.

226. Y. Yang, X. Zhou, R. Johnson and H.N. Wadley, Monte Carlo simulation of hyperthermal physical vapor deposition, *Acta Materialia* 49 (2001), pp. 3321–3332.

227. K.F. Jensen, S.T. Rodgers and R. Venkataramani, Multiscale modeling of thin film growth, *Curr. Opinion in Solid State Materials Science* 3 (1998), pp. 562–569.

228. (a) H. Wadley, A. Zhou, R. Johnson and M. Neurock, Mechanisms, models and methods of vapor deposition, *Progress in Materials Science* 46 (2001), pp. 329–377; (b) A. Dollet, Multiscale modeling of CVD film growth–a review of recent works, *Surface and Coatings Technology* 177–178 (2004), pp. 245–251.

229. F.H. Baumann, D.L. Chopp, T.D. de la Rubia, G.H. Gilmer, J.E. Greene, H. Huang et al., Multiscale modeling of thin-film deposition: Applications to Si device processing, *MRS Bulletin* 26 (2001), pp. 182–189.

230. T.E. Karakasidis and C.A. Charitidis, Multiscale modeling in nanomaterials science, *Materials Science & Engineering C-Biomimetic and Supramolecular Systems* 27 (2007), pp. 1082–1089.

231. (a) H. Ma, L. Pan and Y. Nakayama, Modelling the growth of carbon nanotubes produced by chemical vapor deposition, *Carbon* 49 (2011), pp. 854–861; (b) M. Moraveji et al., CFD modeling and experimental study of multi-walled carbon nanotubes production by fluidized bed catalytic chemical vapor deposition, *International Communications in Heat and Mass Transfer* 38 (7) (2011), pp. 984–989.

232. (a) I. Zahi, P. Mur, P. Blaise, A. Estève, M.D. Rouhani, H. Vergnes et al., Multi-scale modelling of silicon nanocrystal synthesis by low pressure chemical vapor deposition, *Thin Solid Films* 519 (2011), pp. 7650–7658; (b) V. Cocheteau et al., Development of an original model for the synthesis of silicon nanodots by low pressure chemical vapor deposition, *Chemical Engineering Journal* 140 (1–3) (2008), pp. 600–608.

233. Y. Wen, D. Xia and W. Xuan, Modeling for particle size prediction and mechanism of silicon nitride nanoparticle synthesis by chemical vapor deposition, *Aerosol Science and Technology* 51 (2017), pp. 845–855.

234. H. Kim, E. Saiz, M. Chhowalla and C. Mattevi, Modeling of the self-limited growth in catalytic chemical vapor deposition of graphene, *New Journal of Physics* 15 (2013), p. 053012.

235. H. Ni, S. Lu and C. Chen, Modeling and simulation of silicon epitaxial growth in Siemens CVD reactor, *Journal of Crystal Growth* 404 (2014), pp. 89–99.

236. A. Ramos, A. Rodriguez, C. del Canizo, J. Valdehita, J.C. Zamorano and A. Luque, Heat losses in a CVD reactor for polysilicon

production: Comprehensive model and experimental validation, *Journal of Crystal Growth* 402 (2014), pp. 138–146.

237. S.-S. Liu and W.-D. Xiao, Numerical simulations of particle growth in a silicon-CVD fluidized bed reactor via a CFD-PBM coupled model, *Chemical Engineering Science* 111 (2014), pp. 112–125.

238. K. Raji, S. Thomas and C.B. Sobhan, A chemical kinetic model for chemical vapor deposition of carbon nanotubes, *Applied Surface Science* 257 (2011), pp. 10562–10570.

239. J.A. Elliott, Y. Shibuta, H. Amara, C. Bichara and E.C. Neyts, Atomistic modelling of CVD synthesis of carbon nanotubes and graphene, *Nanoscale* 5 (2013), pp. 6662–6676.

240. J. Geiser, Multiscale modeling of chemical vapor deposition (CVD) apparatus: Simulations and approximations, *Polymers* 5 (2013), pp. 142–160.

241. K. Raji and C.B. Sobhan, A computational model for predicting the temperature distribution inside a CVD reactor for carbon nanotube synthesis, in *ASME 2013 International Mechanical Engineering Congress and Exposition*, 2013, p. 87931N.

242. C. Battaile, D. Srolovitz and J. Butler, Atomic-scale simulations of chemical vapor deposition on flat and vicinal diamond substrates, *Journal of Crystal Growth* 194 (1998), pp. 353–368.

243. L. Kadinski, V. Merai, A. Parekh, J. Ramer, E.A. Armour, R. Stall et al., Computational analysis of GaN/InGaN deposition in MOCVD vertical rotating disk reactors, *Journal of Crystal Growth* 261 (2004), pp. 175–181.

244. B. Mitrovic, A. Gurary and W. Quinn, Process conditions optimization for the maximum deposition rate and uniformity in vertical rotating disc MOCVD reactors based on CFD modeling, *Journal of Crystal Growth* 303 (2007), pp. 323–329.

245. C.-F. Tseng, T.-Y. Tsai, Y.-H. Huang, M.-T. Lee and R.-H. Horng, Transport phenomena and the effects of reactor geometry for epitaxial GaN growth in a vertical MOCVD reactor, *Journal of Crystal Growth* 432 (2015), pp. 54–63.

246. Y.-C. Chuang and C.-T. Chen, Mathematical modeling and optimal design of an MOCVD reactor for GaAs film growth, *Journal of the Taiwan Institute of Chemical Engineers* 45 (2014), pp. 254–267.

247. S. Nakamura, M. Senoh, N. Iwasa and S. Nagahama, High-brightness InGaN blue, green and yellow light-emitting diodes with quantum well structures, *Japanese Journal of Applied Physics Part 2–Letters* 34 (1995), pp. L797–L799.

248. M.A. Green, K. Emery, Y. Hishikawa, W. Warta, E.D. Dunlop, D.H. Levi et al., Solar cell efficiency tables (version 49): Solar cell efficiency tables (version 49), *Progress in Photovoltaics: Research and Applications* 25 (2017), pp. 3–13.

249. S. Ruhle, Tabulated values of the Shockley–Queisser limit for single junction solar cells, *Solar Energy* 130 (2016), pp. 139–147.

250. K. Evanoff, J. Khan, A.A. Balandin, A. Magasinski, W.J. Ready, T.F. Fuller et al., Towards ultrathick battery electrodes: Aligned carbon nanotube–enabled architecture, *Advanced Materials* 24 (2012), pp. 533–537.

251. W. Ren and H.-M. Cheng, The global growth of graphene, *Nature Nanotechnology* 9 (2014), pp. 726–730.

252. J. Robertson, Ultrathin carbon coatings for magnetic storage technology, *Thin Solid Films* 383 (2001), pp. 81–88.

253. G. Dearnaley and J.H. Arps, Biomedical applications of diamond-like carbon (DLC) coatings: A review, *Surface & Coatings Technology* 200 (2005), pp. 2518–2524.

254. (a) R.K. Roy and K.R. Lee, Biomedical applications of diamond-like carbon coatings: A review, *Journal of Biomedical Materials Research Part B Applied Biomaterials* 83 (2007), pp. 72–84; (b) J.I. Oñate, M. Comin, I. Braceras, A. Garcia, J.L. Viviente, M. Brizuela et al., Wear reduction effect on ultra-high-molecular-weight polyethylene by application of hard coatings and ion implantation on cobalt chromium alloy, as measured in a knee wear simulation machine, *Surface and Coatings Technology* 142–144 (2001), pp. 1056–1062.

255. K. Reichelt and X. Jiang, The preparation of thin films by physical vapour deposition methods, *Thin Solid Films* 191 (1990), pp. 91–126.

256. U. Betz, M. Kharrazi Olsson, J. Marthy, M.F. Escolá and F. Atamny, Thin films engineering of indium tin oxide: Large area flat panel displays application, *Surface and Coatings Technology* 200 (2006), pp. 5751–5759.

257. F. Faggin and T. Klein, Silicon gate technology, *Solid-State Electronics* 13 (1970), pp. 1125–1144.

258. M. Swihart, Vapor-phase synthesis of nanoparticles, *Current Opinion in Colloid & Interface Science* 8 (2003), pp. 127–133.

259. K. Petherbridge, P. Evans and D. Harrison, The origins and evolution of the PCB: A review, *Circuit World* 31 (2005), pp. 41–45.

260. J. Natoli, I. Berbezier, A. Ronda and J. Derrien, Chemical beam epitaxy of iron disilicide on silicon, *Journal of Crystal Growth* 146 (1995), pp. 444–448.

261. A. Robertson, U. Erb and G. Palumbo, Practical applications for electrodeposited nanocrystalline materials, *Nanostructured Materials* 12 (1999), pp. 1035–1040.

262. R.C. Mehrotra, Present status and future potential of the Sol-Gel process, in *Chemistry, Spectroscopy and Applications of Sol-Gel Glasses*, Springer-Verlag, Berlin, Germany, 2005, pp. 1–36.

263. H. Dislich and E. Hussmann, Amorphous and crystalline dip coatings obtained from organometallic solutions: Procedures, chemical processes and products, *Thin Solid Films* 77 (1981), pp. 129–140.

264. R. Ciriminna, A. Fidalgo, V. Pandarus, F. Béland, L.M. Ilharco and M. Pagliaro, The sol–gel route to advanced silica-based materials and recent applications, *Chemical Reviews* 113 (2013), pp. 6592–6620.

265. H. Dislich, Glassy and crystalline systems from gels: Chemical basis and technical application, *Journal of Non-Crystalline Solids* 57 (1983), pp. 371–378.

266. H. Dislich, Sol-gel: Science, processes and products, *Journal of Non-Crystalline Solids* 80 (1986), pp. 115–121.

267. G. Schottner, Hybrid sol–gel-derived polymers: Applications of multifunctional materials, *Chemistry of Materials* 13 (2001), pp. 3422–3435.

268. G. Schottner, J. Kron and A. Deichmann, Industrial application of hybride SOl-Gel coatings for the decoration of crystal glassware, *Journal of Sol-Gel Science and Technology* 13 (1998), pp. 183–187.

Chemical vapour deposition of ultrafine particles

CIGANG XU AND KWANG LEONG CHOY

2.1 INTRODUCTION

Particles are materials with diameters from nanometres to micrometres. Nanoscale refers to dimensions between 1 and 100 nm, while microscale typically implies dimensions between 100 nm and 100 μm. In this chapter, 'ultrafine particles' describe nanoparticles and submicron-sized particles. Ultrafine particles have been used since the early days of human civilisation; therefore, there is substantial knowledge and long history. Back in 1500 BC, the Chinese used carbon black from combustion processes for painting applications, while carbon black is now used in fabricating car tires. The most impressive uses of nanoscaled objects in history is the Lycurgus Cup, which dates back to Roman times (fourth century AD). The extraordinary cup benefits from the dichroic glass which contains colloidal gold and silver nanoparticles that give the spectacular optical properties; it shows red when lit from behind and green when lit from the front [1]. In 1857, Faraday was the first investigator to report the existence of metallic nanoparticles in solutions [2], while Mie provided a quantitative explanation of their colour back in 1908 [3].

In perspective, there has been considerable research in developing a variety of processes for the preparation of ultrafine particles, including liquid processes, such as wet chemical process and sol-gel process. Unlike the liquid process, chemical vapour deposition (CVD)-based processes benefit from using small volumes of liquid precursors, surfactants, or precipitation agents.

The CVD process has been used to produce ultrafine particles in industrial settings, such as the flame synthesis by companies like Cabot, Evonik (former Degussa), and Du Pont [4a].

The principle of the CVD process is explained in Chapter 1. When the CVD process is used to prepare ultrafine particles, gas phase chemical reactions of precursors would lead to the formation of ultrafine particles; depending on the CVD processes, the particle formation sequence varies. For example, in a high-temperature flame-synthesis process, the particle formation can involve homogeneous reactions of monomers, nucleation and formation of cluster or primary particles, and coalescence/agglomeration. Figure 2.1 shows a schematic diagram of the gas-to-particle conversion process [4b]. Conventional chemical vapour-based processes may be referred to as chemical vapour synthesis (CVS) or chemical vapour condensation (CVC) as they do not deposit films in the same approach as the conventional CVD process [5a]. Chemical vapour reaction (CVR), CVC, and chemical vapour precipitation (CVP) are synonyms used frequently in the literature as CVS. CVS process refers to the process for the formation of nanoparticles in the vapour phase rather than deposition of film, as shown in Figure 2.2 [5b]; CVC refers to the process in which an evaporative source is used to generate the ultrafine particles, which are collectively transported to and collected on a cold substrate, as shown in Figure 2.3 [5c]. In this chapter, 'CVD' refers to all forms of vapour process where chemical reaction occurs [5].

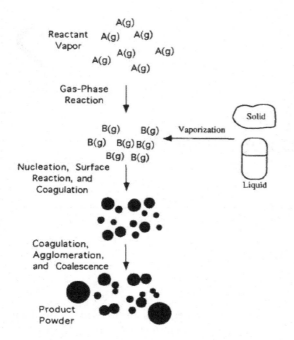

Figure 2.1 Schematic-to-particle conversion process. (From Gurav, A. et al., *Aerosol Sci. Tech.*, 19, 411–452, 1993.)

are also appropriate for large-scale separation with high effectiveness and for continuous processing. Generally, the technique varies for the process of random particles and particles on the surface of substrate or support, and even for the random particles. For example, the production process of carbon black includes the channel process, thermal process, and furnace process. In the channel process, carbon black is mechanically removed from the channels into hoppers and then conveyed and prepared for shipment; in the thermal process, carbon black is carried off in the gas stream to be collected in a bag filter; in the furnace process, carbon black is generated in the flame and quenched while still suspended in the gas stream, then collected by a collector such as a fabric filter. The collection techniques will be discussed together with the related processes.

In addition to conventional thermal CVD technologies, various CVD techniques are also used for the processing of ultrafine particles. This chapter gives an overview on the preparation of ultrafine particles by CVD technologies in various formats, instead of solely bulk powder synthesis that is most frequently the main focus of previous reviews. In addition, a broad range of methodologies will be discussed as well as the current progress on the scale-up production of ultrafine particles.

The collection of ultrafine particles depends on CVD processes [5c]. A simple method is to use a mechanical filter with small pore size to collect particles. Corona discharge and electrostatic devices

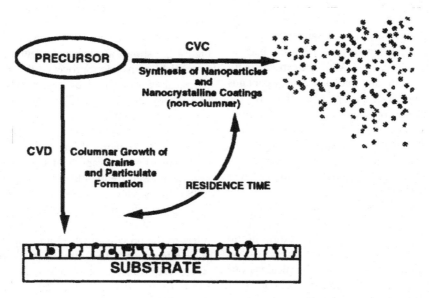

Figure 2.2 Comparison of chemical vapour deposition (CVD) for formation of thin films and chemical vapor condensation (CVC) for synthesis of nanoparticles. (From Hahn, H., *Nanostruct. Mater.*, 9, 3–12, 1997.)

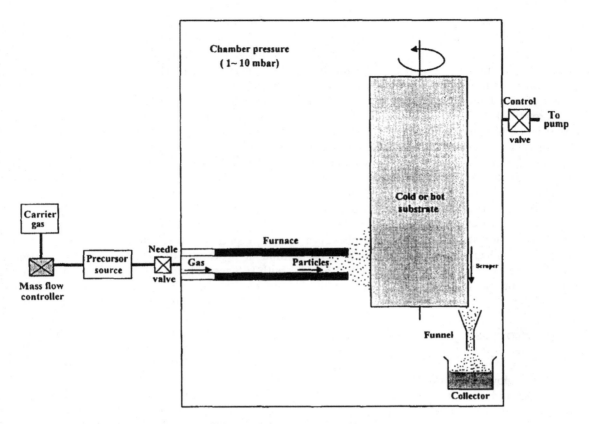

Figure 2.3 Schematic of the conventional gas condensation (CVC) processing apparatus. (From Chang, W. et al., *Nanostruct. Mater,* 4, 345–351, 1994.)

2.2 PROCESSING, FORMATION MECHANISM, STRUCTURE, AND PROPERTIES

Although ultrafine particles, such as carbon black and colloidal gold, were reported to be used since ancient times, it was only in the twentieth century that the research field for further studies has emerged. Since Richard Feynman's famous speech 'There's plenty of room at the bottom' in 1959 [6], effort has been made to explore structural features and devices in the sub-micron range. In 1974, Norio Taniguchi used the term 'nanotechnology' to describe tiny things that would be measured in nanometre scale [7]. Since then, the prefix 'nano' has been applied in many areas. Nanomaterial refers to the material with at least one dimension (1D) (diameter, thickness, height, etc.) in the range of 1–100 nm [8]. As mentioned earlier, in this chapter the term 'ultrafine particle' will be used to discuss nanoparticles and submicron-sized-particles.

Thus far, the research on ultrafine particles has spread to many areas covering physics, chemistry, biology, materials science, and engineering [9].

Ultrafine particles can be prepared in top-down and bottom-up approaches [9]. The top-down approach is through size reduction from bulk materials like milling method, or converting films into nanoparticles, such as annealing treatment of metal films to nanoparticles for the catalytic growth of 1D nanomaterials [10]. The drawbacks of milling method include the low-energy efficiency, the necessity of a liquid medium for producing materials in the nanoscale range, the susceptibility to impurities from abrasion, and a broader particle-size distribution. The bottom-up approaches include liquid-phase and gas-phase synthesis methods [11]. These methods generally result in well-controlled nanoparticles.

Ultrafine particles consist of binary, ternary, or doped composition, for example, TiO_2, $SrMnO_3$, and Y_2O_3:Eu, respectively [12]. Moreover, ultrafine particles can be prepared in various for-mats, including: (1) bulk ultrafine particles, such as the above-mentioned ultrafine particles obtained through milling method [9]; (2) ultrafine par-ticles on the surface of a substrate, for example, those obtained through annealing treatment for the growth of 1D nanomaterial [10]; (3) ultrafine particles on the surface of a support, such as those for the catalysis applications [13]; and (4) ultrafine particles inside a matrix of another material, for instance, Si ultrafine particle in SiOx matrix [14].

There are substantial literatures on the synthe-sis of ultrafine particles in the liquid phase [11]. Although such wet-chemical method is low cost, there is a purity issue with such liquid-phase pro-cessing. This chapter is concerned with gas-phase synthesis methods, in particular, CVD-based methods for the synthesis of ultrafine particles with high purity.

2.2.1 Processing methods using solid precursors

Physical vapour deposition (PVD) involves con-densation and formation of ultrafine particles from the vapour phase with no chemical reaction during the fabrication process. On the other hand, CVD mode is when a chemical reaction occurs, as shown in Figure 2.4, which is the key focus in this chapter.

2.2.1.1 CONDENSATION METHOD

In the condensation method, the solid (con-densed) precursor is heated to above its boiling temperature when it is evaporated into a back-ground gas. The gaseous precursor is then mixed with the cold background gas in order to pro-mote condensation and subsequent formation of solid particles. This method can be in PVD or CVD mode, while in CVD mode, chemical reac-tion takes place where the gas phase consists of reactive gas composition [15].

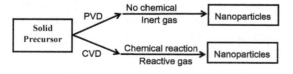

Figure 2.4 The use of solid precursors for the processing of nanoparticles.

This method dates back to 1930 where Bi ultrafine particles (Bi black) were prepared for radiometric devices [16]. Substantial efforts have been performed to understand the mechanism through detailed experimental and theoretical studies, which can differentiate between different sintering mechanisms, or even provide control on the synthesis of nanoparticles. When the method is used in the CVD mode, it generally includes the following steps: 1. evaporation of powder, 2. condensation of vapour to produce nanoparti-cles, and 3. chemical between the reactive gas and nanoparticles. For example, MoS_2 nanoparticles with the size ranging 15($\pm$10nm)–250($\pm$50nm) can be prepared through the thermal evapora-tion of MoO_3 at 710°C–810°C and its reaction with H_2S gas [15] at 800°C $\pm$ 10°C at atmospheric pressure or reduced pressure. Figure 2.5a shows a schematic diagram of the CVD reactor, and Figure 2.5b shows a typical transmission elec-tron microscopy (TEM) image of MoS_2 nanopar-ticles [15a]. This approach has been scaled-up to 20–100 g/batch. MoS_2 nanoparticles have poten-tial for solid-lubricant applications [15b].

This condensation process can also be con-ducted using a solid precursor to prepare nanopar-ticles in a conventional tube reactor as employed in conventional CVD processes. For instance, MnO nanoparticles are synthesized from the reaction of $MnCl_2$ vapour (in Ar or a mixture of Ar/H_2), with O_2 from the system leak at 778°C and pressure of 0.05 MPa [17a].

Moreover, the work on CeO_2 and $Ce_xY_{1-x}O_{2-\delta}$ nanoparticles has shown that in a thermophoresis-assisted CVD tube reactor, the way to collect nanopar-ticles (on a cooled quartz susceptor in the furnace (set-up A), in a conventional cold-wall container out-side the furnace (set-up B)) can affect the particle size and distribution. Typical synthesis conditions were: oxygen flow 0.5–3.2 l/h, argon flow 1.25–4.5 l/h, CeCl evaporation temperature 973–1253 K, and YCl evaporation temperature 873–1053 K, typically in the pressure B). The temperature of the susceptor in set-up A is range of 1–20 mbar. TEM characterisation shows that the particles (for CeO_2 and $Ce_xY_{1-x}O_{2-\delta}$) are 30–80 nm for the set-up (a), and 20–30 nm for the set-up (c), as shown in Figure 2.6 [17b].

Generally, this method is suitable for synthesiz-ing materials that can be thermally evaporated and undergo chemical reactions (e.g., reduction, oxida-tion etc.), leading to the formation of nanoparticles

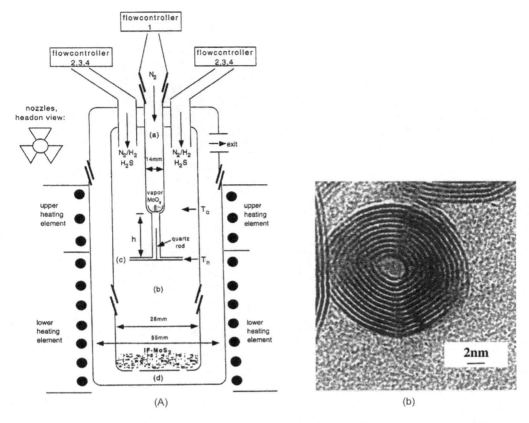

Figure 2.5 **(A)** Schematic representation of the experimental vertical gas phase reactor: **(a)** inner tube; **(b)** middle tube; **(c)** nozzles; **(d)** external tube. **(B)** A typical TEM micrograph of inorganic fullerene-like (hollow onion-like) nanoparticles of 20 nm. (From Zak, A. et al., *J. Am. Chem. Soc.*, 122, 11108–11116, 2000; Feldman, Y. et al., *Solid State Sci.*, 2, 663–672, 2000.)

of metal, oxide, sulphides, carbides, borides, and nitrides. The composition of nanoparticles can be adjusted through the use of relevant gas atmosphere. The issues of this method include: chemical reactions between materials with the refractory crucibles; difficulty to produce multicomponent materials, alloys, intermetallic compounds, and composites due to difficulties to control compositions resulting from the difference of thermodynamic activities and vapour pressures of the constituents; and the technical difficulty for the materials with high melting points and low vapour pressure. Moreover, except thermal evaporation, other methods like sputtering or arc-discharge approach have also been used to evaporate materials, and some materials (like ZnO, Al_2O_3 nanoparticles) are now produced through the arc-discharge process for the applications like personal care and coating sectors [4c]. This will be covered in the related section.

2.2.1.2 LASER-ABLATION METHOD

Instead of thermal heating, a laser can be used to heat and evaporate a solid target. Laser-ablation method involves exposing the material to laser irradiation. This can cause heating, melting, vaporisation, ejection of atoms, ions and molecules, shock waves, plasma initiation, and plasma expansion. The process can be used for the preparation of either aerosol nanoparticles or colloidal nanoparticles [18]. The ablated substance can be modified accordingly through varying the energy, intensity output, and wavelength of the laser. Figure 2.7 illustrates an example of the laser-ablation apparatus used in the preparation of aerosol nanoparticles [19].

The wavelength of the laser can range from the mid infrared to the ultraviolet region, i.e., excimer laser of different UV wavelengths such as 193 nm. The laser-ablation mechanism is illustrated in Figure 2.8. The material absorbs the laser light and

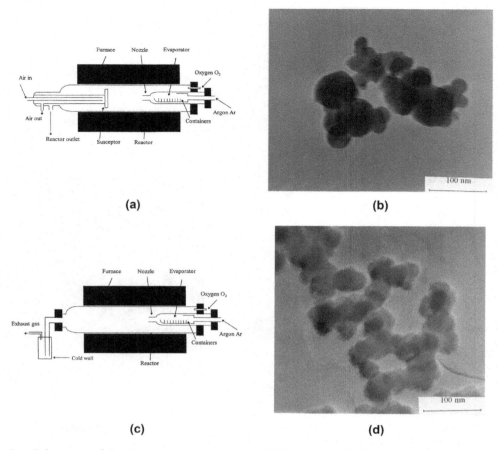

(a)

(b)

(c)

(d)

Figure 2.6 Schematic of the thermophoresis-assisted CVD reactors. (a) Set-up of a cooled susceptor, (b) TEM micrographs of cerium oxide powders collected from set-up (a), (c) set-up (b) with cold-wall collector, (d) TEM micrographs of cerium oxide powders collected from set-up (c). (From Bai, W. et al., *Solid State Ionics*, 116, 225–228, 1999.)

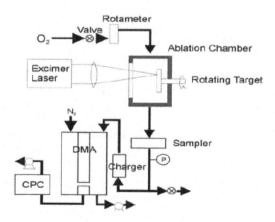

Figure 2.7 Schematic of the laser ablation set-up (DMA, differential mobility analyser; CPC, condensation particle counter). (From Ullmann, M. et al., *J. Nanoparticle Res.*, 4, 499–509.)

forms the melt, which propagates into the remaining solid. The melt then absorbs the laser energy and evaporates. The vaporisation continues to form the plume which absorbs the laser, leading to the formation of a plasma. The plume will expand away from the interaction volume, the target would cool and become solid after cessation of laser pulse [20].

Laser ablation can be in PVD mode, when the atmosphere is non-reactive gas in the absence of chemical reaction, or in CVD mode when the reactive gas is used. Laser-ablation method can be used to produce metal nanoparticles, refractory oxide, and other materials through PVD mode, or produce oxides in CVD mode [21]. Figure 2.9 shows TEM and HRTEM images of NiO nanoparticles produced through the laser ablation of Ni in an oxygen atmosphere [22]. The TEM image in Figure 2.9a shows that the shape of most nanoparticles is cubic

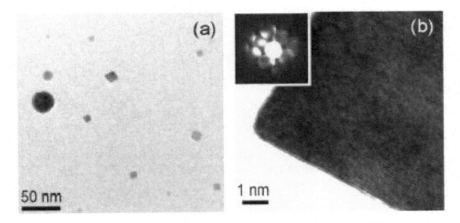

Figure 2.8 Schematic illustration of the key elements of the pulse laser ablation event. **(a)** Initial absorption of laser radiation (indicated by long arrows), melting and vaporisation begin (shaded area indicates melted material, short arrows indicate motion of solid–liquid interface). **(b)** Melt front propagates into the solid, vaporisation continues and laser-plume interactions start to become important. **(c)** Absorption of incident laser radiation by the plume, and plasma formation. **(d)** Melt front recedes leading to eventual re-solidification. (From Ashfold, M. et al., *Chem. Soc. Rev.*, 33, 23–31, 2004.)

Figure 2.9 **(a)** TEM and **(b)** HRTEM micrographs of the nanoparticles obtained by ultrafast laser ablation of nickel in 40 Pa oxygen. Inset of **(b)** is a nanobeam electron diffraction pattern, which can be indexed as the 100 zone axis of NiO. (From Liu, B. et al., *Appl. Phys. Lett.*, 90, 44103, 2007.)

and the shape of large mesoparticles remains round, while the HRTEM image in Figure 2.9b reveals that the cubic nanoparticles are single crystals.

One of the advantages of the laser-ablation method is that it can be used to vaporise materials that are difficult to thermally evaporate, such as refractory oxides. However, this method can only produce a small number of ultrafine particles (~10 mg/h) and has a high production cost; therefore, it is not suitable for large-volume production of ultrafine particles. It is worth mentioning that C_{60} was initially observed through this method in 1985 [23].

2.2.1.3 ARC-DISCHARGE METHOD

This process uses arc-discharge method to evaporate materials. The conventional arc-discharge method

uses two electrodes, where one electrode has the evaporant material and the other is a counter electrode, and the arc is established between the evaporant material and the counter electrode, as shown in Figure 2.10 [24]. The evaporation process is in the atmosphere of a flowing gas. When the inert or non-reactive gas is used, no chemical reaction occurs; somehow it is like the PVD process, and the evaporated material vapour cools down by the gas to form ultrafine particles. When a reactive gas is used, such as oxygen, arc-discharge method can be treated as CVD process. The material and gas selection can give a route to different types of materials.

A broad range of materials have been produced with this method. The solid material can be pure metal, nitride, carbide, or oxide [25], as shown in

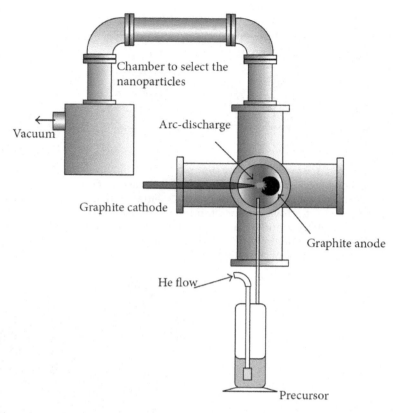

Figure 2.10 A schematic of a typical arc-discharge tool used for the synthesis of the iron encapsulated in carbon nanoparticles. (From Chaitoglou, S. et al., *J. Nanomater.*, 9, 2014.)

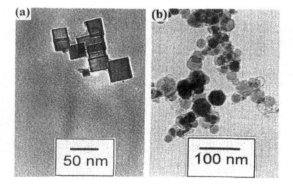

Figure 2.11 TEM micrographs of **(a)** TiN and **(b)** TiC nanoparticles. (From Mahoney, W. et al., *Mat. Res. Soc. Symp. Proc.*, 400, 65–70, 1996.)

Table 2.1 Examples of nanoparticles manufacturing using the arc discharge method

Material	Atmosphere gas kind	Obtained particles
Pure metal	Hydrogen	Fe, Cu, Ni, Co, Cr, Ti, In, ...
Alloy	Hydrogen	Fe–Ni, Fe–Co, Ag–Cu, Ag–Pd, ...
Metal or nitride	Nitrogen	TiN, ZrN, AlN, AlN+Al, ...
Metal or carbide	Hydrogen	SiC, TiC, ...
Boronide	Hydrogen	LaB6
Metal or oxide	Oxygen	WO_3, MoO_3, Nb_2O_5, ...

Source: Nakajima, N., Application 2–Generation of metal nanoparticles using reactive plasma arc evaporation, in *Nanoparticle Technology Handbook*, Elsevier, Amsterdam, the Netherlands, pp. 428–431, 2012.

Table 2.1. Figure 2.11 shows TEM images of TiN and TiC nanoparticles produced by arc-discharge method; typical arc voltage for evaporation into Ar gas was approximately 11 Volts. For the production of metal nitrides and metal carbide nanoparticles, either nitrogen or methane, respectively, was added into the higher-temperature arc region, and they were both added as a small fraction (up to 10%) in an argon carrier gas used to maintain arc stability. Both TiN and TiC nanoparticles have the NaCl crystal structure indicated through electron diffraction measurements [26].

Effort has also been made to produce nanoparticles on a large scale using this method, as shown in Figure 2.12. Since a single arc unit can only produce a small amount of materials, one way to scale-up the process is to use multiple optimized single units. This approach is the scheme of EU-FP7 BUONAPART-E's project for the R&D of large-scale production of metal nanoparticles using electrical discharge methods [9]. The reported production rate is 0.6 Kg/day for the system with eight optimized single units. This type of tool can be operated under PVD mode with Ar as process gas, or CVD mode if O_2 (through air) is added for the processing of SnO/SnO_2 nanoparticles, for example [9b]. It is possible to further increase production rate by increasing the number of optimised single units [9], or the process can be performed in a continuous way. More detail will be covered [17d].

Another type of arc-discharge process is pulsed-wire discharge as shown in Figure 2.13 [27].

Figure 2.12 A multi-unit tool for the scale-up production of nanoparticles. (From Charitidis, C.A. et al., *Manufacturing Rev*, 1, 11–19, 2014;Chen, J. et al., *J. Nanoparticle Res.*, 9, 203–213, 2007.)

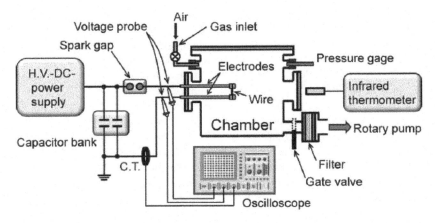

Figure 2.13 Experiment set-up. (From Ishihara, S. et al., *Ceram. Int.*, 38, 4477–4484; Jiang, W.H. and Yatsui, K., *IEEE Trans. Plasma Sci.*, 26, 1498–1501, 1998.)

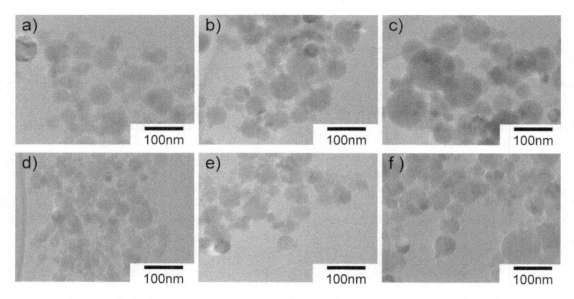

Figure 2.14 Bright-field TEM images of the particles synthesised by PWD at the various conditions of wire diameter and atmosphere (wire diameters of **(a)** and **(d)**: f 0.3; **(b)** and **(e)**: f 0.4; **(c)** and **(f)**: f 0.5, and atmosphere of **(a)**, **(b)**, and **(c)**: closed; **(d)**, **(e)**, and **(f)**: air flow of 33 L/min). (From Ishihara, S. et al., *Ceram. Int.*, 38, 4477–4484; Jiang, W.H. and Yatsui, K., *IEEE Trans. Plasma Sci.*, 26, 1498–1501, 1998.)

The choice of wire and process atmosphere depends on the particle composition. An example experiment setup is shown in Figure 2.14. For the preparation of Al_2O_3 nanoparticles, pure Al wire with the diameter of 0.3 mm, 0.4 mm, or 0.5 mm was used and connected between the electrodes. Air was used as the process gas, pulsed large current from the capacitors was applied to the set wire to result in wire evaporation, and the formed particles were evacuated and collected on a membrane filter of 0.1 μm pore size. The product of Al_2O_3 ultrafine particles were dominantly composed of γ-Al_2O_3 and δ-Al_2O_3 phases. The δ/γ ratios of the Al_2O_3 phases increased with increasing air flow rate and wire diameter, small amount of metallic Al phase was observed when the air flow atmosphere of 33 L/min was used [27a]. Similarly, for this type of process, when the gas is inert gas, like Ar, the metal nanoparticles such as Al, Cu nanoparticles can be obtained in PVD mode, when the gas is reactive gas, O_2 or N_2, the oxide or nitride nanoparticles can be produced in CVD mode, respectively [27b].

It is worth mentioning that C_{60} particles were observed from the laser-ablation process, but the quantity is too low to characterise the material. Arc-discharge method was used for the synthesis of C_{60} particles on the scale of grams from the process of arc discharging graphite rod, and from the resulting deposit of the process, carbon nanotubes were found and characterized [28]. This work on carbon nanotubes subsequently stimulated intensive research on carbon nanotubes and other 1D materials.

2.2.1.4 REACTIVE SPUTTERING METHOD

Instead of using thermal or laser energy to evaporate the solid materials as mentioned earlier, sputtering method uses energetic ions to bombard the solid surfaces so that the material is evaporated from solid surfaces; this can provide a route to prepare particles for the materials of low vapour pressure that are challenging for the evaporation/condensation processes, but the process is difficult to scale-up. It is a momentum-transfer process rather than an evaporation process [29].

Sputtering method can also be used in PVD and CVD mode. In PVD mode, no chemical reaction occurs, and the cathode is generally not reacted; in CVD mode, the atmosphere gas is used to provide the chemical reaction and it is also called reactive sputtering. In this case, the cathode can be oxidized, which is referred to as poisoning. However, cathode poisoning can be reduced by passing a high flow rate of inert gas through a hollow cathode to suppress the reactive gas from entering into the

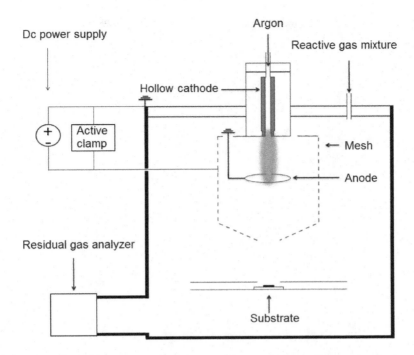

Figure 2.15 A sketch of the experimental sputtering setup with the hollow cathode arrangement (Figure not to scale). (From Gunnarsson, R. et al., *J. Nanopart. Res.*, 17, 353–367, 2015.)

cathode [30]. Figure 2.15 shows a set-up for such type of reactive sputtering tool. The setup consists of a hollow cathode driven by high-power pulses as the sputter cathode, a grounded anode ring, and a mesh that can be biased. The nanoparticles are steered by the mesh through the opening at the bottom and deposited onto a substrate that is biased positively in order to attract them [30]. Figure 2.16 shows TiO$_2$ nanoparticles synthesized under various CVD conditions.

Sputtering method can be combined with other methods by generating a cluster beam. For instance, the combination of sputtering and gas-aggregation techniques has been developed to prepare Fe nanoparticles for applications such as biomedical imaging or zero-valent iron (ZVI) for environmental cleaning. The cluster beam system consists of three parts: (1) a cluster source, (2) an e-beam evaporation chamber, and (3) a deposition chamber, as shown in Figure 2.17. High-pressure magnetron-sputtering gun is used to produce Fe^0 atoms that are decelerated by collisions with continuously injected Ar gas in the chamber cooled by chilled water.

The cluster with a size of 1–100 nm can be produced by varying the aggregation distance, the sputtering power, the pressure in the aggregation tube, and the ratio of He-to-Ar gas flow rate.

The clusters are ejected, part of it is intercepted, and the rest is then deposited onto a sample holder in the deposition chamber. The major advantage of this approach is that the clusters have much smaller size dispersion as compared with other typical vapour-deposition systems. The log normal size distribution can have a standard deviation of about 3%. To form oxides, the four sccm reactive gases such as O$_2$ can be introduced into the chamber.

In the case of Fe, a layer of oxide (2.5 nm thick) can form on the outside layer of iron nanoparticles (with the diameter of 3–85 nm) to avoid oxidation and retain the high magnetic moments of pure iron and enhance their applications, as shown in Figure 2.18 [31].

This cluster-style sputtering method can have different styles of tool. For example, as shown in Figure 2.19 [32], the cluster is produced through sputtering, and the gas can be the reactive gas, such as oxygen, to form oxide nanoparticles, like the diameter of 2.5–12.5 nm TiO$_2$ nanoparticles as shown in Figure 2.20 [33].

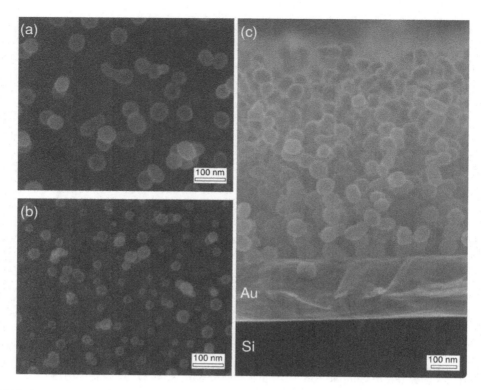

Figure 2.16 SEM images of TiO$_2$ nanoparticles produced at O$_2$ flow rate of 0.050 sccm and −50 V mesh bias **(a)** and −1 V mesh bias **(b)**. The nanoparticle size is smaller for less negative mesh bias. A weak contrast can be seen from the underlaying gold-coated substrate. Cross-sectional SEM image of a sample with nanoparticles produced at a O$_2$ flow rate of 0.15 sccm and a deposition time of 20 min **(c)**. The abrupt changes from gold coating to nanoparticles are clear. (From Gunnarsson, R. et al., *J. Nanopart. Res.*, 17, 353–367, 2015.)

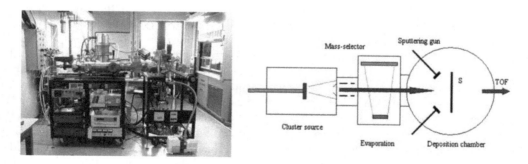

Figure 2.17 Nanocluster deposition system. (From Qiang, Y. et al., *J. Nanopart. Res.*, 8, 489–496, 2006.)

2.2.1.5 ION-BEAM-BASED METHOD

Like the sputtering cluster beam method, ion-beam-based technology is another type of energetic beam-based method for preparing nanoparticles. The process can be performed in various ways. In one case, an ion beam is used to bombard a target, like sputtering method. The resulting species can then be converted to the films or nanoparticles. The atmosphere can be the inert gas or the reactive gas atmosphere. When the gas is inert, the process is in

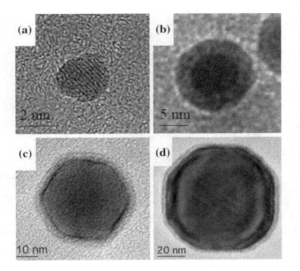

Figure 2.18 HRTEM micrographs of the oxide-coated Fe clusters with diameter from around 3 to 85 nm prepared on carbon microgrids. **(a)** Fe cluster with diameter around 3 nm, no core-shell structure, **(b)** Fe cluster with diameter circa. 10 nm, with core-shell structure, **(c)** Fe cluster with diameter around 40 nm, with core-shell structure, **(d)** Fe cluster with diameter circa. 85 nm, with core-shell structure. (From Qiang, Y. et al., *J. Nanopart. Res.*, 8, 489–496, 2006.)

PVD mode. If the gas is a reactive gas, it is in CVD mode as there is chemical reaction occurring.

Figure 2.21 is one example of ion beam evaporation tool for the preparation of nanoparticles [34].

The ion beam is generated in the vacuum (10^{-4} Torr) and the target chamber is filled with gas (reactive gas like O_2 or N_2) at high pressure (1–100 Torr). A film is used to separate ion beam generation from the sample preparation chamber, no substrate is used, and nanoparticles are collected in the mesh. When ion beam hits the target, the target material is evaporated to form plasma and the vaporised material condenses in the surrounding gas atmosphere to form nanoparticles, as shown in Figure 2.22a. When the surrounding gas is O_2 or N_2, corresponding oxide or nitride nanoparticles are formed as shown in Figure 2.22b, which shows TEM image of TiN nanoparticles (mostly in the diameter of 5–20 nm) produced by use of 20 shots with Ti (99.5% in purity) and N_2 (1–100 Torr) as the target and the gas, respectively.

Ion beam can be used for ion implantation or ion irradiation. This process has been used for almost four decades to fabricate nanoparticles of dielectric materials (e.g., SiO_2) through the control of ion energy, mass, fluence, current density to be compatible with industrial device fabrication, as shown in Figure 2.23. It is worth mentioning that there is no chemical reaction during the process [35].

There are various other ion beam processing tools, such as cluster beam sputtering [32,36]. However, this method generally operates in low pressure for fundamental studies, and it is not intended for bulk synthesis.

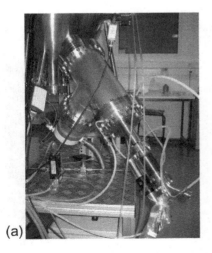

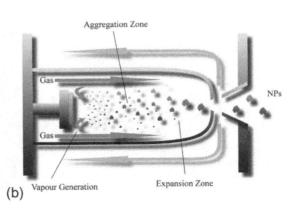

(a) (b)

Figure 2.19 **(a)** The photo of the hardware of sputtering tool and **(b)** the associated schematic to show the principle of producing nanoparticles. (From Verrelli, E. and Tsoukalas, D., *Solid-State Electron.*, 101, 95–105, 2014.)

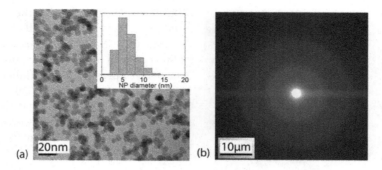

Figure 2.20 **(a)** TEM plan view image of titanium oxide nanoparticles deposited on carbon-coated TEM copper grids. The nanoparticles have a mean diameter of 5 nm as evidenced by the NP size distribution graph in the inset. **(b)** Diffraction pattern taken on a specimen similar to the one shown in **(a)** but with higher density of nanoparticles, showing that the TiO_2 nanoparticles are mainly amorphous. (From Verrelli, E. et al., *Appl. Phys. Lett.*, 102, 22909, 2013.)

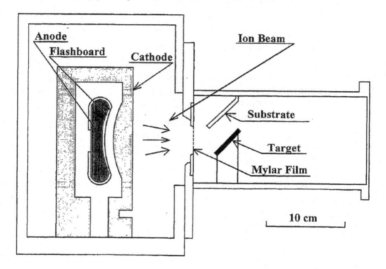

Figure 2.21 Experimental set-up for an ion beam evaporation tool. (From Yatsui, K. et al., *Jpn. J. Appl. Phys.*, 36, 4928–4934, 1997.)

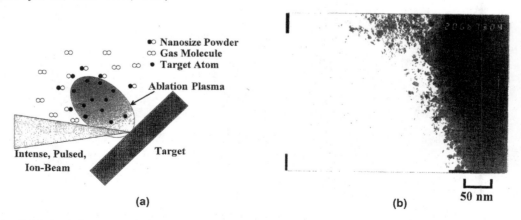

Figure 2.22 **(a)** Schematic of ion beam process mechanism; **(b)** TEM image of nanoparticles TiN (mostly 5–20 nm in diameter). (From Yatsui, K. et al., *Jpn. J. Appl. Phys.*, 36, 4928–4934, 1997.)

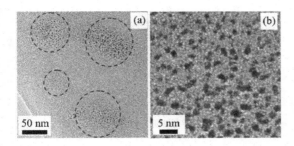

Figure 2.23 **(a)** Bright field TEM micrograph of four completely dissolved nanocrystals at a fluence of 8 × 1016 cm−2. **(b)** Detailed representation of a region taken from the micrograph **(a)**, showing monodispersed Au nanoparticles (appr. 2.0 nm). (From Rizza, G. et al., *J. Appl. Phys.*, 101, 14321, 2007.)

2.2.2 Processing methods using liquid or vapour precursors

Instead of solid precursors, liquid or gaseous precursors can be used for the preparation of ultrafine particles. The process depends on the type of precursor, energy input method, and the morphology of the resulting nanoparticles. As mentioned earlier, the discussion will also cover processes referred to as CVS or CVC in the literatures.

2.2.2.1 THERMAL CVD METHOD

Conventional CVD process for the deposition of films is well developed, and it can also be used for the preparation of nanoparticles. The liquid or gaseous precursor is delivered to the reactor as a vapour. The common precursors used are inorganic precursors based on halides. The reactor can be based on hot-wall reactor or fluid-bed-style reactor.

Figure 2.24 shows an example of a hot-wall style reactor. $FeCl_2$ is vaporised by heating at 900°C, and then carried into the reaction zone by Ar to react with H_2 at 900°C for the formation of Fe nanoparticles [37a] under atmospheric pressure; similarly, liquid precursor like $Zn(NO_3)_2$ can also be used through mist-CVD method to prepare ZnO nanoparticles, with air as the reactive gas. Figure 2.25 shows a TEM micrograph of ZnO nanoparticles synthesised via CVD [37b].

Figure 2.26 shows a fluid-bed style reactor; the gases are fed from two different directions. Nitrogen gas from the bottom is necessary to fluidize the particle bed; the upper tube feeds ammonia gas and contains a thermocouple to measure the temperature inside the particle bed. This provides a good route to form coated nanoparticles [38].

This process can take advantage of precursor chemistry or delivery method of the conventional CVD process, and can be used to produce various types of nanoparticles, including single-composition, multi-component nanoparticles. The properties of nanoparticles depend on the composition and structures. The process parameters, like effect of reaction temperature and precursor concentration, can affect the particle size [37a].

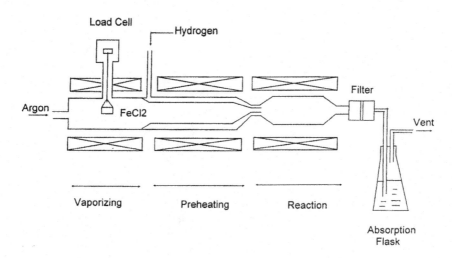

Figure 2.24 Schematic representation of experimental apparatus of CVD process. (From Park, K.Y. et al., *Aerosol Sci. Technol.*, 28, 215–223, 1998;)

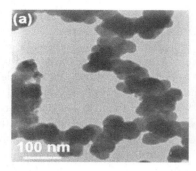
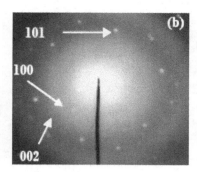

Figure 2.25 **(a)** Bright field TEM image and **(b)** the corresponding diffraction pattern for ZnO nanopowder. (From Singh, P. et al. *Optical Mater.*, 30, 1316–1322, 2008.)

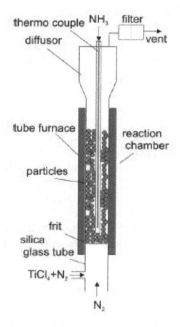

Figure 2.26 Schematic of the vertical fluid-bed reactor. (From Brust, S. et al., *Open J. Appl. Sci.*, 6, 260–269, 2016.)

2.2.2.2 MOCVD METHOD

As mentioned in the previous chapter, MOCVD method uses metal organic precursor(s). MOCVD process can be a thermal process, in which a metal organic precursor decomposes to form nanoparticles [39], or via flame-assisted aerosol process. This part focuses on the former, and the latter will be discussed in the flame synthesis section.

A typical example is the growth of InAs III-V quantum dots (QDs), in which tertiarybutylarsine

(TBAs) was used as (V) precursor to replace AsH_3 due to the high toxicity of AsH_3, and trimethylindium (TMIn) was used as the group III source. The growth was performed using a low-pressure MOCVD system using GaAs(100) substrate with 2° off towards (011). The growth of InAs QDs was affected by the V/III ratio and growth temperature, Figure 2.27 shows the AFM images of InAs QDs grown at 500°C with three different V/III ratios of 21, 42, and 84, respectively, showing the increased size of QDs with the increase of V/III ratio [40].

MOCVD can be used for the synthesis of various III-V materials, even on a large scale [41].

A modified MOCVD process is the aerotaxy process, as shown in the schematic process in Figure 2.28. The low-melting-point group III metal (Ga, or In) is heated in the first furnace with a temperature between 1090°C–1130°C. H_2 is used as a carrier gas to form the metallic aerosol consisting of ≤10 nm metallic nanoparticles, which are then charged by a diffusion charger placed after the furnace. The particle size selection takes place in a differential mobility analyser (DMA). Subsequently, the size-selected metal aerosol is mixed with a group-V containing gas and sent into a second furnace for reaction. The group-V gas flow is low as compared with the carrier gas flow to avoid altering the aerosol flow conditions. The ratio of the number of group-V molecules to the total number of metal atoms in the reaction zone is controlled at the order of 10^6 and 10^7. The group V gas thermally decomposes to react with the metal aerosol to form nanoparticles at the temperature between room temperature to the desired temperature depending on the group-V gas. A second DMA is used to

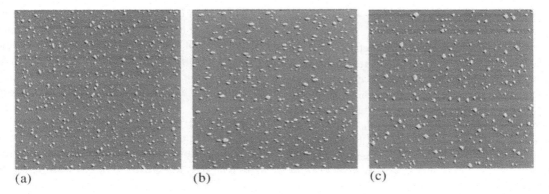

Figure 2.27 AFM images of self-assembled InAs QDs grown at 500°C with V/III ratio of **(a)** 21, **(b)** 42 and **(c)** 84. Scan area is 4 mm 4 mm . (From Zhang, X. et al., *J. Cryst. Growth*, 268, 420–425, 2004.)

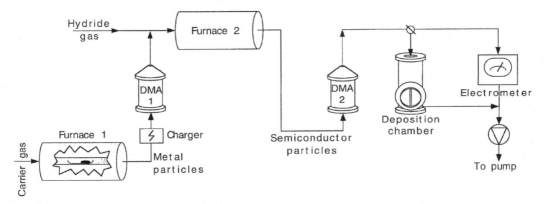

Figure 2.28 Schematic of the aerotaxy process for the processing of GaN. (From Deppert, K. et al., *J. Aerosol Sci.*, 29, 737–748, 1998.)

measure the size distribution of nanoparticles to monitor the reaction process, and an electrometer can be used to determine the particle concentrations, while the mean particle diameter can be controlled through a mobility scanner. The resulting nanoparticles may be deposited on a substrate by means of an electric field. Figure 2.29 shows an example TEM micrograph of InP nanoparticles [42]. As the aerotaxy approach uses the Hydride (AsH_3 or PH_3) to react with metal droplet without contacting substrate, and it can choose the intended size, compared with other methods (like MOCVD method), this approach presents a simple, reliable, and efficient aerosol route for size-selected, contamination-free nanocrystals of III-V semiconductor material in the size range below 20 nm. As the process needs charging nanoparticles, and the diffusion charging is inefficient for nanoscale particles, the process is still mainly for research purposes.

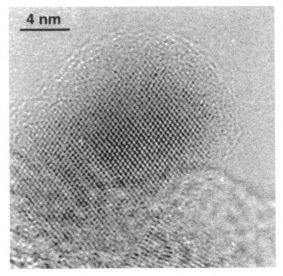

Figure 2.29 TEM micrograph of an InP nanoparticle produced at 512°C. (From Deppert, K. et al., *J. Aerosol Sci.*, 29, 737–748, 1998.)

2.2.2.3 SPRAY PYROLYSIS/AEROSOL-ASSISTED CVD

In this process, small aerosol droplets of the liquid precursor are delivered into a hot reactor to form nanoparticles. It can also be called scraper releases or aerosol decomposing synthesis, droplet-to-particle conversion. This process expands the range of precursors from the gases, providing flexibility on the selection of precursors, and it can be thermal or plasma-based depending on the energy used. The process involves droplet-generation, solvent evaporation, precursor precipitation, decomposition of precursors within the particles, and/or agglomeration steps. The approach to generate droplets from precursor solution includes rotating disk and rotating cup sprayers, air-assisted generators, ultrasonic droplet generators, and electrostatic process. Figure 2.30 shows an example of spray pyrolysis apparatus used [43a]. It consists of two parts: one part is an atomiser; the other part is a furnace to decompose aerosol [43a].

During the process, the precursor solution, like nitrate in methanol with various concentrations, is atomized at a constant flow rate with an air pressure. The oxide nanoparticles can be single composition or multi-component by choosing appropriate precursors [43]. Figure 2.31 shows TEM micrograph of ultrafine particles of Zn and Co oxide; zinc oxide powders obtained from acetate solutions (0.0067 mol/am^3) give aggregate-free particles of about 15 nm of diameter, and zinc oxide powders from nitrate solutions (0.067 mol/dm^3) consist of particles of about 20 nm, which form spherical aggregates with a medium dieter of 200 nm [43b].

Besides thermal CVD, other types of spray using different energy sources are also developed. Figure 2.32 shows a diagram of electrospray process. It uses electrostatic hydrodynamic forces to break up the liquid that is fed into a capillary into small droplets with the system. An 80-mm inside-diameter platinum capillary tube is set in a point-to-plane configuration with 5-mm gap spacing. High-voltage-induced electrospray produces droplets in the range of 40 nm–1 mm, which can be further reduced by evaporation, yielding 10–30 nm Ag nanoparticles. In this case, the liquid droplets evaporate to form nanoparticles [44a]. This method may have difficulty to produce large-volume ultrafine particles. Another type of electrospray process is electrostatic-assisted aerosol jet decomposition/deposition (EAAJD) process; the principle and hardware of EAAJD is similar to that of electrostatic spray-assisted vapour deposition (ESAVD) described in Chapter 1. EAAJD process can be utilised for the preparation of nanoparticles

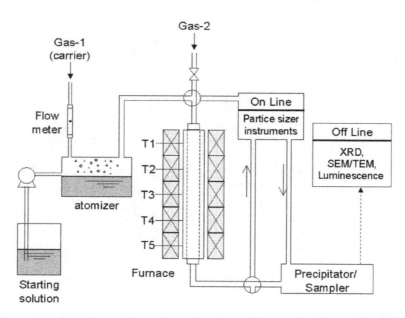

Figure 2.30 Representative spray pyrolysis system used for particle preparation. (From Okuyama, K. and Wuled Lenggoro, I., *Chem. Eng. Sci.*, 58, 537–547, 2003.)

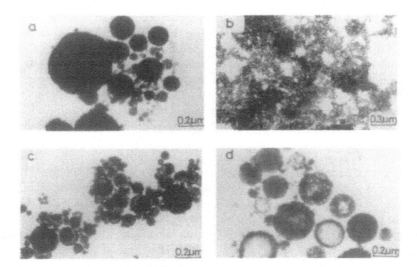

Figure 2.31 TEM micrographs of the powders produced by spray pyrolysis of methanolic solutions: **(a)** 0.067 mol dmm³ zinc nitrate; **(b)** 0.0067 mol dmm³ zinc acetate; **(c)** mixed solution 0.067 mol dmm³ zinc nitrate and acetate; **(d)** 0.067 mol dmm³ cobalt nitrate. (From Carreño, T.G. et al., *Mater. Chem. Phys.*, 27, 287–296, 1991.)

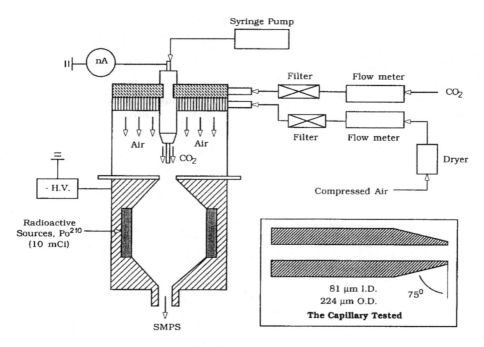

Figure 2.32 A diagram of electrospray process. It consists of an 80-mm inside-diameter platinum capillary tube that is set in a point-to-plane configuration with 5-mm gap spacing. High-voltage-induced electrospray produces droplets in the 40-nm to 1-mm range, which can be further reduced by evaporation, yielding 10–30-nm Ag nanoparticles to be used as quantum dots. In this case, the liquid droplets evaporate to form nanoparticles. (From Chen, D. et al., *J. Aerosol. Sci.*, 26, 963, 1995; Su, B. et al. *Mater. Lett.*, 47, 83, 2001.)

or thin films, for example, for the preparation of CdS semiconducting nanoparticles. Stoichiometric $CdCl_2$ and $(NH_2)_2CS$ aqueous solution with concentrations of 0.005 to 0.01 M were used as the precursors, and aerosols were generated using an ultrasonic aerosol generator at a frequency of 1.7 MHz with nitrogen as a carrier gas (2 l/min). An electrostatic nozzle with electric potential of 5 to 15 kV was used to discharge the aerosol. The charged aerosol was then passed vertically through a heated zone to produce nanopowder. The resulting CdS nanoparticles deposited at 350°C have average diameter of 9.9 nm with a standard deviation of 3.3 nm. This process has been developed on a pilot scale [44b].

Plasma spray is another type of spray pyrolysis method; a schematic diagram is shown in Figure 2.33 [45].

2.2.2.4 FLAME SYNTHESIS

Flame aerosol process was used in prehistoric and ancient times for producing carbon black for painting on cave walls and Chinese ink artwork. This process involves *in-situ* synthesis of particles in a flame produced by the combustion reactions, instead of externally supplied energy.

Figure 2.34a shows an example of the reactor; different reactant gas mixture configurations can be found in Figure 2.34b [46a]. The precursors are directly delivered to the flame, and the

vapour becomes sufficiently supersaturated during the reaction. This leads to nucleation of thermodynamically stable clusters. One example is shown for flame synthesis of TiO_2 nanoparticles in Figure 2.35 [46b].

The morphology of the products depends on the competition between the times for particle-particle collisions governed by Brownian coagulation and inter-particle coalescence, as shown in Figure 2.36 [12]. The process can give nanoparticles with high purity, well-controlled crystallinity, ultrafine diameter, and narrow size distribution. Various materials, such as oxide SiO_2, Al_2O_3, TiO_2, V_2O_5, Fe_2O_3, ZrO_2, and SnO_2, or their mixed composites, and some non-oxide ceramic particles TiN, TiC, TiB_2, $TiSi_2$, and SiC, have been prepared [12]. The advantage of this process would involve low capital investment with proven scalability. It is used for the production of commodity nanoparticles such as carbon black, fumed silica, alumina, and titania due to its simplicity and high throughput capabilities; however, the poor availability of low-cost volatile precursors limits the process to a few materials (e.g., TiO_2 from $TiCl_4$). More discussion on the industrial production of nanoparticles using this method will be discussed in a later part [4].

2.2.2.5 FLAME SPRAY PYROLYSIS

Different from flame synthesis, in which vapour precursors are delivered to the flame, flame spray pyrolysis (FSP) process uses liquid precursors, as

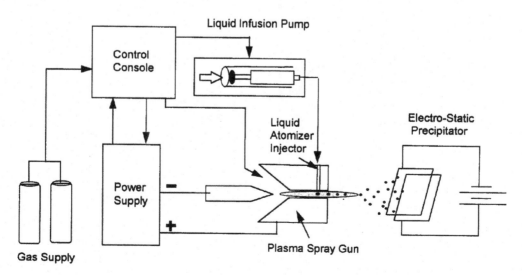

Figure 2.33 A schematic of plasma spray process. (From Karthikeyan, J. et al., *Mater. Sci. Eng-Struct. Mater. Prop. Microstruct. Process.*, 238, 275–286, 1997.)

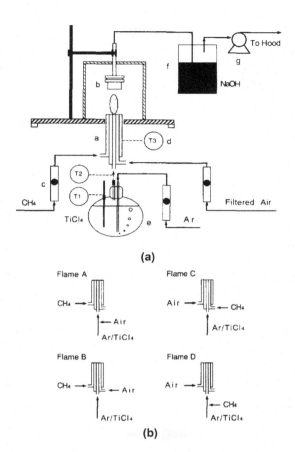

(a)

(b)

Figure 2.34 **(a)** Schematic representation of the experimental flame aerosol set-up and **(b)** reactant gas mixture configuration. (From Pratsinis, S.E. et al., *Powder Technol.*, 86, 87–93, 1996;)

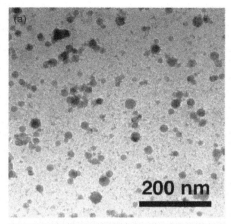

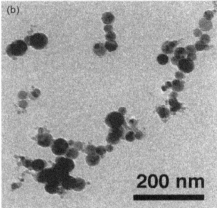

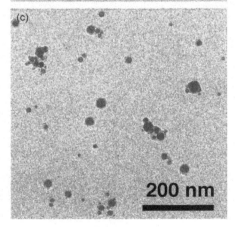

Figure 2.35 TEM pictures of titania particles sampled from the gas phase, **(a)** 10 mm above the burner, **(b)** 250 mm above the burner (in front of the filter) without nozzle and **(c)** 250 mm above the burner with quenching nozzle (oxygen flow rate: 6 l/min, burner–nozzle distance: 15 mm). Flame quenching with the nozzle stops particle growth and preserves particle size and morphology. (From Wegner, K. et al., *Mater. Lett.*, 55, 318–321, 2002.)

shown in Figure 2.37 [12]. It is also called liquid-fed aerosol flame synthesis (LAFS), which is usually called FSP when energy contribution of the liquid precursor is larger than 50%. Otherwise, it is recognized as spray flame synthesis, also known as flame-assisted spray pyrolysis (FASP).

Figure 2.37 is an example of the experimental setup. Liquid precursor and fuel are fed into the torch to generate spray flame. Figure 2.38 shows TEM micrographs of flame-synthesised alumina [47]. As compared with the previously mentioned flame synthesis, this process extends the availability of precursor, as shown in Figure 2.36. The precursors without sufficiently high vapour pressures can now be used, such as hexamethyldisiloane for the production of the silica particles [12]. The preparation of nanoparticles depends on many parameters like the reactor,

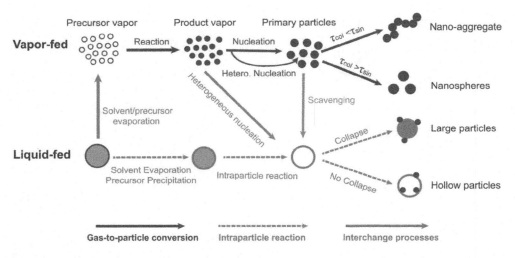

Figure 2.36 Schematic of particle formation mechanism during vapour-fed or liquid-fed flame synthesis, involving gas-to-particle conversion, intraparticle reaction and interchange processes. (From Li, S. et al., *Prog. Energy Combust. Sci.*, 55, 1–59, 2016.)

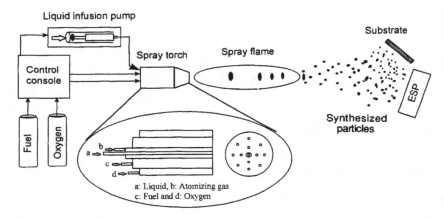

Figure 2.37 A schematic of the Liquid Flame Spray system. (From Karthikeyan, J. et al., *Nanostruct. Mater.*, 8, 61–74, 1997.)

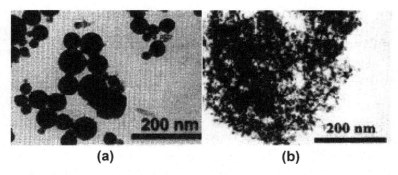

Figure 2.38 Flame spray synthesised alumina (a) and zirconia (b) powders. (From Karthikeyan, J. et al., *Nanostruct. Mater.*, 8, 61–74, 1997.)

precursor, and temperature. For example, conventional spray pyrolysis with furnace reactor often uses acetate salt as the precursors to prepare nanoparticles like MgO and ZnO, but NiO nanoparticles were not reported. In a propane–oxygen diffusion flame, at low flame temperature, nanoparticles were not formed; at high flame temperature (1900°C), either acetate or nitrate salt can give nanoparticles, MgO, ZnO, and even NiO nanoparticles around 30 nm in diameters were prepared [48a]. Many types of materials based on the elements shown in Table 2.2 have been reported [12]. Due to the flexibility and low investment, a broad range of burners/reactors have been developed, as shown in Figure 2.39 [12].

Flame spray process has now been developed for the preparation of special chemicals (like Pt) for the catalyst, electronics, and other applications [48b]. More detail will be covered in the industrial production section.

Table 2.2 The list of element contained in the materials that have now been made through flame spray pyrolysis

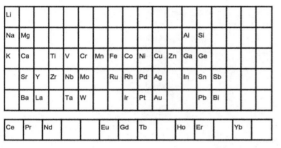

Li														
Na	Mg											Al	Si	
K	Ca		Ti	V	Cr	Mn	Fe	Co	Ni	Cu	Zn	Ga	Ge	
	Sr	Y	Zr	Nb	Mo		Ru	Rh	Pd	Ag		In	Sn	Sb
	Ba	La		Ta	W			Ir	Pt	Au			Pb	Bi
Ce	Pr	Nd			Eu	Gd	Tb			Ho	Er		Yb	

Source: Li, S. et al., *Prog. Energy Combust. Sci.*, 55, 1–59, 2016.

2.2.2.6 LASER PYROLYSIS

Instead of the thermal heat, laser can be absorbed by the reactant or reactant mixture, which is initially excited into non-dissociate states. After energy relaxation, the reactant gas can become hot and decompose as in conventional thermal CVD. This is referred to as laser pyrolytic process. It can use gaseous precursor or liquid precursor or the mixture of gaseous and liquid precursors; the latter can also be named as laser-assisted spray pyrolysis [49].

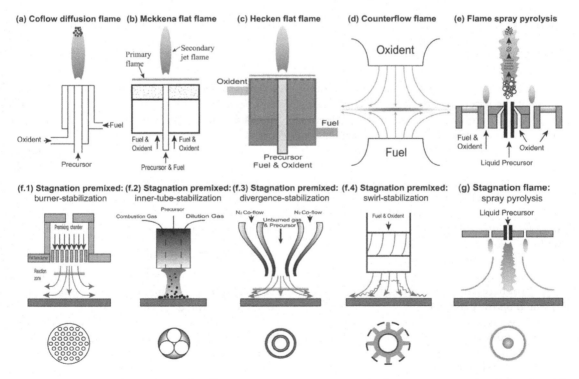

Figure 2.39 Various burners used in flame synthesis. (From Li, S. et al., *Prog. Energy Combust. Sci.*, 55, 1–59, 2016.)

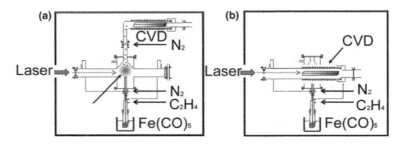

Figure 2.40 Experimental set-ups for the production of iron oxide nanoparticulate films: **(a)** indirect set-up and **(b)** direct set-up. (From de Castro, V. et al., *Thin Solid Films*, 519, 7677–7682, 2011.)

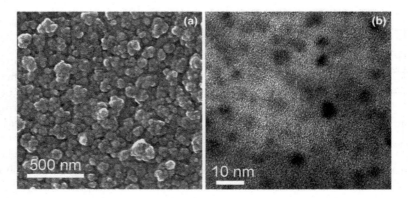

Figure 2.41 Electron microscopy micrographs of coated glasses obtained from the indirect set-up: **(a)** sample IS-M20 (SEM) and **(b)** sample IS-M20 (TEM). (From de Castro, V. et al., *Thin Solid Films*, 519, 7677–7682, 2011.)

During the process, the laser energy is absorbed by one of the precursors or by an inert photosensitizer. Laser pyrolysis process can be performed using the laser only to make nanoparticles, or combined with CVD to form nanoparticles on the surface. An infrared (CO_2) laser is often used.

The set-up for the apparatus varies. Figure 2.40 shows an indirect set-up and direct set-up for the production iron oxide nanoparticulate films, for examples, Figure 2.41 shows SEM and TEM micrographs of iron oxide nanoparticulate films produced by the continuous-wave CO_2 laser-induced pyrolysis of iron pentacarbonyl vapours under N_2 atmosphere and subsequently carried and deposited on the glass [49].

In some cases, UV or vacuum UV (VUV) light is used to irradiate the vapour. Gas discharge lamps, including mercury, deuterium, xenon and microwave-excited krypton, and excimer lasers have been used as the light source, the process is also classified as photothermal synthesis or photo-assisted CVD [50].

This process can be used to produce bulk nanoparticles, composite nanoparticles [51], or the particles in the support [52] or on the surface [53]. It is worth mentioning that laser-assisted process is now also used in the liquid medium to prepare nanoparticles in the liquid in the lab. For example, Ag particles with the diameter ranging from 5 nm to 1.2 μm were prepared though pulsed-laser ablation in supercritical CO_2 medium [18]; this opens another route to produce ultrafine particles.

2.2.2.7 PLASMA-BASED SYNTHESIS

Plasma is another way to provide energy for the reaction. There are many types of plasma processes, including dc plasma jet, dc arc plasma, RF induction plasma, microwave plasma source, and ICP plasma [29]. The previously mentioned sputtering, ion beam, pulsed-laser ablation, arc discharge, and plasma spray are also plasma-based processes as the plasma state is involved. In this part, the focus is on gas-based plasma processes, in which precursor reactions occur in the plasma to lead to the formation of particles [54].

Figure 2.42 shows a type of plasma tool for the preparation of WC using CH_4 as the source of carbon, and Figure 2.43 shows TEM micrograph of the WC nanopowders synthesised at different power level [55]. As-prepared particles from the thermal plasma process contain a small amount of W2C phase, less than about 20 nm in diameter. After posttreatment by H_2, particles become fully carburized WC phase, and the particle size is less

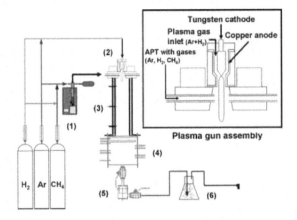

Figure 2.42 Schematic of the plasma reactor system: (1) entrained flow powder feeder for ammonium paratungstate (APT), (2) plasma gun, (3) cylindrical reactor, (4) cooling chamber, (5) powder collector, (6) scrubber and (7) off-gas exhaust system. (From Ryu, T. et al., *J. Am. Ceram. Soc.*, 92, 655–660, 2009.)

than 100 nm in diameter after the posttreatment under the conditions tested. The work shows the advantage of thermal plasma process, such as high processing temperature to vaporise all reactants, high quench rate to form ultrafine powders, and a wide choice of reactants. The process has potential to be scaled up [55].

Plasma processes can be used for the preparation of nitride, carbide, and oxide nanoparticles; carbon nanotubes; and fullerene [28]. However, technological tools and processing conditions vary for individual plasma-chemical synthesis. The plasma process has been used to produce advanced ceramic materials possessing hardness, strength, and high-temperature capability for many applications, pilot plant for the production of TiB_2 ($TiCl_4$ + BCl_3 + H_2 precursor) at the scale of 10 Kg/h. Other particles including SiC and B_4C can be produced from vaporous reactants with plasma processes; these materials are produced at relatively smaller scale compared with large-scale commercial production of TiO_2 and SiO_2, but they are high-value materials needed for some market sectors like composites [56].

2.2.2.8 MBE METHOD

As mentioned in the previous chapter, MBE method is a method for film deposition. With the same operating principles, MBE methods have been used for the productions of nanoparticles on the substrate, on the surface of a support, or nanoparticles in a matrix (as a nanocomposite).

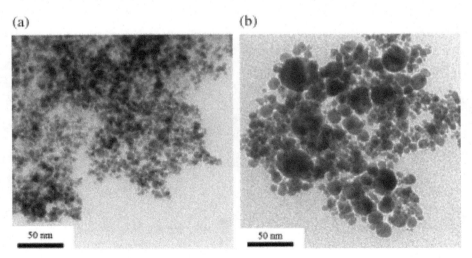

Figure 2.43 TEM micrographs of WC_{1-x} nanopowders synthesised at various power levels of plasma torch: **(a)** 7 kW and **(b)** 29 kW. (From Ryu, T. et al., *J. Am. Ceram. Soc.*, 92, 655–660, 2009.)

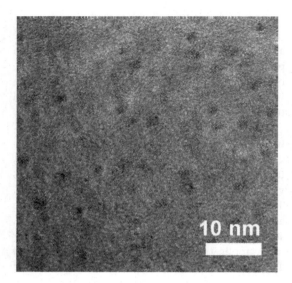

Figure 2.44 TEM picture of 3% ErAs in In0.53Ga0.47As. (From Kim, W. et al., *Nano Lett.*, 8, 2097–2099, 2008.)

The process has a good control for the formation of nanoparticles, as shown in Figure 2.44 [57]. For example, the structure of nanoparticles is different from that of bulk structure, for which, MBE provides a valuable process to grow well-controlled nanoparticles for detailed studies of unusual structures different from bulk structures, as shown in Figure 2.45 [58]. This is mainly from the perspective of fundamental studies. Similarly, MBE method has also been explored for the growing of 1D nanomaterials, for which, catalyst nanoparticles are needed, and MBE method can be used to synthesize nanoparticles with firm control of the size and density. For example, 1nm Fe was evaporated on ZnSe (90nm)/GaAs (001) at room temperature, and followed by heating to 300°C for a few minutes to generate Fe nanoparticles, which were then used to catalyse the growth of ZnSe nanowires through the exposition of Zn and Se fluxes (the flow ratio of 1:2). Simultaneously, for 1 h in an

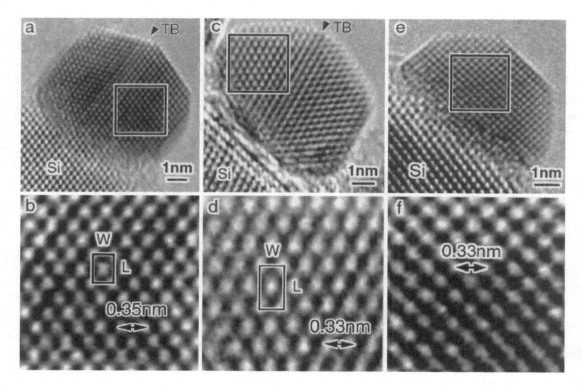

Figure 2.45 HRTEM images of three lead nanoparticles with: **(a)** FCC structure; **(c)** and **(e)** BCT structure. Images **(b)**, **(d)** and **(f)** are enlarged sections of **(a)**, **(c)** and **(e)**, respectively. (From Wu, Y. et al., *Surf. Sci.*, 462, 203–210, 2000.)

MBE chamber, the resulted ZnSe nanowires have the diameters at the tops of the nanowires ranged from 8 to 20 nm, with the typical length about 200 nm and the number density of the nanowires at the order of 10^9 cm^{-2}. The optical property (photoluminescence)of the nanowires differs from that of the bulk crystals [59].

2.2.2.9 IONISATION METHOD

Nanoparticles prepared by CVD method tend to agglomerate due to their Brownian coagulation in the reactor. However, ionisation CVD process has addressed these issues by introducing charged particles, and the size of the particles can be lowered to the size of the ions through controlling the process. One example is the production of the silicon nanoparticles on a Si substrate. The process monitoring shows the negative cluster islands. Negative TEOS and/or TEOS cluster ions were formed by a corona discharge of the mixture of TEOS and O_2 (the mixture flow rate: 2.4 L/min, TEOS concentration: 3.51×10^{-5} mol/L) in the ionizer at 0.3 MPa with the tube furnace temperature at 900°C [60].

Figure 2.46 shows a diagram of Ionization CVD nanoparticle reactor and Figure 2.47 shows a possible mechanism for the formation of nanoparticles. Figure 2.48 shows SEM micrographs of the resulting SiO$_2$ nanoparticles [61]. There are various methods to generate ionised species to form nanoparticles, as shown in Figures 2.49 [62] and 2.50 [63]. These methods have been developed

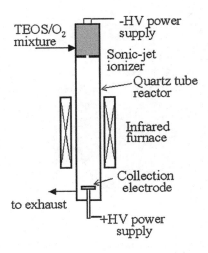

Figure 2.46 Ionisation CVD nanoparticle generator. (From Adachi, M. et al., *Aerosol Sci. Technol.*, 38, 496–505, 2004.)

to generate charged nanoparticles intended to obtain control over the movement of nanoparticles through manipulation of charged nanoparticles, for which the optimisation is needed to give the low particle loss, predictable or high charging efficiency. These will enhance the fundamental research of nanoparticles.

2.2.2.10 ATOMIC LAYER DEPOSITION

The atomic layer deposition (ALD) method was developed to deposit thin films in the 1970s. With the development of the semiconductor industry for high-k dielectric materials, more processes, like deposition of GaAs film, have been developed through ALD method [64]. These processes are also used for the preparation of nanoparticles, such as Ir [65], Pt [66], Pd and Ru nanoparticles [67] on silica or alumina supports for various catalytic applications, and 2–6 nm Cu/Pd alloy [68].

ALD process is a multi-step gas-phase chemical process based on self-limiting surface chemistry, as shown in Figure 2.51 [69]. The deposition of Pt on graphene involves several steps in a cycle; the absorption of precursor, purge, reaction of the precursor, purge, and then more reaction can occur through more cycles. ALD process is a layer-by-layer film deposition process, but it is observed that the deposited Pt forms dispersed particles, growing from individual single atoms and sub-nanometre clusters via the agglomeration in a cubic closely pack manner, instead of layer-by-layer growth.

When ALD method is used to prepare bimetallic nanoparticles, each precursor is deposited through its own cycle. This approach is different from other methods, such as flame synthesis, in which both precursors will be added in the same time. For example, to prepare Pt/Pd bimetallic nanoparticles, Pt layer will be deposited in certain cycles, then Pd layer will be deposited in a desired cycle. In this way, it is easy to control the layer of deposition of individual metal. The composition of resulting nanoparticles depends on the number of each layer [70]. Figure 2.52 shows an example of the Pt-Pd bimetallic nanoparticles [70], which have five core layers of Pd deposition and 15 shell layers of Pt deposition. The diameter of nanoparticles is below 8 nm. Another bimetal nanoparticle is Ru-Pt [13], with diameter of around 1–2 nm developed for the catalysis of CH$_3$OH decomposition reaction. This shows that the Ru-Pt mixed

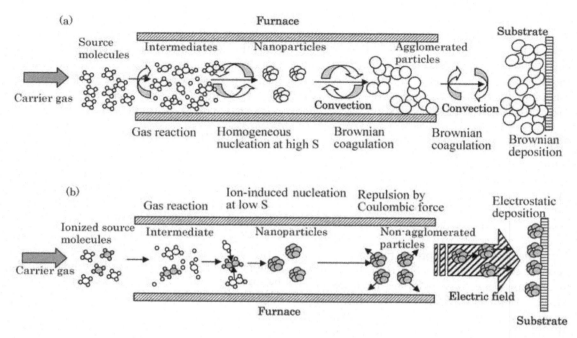

Figure 2.47 Particle formation in **(a)** a common CVD and **(b)** an ionisation CVD reactors. (From Adachi, M. et al., *J. Nanoparticle Res.*, 5, 31–37, 2003.)

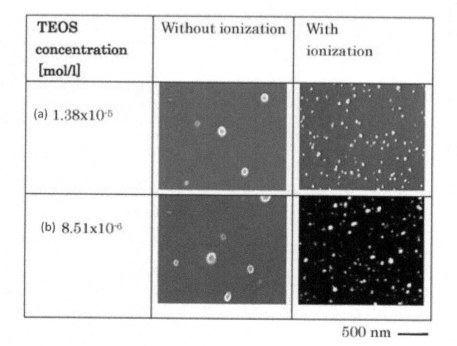

Figure 2.48 SEM micrographs of SiO_2 particles prepared with a reactor temperature of 873 K and reaction time of 5 s. Particles were deposited onto Si wafers for 1 min. For these conditions, the CVD reactor was operated without and with ionisation for TEOS concentrations of **(a)** 1.38×10^{-5} and **(b)** 8.51×10^{-6} mol/l. (From Adachi, M. et al., *J. Nanoparticle Res.*, 5, 31–37, 2003.)

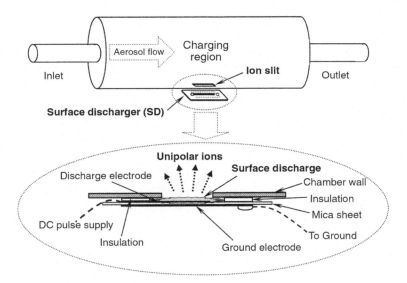

Figure 2.49 Dielectric barrier discharge source emits ions via a slit into particle charging zone. (From Kwon, S.B. et al., *J. Nanopart. Res.*, 9, 621–630, 2007.)

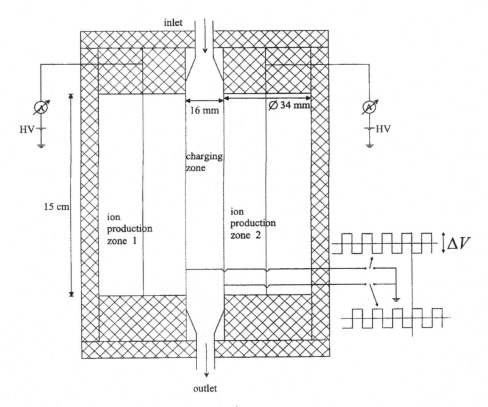

Figure 2.50 Schematic of the twin Hewitt charger. Ions are produced in two cylindrical sections with a central Au wire with a diameter of 25 μm and move through a slit covered with wire mesh into the charging zone. A square-wave electric potential is established between both wire meshes. (From Kruis, F.E. and Fissan, H., *J. Nanopart. Res.*, 3, 39–50, 2001.)

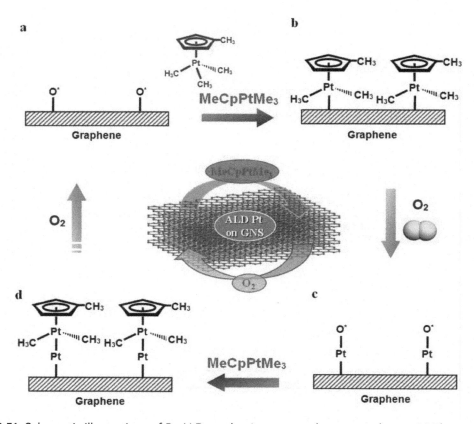

Figure 2.51 Schematic illustrations of Pt ALD mechanism on graphene nanosheets. **(a)** There exists a monolayer of oxygen containing function group on the surface of graphene nanosheets. **(b)** During the MeCpPtMe$_3$ exposure, some of the precursor ligands react with the adsorbed oxygen, and the limited supply of surface oxygen provides the self-limiting growth necessary for ALD, creating a Pt containing monolayer. **(c)** The subsequent oxygen exposure forms a new adsorbed oxygen layer on the Pt surface. These two processes (**b** and **c**) form a complete ALD cycle, producing one atomic layer of Pt atoms. By tuning the number of ALD cycles **(d)**, the Pt deposition can be precisely controlled. Here, '*' represents an active surface species, and Pt-O*represents oxygen molecules (or dissociated oxygen ions) that are adsorbed on the Pt surface. (From Sun, S. et al., *Sci. Rep.*, 3, 1775, 2013.)

metal ALD catalysts yield a higher conversion compared to a physical mixture of monometallic Ru and Pt catalysts, which indicates the Ru-Pt mixed metal ALD yields bimetallics. These results are still in R&D status, but they are encouraging for the development of multicomponent catalysts through ALD method.

ALD method has the advantage of good control on the deposition rate. This is valuable for the deposition of catalyst, particularly when the catalyst is very expensive, such as Pt and Pd metals. The good control on the deposition rate can reduce the excess loading of the catalyst and make use of catalyst more efficiently as they can have good dispersion of the active particles on

substrate. It is then possible to achieve the same activity with less catalyst loading as compared with other methods such as liquid processes with the steps of reflux, centrifugation, and mixing with solvent before pasting on the gas-diffusion layer to thin-film supported catalyst, electrodeposition, and sputtering method for the proton-exchange membrane fuel cells (PEMFCs). Platinum nanocatalyst deposited by ALD on carbon nanotubes has been demonstrated to have a higher utilization efficiency in PEMFCs than commercial E-TEK electrodes [66].

Another related application is to use ALD method to coat nanoparticles to improve their performance. ALD coating of nanoparticles is often

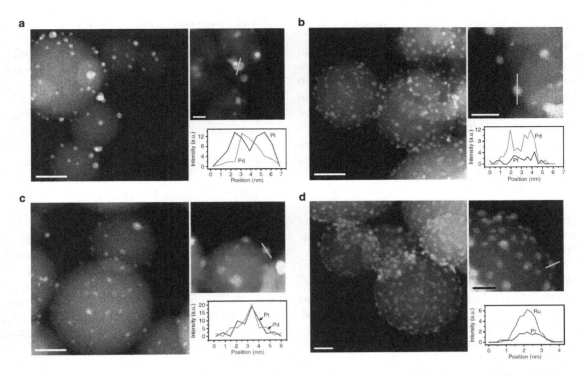

Figure 2.52 Structures of ALD bimetallic nanoparticles. Representative aberration-corrected HAADF-STEM images and corresponding EDS line profiles of **(a)** 5Pd-core 15Pt-shell, **(b)** 12Pt-core 20Pd-rich-shell, **(c)** 12Pt10Pd alloy and **(d)** 1Pt-core 35Ru-rich-shell bimetallic nanoparticles on spherical alumina support; scale bar, 20 nm. (From Lu, J. et al., *Nat. Commun.*, 5, 3264, 2014.)

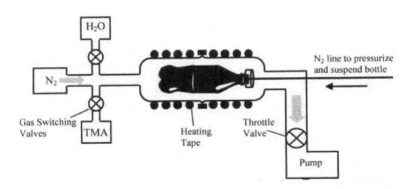

Figure 2.53 Diagram of the viscous-flow reactor used for low-temperature Al2O3 ALD. A special flow tube was employed to accommodate a 510-mL PET bottle. A nitrogen purge line was used to suspend and pressurise the PET bottle during Al2O3 ALD. (From Groner, M.D. et al., *Chem. Mater.*, 16, 639–645, 2004.)

performed using viscous flow reactor; a schematic diagram is shown in Figure 2.53. This type of reactor is different to the one used for thin-film deposition that is shown in Chapter 1. For example, ALD alumina (Al_2O_3) overcoating of supported metal nanoparticles (NPs) effectively reduced deactivation by coking and sintering in high-temperature applications of heterogeneous catalysts [71b]. ALD process is a slow process, and the production of nanoparticles can be a long process [72]. Therefore, ALD process is not suitable for large-scale production of nanoparticles.

2.2.3 Scale-up production

The scale-up production of nanoparticles is an evolved process with the development of technology and the demand from the real world. Carbon black was produced as the filler for rubber used in automobile tires in the late nineteenth century by Cabot in USA by pyrolyzing undesirable by-products of natural gas from oil fields. The more efficient and environmentally friendlier 'furnace process' of flame process in the mid-twentieth century is now a more dominant process utilized by companies such as Cabot, Degussa (currently Evonik), Columbia, Ashi, etc. It is estimated that more than 8 Mt/year of carbon black is produced, of which 70% is for tires and 20% is for rubber application [4].

TiO_2 ultrafine particles have been produced since the 1920s due to the demand from the pigment industry. In the 1930s, sulphate process was used to produce anatase to compete with white lead. In the 1940s, rutile particles were produced for the coating applications to replace anatase for the white colour. In the 1950s, the chloride process was developed to produce rutile as it is whiter and with less waste. Since then, the scale of TiO_2 production has increased tremendously and it has become one of the most important commodities for many industries (e.g., painting). Figure 2.54 shows a schematic production process of TiO_2 from the ores [73].

Figure 2.55 shows SEM images of SiO_2 produced through FSP; particle size depends on the process conditions [74]. Fumed silica (SiO_2) was produced through flame synthesis in the 1940s to function as the 'white filler'. Its success has driven the development of the chloride process for TiO_2 ultrafine particles. Figure 2.56 shows a photo of large flame synthesis reactor used in Degussa (now Evonik) factory [75].

Figure 2.57 shows the schematic production route for fumed silica from $SiCl_4$ [76].

Flame synthesis has been used for the commercial production of many nanoparticles, such as fumed silica, TiO_2, Al_2O_3, carbon black, fine metal (Ni, Pt), ceramics (AlN, SiC, et al.), and cermets (Zn_2SiO_4:Mn for phosphors) [4]. For the industrial production process, a single nanoparticle product is produced using a specific reactor. Other materials such as nano-ZnO, CeOx, Nano-Ag, and fullerene are also produced on the industrial scale, these particles are in many different sectors, including cosmetics (incl. sunscreens), coating and cleaning

agents, paints, cement, fuel catalyst, imaging, textiles, and light conversion for LED/OLED [78].

Ultrafine particles can be produced as commodities on an industrial scale through flame synthesis. Flame-spray process with the flexibility on the selection of precursors is also a scalable process, but the scale of flame-spray process is not as large as flame synthesis. Figure 2.58a shows a schematic diagram of a pilot plant [77a], Figure 2.58b shows a photo for a FSP unit for pilot-scale production of nanoparticles [77b].

Figure 2.59 shows the pilot production plant of nanoparticles at Johnson Matthey Technology Centre. The nanoparticles are produced for the use in catalysis, electronics, and thin-film applications [48].

Another type of scale-up process is arc-discharge-based process for the production of oxide nanoparticles, including ZnO, Al_2O_3, TiO_2, and Fe_2O_3. The process is to use arc-discharge process to evaporate metal; the oxidant is then added during the quench stage for the chemical reaction to form oxide nanoparticles. The dimension of particles can be controlled by adjusting process parameters, such as working gas and quench gas injection position in the system. A schematic process diagram is shown in Figure 2.60 [79a]. For example, 20 nm, 40 nm, and 60 nm non-porous, non-agglomerated ZnO particles can be produced. The scale for various nanoparticles is hundreds of tons per year [79b].

Figure 2.61 shows the current commercial-scale production methods used to produce ultrafine particles extracted from the survey performed with many companies. The choice of particular methods can be related to historical reasons and accessibility to raw materials, not necessarily indicating the superiority or versatility of the techniques. Vapour phase techniques are widely used for the production of metal oxide nanoparticles, like TiO_2, Al_2O_3, and SiO_2, due to the scalability of the process. The fewer sources of contaminants and high temperature involved, the disadvantage is the difficulty in controlling particle growth and preventing agglomeration. Moreover, for the same material, like ZnO nanoparticles, there is difference in the particle size and surface area [79c].

The previous section describes various methods for the preparation of nanoparticles. The choice of method depends on many factors, including: (1) the desired morphology of nanoparticles (bulk nanoparticles, nanoparticles on surface or a support,

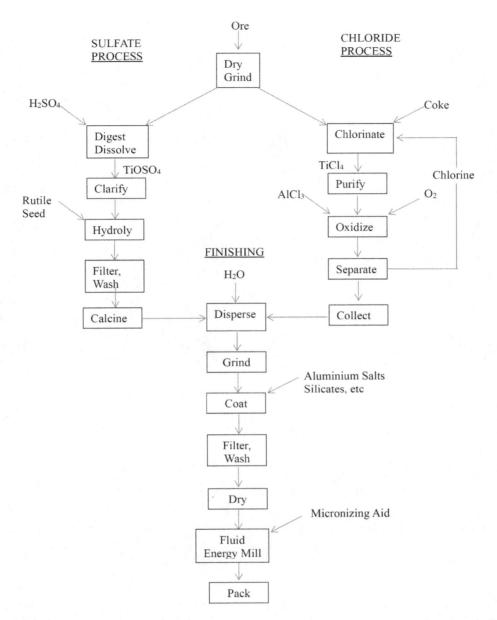

Figure 2.54 Flow charts of titanium dioxide manufacturing processes. (From Braun, J.H. et al., *Prog. Org. Coatings*, 20, 105–138, 1992.)

nanoparticles in a matrix); (2) the availability of the precursor for the material; (3) the compatibility of material and process; (4) production rate; (5) scalability of process; and (6) cost and equipment investment.

There are additional factors, or even sometimes solely one factor, or a combination of factors that lead to a decision of whether to choose one process over another for the synthesis of nanoparticles. Importantly, the first production of ultrafine particles was in the 1880s, like carbon black. While action on the regulation on the health and safety for handling ultrafine particles was taken in the 1970s as reports for safety issues were raised [80a], the related regulation continued to be optimised, and there is research on the safety issue of particles [80b]. More regulation is needed on ultrafine particles; for example, Ultrafine and nanoparticles in Europe are regulated by REACH [80c].

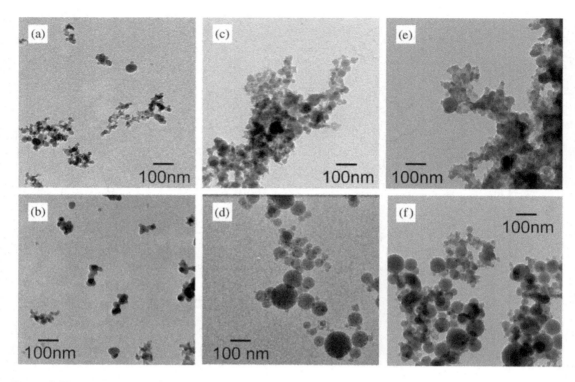

Figure 2.55 Transmission electron micrographs of silica nanoparticles at production rates of 150 g = h (top row) and 300 g = h (bottom row) using 12:5 l = min air (**a, b**) or O_2 as dispersion gas without (**c, d**) and with (**e, f**) additional 25 l = min of O_2 sheath using 1:26 M HMDSO in EtOH. (From Mueller, R. et al., *Chem. Eng. Sci.*, 58, 1969–1976, 2003.)

Figure 2.56 Degussa (Currently Evonik) large-scale industrial flame reactor. (From Roth, P., *Proc. Combust. Inst.*, 31, 1773–1788, 2007.)

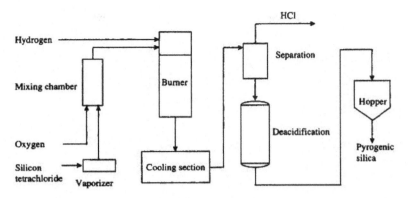

Figure 2.57 Schematic of the fumed silica synthesis by flame hydrolysis. (From Pratsinis, S.E., *Prog. Energy Combust. Sci.*, 24, 197–219, 1998.)

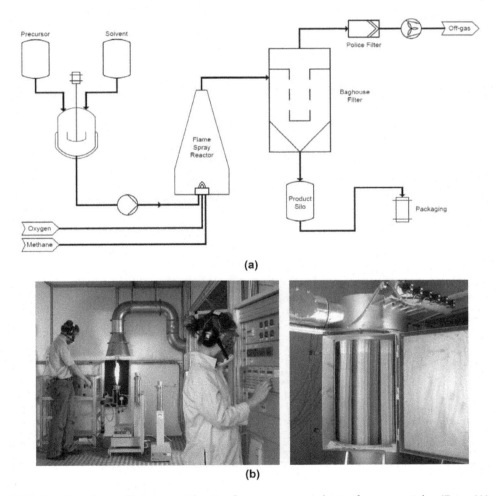

Figure 2.58 **(a)** Flow chart of pilot plant for the flame spray pyrolysis of nanoparticles (From Wegner, K. et al., *KONA Powder Part J.*, 29, 251–265, 2011.); **(b)** flame spray pyrolysis unit for pilot-scale production of nanoparticles up to 1 kg h21: (left) flame reactor and control unit (right) baghouse filter for particle collection. (From Strobel, R. and Pratsinis, S.E., *J. Mater. Chem.*, 17, 4743–4756, 2007.)

Figure 2.59 Pilot plant inside a fully enclosed production space at Johnson Matthey Technology Centre. (From Thiebaut, B., *Platin. Met. Rev.*, 55, 149–151, 2011.)

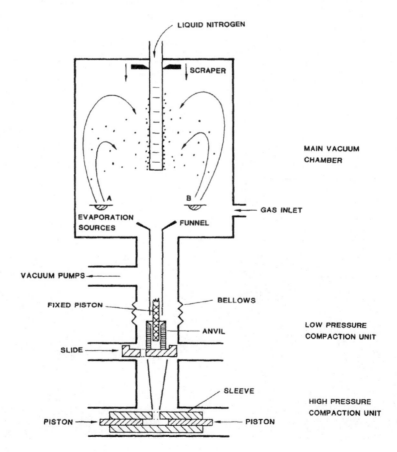

Figure 2.60 Schematic drawing of a gas-condensation chamber for the synthesis of nanophase materials. The material evaporated from sources A and/or B condenses in the gas and is transported via convection to the liquid N_2–filled cold finger. The powders are subsequently scraped from the cold finger, collected via the funnel and consolidated first in the low-pressure compaction device and then in the high-pressure compaction device, all in vacuum. (From Siegel, R.W. and Eastman, J.A., *MRS Proc.*, 132, 3, 1988.)

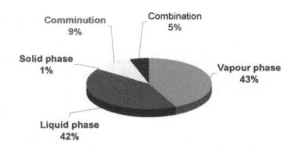

Figure 2.61 Techniques used for the commercial scale production of inorganic nanoparticles. Top-down approach is "Comminution" approach, and bottom-up approaches are "Solid phase", "Liquid phase" and "Vapour phase" approaches, and the "Combination" approach means the combination of those approaches. (From Tsuzuki, T., *Int. J. Nanotechnol.*, 6, 567–578, 2009.)

2.3 SCIENTIFIC AND TECHNICAL ISSUES AND PROCESS CONTROL FOR PRODUCING HIGH-PERFORMANCE NANOPARTICLES FOR APPLICATIONS IN STRUCTURAL, CLEAN ENERGY, BIOMEDICAL, AND ENVIRONMENTAL SECTORS

Ultrafine particles have been prepared via various CVD processes. Some (e.g., TiO_2) are produced on an industrial scale and applied in many sectors including cosmetics (incl. sunscreens), coating and cleaning agents, paints, and cement. This section will focus on the applications of ultrafine particles in structural, clean energy, biomedical, and environmental sectors.

2.3.1 Structural applications

The composition and properties of ultrafine particles vary depending on the structural applications. These include:

1. Ultrafine particles used as fillers in a matrix of materials. A typical example is carbon black used as a filler in the rubber to make tires for automotive and aerospace sectors. There are many examples of ultrafine particles in the matrix of various materials as reinforcements,

such as SiC nanoscale reinforcement in the matrix of alumina and aluminium nitride [80].
2. Ultrafine particles used as a part of composite to improve the functions. For example, TiO_2 nanoparticles, as mentioned earlier that flame synthesis process is used for commercial production, were added to the building materials (e.g., cement, tiles, grouts, sealants, windscreen glass) to obtain advanced functions such as self-cleaning, antibacterial, and anti-fouling [78]. ZnO nanoparticles are introduced in glass coating as UV-barrier for glass [78]. Nanoparticle additives (e.g., metal oxides) are found in polymer nanocomposites to enhance mechanical strength and wear resistance of structural materials [9]. Similarly, CVD synthesis of WS_2 and MoS_2 nanoparticles can be added in the structural nanocomposites to give high impact resistance and they can add extra function such as solid lubricants [81]. As shown in Table 2.3, ultrafine particles of Ni from the

Table 2.3 System/devices with ultrafine particles added as filler and the technical focus

System type	Technical focus
Bulk PZT with Ag	Mechanical reinforcement
$BaTiO_3$ with Ni or Ag	Mechanical reinforcement
Bulk PZT with nanoscale oxide additions	Mechanical reinforcement with retained dielectric properties
$BaTiO_3$ with SiC	Reduction of slow crack growth in bulk ceramics
MgO with $BaTiO_3$	To enhance sinterability of magnesia and to realise a composite possessing ferroelectric properties
Relax or ferroelectrics	Enhanced dielectric and piezoelectric properties arising from nanostructured phase inhomogeneities
PLZT, etc.	Ferroelectric memories and high dielectric constant materials from nanostructured thin films for silicon integration

Source: Cain, M. and Morrell, R., *Appl. Organomet. Chem.*, 15, 321–330, 2001.

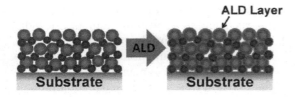

ALD Layer

Figure 2.62 Nanoparticle film reinforced by atomic layer deposition film. (From Dafinone, M. et al., *ACS Nano*, 5, 2078–2087, 2011.)

reduction of NiO, for example, can also be added in the electro-active materials such as piezoelectrics and thin-film devices. The study shows that Ni nanoparticles from the reduction of NiO nanoparticle decorates the grain boundaries of the $BaTiO_3$ grains, 5vol.% Ni additions can give the fracture strength of the composite an increase by a factor of 2.5 times that of the monolithic $BaTiO_3$ [80].

3. Ultrafine particles are part of components for hierarchy structures—for example, nanoparticle film of SiO_2/TiO_2 reinforced by Al_2O_3 film deposited through atomic layer deposition process [82] (Figure 2.62)—or part of porous structures [83a]. Another example is that Al_2O_3 coating (4 monolayer, ~0.8 nm) through ALD process of TMA and H_2O at 180°C on nano MoO_3-based composite anode shows improved mechanical integrity as adhesion to the current collector is significantly improved, which enables enhanced electrical conductivity, and the electrode hardness increases by close to 50% [83b].

2.3.2 Clean energy

There is an increasing demand for clean energy in the modern society, since it covers energy generation and energy storage. In relation to the energy generation, ultrafine particles have been used in various energy devices: For the application of solar cells, TiO_2 nanoparticles have been used in dye-sensitized solar cells (Figure 2.63) [84]. For example, P25 (Evonik-Degussa Aerosil process) TiO_2 nanoparticles have been used to make porous TiO_2 layers to obtain high surface area layer with the possibility of tuning the surface areas and band gap to increase the efficiency of solar cells. From the perspective of manufacturing, P25 TiO_2 nanoparticles have been made through flame aerosol synthesis by Degussa (currently Evonik), as it is an established production process; thus, it is challenging to modify the process to tune the composition of nanoparticles towards the band-gap modification.

For the application of fuel cells, proton exchange membrane fuel cells, also known as polymer electrolyte membrane fuel cells (PEMFCs), are a type of fuel cell being developed for transport applications as well as for stationary and portable fuel cell applications. Currently, Pt nanoparticles are used as catalysts, as nanoparticles exhibit high surface-to-weight ratios, thereby reducing the amount of the costly platinum. For this type of applications, ALD method has an advantage on the controllable deposition as compared with thin-film method and electrodeposition method in terms of simple processing, large area, batch production,

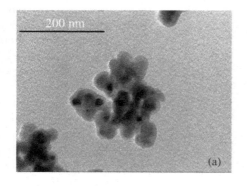

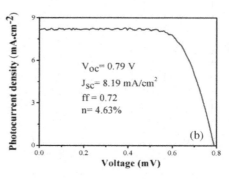

Figure 2.63 TEM micrograph of the core/shell-structured $P25/SiO_2$ nanoparticles **(a)** and photocurrent density-voltage curve of the DSSCs constructed by the core/shell-structured $P25/SiO_2$ nanoparticles **(b)**. (From Niu, H. et al., *Electrochimica Acta*, 81, 246–253, 2012.)

small particle size, lower loading, good uniformity, and conformality. In particular, conformal coating of Pt with controlled particle size is required in the case of micro fuel cells, for lower loading (e.g., 0.016 mg cm^{-2} by ALD as compared to >0.1 mg cm^{-2} in liquid process) and uniformity [66]. As ALD has not been used to make large volume production of ultrafine particles, more research is needed for this application.

Ultrafine particles are also used for energy storage in different formats. A typical example is carbon black, which is used in the Li-ion battery that powers mobile phones [85]. Other examples include Sn-C core-shell particles (~50 nm diameter) prepared from aerosol-through plasma process, which have been used as anodes for Li-ion battery, as-prepared Sn-C nanoparticles cannot give reversable capacity due to Sn volume expansion during lithiation and consequent shell breakage. After acid treatment to partly remove Sn and provide space for expansion, the resulting Sn-C nanoparticles (about 50 nm in diameter) and 6 wt% Sn content show improved cycling performance and initial capacity up to 2000 mAh/g [86], and Co_3O_4 film consisting of nanoparticle deposited on 304 stainless steel substrate (25 μm in thickness) by spray pyrolysis for supercapacitor with a specific capacitance of ~162 F g–1 with a retention capacity of 72.2% after 1000 cycles for a specific current rate of 2.75 A g^{-1} in 6 M KOH electrolyte [87]. Another type of energy storage application is for H_2 storage with the materials composed of Mg nanoparticles and Ni nanoparticles fabricated by gas evaporation method [88].

2.3.3 Biomedical applications

The application of ultrafine particles in the field of biology and medicine is related to the composition. For example, silver is widely used in antimicrobial formulations and dressings. Titania nanoparticles are used for bactericidal purposes in filters, and platinum nanoparticles are used to destroy dangerous toxins and other hazardous organic materials [89].

Different types of nanomaterials (e.g., polymeric nanoparticles, magnetic particles) have been studied for the fields of biology or medicine, including molecular tagging, i.e., fluorescent biological labels and DNA probing, detection of pathogens and proteins, separation and purification of biological molecules and cells, MRI contrast enhancement, drug and gene delivery, tissue

engineering, and tumour destruction via heating (hyperthermia) [89]. Some of the nanoparticle-based medicinal classes approved for clinical use include liposomal drugs and polymer-drug conjugates. Many other nanoparticle-based therapeutics are currently in clinical trials and/or preclinical development, which include metallic nanoparticles (Au/Si, Fe_2O_3) and ceramic particles (silica) and also polymer-based nanoparticles (e.g., PEG-anti TNF-a antibody fragment in Phase III). The applications of metallic nanoparticles include thermo/photothermo therapy and magnetically guided mitoxantrone, and ceramic particles are developed for photodynamic therapy, imaging agents and chemotherapies [90]. Generally, metallic and ceramic nanoparticles towards diagnostics and therapeutic applications can be classified as inside (in vivo) or outside (in vitro) the body, and these applications require the nanoparticles to be stable in water at neutral pH and physiological salinity. Therefore, the dimensions and size of nanoparticles are important to avoid precipitation due to gravitational forces, surface charge, and chemistry. Toxicity is also an important factor to consider when deciding whether nanoparticles are used for in vivo or in vitro applications. Moreover, depending on the applications, nanoparticles can have the advantage over microparticles as nanoparticles can absorb higher power at tolerable AC magnetic fields. This requires uniform particle size and shape for biomedical applications.

The nanoparticles for biomedical applications are mainly prepared through liquid process [89]. CVD-based process mainly focuses on metal and/or ceramic nanoparticles. The particle size and shape uniformity are important factors to be consider when producing nanoparticles for biomedical applications.

For magnetically metallic nanoparticles, spray pyrolysis and laser pyrolysis can be both used for the preparation of Fe_2O_3. The main difference between the two procedures is that spray pyrolysis results in larger particles due to the aggregation of ultrafine particles, while laser pyrolysis results in finer particles as the reaction time is shorter and particles are less aggregated. The aggregation may cause issues for further functionalisation, as shown in Figure 2.64 [91].

Ultrafine particles, in particular, magnetic particles (e.g., Fe_3O_4) have great potential for the biomedical applications, such as Gastromark (i.e., commercial MRI contrast agent comprising

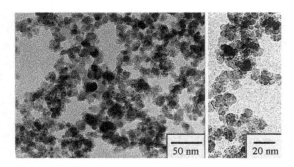

Figure 2.64 TEM micrographs of the synthesised g-Fe$_2$O$_3$ particles. (From Veintemillas-Verdaguer, S. et al., *Mater. Lett.*, 35, 227–231, 1998.)

silicone-coated superparamagnetic iron oxide nanoparticles). Five magnetic particle-based contrast agents have been clinically approved, as listed in Table 2.4 [92]. As there are various methods to prepare ultrafine particles, it is not easy to track the methods used to prepare magnetic particles due to trade secrets. Overall, further research is still needed to understand the physicochemical properties of magnetic particles and *in vivo* performance.

2.3.4 Environmental

Nanoparticles can be used for environmental applications. For each function, a different material would be needed:

1. Environmentally-benign and/or pollution prevention. TiO$_2$ and ZnO nanoparticles are widely used as UV-protection and paint additives. As shown in Figures 2.65 and 2.66 [93], all the coatings were subjected to static immersion conditions and therefore differences in the microbial counts reflected differences in the coatings. An increase in colony counts was observed in bacterial density on control surfaces as, compared to the nanoparticle-based metal oxide coatings, except TiO$_2$ coating at 300°C, on average the attachment of colonies was lower on Nb$_2$O$_5$.

2. Remediation of materials contaminated with hazardous substances. For this application, ZVI has been used to remove pollutants in the soil and water, as shown in Figure 2.67 [94]. An example shows that commercial Fe powder (<10 μm, a specific surface area of 0.9 m^2/g) shows the effect of trichloroethene (TCE) removal (initial concentration 20 mg/L, metal to solution ratio was 2 g/100 mL), but the nanoscale Fe nanoparticles (1–100 nm, surface area of 33.5 m^2/g) show better effect. TCE was completely dechlorinated by the synthesized nanoscale Fe particles within 1.7 h, whereas it still has about 90% TCE after 3h with commercial Fe powder. It must be noted that zero-valent metal can be made through vapour phase or liquid phase process [95].

3. Sensors for environmental agents. There are various types of sensors, such as gas sensors and chemical sensor, for which nanoparticles need to be part of devices to function. For example, In$_2$O$_3$ nanoparticle (2–10 nm in diameter) layer produced through MOCVD process (trimethylindium, H$_2$O as precursor) at 200°C, has shown the lowest detectable air

Table 2.4 The main clinically approved magnetic particles are used in drug delivery

Agent	Commercial name	Particle size (nm)	Target
AM125	AMAG Pharma: Feridex®/ferumoxides; Guerbet: Endorem®; Berlex laboratories: Feridex I.V.® (30/8/1996)	120–180	Liver
Code 7228	AMAG Pharma: Feridex®/ferumoxol (30/6/2009)	~30	Vasculature
SHU555A	Bayer healthcare: Resovist®; Schering AG: Ferucarbotran (2001, Europe, Japan, Australia and China, not in US)	~60	Liver
OMP	Nycomed Imaging: Ferristene/Abdoscan®	>300	Bowel
AM121	AMAG Pharma: GastroMARK®; Guerbet: Ferumoxsil/ Lumirem®	~300–400	Bowel

Source: Mohammed, L. et al., *Particuology*, 30, 1–14, 2017.

Figure 2.65 SEM micrographs of metal oxide coatings on glass substrate **(a)** titania coating on glass HT at 400°C, scale bar is 500 nm; **(b)** niobia coating on glass HT at 400°C, scale bar is 10 μm. (From Dineshram, R. et al., *Colloids Surf. B Biointerfaces*, 74, 75–83, 2009.)

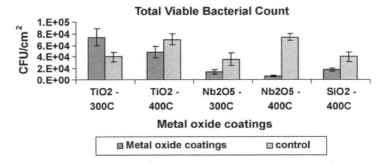

Figure 2.66 Total viable bacterial count (in colony forming units) on metal oxide coatings (1 cm^{-2}) relative to the control (without metal oxide coatings) after static immersion for 15 days. Each value is the mean of duplicate counts (n = 2), and the error bar is the standard deviation. (From Dineshram, R. et al., *Colloids Surf. B Biointerfaces*, 74, 75–83, 2009.)

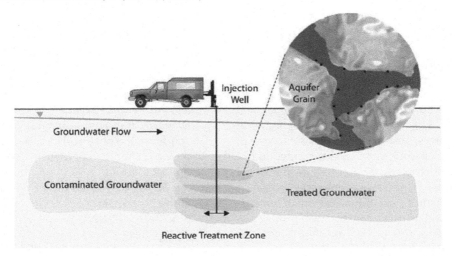

Figure 2.67 An approach to the application of Fe particles for groundwater remediation: a 'reactive treatment zonediation: a 'reactive treatment zone' formed by sequential injection of nanosized Fe to form overlapping zones of particles adsorbed to the grains of native aquifer material. (From Tratnyek, P.G. and Johnson, R.L., *Nano Today*, 1,. 44–48, 2006.)

pollutant NOx concentration, ~200 ppb, and the sensor response towards 200 ppm of NOx is found to be above 10^4. Moreover, the cross-sensitivity against O_2 is very low, indicating that the In_2O_3 nanoparticles are very suitable for the selective NOx detection [96]. As there are many types of environmentally hazardous gases including NO_2, NO, N_2O, H_2S, CO, NH_3, CH_4, SO_2, and CO_2, many semiconducting oxide nanoparticles have been investigated for sensing application. There is good review on this type of sensoring application [97]; one example is WO_3 nanoparticles (smaller than 100nm in diameter) prepared through W filament evaporation under O_2 atmosphere, and the highest sensitivity of 4700 to NO_2 at 1 ppm observed was measured at a relatively low operating temperature of 50°C; with WO_3 nanoparticles as small as 36 nm [98].

In order to explore the applications of ultrafine particles in these sectors, from the scientific and technical perspective, it is required to understand the function of nanoparticles, by which the requirement of nanoparticles can be derived, for example, whether the nanoparticles need to be functionalised and modified in size. Due to the wide variety of nanoparticle type and functions, it is not feasible to discuss all the aspects; however, to obtain high performance of nanoparticles, it is essential to control the process of nanoparticle production with the desired dimension, size, uniformity, purity, and surface function.

2.4 SUMMARY

This chapter has provided an overview on the preparation of ultrafine particles by various CVD technologies. Ultrafine particles can be made in different forms such as particles on a surface or in bulk form. In addition to scalable production of ultrafine particles (e.g., TiO_2, SiO_2) on industrial scale by flame synthesis, flame-spray synthesis with the flexibility on the selection of precursors is a good process for scalable production of nanoparticles that are not produced by flame synthesis such as Pt particles. The combination of scalability, applications and properties means that only a few types of nanoparticles (e.g., TiO_2, SiO_2, Al_2O_3, carbon black) have now been made

at large scale as bulk commodities, and some ultrafine particles can now be made as high-value products on small to medium scales, such as Pt used as catalyst. Considering the progress in the research on the preparation and application of ultrafine particles, it is highly possible that more types of nanoparticles will be produced commercially in the future.

REFERENCES

1. S. Pratsinis, History of manufacture of fine particles in high-temperature aerosol reactor, in *Aerosol Science and Technology: History and Reviews*, Ensor, D.S. (Ed.), RTI International Research Triangle Park, 2011.
2. M. Faraday, The Bakerian lecture: experimental relations of gold (and other metals) to light, *Phil Trans* 147 (1857), pp. 145–181.
3. G. Mie, Contribution to the optical properties of turbid media, in particular of colloidal suspensions of metals, *Ann Phys* 25 (1908), pp. 377–452.
4. (a) W. Stark and S. Pratsinis, Aerosol flame reactors for manufacture of nanoparticles, *Powder Technol* 126 (2002), pp. 103–108; (b) A. Gurav, T. Kodas, T. Pluym and Y. Xiong, Aerosol processing of materials, *Aerosol Sci Tech* 19 (1993), pp. 411–452; (c) H. Hahn, Gas phase synthesis of nanocrystalline materials, *Nanostruct Mater* 9 (1–8) (1997), pp. 3–12.
5. (a) M. Swihart, Vapor-phase synthesis of nanoparticles, *Curr Opin Colloid Interface Sci* 8 (2003), pp. 127–133; (b) H. Hahn, *Gas phase synthesis of nanocrystalline materials*, Nanostructured Mater 9 (1997), pp. 3–12; (c) W. Chang, G. Skandan, H. Hahn, S.C. Danforth, B.H. Kear, Chemical vapor condensation of nanostructured ceramic powders, *Nanostruct Mater* 4 (3) (1994), pp. 345–351.
6. R.P. Feynman, There's plenty of room at the bottom, *Eng Sci* 23 (1960), pp. 22–26,30,34,36.
7. N. Taniguchi, On the basic concept of "nano-technology". *Proceedings of the International Conference on Production Engineering, Part II*, Society for Precision Engineering (1974).

8. M.N. Horenstein, Electrostatics and nanoparticles: What's the same, what's different? *J Electrostat* 67 (2009), pp. 384–393.

9. (a) C.A. Charitidis, P. Georgiou, M.A. Koklioti, A.-F. Trompeta and V. Markakis, Manufacturing nanomaterials: From research to industry, *Manufacturing Rev* 1 (2014), pp. 11–19; (b) J. Chen, G. Lu, L. Zhu and R.C. Flagan, A simple and versatile mini-arc plasma source for nanocrystal synthesis, *J Nanoparticle Res* 9 (2007), pp. 203–213.

10. M. Chhowalla, K.B.K. Teo, C. Ducati, N.L. Rupesinghe, G.A.J. Amaratunga, A.C. Ferrari et al., Growth process conditions of vertically aligned carbon nanotubes using plasma enhanced chemical vapor deposition, *J Appl Phys* 90 (2001), pp. 5308–5317.

11. C. Burda, X. Chen, R. Narayanan and M.A. El-Sayed, Chemistry and properties of nanocrystals of different shapes, *Chem Rev* 105 (2005), pp. 1025–1102.

12. S. Li, Y. Ren, P. Biswas and S. Tse, Flame aerosol synthesis of nanostructured materials and functional devices: Processing, modeling, and diagnostics, *Prog Energy Combust Sci* 55 (2016), pp. 1–59.

13. S.T. Christensen, H. Feng, J.L. Libera, N. Guo, J.T. Miller, P.C. Stair et al., Supported Ru-Pt bimetallic nanoparticle catalysts prepared by atomic layer deposition, *Nano Lett* 10 (2010), pp. 3047–3051.

14. J. Heitmann, F. Müller, M. Zacharias and U. Gösele, Silicon nanocrystals: Size matters, *Adv Mater* 17 (2005), pp. 795–803.

15. (a) A. Zak, Y. Feldman, V. Alperovich, R. Rosentsveig and R. Tenne, Growth mechanism of MoS_2 fullerene-like nanoparticles by gas-phase synthesis, *J Am Chem Soc* 122 (2000), pp. 11108–11116; (b) Y. Feldman, A. Zak, R. Popovitz-Biro, R. Tenne, New reactor for production of tungsten disulfide hollow onion-like (inorganic fullerene-like) nanoparticles, *Solid State Sci* 2 (2000), pp. 663–672.

16. A.H. Pfund, Bismuth black and its applications, *Rev Sci Instrum* 1 (1930), pp. 397–399.

17. (a) Y. Chang, D. Yu, Z. Wang, Y. Long, H. Zhang and R. Ye, Fabrication and abnormal magnetic properties of MnO nanoparticles via vapor phase growth, *J Cryst Growth* 281 (2005), pp. 678–682; (b) W. Bai, K.L. Choy, N.H.J. Stelzer, J. Schoonman, Thermophoresis-assisted vapour phase synthesis of CeO_2 and $Ce_xY_{1-x}O_{2-\delta}$ nanoparticles, *Solid State Ionics* 116 (1999), pp. 225–228.

18. S. Machmudah, T. Sato, Wahyudiono, M. Sasaki and M. Goto, Silver nanoparticles generated by pulsed laser ablation in supercritical CO_2 medium, *High Press Res* 32 (2012), pp. 60–66.

19. M. Ullmann, S. Friedlander and A. Schmidt-Ott, Nanoparticle formation by laser ablation, *J Nanoparticle Res* 4 (2002), pp. 499–509.

20. M. Ashfold, F. Claeyssens, G. Fuge and S. Henley, Pulsed laser ablation and deposition of thin films, *Chem Soc Rev* 33 (2004), pp. 23–31.

21. M. Kim, S. Osone, T. Kim, H. Higashi and T. Seto, Synthesis of nanoparticles by laser ablation: A review, *KONA Powder Part J* (2017), pp. 80–90.

22. B. Liu, Z. Hu, Y. Che, Y. Chen and X. Pan, Nanoparticle generation in ultrafast pulsed laser ablation of nickel, *Appl Phys Lett* 90 (2007), p. 44103.

23. H.W. Kroto, J.R. Heath, S.C. O'Brien, R.F. Curl and R.E. Smalley, C_{60}: Buckminsterfullerene, *Nature* 318 (1985), pp. 162–163.

24. S. Chaitoglou, M. Sanaee, N. Aguilo-Aguayo and E. Bertran, Arc-discharge synthesis of iron encapsulated in carbon nanoparticles for biomedical applications, *J Nanomater* (2014), 9.

25. N. Nakajima, Application 2 – Generation of metal nanoparticles using reactive plasma arc evaporation, in *Nanoparticle Technology Handbook*, Elsevier, Amsterdam, the Netherlands, 2012, pp. 428–431.

26. W. Mahoney, M.D. Kempe and R.P. Andres, Aerosol synthesis of metal and metal oxide nitride and carbide nanoparticles using an arc evaporation source, *Mat Res Soc Symp Proc* 400 (1996), pp. 65–70.

27. (a) S. Ishihara, H. Suematsu, T. Nakayama, T. Suzuki and K. Niihara, Synthesis of nanosized alumina powders by pulsed wire discharge in air flow atmosphere, *Ceram Int*

38 (2012), pp. 4477–4484; (b) W.H. Jiang and K. Yatsui, Pulsed wire discharge for nanosize powder synthesis, *IEEE Trans Plasma Sci* 26 (1998), pp. 1498–1501.

28. S. Iijima, Helical microtubules of graphitic carbon, *Nature* 354 (1991), pp. 56–58.

29. I.S. Tsong and D.J. Barber, Review: Sputtering mechanisms for amorphous and polycrystalline solids, *J Mater Sci* 8 (1973), pp. 123–135.

30. R. Gunnarsson, U. Helmersson and I. Pilch, Synthesis of titanium-oxide nanoparticles with size and stoichiometry control, *J Nanopart Res* 17 (2015), pp. 353–367.

31. Y. Qiang, J. Antony, A. Sharma, J. Nutting, D. Sikes and D. Meyer, Iron/iron oxide core-shell nanoclusters for biomedical applications, *J Nanopart Res* 8 (2006), pp. 489–496.

32. E. Verrelli and D. Tsoukalas, Cluster beam synthesis of metal and metal-oxide nanoparticles for emerging memories, *Solid-State Electron* 101 (2014), pp. 95–105.

33. E. Verrelli, D. Tsoukalas, P. Normand, A.H. Kean and N. Boukos, Forming-free resistive switching memories based on titanium-oxide nanoparticles fabricated at room temperature, *Appl Phys Lett* 102 (2013), p. 22909.

34. K. Yatsui, C. Grigoriu, K. Masugata, W. Jiang and T. Sonegawa, Preparation of thin films and nanosize powders by intense, pulsed ion beam evaporation, *Jpn J Appl Phys* 36 (1997), pp. 4928–4934.

35. G. Rizza, H. Cheverry, T. Gacoin, A. Lamasson and S. Henry, Ion beam irradiation of embedded nanoparticles: Toward an in situ control of size and spatial distribution, *J Appl Phys* 101 (2007), p. 14321.

36. S. Dhara, Formation, dynamics, and characterization of nanostructures by ion beam irradiation, *Crit Rev Solid State Mater Sci* 32 (2007), pp. 1–50.

37. (a) K.Y. Park, H.D. Jang and C.S. Choi, Vapor-phase synthesis and characterization of ultrafine iron powders, *Aerosol Sci Technol* 28 (1998), pp. 215–223; (b) Singh P., Kumar A., Kaur D., ZnO nanocrystalline powder synthesized by ultrasonic mist-chemical vapour deposition, Optical Mater 30 (2008), pp. 1316–1322.

38. S. Brust, A. Röttger and W. Theisen, CVD coating of oxide particles for the production of novel particle-reinforced iron-based metal matrix composites, *Open J Appl Sci* 6 (2016), pp. 260–269.

39. K. Okuyama, Y. Kousaka, N. Tohge, S. Yamamoto, J.J. Wu, R.C. Flagan et al., Production of ultrafine metal oxide aerosol particles by thermal decomposition of metal alkoxide vapors, *AIChE J* 32 (1986), pp. 2010–2019.

40. X. Zhang, J. Dong, S. Chua, J. Zhang and A. Yong, Metalorganic chemical vapor deposition and spontaneous emission of self-assembled InAs quantum dots in open space and in a planar microcavity, *J Cryst Growth* 268 (2004), pp. 420–425.

41. A. Yella, H.A. Therese, N. Zink, M. Panthöfer and W. Tremel, Large scale MOCVD synthesis of hollow ReS2 nanoparticles with nested fullerene-like structure, *Chem Mat* 20 (2008), pp. 3587–3593.

42. K. Deppert, M.H. Magnusson, L. Samuelson, J.-O. Malm, C. Svensson and J.-O. Bovin, Size-selected nanocrystals of III-v semiconductor materials by the aerotaxy method, *J Aerosol Sci* 29 (1998), pp. 737–748.

43. (a) K. Okuyama and I. Wuled Lenggoro, Preparation of nanoparticles via spray route, *Chem Eng Sci* 58 (2003), pp. 537–547; (b) T.G. Carreño, A. Mifsud, C.J. Serna and J.M. Palacios, Preparation of homogeneous ZnCo mixed oxides by spray pyrolysis, *Mater Chem Phys* 27 (1991), pp. 287–296.

44. (a) D. Chen, D.Y. Pui and S.L. Kaufman, Electrospraying of conducting liquids for monodisperse aerosol generation in the 4 nm to 1.8 m diameter range, *J Aerosol Sci* 26 (1995), p. 963; (b) B. Su, M. Wei and K.L. Choy, Microstructure of nanocrystalline CdS powders and thin films by electrostatic assisted aerosol jet decomposition/deposition method, *Mater Lett* 47 (1–2) (2001), p. 83.

45. J. Karthikeyan, C. Berndt, J. Tikkanen, S. Reddy and H. Herman, Plasma spray synthesis of nanomaterial powders and deposits, *Mater Sci Eng-Struct Mater Prop Microstruct Process* 238 (1997), pp. 275–286.

46. (a) S.E. Pratsinis, W. Zhu and S. Vemury, The role of gas mixing in flame synthesis of titania powders, *Powder Technol* 86 (1996), pp. 87–93; (b) K. Wegner, W.J. Stark, S.E. Pratsinis, Flame-nozzle synthesis of nanoparticles with closely controlled size, morphology and crystallinity, *Mater Lett* 2002, 55, pp. 318–321.

47. J. Karthikeyan, C. Berndt, J. Tikkanen, J. Wang, A. King and H. Herman, Nanomaterial powders and deposits prepared by flame spray processing of liquid precursors, *Nanostruct Mater* 8 (1997), pp. 61–74.

48. (a) D.J. Seo, S.B. Sark, Y.C. Kang, K.L. Choy, Formation of ZnO, MgO and NiO nanoparticles from aqueous droplets in flame reactor, *J Nanopart Res* 5 (3–4) (2003), p. 199; (b) B. Thiebaut, Flame spray pyrolysis: A unique facility for the production of nanopowders, *Platin Met Rev* 55 (2011), pp. 149–151.

49. V. de Castro, G. Benito, S. Hurst, C.J. Serna, M.P. Morales and S. Veintemillas-Verdaguer, One step production of magnetic nanoparticle films by laser pyrolysis inside a chemical vapour deposition reactor, *Thin Solid Films* 519 (2011), pp. 7677–7682.

50. S.C. Chan and M.A. Barteau, Preparation of highly uniform Ag/TiO_2 and Au/TiO_2 supported nanoparticle catalysts by photodeposition, *Langmuir* 21 (2005), pp. 5588–5595.

51. H. Maskrot, N. Herlin-Boime, Y. Leconte, K. Jursikova, C. Reynaud and J. Vicens, Blue TiO_2-x/SiO_2 nanoparticles by laser pyrolysis, *J Nanoparticle Res* 8 (2006), pp. 351–360.

52. R. Fernandes, N. Patel, R. Edla, N. Bazzanella, D. Kothari and A. Miotello, Ruthenium nanoparticles supported over carbon thin film catalyst synthesized by pulsed laser deposition for hydrogen production from ammonia borane, *Appl Catal-Gen* 495 (2015), pp. 23–29.

53. A. Honda, T. Kimura, A. Ito and T. Goto, Rh-nanoparticle-dispersed ZrO_2 films prepared by laser chemical vapor deposition, *Surf Coat Technol* 206 (2012), pp. 3006–3010.

54. K. Zaharieva, G. Vissokov, J. Grabis and S. Rakovsky, Plasma-chemical synthesis of nanosized powders-nitrides, carbides, oxides, carbon nanotubes and fullerenes, *Plasma Sci Technol* 14 (2012), pp. 980–995.

55. T. Ryu, H.Y. Sohn, K.S. Hwang and Z.Z. Fang, Plasma synthesis of tungsten carbide nanopowder from ammonium paratungstate, *J Am Ceram Soc* 92 (2009), pp. 655–660.

56. T. Meyer, A.J. Becker, J.F. Edd, F.N. Smith and J. Liu, Plasma synthesis of ceramic powders, *Proceedings of the 8th International Symposium on Plasma Chemistry*, Tokyo 4 (1987), pp. 2006–2011.

57. W. Kim, S.L. Singer, A. Majumdar, J.M.O. Zide, D. Klenov, A.C. Gossard et al., Reducing thermal conductivity of crystalline solids at high temperature using embedded nanostructures, *Nano Lett* 8 (2008), pp. 2097–2099.

58. Y. Wu, Q. Chen, M. Takeguchi and K. Furuya, High-resolution transmission electron microscopy study on the anomalous structure of lead nanoparticles with UHV-MBE-TEM system, *Surf Sci* 462 (2000), pp. 203–210.

59. Y. Ohno, T. Shirahama, S. Takeda, A. Ishizumi and Y. Kanemitsu, Fe-catalytic growth of ZnSe nanowires on a ZnSe,001…surface at low temperatures by molecular-beam epitaxy, *Appl Phys Lett* 87 (2005), p. 43105.

60. M. Adachi, M. Kusumi and S. Tsukui, Ion-induced nucleation in nanoparticle synthesis by ionization chemical vapor deposition, *Aerosol Sci Technol* 38 (2004), pp. 496–505.

61. M. Adachi, S. Tsukui and K. Okuyama, Nanoparticle formation mechanism in CVD reactor with ionization of source vapor, *J Nanoparticle Res* 5 (2003), pp. 31–37.

62. S.B. Kwon, H. Sakurai and T. Seto, Unipolar charging of nanoparticles by the surface-discharge microplasma aerosol charger (SMAC), *J Nanopart Res* 9 (2007), pp. 621–630.

63. F.E. Kruis and H. Fissan, Nanoparticle charging in a twin Hewitt charger, *J Nanopart Res* 3 (2001), pp. 39–50.

64. R.L. Puurunen, A short history of atomic layer deposition: Tuomo Suntola's atomic layer epitaxy, *Chem Vap Depos* 20 (2014), pp. 332–344.

65. R.J. Silvennoinen, O.J.T. Jylha, M. Lindblad, H. Osterholm and A.O.I. Krause, Supported iridium catalysts prepared by atomic layer

deposition: Effect of reduction and calcination on activity in toluene hydrogenation, *Catal Lett* 114 (2007), p. 135.

66. C. Liu, C. Wang, C. Kei, Y. Hsueh and T. Perng, Atomic layer deposition of platinum nanoparticles on carbon nanotubes for application in proton-exchange membrane fuel cells, *Small* 5 (2009), pp. 1535–1538.

67. M. Lashdaf, T. Hatanpaa, A.O.I. Krause, J. Lahtinen, M. Lindblad and M. Tiitta, Deposition of palladium and ruthenium β-diketonates on alumina and silica supports in gas and liquid phase, *Appl Catal A* 24 (2003), p. 51.

68. A.M. Molenbroek, S. Haukka and B.S. Clausen, Alloying in Cu/Pd nanoparticle catalysts, *J Phys Chem B* 102 (1998), pp. 10680–10689.

69. S. Sun, G. Zhang, N. Gauquelin, N. Chen, J. Zhou, S. Yang et al., Single-atom catalysis using Pt/Graphene achieved through atomic layer deposition, *Sci Rep* 3 (2013), p. 1775.

70. J. Lu, K. Low, Y. Lei, J.A. Libera, A. Nicholls, P.C. Stair et al., Toward atomically-precise synthesis of supported bimetallic nanoparticles using atomic layer deposition, *Nat Commun* 5 (2014), p. 3264.

71. (a) M.D. Groner, F.H. Fabreguette, J.W. Elam and S.M. George, Low-temperature Al2O3 atomic layer deposition, *Chem Mater* 16 (2004), pp. 639–645; (b) J. Lu, B. Fu, M.C. Kung, G. Xiao, J.W. Elam, H.H. Kung et al., Coking- and sintering-resistant palladium catalysts achieved through atomic layer deposition, *Science* 335 (2012), pp. 1205–1207.

72. P. Chalker, S. Romani, P. Marshall, M. Rosseinsky, S. Rushworth and P. Williams, Liquid injection atomic layer deposition of silver nanoparticles, *Nanotechnology* 21 (2010).

73. J.H. Braun, A. Baidins and R.E. Marganski, TiO_2 pigment technology: A review, *Prog Org Coatings* 20 (1992), pp. 105–138.

74. R. Mueller, L. Mädler and S.E. Pratsinis, Nanoparticle synthesis at high production rates by flame spray pyrolysis, *Chem Eng Sci* 58 (2003), pp. 1969–1976.

75. P. Roth, Particle synthesis in flames, *Proc Combust Inst* 31 (2007), pp. 1773–1788.

76. S.E. Pratsinis, Flame aerosol synthesis of ceramic powders, *Prog Energy Combust Sci* 24 (1998), pp. 197–219.

77. (a) K. Wegner, B. Schimmoeller, B. Thiebaut, C. Fernandez and T. Rao, Pilot plants for industrial nanoparticle production by flame spray pyrolysis, *KONA Powder Part J* 29 (2011), pp. 251–265; (b) R. Strobel and S.E. Pratsinis, Flame aerosol synthesis of smart nanostructured materials, *J Mater Chem* 17 (2007), pp. 4743–4756.

78. F. Piccinno, F. Gottschalk and S. Seeger, Industrial production quantities and uses of ten engineered nanomaterials in Europe and the world, *J Nanopart Res* 14 (2012), p. 1109.

79. (a) R.W. Siegel and J.A. Eastman, Synthesis, characterization, and properties of nanophase ceramics, *MRS Proc* 132 (1988), p. 3, doi:10.1557/PROC-132-3; (b) Nanophase Technologies, Nanophase Technologies Corporate History, http://nanophase.com/about-us/history, 2017; (c) T. Tsuzuki, Commercial scale production of inorganic nanoparticles, *Int J Nanotechnol* 6 (2009), pp. 567–578.

80. (a) National Institute for Occupational Safety and Health, Current Intelligence Bulletin 65, Occupational Exposure to Carbon Nanotubes and Nanofibers, 2013; (b) K. Donaldson, V. Stone, A. Clouter, L. Renwick, W. MacNee, Ultrafine particles, *Occup Environ Med* 58 (2001), pp. 211–216; (c) What is REACH?, http://www.hse.gov.uk/reach/whatisreach.htm, http://www.hse.gov.uk/nanotechnology/index.htm, 2017; (d) M. Cain and R. Morrell, Nanostructured ceramics: A review of their potential, *Appl Organomet Chem* 15 (2001), pp. 321–330.

81. L. Rapoport, N. Fleischer and R. Tenne, Applications of WS2 (MoS_2) inorganic nanotubes and fullerene-like nanoparticles for solid lubrication and for structural nanocomposites, *J Mater Chem* 15 (2005), pp. 1782–1788.

82. M. Dafinone, G. Feng, T. Brugarolas, K.E. Tettey and D. Lee, Mechanical reinforcement of nanoparticle thin films using atomic layer deposition, *ACS Nano* 5 (2011), pp. 2078–2087.

83. K. Ariga, Q. Ji, J.P. Hill, Y. Bando and M. Aono, Forming nanomaterials as layered functional structures toward materials nano-architectonics, *NPG Asia Mater* 4 (2012), p. e17. doi:10.1038/am.2012.30.

84. H. Niu, L. Liu, H. Wang, S. Zhang, Q. Ma, X. Mao et al., Significant influence of nano-SiO_2 on the performance of dye-sensitized solar cells based on P25, *Electrochimica Acta* 81 (2012), pp. 246–253.

85. K. Brandt, *Historical development of secondary lithium batteries, Solid State Ion* 69 (1994), pp. 173–183.

86. S. Wakeland, Y. Cui, A. Knapp, M. Richard, J. Phillips and C. Luhrs, Multilayered nanoparticles generated by plasma methods for energy storage applications, *Nanosci Nanotechnol Lett* 4 (2012), pp. 316–322.

87. R. Tummala, R. Guduru and P. Mohanty, Nanostructured Co_3O_4 electrodes for super-capacitor applications from plasma spray technique, *J Power Sources* 209 (2012), pp. 44–51.

88. T. Fujimoto, S. Ogawa, T. Kanai, N. Uchiyama, T. Yoshida and S. Yagi, Hydrogen storage property of materials composed of Mg nanoparticles and Ni nanoparticles fabricated by gas evaporation method, *Int J Hydrog Energy* 40 (2015), pp. 11890–11894.

89. O. Salata, Applications of nanoparticles in biology and medicine, *J Nanobiotech* 2 (2004), p. 3.

90. L. Zhang, F. Gu, J. Chan, A. Wang, R. Langer and O. Farokhzad, Nanoparticles in medicine: Therapeutic applications and developments, *Clin Pharmacol Ther* 83 (2008), pp. 761–769.

91. S. Veintemillas-Verdaguer, M. Morales and C. Serna, Continuous production of γ-Fe_2O_3 ultrafine powders by laser pyrolysis, *Mater Lett* 35 (1998), pp. 227–231.

92. L. Mohammed, H.G. Gomaa, D. Ragab and J. Zhu, Magnetic nanoparticles for environmental and biomedical applications: A review, *Particuology* 30 (2017), pp. 1–14.

93. R. Dineshram, R. Subasri, K.R.C. Somaraju, K. Jayaraj, L. Vedaprakash, K. Ratnam et al., Biofouling studies on nanoparticle-based metal oxide coatings on glass coupons exposed to marine environment, *Colloids Surf B Biointerfaces* 74 (2009), pp. 75–83.

94. P.G. Tratnyek and R.L. Johnson, Nanotechnologies for environmental cleanup, *Nano Today* 1 (2006), pp. 44–48.

95. C. B. Wang and W.X. Zhang, Synthesizing nanoscale iron particles for rapid and complete dechlorination of TCE and PCBs, *Environ Sci Technol* 31 (7) (1997), pp. 2154–2156.

96. C.Y. Wang, M. Ali, T. Kups, C. Rohlig, V. Cimalla, T. Stauden, O. Ambacher, NOx sensing properties of In_2O_3 nanoparticles prepared by metal organic chemical vapor deposition, *Sensor Actuat B* 130 (2008), pp. 589–593.

97. K. Wetchakun, T. Samerjai, N. Tamaekong, C. Liewhiran, C. Siriwong, V. Kruefu, A. Wisitsoraat, A. Tuantranont, S. Phanichphant, Semiconducting metal oxides as sensors for environmentally hazardous gases, *Sensor Actuat B* 160 (2011), pp. 580–591.

98. D. Meng, T. Yamazaki, Y. Shen, Z. Liu, T. Kikuta, Preparation of WO_3 nanoparticles and application to NO_2 sensor, *Appl Surf Sci* 256 (2009), pp. 1050–1053.

3

CVD of nanowires and nanotubes, mass production, and industrialization

CIGANG XU AND KWANG LEONG CHOY

3.1 INTRODUCTION

Nanostructures are referred to the structures with at least one dimension in the range between 1 and 100 nm. Nanotube is used to describe a tube-like structure that has at least 1 dimension (e.g., diameter) in the range of 1–100 nm. In addition, other types of structures like nanowires, nanofibres, nanowires, nanorods, nanobelts, and nanoribbons are also reported as indicated in Figure 3.1 [1]. Nanostructures as 1D nanomaterials have attracted considerable attention due to their peculiar properties and applications [1].

Filament, fibre, whisker, were referred to as the 1-dimensional materials. However, most of these materials are on the micrometre scale, rather than nanoscale. The description of these materials can provide the background information on the development of the field. Recent characterisation of an ancient Damascus sabre has demonstrated the presence of carbon nanotube structures; therefore, it is possible that carbon nanotubes were formed during the thermal cycling and cyclic forging process for making the sabre more than 400 years ago, despite the fact that the craftsman did not have the necessary equipment to verify carbon nanotubes at that time [2]. The knowledge of filamentary growth on non-metal was dated back to 1720, and there was a report on silver filament in 1722. Hair silver has been found on proustite ($3Ag_2S$, As_2S_3) and other related minerals, and there is record on these in the British Museum (Natural History). Many researchers performed the

work to understand how the filamentary material formed in the following years in eighteenth century [3]. In 1877, metals in the form of fine filaments or whiskers grown by the reduction of some of their salts, such as silver from silver sulphide, were reported [4]. In 1880, chemical vapour deposition (CVD) process was developed to deposit carbon for the fabrication of carbon fibres [5]. Further work and review on the filamentary growth of silver was reported in the end of nineteenth century and the early twentieth century [3]. In 1921, the formation of Hg nanofibre of 20 nm was determined by measuring the Brownian motion under optical microscope, and length up to 1 mm was observed through mercury vapour condensation on a glass surface cooled below the melting point of mercury [6]. Additional work on the metal whisker was performed to cause short-circuit in the component of the channel filters used in the carrier telephone systems at Bell Telephone Laboratories in 1948 [7]. Vapour condensation and reduction of metal halide was employed for the formation of metal whisker in the 1950s and 1960s [4], the vapour-liquid-solid (VLS) growth mechanism was proposed for the material that was referred as Si whiskers [8]. During the 1960–1970s, the formation of carbon deposits on nuclear fuel-pins was observed and its investigation has led to the observation of the growth of carbon filaments on various metal surfaces. The growth mechanism was then proposed [9]. The CVD growth of carbon nanotubes was also reported [10].

The emergence of the aerospace industry in the 1960s–1970s has prompted the development

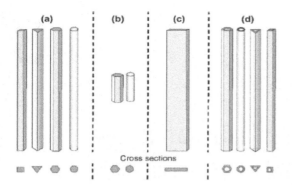

Figure 3.1 Schematic illustration of different 1D nanostructure morphologies and the terms typically used to describe them: **(a)** nanowires (NWs), nanofibres or whiskers; **(b)** nanorods (NRs); **(c)** nanobelts (NBs) or nanoribbons and **(d)** nanotubes (NTs). (From Barth, S. et al., *Prog. Mater. Sci.*, 55, 563–627, 2010.)

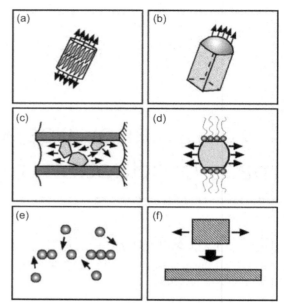

Figure 3.2 Schematic illustration of six different strategies for achieving 1D growth: **(a)** diction by the anisotropic crystallographic structure of a solid; **(b)** confinement by a liquid droplet as in the vapour–liquid–solid process; **(c)** direction through the use of a template; **(d)** kinetic control provided by a capping reagent; **(e)** self-assembly of 0D nanostructures and **(f)** size reduction of 1D microstructure. (From Xia, Y.N. et al., *Adv. Mater.*, 15, 353–389, 2003.)

of composites, which requires light weight, high strength, and resistance to high temperatures. This has driven the research on many types of fibres and commercial production of SiC monofilament fibres (e.g., 100 μm–142 μm in diameter), boron fibres, and the development of many different processes for the growth of various kinds of micron sized fibres [11]. This research and development have provided good foundation for the further work on 1D nanomaterials, such as the VLS growth mechanism that is regarded as the growth mechanism for many nanowires. In the modern literatures, 'whisker' is still used to describe nanowires.

In 1959, Feynman made his famous speech on the nanoscience, and in 1974, Taniguchi started using the 'nano-' prefix [12]. The discovery of C_{60} in 1985 by laser ablation [13], and the development of arc-discharge method for the large-scale production of C_{60} and Iijima's transmission electron microscopy (TEM) work on carbon nanotubes in the 1990s [14,15] accelerated the studies of these structures and other 1D materials in the decades to come, because of distinguished properties and applications that are different from or even superior to their bulk counterparts [6].

Nanostructures can be prepared by top-down or bottom-up approaches as shown in Figure 3.2 using different methods [6]. Among them, CVD method is a very versatile method that is the focus of this chapter; many CVD approaches have been reported, including thermal CVD, plasma-enhanced CVD (PECVD), metal organic

CVD (MOCVD), laser-ablation method, atomic layer deposition (ALD), molecular beam epitaxy (MBE) and chemical beam epitaxy (CBE) [6]. It is worth mentioning that ALD, MBE, and CBE methods and processes were not developed until after 1960s.

The selection of CVD method depends on many factors covering 1D nanomaterial, chemical precursor, substrate material, and the desired growth orientation. In some cases, one type of 1D material can be made by different methods. For example, Si nanowire can be grown via solution method (by controlling the diameter and orientation of solution-grown silicon nanowires) [16], laser-vaporization method [17], CVD [8], and PECVD method [18].

Currently, the proposed growth mechanism of 1D nanomaterials consists of direct vapour, indirect vapour, VLS, and solution-liquid-solid mechanism [6].

3.2 PREPARATION, STRUCTURE, PROPERTIES, AND APPLICATIONS OF NANOSTRUCTURES

3.2.1 Carbon nanotubes

As mentioned, previous work has been reported from the 1950s, including Radushkevich and Lukyanovich in 1952 [19]; Hofer, Sterling, and McCarney in 1955 [20]; Roger Bacon and Bowmann in 1957 [21]; and Lieberman and Morinobu Endo in 1970s [10,22]. Since Iijima's TEM work in 1991 [15], great progress has been made in the synthesis, characterisation, and applications of carbon nanotubes, such as the discovery and synthesis of single-walled carbon nanotubes (SWNTs), first reported in 1993 through arc-discharge process [23,24].

Following the development of carbon nanotube research, carbon nanotubes are generally named according to the number of walls. If only one wall is present, it is named as SWNTs. It is called double-walled carbon nanotubes (DWNTs) or multi-walled carbon nanotubes (MWNTs) when two, three, or more layers (i.e., multi-walled) are present, respectively [15].

Figure 3.3 shows the time line of the important progress in the synthesis of carbon nanotubes since Iijima's TEM work in 1991, and Table 3.1 provides information on various types of carbon nanotubes, their morphology and preparation methods.

Generally, MWNTs can now be prepared with various morphologies, DWNTs can be made in the morphology of bulk materials and bush-style, and SWNTs are in the morphology of bulk materials and bush-style. From the alignment perspective, MWNTs can be vertically aligned due

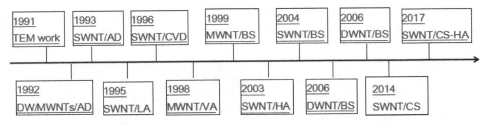

Figure 3.3 The time line of the important progress in the synthesis of carbon nanotubes.

Table 3.1 Different types of carbon nanotubes and their morphology and related preparation processes

Morphology / Nanotube type	Random/Bulk	Alignment		Bushy style
MWNTs	Arc Discharge[25] Laser[26] CVD [27] PECVD[28] Flame synthesis[29]	Vertically	PECVD[30]	PECVD [31] CVD[32] Flame synthesis [29]
DWNTs	Arc Discharge [25] CVD[33]	Horizontally	CVD[34]	CVD[35]
SWNTs	Arc Discharge[23] CVD[36] Laser[26] PECVD[37] Flame synthesis[38]	Horizontally	CVD[40]	CVD[41] PECVD[42] Molecular beam[39]

to a large diameter while DWNTs and SWNTs are horizontally aligned. Figure 3.4 gives representative morphologies of carbon nanotubes [30a,41b,44,45].

As shown in Table 3.1 [25–42], there are many methods to grow different types of carbon nanotubes with various morphologies. These include arc discharge [23], laser-ablation [26], CVD [43], and PECVD [30]. The growth mechanism of carbon nanotubes depends on the growth process and type of carbon nanotubes. For arc discharge method, no catalyst is needed for the growth of MWNTs [25]; however, for the growth of SWNTs [23], metal catalyst (e.g., Fe) is usually required. Different types of carbon materials can be produced by laser ablation, including fullerenes, nanoparticles, and nanotubes depending on the processing conditions [46]. Figure 3.5 shows a proposed mechanism for the growth of various carbon materials [26].

For the CVD-based methods, the use of catalyst is usually needed for the synthesis of carbon nanotubes, in particular, for the synthesis of SWNTs. The mechanism based on the VLS model that was developed to explain the growth of Si whiskers [8] was first used to explain the growth mechanism of carbon filaments (see Figure 3.6) involving a few stages [47–49]. Such VLS mechanism is well received and supported, but some other observations cannot be explained when the dissociation of the carbon precursor is endothermic instead of exothermic, such as alkanes [50–53]. There are factors that limit the indefinite growth of carbon nanotube assemblies in relation to carbon nanotubes with the high density (see Figure 3.7) [54]. The carbon nanotube growth involves the sequential stages of catalyst formation, nucleation, self-organization, evolution, and termination. The specific and underlying causes of growth termination of individual carbon nanotube is still not resolved as it is difficult to monitor through metrology approaches. However, experiments thus far indicate that the growth of carbon nanotubes self-terminates. Possibilities

Figure 3.4 SEM images of different types of carbon nanotubes: **(a)** Random/bulk SWNTs (From Journet, C. et al., *Nature*, 388, 756–758, 1997.); **(b)** vertically aligned MWNTs (From AuBuchon, J.F. et al., *Nano Lett.*, 4, 1781–1784, 2004.); **(c)** horizontally aligned SWNTs on the substrate surface (From Zhou, W. et al., *Nano Lett.*, 6, 2987–2990, 2006; Ding, L. et al., *J. Am. Chem. Soc.*, 130, 5428–5429, 2008.) and **(d)** bush-style SWNTs (From Einarsson, E. et al., *Carbon*, 462008, 923–930, 2008.).

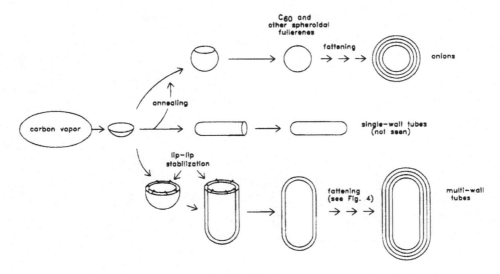

Figure 3.5 Proposed mechanism for various carbon materials. (From Guo, T. et al., *J. Phys. Chem.*, 99, 10694–10697, 1995.)

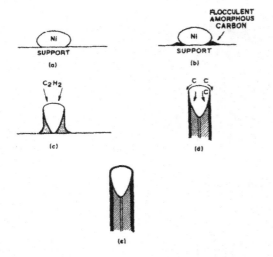

Figure 3.6 Stages in the growth of filaments: **(a)** Ni particle on the surface of the support, **(b)** carbon source gas decomposes on the exposed surfaces of the nickel particle, **(c)** carbon from the decomposed carbon source gas is taken into solution, diffuses down the thermal gradient to be deposited predominantly in the protected regions, and the precipitation of carbon at the rear of the particle builds up a deposit of carbon which forces the particle away from the support, **(d)** the bulk material arises from carbon transported through the particle and the skin by carbon transport around its peripheral surfaces, and **(e)** the particle is completely encapsulated by surface carbon and filament growth ceases as observed on particles which lost their activity. (From Baker, R.T.K. et al., *J. Catal.*, 26, 51–62, 1972.)

include the change of the status of catalyst particles (e.g., coalescence, evaporation), carbonaceous overcoat of catalyst, the formation of inactive carbide phase, insufficient supply of carbon in the process atmosphere, and the interaction of catalyst to the substrate. The combination of *in-situ* and *ex-situ* methods shows that interactions among the growing population of carbon nanotubes govern their collective growth behaviour together with the performance of individual catalyst particles, and these interactions limit the indefinite growth of carbon nanotubes. For this reason, the statistical behaviour of carbon nanotubes can be important for the controllable growth in a collective fashion. This understanding can be valuable for future reactor and process design, to overcome existing limits for the growth of long carbon nanotubes [54].

Regarding the growth of vertically aligned carbon nanotubes [30], with a large diameter (e.g., >20 nm), the carbon nanotube is reasonably upright as shown in Figure 3.4b. Whereas for the growth of a smaller diameter (e.g., <20 nm), carbon nanotube is less vertically aligned. The process can involve plasma or electric bias to provide alignment as the movement of catalyst particle is affected by plasma or electric field [55], and the diameter of carbon nanotubes depends on the process conditions, as shown in Figure 3.8 [30c]. Under the similar processing conditions, thermal CVD-based process can only produce curly carbon nanotubes [31]. The plasma-based method can be used to create various

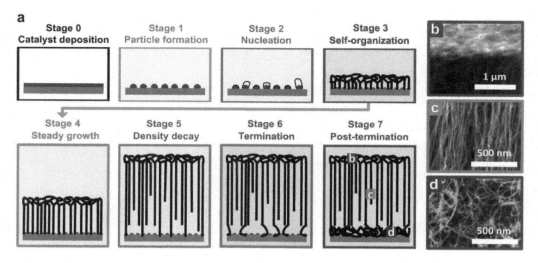

Figure 3.7 Stages for the growth of high-density carbon nanotubes and collective mechanism of CNT forest growth: **(a)** sequential stages of catalyst formation, nucleation, self-organization, evolution and termination; **(b, c, d)** SEM images of representative morphology of top, middle and bottom of a CNT forest. (From Bedewy, M. et al., *J. Phys. Chem. C*, 113, 20576–20582, 2009.)

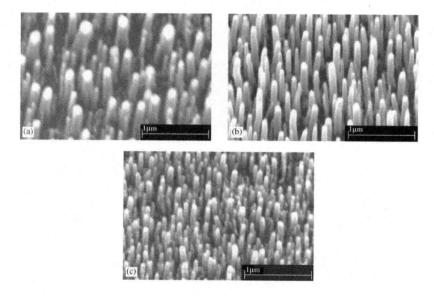

Figure 3.8 SEM images of vertically aligned carbon nanotubes: **(a)** CNT sample with average out-diameter of 104 nm; **(b)** CNT sample with average out-diameter of 90 nm and **(c)** CNT sample with average out-diameter of 55 nm. The samples are tilted by 25° to reveal the forest-like landscapes. (From Qi, H.J. et al., *J. Mech. Phys. Solids*, 51, 2213–2237, 2003.)

morphologies of carbon nanotubes [56], whereas electric field can be applied to produce bent carbon nanotubes [55].

As mentioned earlier, the growth of SWNTs usually needs the use of catalyst particles [24], and the size of particles can play an important role in controlling the diameter of SWNTs. Figure 3.9

shows the mechanism from a molecular study [57]. For the growth of SWNTs without alignment, the catalyst can be in a form of a floating catalyst [58a], or in a support [58b]. The growth mechanism of SWNTs is still not fully understood.

Regarding the growth of DWNTs, its diameter and growth rate is similar to that of SWNTs, except

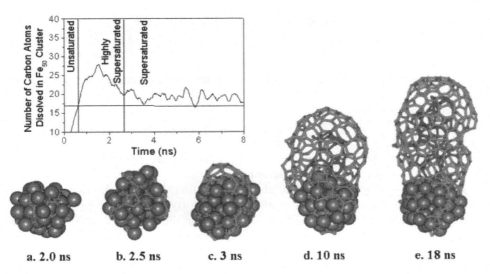

Figure 3.9 Proposed mechanism from molecular study of carbon nanotubes. (From Ding, F. et al., *J. Phys. Chem. B*, 108, 17369–17377, 2004.)

that the presence of H_2 in a certain ratio is required in order to favour the growth of DWNTs rather than SWNTs [33]. In relation to the growth of high-density of bushy-style SWNTs, the presence of H_2O vapour [41], the oxygen in the processing gas [58], or the component of the precursor (e.g., ethanol) can favour their growth [59]. The understanding is that under such conditions, the activity of catalyst particles can be sustained for the growth.

When SWNTs are grown horizontally, as shown in Figure 3.10 [40,60], various phenomena have been observed. High density of SWNTs can

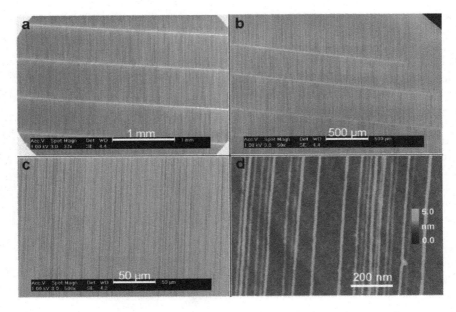

Figure 3.10 SEM and AFM images of high-density and perfectly aligned arrays of long SWNTs along [100] direction on the ST-cut quartz substrate using patterned copper catalyst. The bright stripes in **(a)** and **(b)** correspond to copper catalyst. **(c)** High-magnification SEM image of the arrays of SWNTs. **(d)** AFM image of 1 μm × 0.75 μm area, in which 22 SWNTs have been found. (From Li, Y. et al., *Adv. Mater.*, 22, 1508–1515, 2010.)

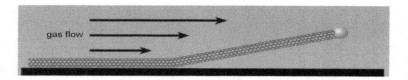

Figure 3.11 Schematic representation of the 'kite mechanism' for the growth of ultralong SWNTs. (From Li, Y. et al., *Adv. Mater.*, 22, 1508–1515, 2010.)

grow on ST-cut quartz substrate, and different types of quartz substrates may give different morphologies of SWNTs [61], indicating that such substrate is essential for the growth of high density of horizontally aligned SWNTs. It is proposed that SWNTs grow in a kite mechanism, as shown in Figure 3.11 [60]. The growth of SWNT can be suspended on the microfabricated catalyst islands (Figure 3.12) [62], or grow across a slit sample with an elevated stage in the middle of the substrate (Figure 3.13) [63], while also as the block with the possible mechanism (Figure 3.14). Recently, it was also shown that SWNT could grow in a flying way such that the crawling style or serpentine shape can be obtained [64,65], whereby the impact of gas flow, gravity, or the interaction of SWNT with substrate could be essential.

From the perspective of structure, carbon nanotubes can be SWNTs, DWNTs, or MWNTs, but it is confirmed that each nanotube is hollow and concentric. The preparation of SWNTs can be more challenging, as individual SWNT can be semiconducting or metallic, different diameters, chiralities, or lengths. Many catalysts have been developed for the growth of SWNTs, Table 3.2 [66–74] gives the list of elements and related materials that are used as the catalysts.

In general, bulk SWNTs are a mixture of semiconducting and metallic SWNTs with different chiralities. Due to the structure-properties relationship of SWNTs, it is desirable to obtain monodisperse SWNTs. For example, SWNTs from high pressure CO (HiPco) process consists of about 60% semiconducting SWNTs, and laser ablation preferentially grows about 70% metallic SWNTs. A remote plasma-based process can give nearly 90% of semiconducting SWNTs in the form of bulk material, indicating highly preferential growth of semiconducting over metallic tubes in the PECVD process [75]. When combining plasma process and fast heating, bush-style SWNTs with about 96%

semiconducting SWNTs can be produced via the pyrolysis of pure C_2H_2 under a low pressure (e.g., ~30 mTorr) [76]. In order to produce horizontally aligned SWNT array, using ST-cut quartz as the substrate, the combination of methanol and ethanol as the process gas, over 95% of the nanotubes in the arrays are semiconducting, and it is proposed that the interaction between the SWNTs and the quartz lattice may play an important role in the

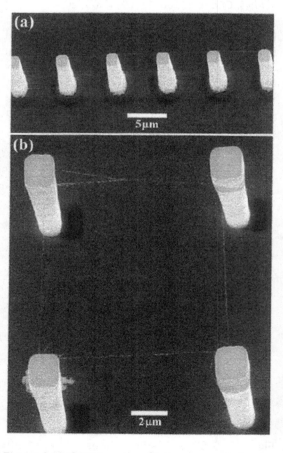

Figure 3.12 SEM images of **(a)** a suspended SWNT power line and **(b)** a square of suspended SWNT bridges. (From Cassell, A.M. et al., *J. Am. Chem. Soc.*, 121, 7975–7976, 1999.)

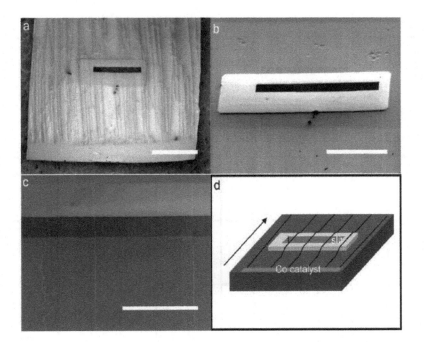

Figure 3.13 SEM images of SWNTs grow across a slit sample with elevated stage in the middle (20 μm in height): **(a)** top view, scale bar 1 mm and **(b)** sample tilted by 45°, scale bar 500 μm. **(c)** High magnification image of the SWNTs in **(b)**, showing the nanotubes clambering over the stage, with the sample tilted by 45°. Scale bar: 50 μm. **(d)** Schematic view of the growth. The arrow shows the gas flow direction. Ethanol with 0.2 wt% water was used as carbon feedstock for the growth. (From Huang, L. et al., *J. Phys. Chem. B*, 110, 11103–11109, 2006.)

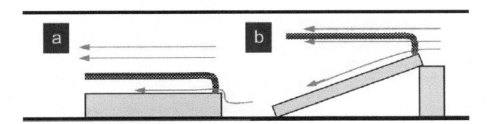

Figure 3.14 Schematic views of a possible mechanism of CVD growth of carbon nanotubes: **(a)** a carbon nanotube first grows upward and gets caught by the stable and ordered laminar flow and then grows in the flow steadily until gravity drags the tube down to the surface; **(b)** by a slight tilting of the sample to raise the catalysts, the carbon nanotube can stay in the gas flow for more time and hence can grow longer. (From Pisana, S. et al., *J. Phys. Chem. C*, 111, 17249–17253, 2007.)

preferential growth of semiconducting SWNTs [77]. In order to obtain a high ratio of metallic SWNTs, Fe nanocatalysts were deposited onto a SiO_2/Si support and *in-situ* annealed in a He or Ar ambient that contains various ratios of H_2 and H_2O, using methane as the carbon source at 860°C. It was found that by tuning the pre-treatment conditions with different noble gases (e.g., He, Ar), the ratio of metallic SWNTs can be increased from one-third to 91%. *In-situ* TEM studies reveal that this variation of pre-treatment would lead to differences in both morphology and coarsening behaviour of the nanoparticles used to nucleate nanotubes, indicating that there is correlation between the catalyst morphology and the resulting nanotube electronic structure [78].

Table 3.2 List of elements and related materials that are used as the catalyst for SWNTs

Mg [66]									Al [66]	Si [67a] SiO$_2$ [68]		
	TiO$_2$ [69]	Cr [66]	Mn [66]	Fe [66]	Co [66]	Ni [66]	Cu [67b]	ZnO [69]		Ge [71]		
	ZrO$_2$ [69]	Mo [66]				Pd [66]	Ag [67c]			Sn [66]		
La$_2$O$_3$ [72]		W [73]	Re [74]			Pt [66]	Au [66]			Pb [60]		

Monodispersed SWNTs may also be obtained via: (1) catalyst design and (2) the templated growth. In the case of catalyst design, the intention is to prepare uniform catalyst particles so that SWNTs with a uniform diameter are prepared. For example, nanoclusters with the formula of $[Hx\text{-}PMo_{12}O_{40}cH_4Mo^{VI}{}_{72}Fe^{III}{}_{30}(CH_3COO)_{15}O_{254}(H_2O)_{98}]\cdot60H_2O$ on Si substrate were reduced to produce Fe-Mo catalyst particles, using CH_4 as process gas, the resulting SWNTs have a narrow diameter distribution, from 0.7 to 1.5 nm with a standard deviation of 17% [79]. Another work is to use the molecular cluster $Na_{15}[Na_3c\{Co(H_2O)_4\}_6\{WO(H_2O)\}_3(P_2W_{12}O_{48})_3]\cdot nH_2O$ (denoted as $\{W_{39}Co_6O_x\}$), where 'c' indicates that the three Na atoms preceding the symbol are encapsulated by the units listed after the symbol, to obtain W-Co alloy nanoparticles used to catalyse SWNT growth on SiO_2/Si substrates via CVD process, SWNTs of a single chirality, (12,6), can be produced directly in abundance (higher than 92%), as shown in Figure 3.15 [73].

Further development has demonstrated that horizontally aligned metallic SWNT arrays can

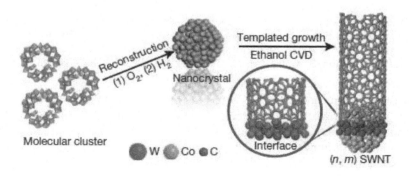

Figure 3.15 Preparation of the W–Co nanocrystal catalyst and the templated growth of a SWNT with specified (n, m). (From Yang, F. et al., *Nature*, 2014, 510, 522–524, 2014.)

be obtained with an average density of more than 20 tubes per micrometre, in which 90% of the tubes had chiral indices of (12,6), and semiconducting SWNT arrays with an average density of more than 10 tubes per micrometre, in which 80% of the nanotubes had chiral indices of (8,4). The nanotubes were grown using uniform-size

Mo_2C and WC solid catalysts by controlling the symmetries of the active catalyst surface, as shown in Figure 3.16 [80].

In the case of templated-growth, the intention is to synthesize organic pieces or use a carbon nanotube to function as the template for the growth of carbon nanotube in the same diameter.

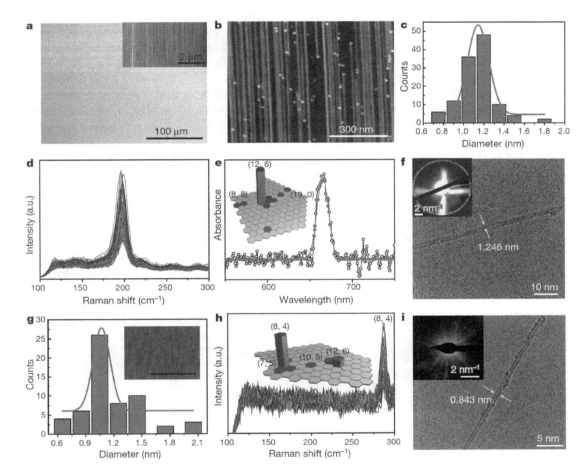

Figure 3.16 Successful growth of (12, 6) and (8, 4) SWNTs arrays using uniform catalysts, and their characterizations. **(a)** SEM image of the as-grown (12, 6) SWNT arrays using Mo_2C catalysts. Inset, a magnified image. **(b, c)** The corresponding AFM image **(b)** and size distribution and Gaussian fit of the (12, 6) SWNT array **(c:** mean, 1.21 nm; standard deviation, 0.26 nm). **(d)** Raman spectra in the RBM region, obtained using line mapping with a scanning step of 1 μm and an excitation of 633 nm. **(e)** Ultraviolet–visible–near infrared absorption spectrum (background subtracted; open circles) of an aqueous dispersion of SWNTs collected from 40 samples and the fitted spectrum (the line with filled circles). Inset, the relative abundances of the various chiralities. **(f)** TEM image of carbon nanotubes. Inset, a typical electron diffraction pattern of a (12, 6) nanotube. **(g)** The size distribution of catalysts using monodispersed tungsten oxide on sapphire (mean, 1.09 nm; standard deviation, 0.46 nm). Inset, the corresponding AFM image; scale bar, 1 μm. **(h)** Raman spectra in the RBM region, obtained using line mapping with a scanning step of 1 μm and an excitation of 633 nm. Inset, the population of the chiral species of the tubes. **(i)** Typical TEM image of the SWNTs. Inset, the corresponding electron diffraction pattern. (From Zhang, S. et al., *Nature*, 543, 234–238, 2017.)

Using a carbon nanotube as the template, early work was performed to cut carbon nanotubes, then through functionalisation and co-ordination steps to deposit catalyst particles at the end of carbon nanotubes, followed by the growth of carbon nanotube with the catalyst particles, and AFM was used to verify the continuous growth phenomenon [81]. When no catalyst is involved, SWNTs were cut and then the carbon nanotube pieces were used as the templates, as shown in Figure 3.17, by tuning the processing conditions, the yield was about 9.3% (56/600). Three observations during this cloning process included the continued growth (the seeds being longer), tube closure (the seeds with no change in length), and SWNT etching (the seeds

becoming shorter). In the second growth, the yield was greatly improved (24/59 = 40.7%) despite all the above three observations were made; however, additional work is needed to understand the origin for the increase in growth efficiency utilizing quartz substrate [82]. Similarly, using purified SWNTs as the template, (7,6), (6,5), (7,7) nanotubes can be obtained [83].

Another route of templating approach is to use organic pieces as the starting template; for example, opened C_{60} can be applied as caps to grow SWNTs, as shown in Figure 3.18, and the cage opening temperature directly affects the diameter distribution of SWNTs. Higher cage-opening temperatures are favourable for growing

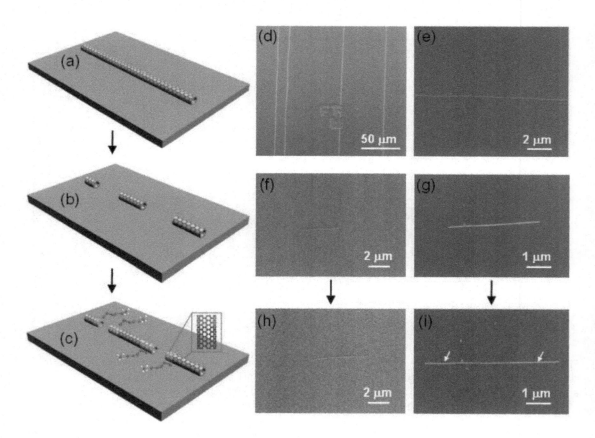

Figure 3.17 **(a–c)** Schematic of as-grown ultralong SWNTs **(a)** on substrate by low feeding gas flow, which could be cut into short segments; open-end SWNTs seeds **(b)**; and the duplicate SWNTs (the section next to the line with the filled circles), which could be continued to grow from the parent SWNTs segments via directly adding Cx (C_2 and/or C_3) radicals to the open-end seed **(c)**. **(d, e)** SEM image and AFM image of ultralong SWNTs used for preparing open-end SWNTs seeds, respectively. **(f, g)** Representative SEM image and AFM image of short parent SWNTs segments for the second growth. **(h, i)** SEM image and AFM image of duplicate SWNTs continued from the SWNTs in panels **(f)** and **(g)**, respectively. (From Yao, Y. et al., *Nano Lett.*, 9, 1673–1677, 2009.)

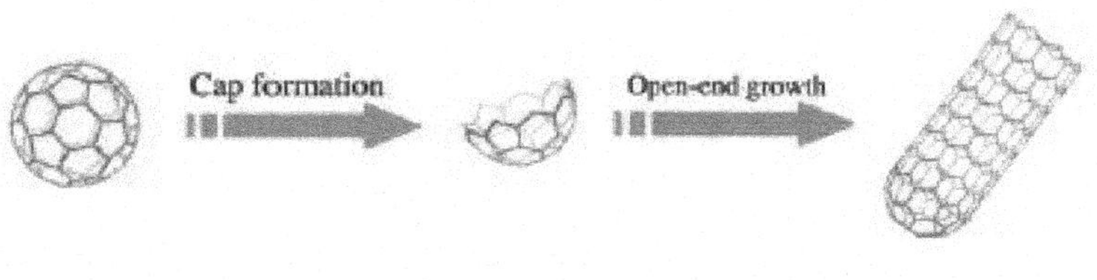

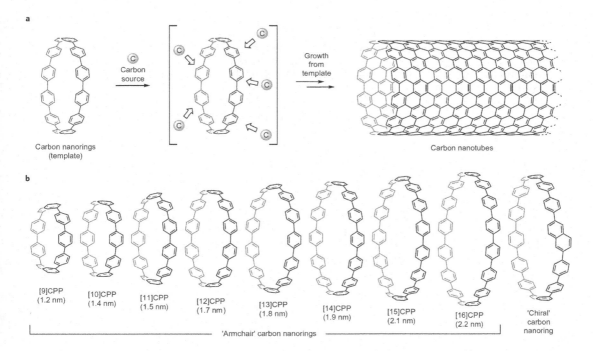

Figure 3.18 Schematic illustration of SWNT growth using fragmented C60 as an initiating cap. (From Yu, X. et al., *Nano Lett.*, 10, 3343–3349, 2010.)

small-diameter SWNTs, while lower cage-opening temperatures are favourable for growing large-diameter SWNTs. In addition, SWNTs showed a step-like diameter distribution as compared to SWNTs grown by Fe catalyst [84].

Another example is to synthesize carbon nanotubes using carbon-ring as the template, as shown in Figure 3.19. Carbon nanotubes can be obtained when the 12-ring Cycloparaphenylene (CPP) is deposited on a C-plane sapphire substrate exposed to a flowing ethanol gas under vacuum (1 torr) at 500°C for 15 minutes. It was observed that: (a) carbon nanotubes were not

formed without CPPs; (b) carbon nanotubes were not formed without ethanol; and (c) the diameters of grown carbon nanotubes matched those of the CPPs employed. These findings indicated that carbon nanotubes could be grown with this approach, although the scale of the process was limited [85].

Similarly, molecular precursors were converted to ultrashort singly capped (6,6) 'armchair' nanotube seeds using surface-catalysed cyclodehydrogenation on a Pt (111) surface using metal surface to convert molecular precursors into the nanotube seeds (which can then be used to grow single-chirality SWNTs), as

Figure 3.19 A 'growth-from-template' strategy for the bottom-up synthesis of structurally uniform CNTs. **(a)** General strategy for CNT growth using a carbon nanoring as a template. Cycloparaphenylene (CPP) is shown as an example. **(b)** The structures of representative carbon nanorings synthesized. The diameters of carbon nanorings are shown in parentheses. (From Omachi, H. et al., *Nat. Chem.*, 5, 572–576, 2013.)

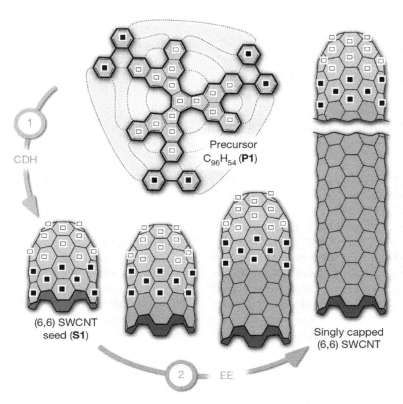

Figure 3.20 Two-step bottom-up synthesis of SWNTs. (1) Formation of singly capped ultrashort (6,6) SWCNT seed S1 via cyclodehydrogenation (CDH) of the suitably designed polyaromatic hydrocarbon precursor $C_{96}H_{54}$ (P1). (2) Nanotube growth via epitaxial elongation (EE). Parts of the precursor P1 involved in the formation of the SWCNT end cap and the ultrashort CNT segment of the seed S1 are highlighted in the sector with C6-rings marked with open square and filled sqaure, respectively. The dashed lines indicate the new C-C bonds formed upon CDH. Epitaxial elongation occurs via the successive addition of carbon species, as indicated in the sector with C6-rings unmarked. (From Sanchez-Valencia, J.R. et al., *Nature*, 512, 61–64, 2014.)

shown in Figure 3.20. SWNTs with lengths up to a few hundred nanometres could be grown through a subsequent growth phase [86]. In addition, a nanotube-end-cap molecule, $C_{50}H_{10}$, prepared by bottom-up organic chemistry synthesis, was used to prepare small-diameter semiconducting SWNTs [87].

Another approach was to obtain pure semiconducting or metallic SWNTs via post-synthesis separation [88a]. Although great progress has been made, it is still challenging to obtain SWNTs with controllable chirality or diameter on a large scale, which is essential in order to open their applications in microelectronics. Carbon nanotubes have special mechanical, electrical, and thermal properties. For example, SWNTs are reported to have high current-carrying capacities (up to 10^9A cm^2), and high thermal conductivities (up to 3500 W $m^{-1}K^{-1}$);

the axial Young's modulus is typically between 1 to 1.8 TPa. The fracture stress of SWCNT bundles can reach 50 GPa, corresponding to a density-normalized strength 50 times larger than that of steel wires [88b], which have potential applications in many different fields. More detail will be provided in the later part.

3.2.2 Si and Ge Nanowires

The preparation of Si nanowires was reported in the 1950s, and the VLS model was developed [8]. Si nanowires can be produced through the top-down etching method, bottom-up solution method, and CVD methods [89]. Various CVD methods including thermal CVD, laser ablation, PECVD, sputtering method, and MBE method have been explored for the growth of Si-based nanowires [89,90].

Table 3.3 List of elemental catalyst used for the growth of Si nanowires

Mg [8]											Al [91]			
		Ti [92]			Mn [8]	Fe [93]		Ni [8]	Cu [8]	Zn [94]	Ga [95]			
								Pd [8]	Ag [8]	Cd [8]	In [96]			Te [96]
	Gd [8]					Os [8]		Pt [8]	Au [8]			Pb [96]	Bi [96]	

Thermal CVD methods include the use of process with and without catalysts. When there is no catalyst being used, Si nanowires could be prepared via the dispropoation of SiO [89]. Gas precursors could be used for the preparation in the presence of catalysts (e.g., Au) as summarised in Table 3.3 [8,91–96].

It is worth mentioning that instead of gas precursor, Si nanowires were grown on Si substrate with and without a thermal oxide layer through RF sputtering process. This approach has the advantages to use pure Si as target without the need of toxic gas like SiH_4. The process uses thin Au film (2–20 nm) on the substrate as the catalyst; the substrate was heated to 700°C through remote radiative heating. It shows that 75% H_2/Ar gives highest nanowire density for the same catalyst and gas composition. The density of Si nanowires increases with the increase of process time, but the substrate with or without thermal oxide layer does result in high difference of the density [97]. This process is not a conventional CVD process.

Depending on the growth process, Si nanowires can be grown randomly or well aligned (either vertically or horizontally) [91,98]. Vertically aligned Si nanowires normally need a good substrate (e.g., Si (111)) in order to provide the epitaxy style growth, as shown in Figure 3.21 [91].

Considering the horizontal growth of Si NWs, this can be achieved through the use of Ni catalyst

Figure 3.21 SEM cross-sectional images (tilt 108) of Si nanowires grown at 490°C. (From Wang, Y. et al., *Nat. Nanotechnol.*, 1, 186–189, 2006.)

with the Si source on a substrate wafer [99]. If the catalyst is deposited on the vertical side of a surface, horizontal Si NWs could be grown via Si(111) epitaxy growth, as shown in Figure 3.22 [98]. In most cases, the growth is via VLS mechanism [8]. A recent study indicated that the presence of impurities such as Al, Cu from the processing reactor tube might act as the catalyst for the growth of Si NW, which was via solid-liquid-solid mechanism [99]. The structural study of Si nanowires shows that Si nanowire surface has an amorphous SiO_2 layer, and the catalyst might form a dopant in Si nanowires [89]. Si nanowires can

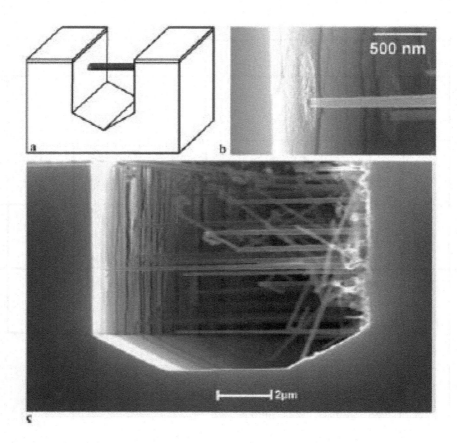

Figure 3.22 **(a)** Schematic and **(c)** scanning electron micrograph of nanowires 'bridging' across 8-μm-wide trench and connecting to opposing side wall. The nanowires grow from right to left in these views. **(b)** Impinging-end disk, showing the details of connection. (From Saif Islam, M. et al., *Appl. Phys.* A, 80, 1133–1140, 2005.)

be used for solar cell, sensors, and field-effect transistors (FET) due to its special electrical (e.g., semiconducting and optical properties [1].

Similarly, random/bulk Ge nanowires can be grown via a top-down approach [100], and various CVD methods based on VLS mechanism, such as vapour transport [101], and thermal CVD [102].

Ge nanowires can be used as photoresistors and visible light detectors because of their electrical (e.g., high carrier mobility) and optical (e.g., photoconductivity) properties [103].

3.2.3 III–V 1D nanomaterials

The development of 1D III–V nanostructures to some extent links with the period of time when the certain techniques was applied as it was used in other fields [1]. 1D III–V nanostructures can be made via top-down or bottom-up methods. The bottom-up CVD methods including arc-discharge, laser-assisted CVD, MOCVD (MOVPE) [104], MBE and CBE [105], and aerotaxy technique [106]. 1D III–V nanostructures were synthesized to follow VLS mechanism. Table 3.4 summarizes the growth of 1D III–V nanomaterials by CVD-based methods [107–132]. It is worth pointing out that aerotaxy includes several stages, as shown in Figure 3.23 [106].

The oriented growth of 1D III–V nanostructures is referred to as vertically aligned nanostructures that depend on the substrate to match the structure of nanostructures; many methods can be used for the oriented growth [133,134]. For example, the epitaxial growth of GaAs

Table 3.4 Growth of 1D III-V nanomaterials by CVD-based method

Material	Structure	Method	Reference
GaAs	Nanowire	Thermal CVD	[107]
GaAs	Nanowire	Laser CVD	[108]
GaAs	Nanowire	MOCVD	[109]
GaAs	Nanowire	MBE	[110]
GaAs	Nanowire	CBE	[111]
GaN	Nanowire	CVD	[112]
GaN	Nanowire	Arc discharge	[113]
GaN	Nanowire	Laser CVD	[114a]
GaN	Nanowire	MOCVD	[114b]
GaN	Nanowire	MBE	[115]
GaP	Nanowire	Laser	[108]
GaP	Nanowire	MOCVD	[116]
GaP	Nanowire	MBE	[117a]
GaP	Nanowire	CBE	[117b]
GaSb	Nanowire	CVD	[118]
GaSb	Nanowire	MOCVD	[119]
GaSb	Nanowire	FIB	[120]
GaAsxSb1-x	Nanowire	MBE	[121]
InAs	Nanowire	Laser	[108]
InAs	Nanowire	MOCVD	[109]
InAs	Nanowire	MBE	[122]
InN	Nanowire	CVD	[123]
InN	Nanowire	Vapour-solid	[124]
InN	Nanowire	PECVD	[125]
InN	Nanowire	Photo CVD	[97]
InN	Nanowire	MBE	[126]
InN	Nanowire	CBE	[127]
InP	Nanowire	Solution LS	[128a]
InP	Nanowire	Laser	[108]
InP	Nanowire	MOCVD	[128b]
InP	Nanowire	CBE	[129]
InSb	Nanowire	Electrodeposition	[130a]
InSb	Nanowire	PVD	[130b]
InSb	Nanowire	CVD	[118]
InSb	Nanowire	MOCVD	[130c]
InSb	Nanowire	MBE	[131]
InSb	Nanowire	CBE	[132]
InAsxSb1-x	Nanowire	MBE	[131]

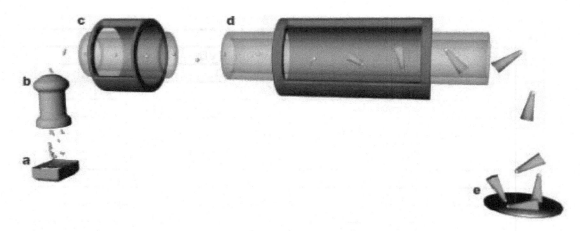

Figure 3.23 Aerotaxy growth of nanowires (a) Formation of Au agglomerate (b) Au agglomerate size sorting using a DMA (c) Au agglomerate compaction into spherical particles in a furnace; (d) nanowire growth and (e) nanowire deposition. (From Heurlin, M. et al., *Nature*, 492, 90–94, 2012.)

Figure 3.24 The 251 tilted view (a) and top view (b) SEM images of GaAs nanowires grown at flux ratio of 2:3 and substrate temperature of 550°C. The length bars indicate 500 nm. (From Plante, M.C. and LaPierre, R.R., *J. Cryst. Growth*, 286, 394–399, 2006.)

nanowires (NWs) on GaAs (111)B substrates by MBE has been systematically investigated as a function of relevant growth parameters, namely, temperature, arsine (AsH₃), and trimethyl-gallium (TMGa) flow rates, growth time, and gold nanoparticle catalyst size, as shown in Figure 3.24 [134].

The growth mechanism of III–V nanowires is mainly through VLS mechanism [8]. III–V nanowires, for example, could also be doped with magnesium (Mg) and silicon (Si) as p-type and n-type dopants, respectively, for the growth of III-Nitride nanowires [135].

III–V nanomaterials have optical and optoelectronic properties with the advantage to tune band gap with diameter, and have potential applications in LEDs, laser, solar cells, and photodetectors [103,135].

3.2.4 II–VI 1D nanomaterials

II–VI materials are promising for optoelectronics applications. Blue-green laser diode was developed from ZnSe in 1991 [136], earlier than GaN-based blue laser diode in 1996, and CdTe for thin-film solar cells. The latter material is now produced on industrial scale by First Solar (USA). 1D II–VI nanostructures (ZnO, ZnS, ZnSe, ZnTe, CdS, CdSe, CdTe) have been studied extensively since the 1990s. This section would highlight the growth of non-oxide II–VI materials while oxide nanomaterials will be reviewed in another chapter. Special attention will be paid to 1D II–VI nanostructures which could be grown via solution-based method, or CVD-based methods, including CVD, MOCVD, laser-ablation, and MBE method.

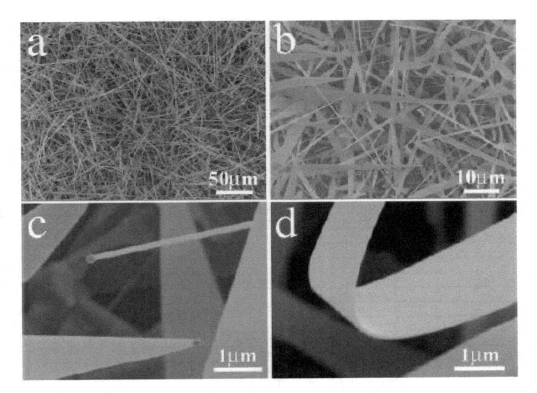

Figure 3.25 SEM images of two types of ZnS nanostructures. **(a)** and **(b)** are low- and high-magnification images, respectively. **(c)** Both types of the nanostructures are terminated by catalytic particles, and the nanobelts show tapering features towards the gold particle. **(d)** Typical bending contour of a ZnS nanobelt. (From Hao, Y. et al., *Nano Lett.*, 6, 1650–1655, 2006.)

Interesting reviews for II–VI 1D nanowires already exist [137]. The growth mechanism of II–VI 1D nanostructures can be via VLS growth, vapour-solid growth, and self-catalytic growth [136]. The morphology of 1D II–VI materials can vary depending on the methods used for the synthesis. As shown in Figure 3.25, both ZnS nanowires and nanobelts were obtained through Au-assisted VLS processes [138]. In Table 3.5 the general methods as well CVD-based methods for the growth of II–VI nanowires are summarised [91,108,114,139–160].

Besides the typical binary 1D II–VI nanostructures, ternary II–VI nanostructures can be produced to tune the band-gap and emission wavelength. Moreover, for the application of II–VI nanostructures, efficient n- and p-doping is needed, but more work is needed to develop the process and characterize the doping level [137]. 1D II–VI nanostructures have distinct electronic and optical properties such as size-confinement

effects, and are expected to have applications in new electronic and optoelectronic devices, such as FET, photodetectors, photoconductors, LEDs, laser diodes, and image sensors [137].

Aligned growth of 1D II–VI nanostructures was mainly vertically growth, in which the substrate structure played an important role [136].

3.2.5 Oxides

As part of the background information for 1D oxide materials, prior to 1969, Al_2O_3 whiskers were of greatest interest. Most of the oxide-based fibres in the commercial production contain a mixture of Al_2O_3 and SiO_2. Such fibres are produced in largest volume (>20,000 tons per year); due to their high melting points, these fibres are mainly used for composite production and insulation applications involving high temperatures [11]. The focus of this part is on metal oxide nanowires. They are versatile materials due to

Table 3.5 The list of II-VI materials and synthesis method

Material	Structure	Method	Reference
CdS	Nanowire	Solution	[139]
CdS	Nanowire	Electrochem	[140]
CdS	Nanowire	Solvothermal	[141]
CdS	Nanowire	PVD	[142]
CdS	Nanowire	Vapour transport	[143]
CdS	Nanowire	CVD	[144]
CdS	Nanowire	Van der Waals epitaxy	[145]
CdS	Nanowire/belt	CVD	[91]
CdSe	Nanowire	Electrochem	[146]
CdSe	Nanowire	Thermal evaporation	[147]
CdSe	Nanowire	Laser CVD	[108]
CdSe	Nanowire	MOCVD	[148]
CdSe	Nanowire	Van der Waals epitaxy	[145]
CdTe	Nanowire	Solution	[149a]
CdTe	Nanowire	Laser	[149b]
CdTe	Nanowire	Gas phase exchange	[150]
CdTe	Nanowire	Van der Waals epitaxy	[145]
ZnS	Nanowire	Thermal evaporation	[151]
ZnS	Nanowire	Laser	[114]
ZnS	Nanowire array	Thermal evaporation	[144a]
ZnS	Nanowire	MOCVD	[144b]
ZnS	Nanowire	MBE	[152]
ZnS	Nanowire/Nanobelt	Thermal evaporation	[138]
ZnSe	Nanowire	Solvothermal	[153]
ZnSe	Nanowire	Self catalysed/CVD	[154]
ZnSe	Nanowire	Laser	[108]
ZnSe	Nanowire	MOCVD	[155]
ZnSe	Nanowire	ALD	[156]
ZnSe	Nanowire	MBE	[157]
ZnTe	Nanowire	Solvothermal	[158a]
ZnTe	Nanowire	AAO, electrochem	[158b]
ZnTe	Nanowire	Thermal evaporation/PVD	[159]
ZnTe	Nanowire	MBE	[160]

their diverse properties (e.g., semiconductor of different type, including p-type CuO, n-type TiO_2, SnO_2 and CdO) and functions (e.g., optical transparency of In_2O_3 and CdO). 1D metal oxide nanostructures have interesting properties, like piezoelectricity, chemical sensing, and photodetection [161].

Oxide 1D nanomaterials can be fabricated via solution phase growth and vapour phase growth. The solution phase growth includes template-assisted synthesis, biochemical deposition, sol-gel deposition, surfactant-assisted growth, sonochemical method, and hydrothermal method [161]. Here, we focus on the vapour phase growth which includes thermal

CVD [162], direct thermal evaporation [163a], laser deposition [163b], MOCVD [164a], and MBE [164b].

Oxide 1D nanomaterials have very broad morphologies, such as nanowire, nanotube, nanoribbon, ring, nanospring, and nanobelt, and this is more diverse than other types of 1D nanomaterials, such as Si nanowire that does not have nanobelt morphology [165]. Besides random 1D material, aligned oxide 1D nanomaterial can be achieved via vertically aligned growth by choosing the appropriate substrate; for example, vertically aligned ZnO nanowire with (001) growth direction has been grown on (110) sapphire substrate [166]. Table 3.6 summarizes the

Table 3.6 List of oxide materials and the synthesis method

Material	Structure	Method	Reference
CdO	Nanowire	VLS	[171]
CeO_2	Nanowire	Sonochemical	[172]
CeO_2	Nanowire	Hydrothermal	[173]
CeO_2	Nanowire	Surfactant assisted	[174]
Cu_2O	Nanowire	Electrochemical	[173]
CuO	Nanowire	Vapour phase growth	[175a]
CuO	Nanowire	Hydrothermal	[175b]
Eu_2O_3	Nanowire	Sonochemical	[176]
Fe_2O_3	Nanowire	Electrochemical	[177]
Ga_2O_3	Nanowire	Hydrothermal	[178a]
Ga_2O_3	Nanowire	VLS	[178b]
Ga_2O_3	Nanowire	Arc discharge	[179a]
Ga_2O_3	Nanowire	MOCVD	[179b]
In_2O_3	Nanowire	VLS	[163a]
In_2O_3	Nanowire	Laser ablation	[163b]
MgO	Nanowire	VLS	[180]
MnO_2	Nanowire	Sol-gel	[181]
MnO_2	Nanowire	hydrothermal	[182]
NiO	Nanowire	Surfactant assisted	[183]
PbO_2	Nanowire	Surfactant assisted	[184]
SiO_2	Nanowire	VLS	[185]
SnO_2	Nanowire	Surfactant assisted	[186]
SnO_2	Nanowire, nanoribbon, nanotube	Vapour transport	[163]
SnO_2	Nanowire	Laser CVD	[171]
TiO_2	Nanowire	VLS	[187]
TiO_2	Nanowire	Sol-gel	[188]
ZnO	Nanowire	Surfactant assisted	[189]
ZnO	Nanowire	Electrochemical	[190]
ZnO	Nanowire	VLS	[191]
ZnO	Nanowire	VLS	[192a]
ZnO	Nanowire	MOCVD	[192b]
ZnO	Nanowire	Laser	[168]
ZnO	Nanowire	MBE	[164b]
ZnO	Nanowire	Epitaxy	[167]
ZrO	Nanowire	Sol-gel	[186]
Ag_2CrO_4	Nanowire	Hydrothermal	[193a]
$BaCrO_4$	Nanowire	Hydrothermal	[193a]
$CaCrO_4$	Nanowire	Hydrothermal	[193a]
$CuCrO_4$	Nanowire	Hydrothermal	[193a]
$La_{0.5}(Ba, Sr)_{0.5}MnO_3$	Nanowire	Hydrothermal	[193b]
$NiCrO_4$	Nanowire	Hydrothermal	[193a]
$PbCrO_4$	Nanowire	Hydrothermal	[193a]
$SrCrO_4$	Nanowire	Hydrothermal	[193a]

list of oxide materials and the synthesis methods [163,164,167,171–193].

From the perspective of 1D material development, doping is also important that would affect the applications of oxide 1D nanomaterials; for example, n-type ZnO 1D nanomaterial can be realised through substituting group-III and IV elements or incorporating excess Zn. However, it is challenging to produce p-type ZnO 1D nanomaterial [161].

Since there are already some reviews on 1D ZnO and other oxide nanostructures [161], here we would give a brief overview on the synthesis methods of typical oxides, as shown in the Table 3.6, and further highlight on ZnO nanowires.

ZnO 1D nanostructure is one of the most studied 1D oxide materials [166]. It can be prepared via top-down or bottom-up approaches. For the bottom-up approaches, besides solution phase growth, many vapour-based methods have been utilized [1], including thermal CVD [162], direct thermal evaporation [192a], pulse-laser deposition [168], MOCVD [164a], and MBE [164b].

The main growth mechanisms are vapour-solid (VS) and VLS mechanism [161], and the growth can be conducted with the use of catalyst [167] or catalyst-free [168]. In some cases, the growth of ZnO nanowires depends on the buffer ZnO film. For example, ALD method was used to deposit ZnO films with preferred orientation (i.e., self-catalysed layer) at different temperatures, as shown in Figure 3.26, which was subsequently used to grow ZnO nanowires via CVD process, and the resulting ZnO nanowires show that the alignment of ZnO nanowires depends on the buffer ZnO film (Figure 3.27) [169].

The aligned growth of ZnO nanowires could be achieved via the selection of a proper substrate; for example, (110) sapphire substrate was used to grow vertically aligned ZnO nanowires [167]. It is interesting that ZnO 1D nanomaterials can have many different morphologies, such as ZnO nanowires shown in

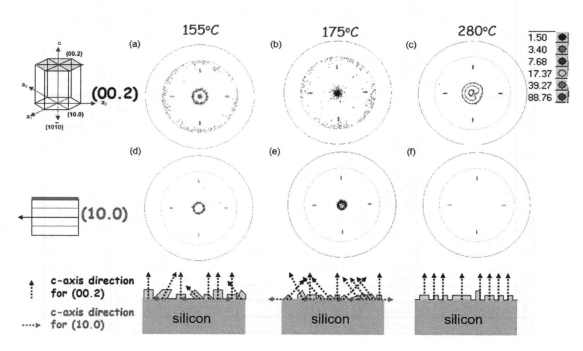

Figure 3.26 Pole figures of XRD texture analysis on as-deposited ZnO thin films produced by ALD at **(a)** 155°C, (100) plane; **(b)** 175°C, (100) plane; **(c)** 280°C, (100) plane; **(d)** 155°C, (002) plane; **(e)** 175°C, (002) plane and **(f)** 280°C, (002) plane, and schematic of the c axis of crystal grains in ZnO thin films. (From Pung, S.-Y. et al., *Nanotechnology*, 19, 435609, 2008.)

Figure 3.27 Homoepitaxial growth of ZnO nanowires by the CVD method on ZnO thin films deposited at **(a)** 155°C, **(b)** 175°C and **(c)** 280°C. The synthesis time of ZnO NWs was 30 min for each type of ZnO film. (From Pung, S.-Y. et al., *Nanotechnology*, 19, 435609, 2008.)

Figure 3.27 [169], and ZnO nanobelts shown in Figure 3.28 [170].

In order to explore broad applications, doping ZnO is an important step. As mentioned earlier, n-type ZnO is easy to be prepared, but p-type ZnO is rather difficult to obtain [161].

Oxide 1D nanomaterials can be used for the development of FET (e.g., ZnO and In_2O_3), emitter (e.g., IrO_2 and RuO_2), laser and waveguide (e.g., ZnO), LED (e.g., ZnO), photodetector (e.g., ZnO, SnO_2, and In_2O_3), solar cells (e.g., ZnO, TiO_2, and CuO), and chemical sensors (e.g., Ga_2O_3 and WO_3), based on their electrical, optical, and mechanical properties [161].

3.2.6 Carbides

As SiC can withstand high temperatures and possesses peculiar mechanical properties such as a standard SiC fibre with a carbon core from Textron Specialty Materials showing a tensile strength of 5.5 GPa, SiC fibres have been developed commercially for the application of composites utilized in aircraft and space vehicles [11]. SiC whiskers were investigated in the 1950–1960s via VLS growth mechanism [8] for potential applications in composites. Since then, other carbide nanostructures have been prepared as shown in Table 3.7 via CVD and carbon-thermal reactions

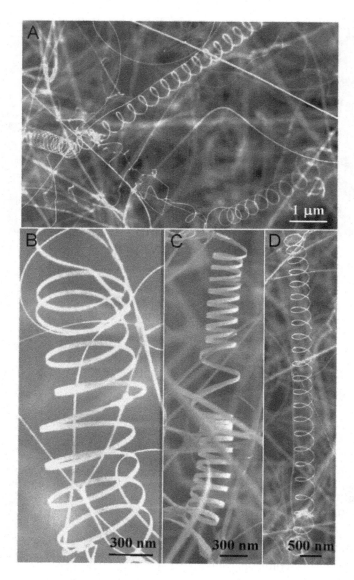

Figure 3.28 SEM images **(a–d)** of the as-synthesized ZnO nanobelts, showing helical nanostructure. The typical width of the nanobelt is 30 nm, and pitch distance is rather uniform. The helixes are right handed. (From Kong, X.Y. and Wang, Z.L., *Nano Lett.*, 3, 1625–1631, 2003.)

[194–205]. For CVD processes, carbon nanotubes could be used as a template for the preparation of carbide nanorods, as shown in Figure 3.29 [194a], or through conventional CVD process [194b]. For example, in Figure 3.30, $TiCl_4$-CH_4-H_2-Ar gas mixture was used as the reactant for the synthesis of TiC nanowires using a pure hollow nickel cylinder as the substrate placed near the gas inlet around the inner reaction tube wall,

at 1125°C, with the reaction time of 75 mins. The resulting TiC nanowires had a diameter of 50 nm–1.5 μm, and in the growth direction of (110) and (111) with 71% yield [194b].

TiC deposits are often different in different locations of the nickel substrate (see Figure 3.30 for the experimental set-up and Figure 3.31 for the surface morphology). From bottom to the top, the inner surface of the nickel substrate cylinder is divided

Table 3.7 List of elements that could form carbide nanomaterials

										B [195a]		N [195b]
										Al [196]	SiC [194a]	
	Ti [197]	V [198]	Cr [199]	Fe [194a]								
	Zr [200]	Nb [201]	Mo [202]									
	Hf [203]	Ta [204]	W [205]									

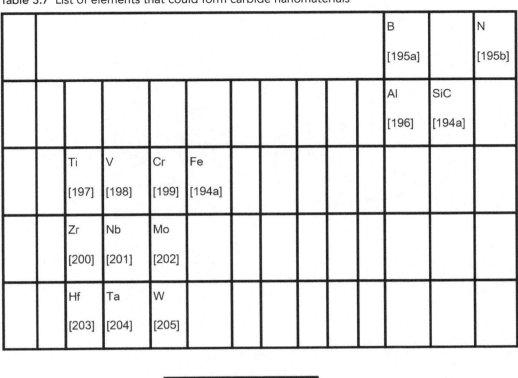

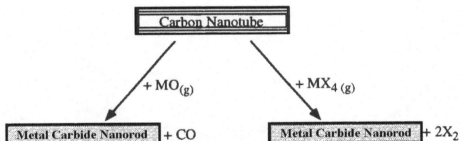

MO = volatile metal or non-metal oxide

MX$_4$ = volatile metal or non-metal halide

Figure 3.29 Reaction scheme used to prepare metal carbide nanorods. (From Dai, H.J. et al., *Nature*, 375, 769–772, 1995.)

into three sections: I, II, and III. Three distinctive morphologies were observed: (1) Type A – whiskers from location I of the substrate were generally thick (0.1–1.0 μm), long (0.05–2.0 mm), straight and smooth (Figure 3.31a). The quality of TiC whiskers with a high yield was obtained in this location; (2) Type B – whiskers from location II were thin (0.02–0.5 μm) and short (5–500 μm), as shown in Figure 3.31b; (3) Type C – whiskers from location III seem to have a similar size as in Type B whiskers, but they grew as a dot-distribution, as shown in Figure 3.31c. It is suggested that the

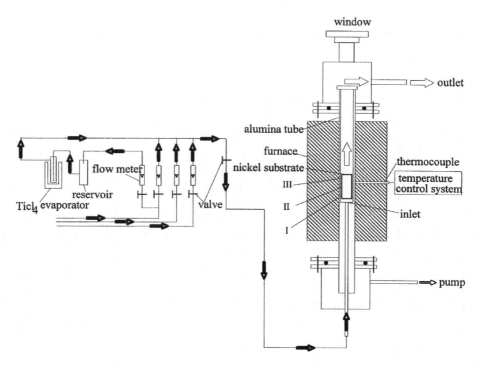

Figure 3.30 Schematic of the whisker deposition apparatus. (From Yuan, Y.W. and Pan, J.S., *J. Cryst. Growth*, 193, 585–591, 1998.)

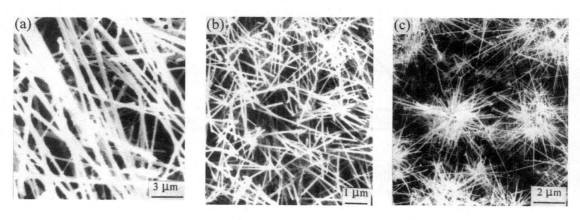

Figure 3.31 Three types of TiC whiskers obtained at different locations of the nickel substrate: **(a)** Type A, **(b)** Type B and **(c)** Type C. (From Yuan, Y.W. and Pan, J.S., *J. Cryst. Growth*, 193, 585–591, 1998.)

different whisker morphologies at different locations might be due to the difference of contacting time and axial effective vapour concentration along the nickel substrate cylinder, and the growth mechanism was VLS mechanism [194b].

Moreover, it is possible to synthesise many carbide nanowires via carbon thermal reduction (CTR) in combination of VLS mechanism, such as TiC, TaC, NbC, and $Ti_xTa_{1-x}(C, N)$ nanowires. For instance, for the preparation of TaC, Ta_2O_5 was used as a source for Ta, while carbon black was used to reduce the oxide, Ni, to catalyse the growth, and NaCl was used as a source of Cl for the vapour-phase transportation of Ta-oxochlorides to the catalyst. The growth rate would depend on process parameters such as temperature, catalyst, and the atmosphere. The yield could be obtained in the range of 70–90 vol.%. The carbide nanowires could be utilized as reinforcing materials in ceramic composites for ceramic cutting tools, similar to conventional carbide materials [201].

3.2.7 Nitrides

Nitrides are refractory materials. The early CVD deposition of nitride 1D materials in 1925 was based on the deposition of TiN layer onto the heated W filament using a mixture of $TiCl_4$-H_2-N_2 to form TiN filaments. Si_3N_4 fibres were produced for the potential application in metal/ceramic composites, and short Si_3N_4 fibres have been explored for the use in ceramic composites for specialty electrical parts, aircraft parts, and microwave windows [11].

BN fibres were also made for applications that require electrical insulation and thermal conduction. Moreover, BN fibres have been used in aluminium composites because they can be easily wet by molten aluminium [11].

BN 1D nanomaterial is one of the nitride materials that attracted considerable interest and it would be used as an example for the discussion of 1D nitride-based nanomaterials. BN 1D nanomaterials are mainly in the form of nanotubes and nanowires [206]. BN nanotubes could be single-walled, double-walled, or multi-walled nanotubes, as shown in Figure 3.32 [207]. Various methods, including non-CVD methods and CVD methods, have been employed. Non-CVD methods include auto-clave and ball-milling methods, while CVD methods include arc discharge [208], laser ablation [207], CVD [209], and PECVD processes [210]. CVD processes can be conducted in different ways; the process can be via a substitution reaction with carbon nanotubes [211], or through vapour deposition without catalyst such as the pressurized vapor/condenser (PVC) method [212], or with the use of catalysts (e.g.,

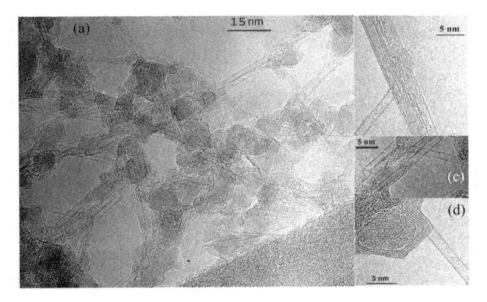

Figure 3.32 (a–d) High-resolution TEM (HRTEM) images of the different objects present in the samples: BN nanotubes (single-walled [individuals or organized in small bundles] and multi-walled [mostly double-walled]), cages and other kinds of nanoparticles. (From Arenal, R. et al., *J. Am. Chem. Soc.*, 129, 16183–16189, 2007.)

Co, Ni, NiB, and Ni$_2$B) [209]. BN can be doped by other elements, such as C or Si [206].

In boric oxide-based CVD (BOCVD), B, MgO, and other metal oxides are used as the precursors while induction heating is applied to activate the precursors at a temperature >1300°C. The volatile BxOy vapour generated was carried by Ar gas to interact with ammonia gas to form BN nanotubes. This process required a specially designed vertical induction furnace with rapid heating and a large temperature gradient [213], and Figure 3.33 shows a proposed growth model for the formation of the tubes. When this process was performed in a horizontal furnace via a growth vapor trapping (GVP) approach (Figure 3.34), then BN nanotubes could be grown at 1200°C with an absorption band edge of about 6.0eV approaching that of single h-BN crystals. Moreover, it was found that the yield of BN nanotubes was also higher when the combustion boat was partially covered by substrates. It was observed that when the closed-end quartz was replaced with an open-end tube there was no BN nanotube growth. This indicated that such set-up

would help to accumulate and trap the growth vapour so as to increase the rate of nucleation of BN nanotubes, which would react with NH$_3$ gas and prevent the reactive growth vapour being carried away before reaching the critical vapor pressure for the effective nucleation of BN nanotubes. The growth mechanism was via VLS mechanism [214]. If the patterned substrate covered the reaction boat, the patterned growth of BN nanotubes could be obtained, as shown in Figure 3.35 [215]. The following reactions may take place during the process

$$2B_{(s)} + 2MgO_{(s)} \rightarrow B_2O_{2(g)} + 2Mg_{(g)}$$

$$B_2O_{2(g)} + 2NH_{3(g)} \rightarrow 2BNNT_{(s)} + 2H_2O_{(g)} + H_{2(g)}$$

The yield of BN nanotubes increased when SnO or FeO was added into the precursor mixtures, and the role of the metal oxides was to function as solvents at a high temperature during the reaction between B and MgO; however, it is challenging to

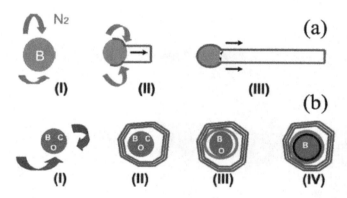

Figure 3.33 **(a)** and of the cages of BN **(b)**. **(a)** Model is as follows: **(a – i)** formation of boron drops from the decomposition of h-BN and from the boron oxide of the target; **(a – ii)** reaction of these drops of boron with the nitrogen injected into the reaction chamber and with nitrogen coming from the h-BN target. Recombination of the boron and nitrogen to form boron nitride; **(a – iii)** incorporation of the nitrogen atoms at the root of the boron particle that achieves the growth of the tube. Concerning the cages and the presence of carbon and boron oxide filling these objects (for the carbon, it is also possible to find it inside some of the BNNTs), the growth mechanism is: **(b – i)** and **(b – ii)**; these steps correspond to the previous ones (**[a – i]** to **[a – iii]**), taking into account the dissolution of oxygen and carbon in the boron drops; **(b – iii)** and **(b – iv)** correspond to the segregation of the carbon and oxygen. For carbon, the segregation occurs at temperature close to 2000°C, whereas the segregation of the oxygen is at lower temperatures, toward 700°C. (From Arenal, R. et al., *J. Am. Chem. Soc.*, 129, 16183–16189, 2007.)

Figure 3.34 **(a)** Experimental layout. **(b)** Typical SEM images of the as-grown sample. Dense BNNTs can be scratched off from the sample. **(c)** High magnification SEM showing the hollow centre channel of a tube structure. **(d)** Dense BNNTs film on a Si substrate after slight compression. The diameter and length of the tubes are estimated to be 15–100 nm and >10 µ1, respectively. **(e)** Cross-sectional view: Partially vertically aligned BNNTs. (Artificial colour was added to the SEM images.) (From Lee, C.H. et al., *Chem. Mater.*, 22, 1782–1787, 2010.)

control the diameter and the number of walls of BN nanotubes, as shown in Figure 3.36 [212a]. It is interesting to know that this process can produce BN nanotubes in a wide temperature range, as shown in Figure 3.37 [212b].

Another method was the PVC method without catalyst, in which the boron vapor was produced at a quasi-point source by the local heating of a target centred in the chamber. The large density difference between the hot boron vapour (over 4000°C) and the surrounding high-pressure N_2 gas (room temperature) generated a strong buoyancy force and a narrow vertical plume of boron vapor with a velocity profile. When a cooled metal wire traversed the boron plume, the wire acted as a condenser and left boron droplets in its wake by the homogeneous nucleation. Subsequently, clusters of BN nanotube would grow, intermingle, interlock, and be shaped by the fluid shear into a dimensionally stable fibril shape,

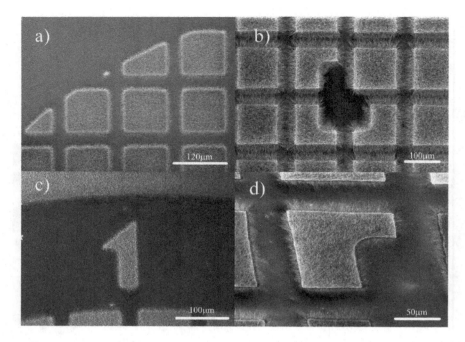

Figure 3.35 SEM images for the well-defined patterned growth of BNNTs in different areas **(a–d)** on a substrate (From Lee, C.H. et al., *Chem. Mater.*, 22, 1782–1787, 2010.)

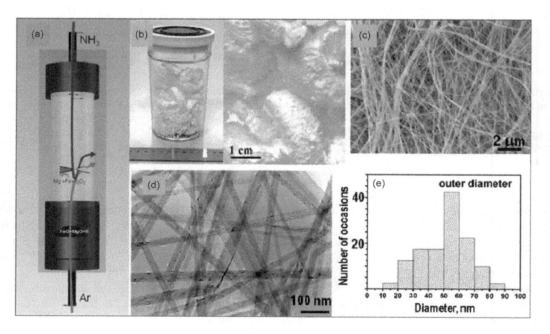

Figure 3.36 **(a)** Illustration of the apparatus for the synthesis of pure snow-white multi-walled BNNTs by the BOCVD method. **(b)** Photo images of as-prepared BNNT products. **(c, d)** Typical scanning (SEM) and transmission electron microscopy (TEM) images, displaying high yield and pure phase of a BNNT product. **(e)** Histograms of the tube diameter distributions. (From Golberg, D. et al., *ACS Nano*, 4, 2979–2993, 2010.)

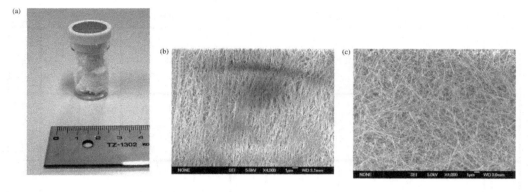

Figure 3.37   **(a)** BNNTs obtained during one synthetic run at 1500°C. **(b)** and **(c)** SEM images of as-grown BNNTs obtained at 1300°C and 1700°C, respectively. (From Zhi, C. et al., *Solid State Commun.*, 135, 67–70, 2005.)

all in about 100 ms. A typical BN fibril could be grown in a continuous assembly line by repeatedly intercepting the B plume with a moving condenser. The fibrils exhibit natural alignment along the axis of growth, giving the raw material the appearance of combed cotton. The process of fibril formation is efficient, as the mass of the BN nanotube raw material is typically equal to circa. 80% of the mass loss of the boron source target. The material has an unusually low density. In the current case 200 mg would fill a 10 cm by 10 cm jar, indicating that the PVC-grown BN nanotube product has a much higher surface area than previously available. Moreover, BN nanotubes can be spun to produce yarn. Various nitride nanomaterials have been prepared for the growth of 1D nitride nanomaterials via CVD process, as shown in Table 3.8 [125,216–226].

The growth mechanism for each process varies. It is proposed that in CVD process with the particulate catalyst, the BN nanotubes grow through a root-growth mechanism. Nanotubes nucleated on a nickel boride particle with often irregular (e.g., bulbous, club-like, flaglike, or other) initiation caps grew by the incorporation of additional BN at the catalyst-nanotube junctions, such that the growth separated the initiation caps from the catalytic surface. At the termination of growth, BN nanotubes remained attached by roots consisting of incomplete BN layers adhering to the large nickel boride particles on the substrate [209]. During the laser ablation

process with the target of h-BN powder pressed with boron oxide (B_2O_3) as a binding material, the composition (in weight percentage) of the target: 91.2% BN; 4.0% O; 4.5% B_2O_3; 0.1% Ca; 0.02% C; and 0.18% other impurities. Figure 3.32 shows the TEM of BN nanotubes with boron particles attached to the end of the nanotubes. It has been suggested that a root growth mechanism would involve three steps, as in a V-L-S model: (1) the decomposition by the laser of boron oxide contained in the binder of the target as well as the h-BN crystallites of the target into nitrogen gas and liquid boron which were finally vaporized in both cases; (2) upon cooling, boron vapour condensed into small boron liquid droplets, which would react with nitrogen gas, either coming from the carried gas or issued from the decomposition of the target, to form BN caps at the surface of the droplets; (3) the growth of the nanotubes from the progressive incorporation of nitrogen and boron at the interface between the cap and the particle. The growth would stop at the solidification of this boron core particle [207]. During the plasma process, BN nanotubes were grown on Fe nanoparticles according to the vapor-liquid-solid mechanism [210].

BN can also be produced in the form of nanosheets as 2D materials such as graphene. Similar to bulk h-BN, BN 1D nanomaterials also have the advantages of low-density, high-thermal conductivity (~350 WmK^{-1}), good electrical insulation, superb oxidation resistance (up to 1100°C in air), excellent

Table 3.8 The list of element that could form nitride nanomaterials

										B [217a]	C [218]		
										Al [216]	Si [219]		
	Ti [220]					Co [221]		Cu [222]	Zn [223]	Ga [224]			
	Zr [225]	Nb [226]								In [125]			
	Hf [203]												

inertness—such as passivity to reactions with acids, alkalis, and melts—and low friction coefficient [217b].

3.2.8 Elemental nanowires

The development of metal whiskers could be dated back to 1720 [3], and the synthesis of Hg nanofibre was observed in 1921 [6]. The use of metal whiskers to cause short-circuiting filter components for telecommunication in 1948 has driven the research on the growth mechanism of metal whiskers [7], and other whiskers which could be grown from the metal solder would not be discussed here as no CVD process is involved.

Elemental nanowires can be prepared via the top-down approach, such as the processing of Au nanowires via scanning tunnelling microscopy (STM) process [227a], or bottom-up approaches depending on the materials, such as a solution-based process that can be used to grow metal nanowires like electrodeposition [227b] as well as the CVD-based processes. CVD-based growth methods, which are the focus of this report,

include vapour-based method for Hg [228], thermal CVD reduction of halide as shown in Figure 3.38 [4], and thermal CVD for Al, Fe [229], ion-beam-induced CVD for W [230,231], MOCVD for Cu [232], MBE for Cu, Au, Ag, Al, Si, [233], arc-discharge [233b], laser-ablation [17], and ALD for Pt [234].

Table 3.9 shows the reported elemental 1D nanostructures [4,15,17,102,185,232,234–247]. Many methods exist for the growth of various elementary nanowires and the growth mechanism depends on the material and process. For example, in MOCVD process, where Cu nanowires were prepared using Cu(etac)[P(OEt)$_3$]$_2$ precursor that was evaporated in a glass cold-wall reactor onto a Si(111) substrate at a temperature range of 200°C–300°C under 0.1–1.0 Torr using argon as a carrier gas. No reducing agent such as hydrogen was used. The resulting Cu nanowires which exhibited high purity and crystallinity with [111] orientation are shown in Figure 3.39. The diameters of the copper nanowires were in the range of 70–100 nm, and the lengths of the copper nanowires could be controlled by the processing

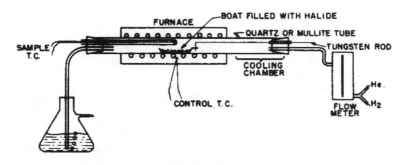

Figure 3.38 Whisker growth apparatus for reduction of halide. (From Brenner, S.S., *Acta Metallurgica*, 4, 62–74, 1956.)

Table 3.9 List of element that has been grown as 1D nanomaterial

										B [185]	C [15]		
										Al [236]	Si [17]		
			Cr [237]		Fe [4]	Co [4]	Ni [4, 238]	Cu [232, 235]	Zn [239]		Ge [102]		
			Mo [240]					Ag [4]	Cd [239]	In [241]	Sn [246]		Te [242]
			W [243, 244]				Pt [234,245]	Au [4]	Hg [228]		Pb [247]		

parameters such as deposition time and precursor feed rate [232]. Further work showed that Cu nanowires could potentially be used in electron emitter [235].

3.2.9 Complex structures of 1D nanomaterial

The heterostructures and superlattices in thin films have been developed previously. 1D nanomaterials can also function as the component of complex structures obtained through CVD processes. There are some reviews on certain types of materials for the formation of heterostructures and superlattices, such as semiconducting and II–VI 1D material-based structures. This chapter gives an overview on heterostructures and superlattices consisting of various types of 1D nanomaterials produced by CVD-based methods, including carbon nanotubes, Si/Ge nanowires, III–V, II–VI, and metal nanowires.

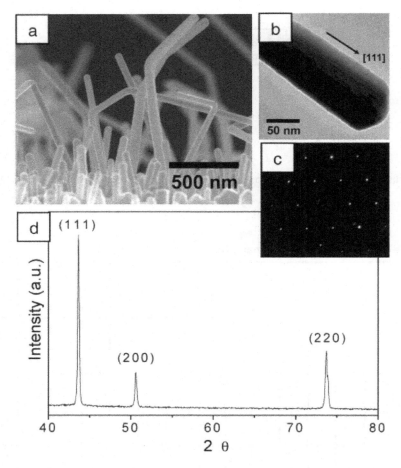

Figure 3.39 **(a)** SEM image, **(b)** TEM image, **(c)** [110] zone diffraction pattern and **(d)** XRD pattern of the copper nanowires grown at 250°C. (From Choi, H. and Park, S., *J. Am. Chem. Soc.*, 126, 6248–6249, 2004.)

Many CVD methods can be used for the preparation of heterostructures and superlattices, including thermal CVD, PECVD, laser-based method, MOCVD, ALD, MBE, and CBE. The complex structures can be divided into two types: 1D-style complex structures and 3D complex structures. For the 1D-style complex structure, it still preserved the morphology of 1D nanomaterials, but the structure or composition would differ from the pure 1D nanomaterials.

As shown in Figure 3.40, there are various types of 1D-style complex structures, including: A-B style axial junction, and it can be intramolecular junction, such as carbon nanotube-based metal-metal, metal-semiconductor, semiconductor-semiconductor junctions [248,249], or various composition p-n junction as shown in p-n Si nanowire structure [254], or various material junction such as Si-Ge junction [255].

The morphology of 1D-style complex structure would depend on the material. It can be randomly grown or aligned ID nanostructures. The growth of 1D-style complex structures also depends on the material of interest. For example, for Si-based radial p-n, p-i-n, or axial p-n, p-i-n, the CVD processes can be used for the growth of the nanostructures [254], while for III-V material, MOCVD method is often preferred [119]. It can also involve various methods for different parts of the structures; for example, for MgO/YBCO core-shell structure, MgO nanowires were grown via CVD method, while the YBCO layer was grown via PLD method. Figure 3.41 shows SEM images of an axial junction of

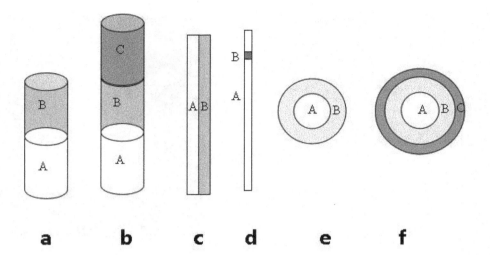

Figure 3.40 Various types of 1D style complex structures; moreover, this type of structure includes the **(a)** A-B style superlattice structure, **(b)** A-B-C style axial junction, like Si p-i-n junction, **(c)** biaxial style structure, like oxide type structure, **(d)** the nanoparticle /nanowire style heterojunction and **(e)** radial A-B style heterostructure; it can be Si p-n junction, or Ge-C style material. Similarly, this can be extended to cover radial superlattice, **(f)** radial A-B-C style heterostructure, like Si p-i-n structure.

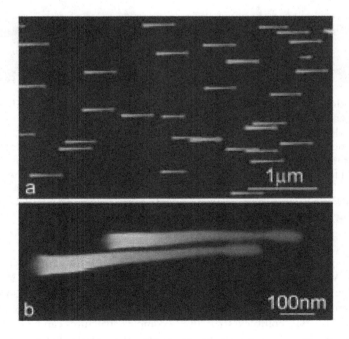

Figure 3.41 SEM image of an axial junction of GaAs/GaSb. **(a)** A low magnification image showing all nanowires being well aligned and **(b)** a high magnification image showing their tapered bodies with thick column-forming heads. SEM images taken from the substrate being tilted off 30°. (From Guo, Y.N. et al., *Appl. Phys. Lett.*, 89, 231917, 2006.)

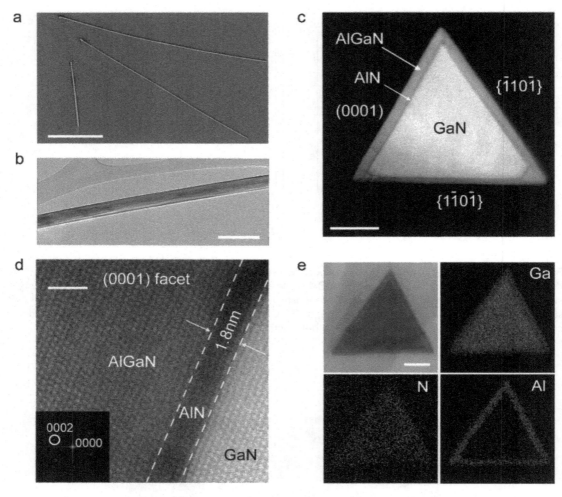

Figure 3.42 The random core-shell style III-V 1D style complex structures. **(a)** SEM image of GaN/AlN/AlGaN nanowires. Scale bar is 5 μm. **(b)** Bright-field low-resolution TEM image of a GaN/AlN/AlGaN nanowire. Scale bar is 500 nm. **(c)** HAADF-STEM image of the nanowire cross section. Scale bar is 50 nm. **(d)** Lattice-resolved HAADF-STEM image recorded at the (0001) facet of the nanowire. Dashed lines highlight the heterointerfaces between layers. Scale bar is 2 nm. Inset: electron diffraction pattern indexed for the [11-20] zone axis. **(e)** Bright-field STEM image and the corresponding EDX elemental mapping of the same nanowire, indicating spatial distribution of Ga, Al, and N, recorded on a GaN/AlN/AlGaN nanowire cross section. Scale bar is 50 nm. (From Li, Y. et al., *Nano Lett.*, 6, 1468–1473, 2006.)

GaAs/GaSb [119]. Figure 3.42 shows images of core-shell style III–V 1D-style complex structures [250], and Figure 3.43 are TEM images of nanowire twinning superlattices [251]. The image of vertically aligned 1D-style complex structure is shown in Figure 3.44 [252]. Table 3.10 lists 1D-style complex structures from various materials [101,113,119,154,248,253–284].

1D nanomaterials can also form 3D-style complex structure, as shown in Figure 3.45. 3D-style complex structures can be from same type of material, for example, bending carbon nanotubes or

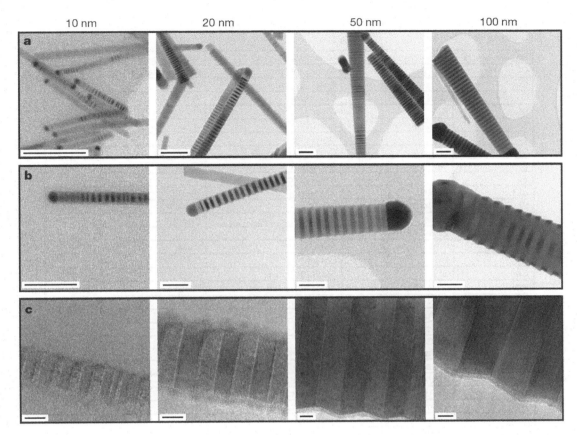

Figure 3.43 TEM images of nanowire twinning superlattices. **(a, b)** Overview and **(c)** high-resolution TEM images of InP nanowires with a diameter of nominally 10, 20, 50, and 100 nm The scale bars correspond to **(a)** 100 nm, **(b)** 50 nm and **(c)** 5 nm. (From Algra, R.E. et al., *Nature*, 456, 369–372, 2008.)

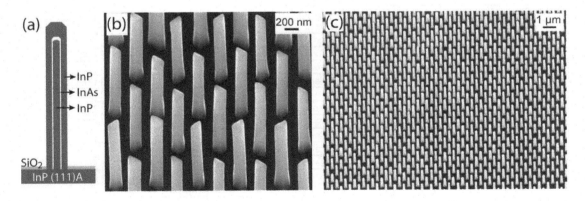

Figure 3.44 Image of vertically aligned 1D style complex structure, **(a)** colour online a schematic cross-sectional image of InP/InAs/InP core-multishell nanowire. **(b)** SEM image of periodically aligned InP/InAs/InP core-multishell nanowire array. **(c)** Low angle–inclined SEM image showing high dense ordered arrays of core-multi-shell nanowires. (From Mohan, P. et al., *Appl. Phys. Lett.*, 88, 133105, 2006.)

Table 3.10 1D style complex structures from various materials to show the base material and other materials

Base material	Other material	Structure	Method	Ref
C	C	Axial junction	Arc discharge	[248]
C	Si	Axial junction	CVD	[253]
C	GaN	Core-shell	Arc discharge	[113]
Si	Si	P-i-n	CVD	[254]
Si	SiO2/Si	p-SiO2-i	CVD	[255]
Si	Ge	Core-shell	CVD	[255]
Si	Ge-Si	Core shell (multi)	CVD	[255]
Si	Ge	Axial junction (multi)	CVD	[256a]
Si	NiSi	Axial junction	CVD	[256b]
Si	CdSe	Core-shell	Vapour	[257]
Si	CdSe	Biaxial	Vapour	[258]
Si	ZnSe	Biaxial	Vapour	[259]
Si	ZnS-Si-ZnS	Triaxial	CVD	[260]
Si	III-V (GaAs, GaP)	Axial	CVD	[261]
Ge	C	Core-shell	CVD	[101]
Ge	Si	Axial junction	CVD	[262]
AlN	GaN	Axial superlattice	MBE	[263]
AlN	GaN	Radial superlattice	MBE	[263]
GaAs	GaAs	Wurtzite-Zinc Blende		[264]
GaAs	AlAs/GaAs	Core-shell (multi)	MBE	[265]
GaN	InxGa1-xN/GaN/p-AlGaN/p-GaN	Core-shell	MOCVD	[266]
GaN	n-GaN/InGaN/p-GaN	Core-shell-shell	MOCVD	[267]
GaN	AlN/AlGaN	Core-shell-shell	MOCVD	[268]
InAs	GaAs	Axial junction	CBE	[269]
InAs	InSb	Axial junction	MOCVD	[270]
InAs	InP	Core shell	CVD	[271]
InAs	InAsP	Axial junction (multi)	MOCVD	[272]
InP	InAs/InP	Core-shell	MOCVD	[252]
InSb	InAs	Axial junction	MOCVD	[273]
InSb	GaSb/InSb	Axial junction	MOCVD	[274]
GaAS	GaSb	Axial junction	MOCVD	[119]
ZnS	Si	Core-shell	Vapour/CVD	[275]
ZnS	Si	Biaxial	CVD	[260]
ZnSe	Si	Biaxial	CVD	[260]
ZnS	Si-ZnS	Triaxial	CVD	[260]
ZnS	C	Core-shell	CVD	[154]
In2Se3	CuInSe2	Core-shell	CVD/solid reaction	[276a]
Ga2O3	ZnO	Core-shell	MOCVD	[276b]
Ga2O3	In2O3	Nanobelt/nanotape	CVD	[276c]
In2O3	Sb	Core-shell	CVD	[277]
MgO	YBCO	Core-shell	CVD/PLD	[278]
MgO	LMCO	Core-shell	CVD/PLD	[278]
MgO	PZT	Core-shell	CVD/PLD	[278]
MgO	Fe3O4	Core-shell	CVD/PLD	[278]
SnO2	TiO2	Nanotape	vapour/PVD	[279]
SnO2	Co0.05Ti0.95O2	Nanotape	vapour/PVD	[279]
SnO2	BaTiO3	Nanotape	vapour/PVD	[279]
ZnO	ZnS	Axial junction	Vapour	[280a]
ZnS	ZnO	Nanobelt	CVD (vapour)	[280b]
ZnO	GaN	Axial junction	MOCVD	[281]
ZnO	ZnMgO	Axial superlattice	MOCVD	[282]
Zn	ZnO	Core-shell	CVD	[283]
Zn	ZnO	Core-shell	MOCVD	[284]

Figure 3.45 Schematic diagrams **(a–e)** of different shapes of the 3D structures.

Y-style carbon nanotubes, as shown in Figure 3.46 [285], or branched Si nanowires [285b], or different materials, like core-shell In-ZnS 3D structures [288]. The complex structures would cover many types of materials. Figure 3.47 shows the 3D structure of carbon nanotube-graphene complex structure [286], while Figure 3.48 shows CdS-based tetrapod style of structure [287a], and different kinds of ZnS-based 3D structure are shown in Figure 3.49 for the tetrapod style of structure, Figure 3.50 for nanoawls

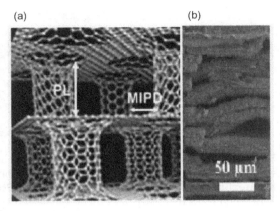

Figure 3.47 Complex structure of carbon nanotube-graphene. **(a)** Schematic of a 3D-pillared VACNT graphene nanostructure. **(b)** Typical SEM images of the 3D-pillared VACNT graphene architectures under different magnifications. (From Du, F. et al., *Chem. Mater.*, 23, 4810–4816, 2011.)

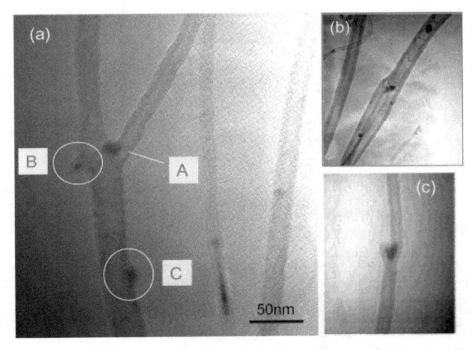

Figure 3.46 Y-style carbon nanotube-based 3D structures. **(a)** TEM image of a Y-junction. A catalyst particle is often present at the junction (marked A) leading to the formation of the two branches. **(b)** Another example of a catalyst particle present at the junction. **(c)** A catalyst particle inducing Y-junction formation at an early stage. (From Gothard, N. et al., *Nano Lett.*, 4, 213–217, 2004; Wang D. et al., *Nano Lett.*, 4, 871–874, 2004.)

Figure 3.48 CdS-based rocket-like tetrapodal structure. **(a)** Low-magnification SEM image of the as-deposited CdS products, **(b)** high-magnification SEM image of the rocket-like tetrapodal nanorods architectures and **(c)** SEM image of two-level RTPs. (From Zhai, T. et al., *Cryst. Growth Design*, 7, 488–491, 2007.)

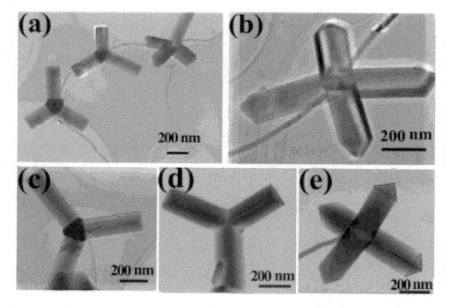

Figure 3.49 Images of nanocable-aligned ZnS tetrapods. **(a)** Aligned ZnS tetrapods. **(b)** Nanocable passing through the centre of the tetrapod. **(c)** Tetrapod image viewed along the [111] direction; **(d)** viewed along the [1h1h1h] direction and **(e)** viewed along the [100] direction. (From Zhu, Y.C. et al., *J. Am. Chem. Soc.*, 125, 16196–16197, 2003.)

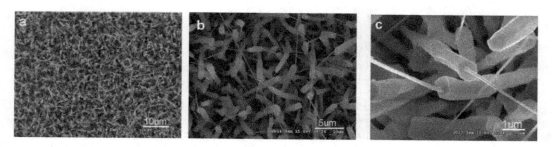

Figure 3.50 FESEM images of the as-synthesized ZnS nanoawls. **(a,b)** High yield and uniform morphology of the nanoawls. **(c)** Enlargement of the region in **(a)**, revealing a typical ZnS nanoawl structure. (From Zhai, T. et al., *Cryst. Growth Design*, 7, 1388–1392, 2007.)

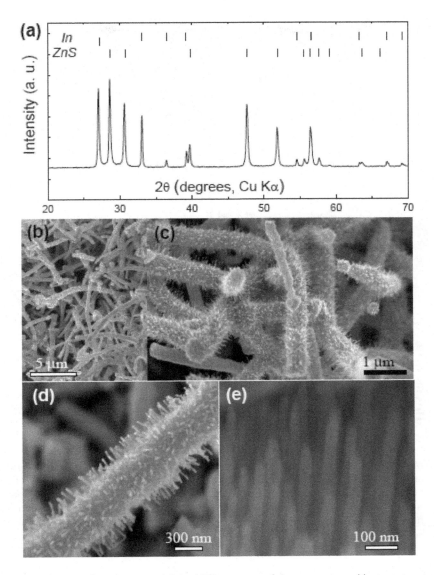

Figure 3.51 Nanorod-nanotube structure. **(a)** An XRD pattern of the as-prepared heterostructures. The standard peak positions for wurtzite ZnS and In are shown at the panel top. **(b, c, d)** SEM images of the heterostructures at different magnifications. **(e)** A high-magnification SEM image showing the ZnS branches quasi-aligned at the heterostructure tip. (From Gautam, U.K. et al., *ACS Nano*, 2, 1015–1021, 2008.)

[287b, c], and Figure 3.51 for the nanorod-nanotube heterostructure [288]. Figure 3.52 gives images for branched Zn_3P_2 nanostructures [289], and Figure 3.53 gives an example of the nanorod-block style complex structure [290].

Table 3.11 lists 3D-style complex structures from various materials [55,156,165,259,285–307].

The growth process of these materials would vary. It can be from CVD processes, like bending or Y-style carbon nanotubes, or from many steps of the same type of process, such as branched Si nanowire via CVD process. As compared with simple 1D nanostructure, the complex structures provide a route to combine the different materials,

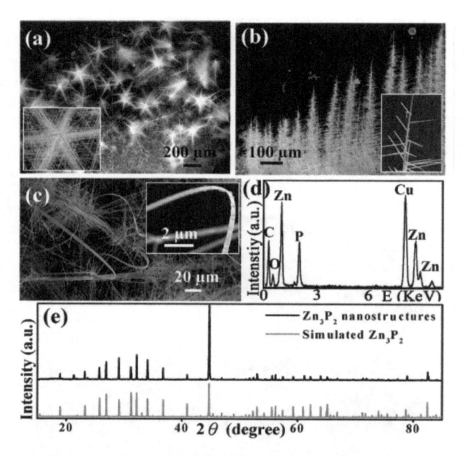

Figure 3.52 Zn_3P_2 branch, SEM, EDS and XRD analysis of Zn_3P_2 nanostructures. (a) Low magnification SEM image and an enlarged one of the as-grown Zn_3P_2 nanostructures on Si substrate, showing 6-fold symmetry. (b) SEM image of the side view of tree-shaped nanostructures from target with an enlarged part in the inset. (c) SEM images of belt-shaped Zn_3P_2. (d) EDS indicating the chemical signature of grown nanostructures. (e) XRD data show a single phase of tetragonal structured Zn_3P_2 with the bottom line displaying a simulated spectrum for the standard Zn_3P_2 powder. (From Yang, R. et al., *Nano Lett.*, 7, 269–275, 2007.)

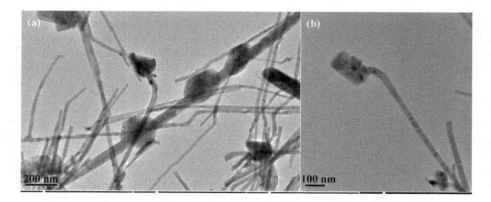

Figure 3.53 Example of nanorod-block style complex structure. (a) The HRTEM image of Zn_2SnO_4/ZnO and (b) an individual Zn_2SnO_4/ZnO HNS. (From Wang, B. et al., *Sens. Actuators B Chem.*, 195, 549–561, 2014.)

Table 3.11 3D style complex structures from various materials to show the base material and other material

Base material	Other material	Structure	Method	Ref
C	C	Bending/curly	PECVD	[55]
C	Graphene	3D	CVD	[286]
C	CNT branch	Branch	CVD	[285]
Si	Si	Branch	CVD	[285b]
Si	SiO_2	Core shell	CVD	[291]
Si	Si	P-n	CVD	[291]
Si	ZnSe/ZnSe	Branch	Vapour	[259]
CdS	CdS/ZnS	Core-Shell	MOCVD	[292]
CdS	CdS/Si	Core-Shell	Vapour transport	[293]
CdSe	CdSe/Si	Core Shell, Biaxial, Sandwich	Vapour transport	[294]
CdSe	CdSe/CdS	3D	Two-step thermal	[295]
CdS	CdS	Tetropod rocket style	MOCVD	[296]
Zn_3P_2	Zn_3P_2	Branch	Vapour Transport	[289]
ZnSe	$ZnSe/SiO_2$	Core-shell	Vapour Transport	[297]
ZnSe	ZnSe/CdSe	Superlattice	ALD	[156]
ZnSe	ZnSe/ZnO	Core-shell	CVD/Laser ablation	[298]
ZnSe	ZnSe/Si	Bi-coaxial	Vapour transport	[259]
ZnSe	ZnSe/ZnCdSe	Bicrystalline nanoribbon	MOCVD	[299]
ZnS	ZnO	Branch	Vapor/solution	[300]
ZnS	CdS	Branch (2 step)	MOCVD	[301]
ZnS	ZnS	Nanoribbon/nanocantilever	Vapour	[302]
ZnS	ZnS/ZnO	Biaxial nanoribbon	Vapour/CVD	[303]
ZnS	ZNS/Zn_3P_2	Core-Shell	CVD	[304]
ZnS	ZnS/SiO_2	Core/Shell	Vapour	[305]
ZnTe	ZnTe/CdTe	Axial junction	MBE	[306]
ZnTe	ZnTe/SiOX	Core-Shell	Thermal evaporation	[307]
ZnO	Zn_2SnO_4	Wire/sheet	Vapour type CVD	[290]
In	ZnS	Core-shell (NT)	CVD	[288]

structures, or functions, although the growth mechanism and properties of the complex structures cannot be fully understood.

These complex structures consisting of 1D nanomaterials can be used as building blocks for functional nanoscale devices, such as LED or environmental water treatment, thermal electric devices. The complex structure of 1D nanomaterials, such as nanowire heterostructure devices, could take advantage of the strain relaxation in nanowires to produce the device components that would not be possible in bulk. Various types of devices, including nanowire heterostructures of single-electron transistors (SETs), nanowire-based resonant tunnelling diodes, and memory devices from nanowire heterostructure superlattices, have been demonstrated, in some case, such as resonant tunnelling diodes, the functionality could be the use of optimal material combination of nanowires [103].

Scientific and technical issues and process control for producing high-performance nanowires and nanotubes for applications in structural, clean energy, biomedical, and environmental sectors are highlighted below.

As reviewed earlier, considerable effort has been made to fabricate 1D nanomaterials of different compositions and morphologies. There are reviews on the synthesis, characterization, and applications of general 1D nanomaterial [6], and specific types of materials, such as carbon nanotubes [308], Si nanowires [89], III–V materials [103], II–VI materials [136,137], oxide [161,309], and semiconductor nanowires [1]. In one aspect, the research and development of 1D nanomaterials is still ongoing, and this review provides an update on the applications of various 1D nanomaterials, as highlighted in Table 3.12. The applications of these materials are still in the stage of low-technology readiness level, i.e., early research and development. In another aspect, only Ag nanowires from solution route and carbon nanotubes from CVD route are now being produced on an industrial scale.

This chapter will mainly discuss CVD processes, describing the applications of carbon nanotubes that have been explored.

3.2.10 Structural applications

Carbon nanotubes as one-dimensional materials have shown remarkable thermal (e.g., up to 3500 W m^{-1}K^{-1} of thermal conductivity for SWNTs) and electrical (e.g., up to 10^9A cm^2 of high current-carrying capacities for SWNTs) properties that are better than copper, and the density-based modulus and strength of highly crystalline SWNTs are 19 times and 56 times that of steel, respectively [310,311]. Carbon nanotubes could potentially be used in structural applications in different forms: (1) as a powder filler in metal/ceramic/polymer composites [310–312]; (2) as a component in coatings and films to form nanocomposite coatings [311]; (3) as a fibre obtained from synthesis process or post-synthesis processing [37,311].

When carbon nanotubes are used as powder fillers and due to currently available length, they cannot reinforce large components while the cost of carbon nanotubes is still high for the commercialization of structural components which would require large quantity. From the perspective of performance/price ratio, carbon nanotubes nanocomposites cannot compete with carbon fibre reinforced plastics (CFRP). Carbon nanotubes are mainly used as secondary fillers in CFRP to reinforce the resin that surrounds the carbon fibres in order to improve the compressive strength and shock-damping characteristics of the composite, so that the amount of resin used can be reduced and the resin itself is reinforced. These carbon nanotube-based composites are already commercialized and used in frames, cranks, handlebars, and front forks of road racing bikes from various manufacturers [310], such as bicycles used in the Tour de France in 2005 [311]. Lightweight carbon nanotube-fibre composites have also been used to make boat hulls for maritime security boats; it is expected that carbon nanotubes, usually MWNTs with increased compressive strength and carbon fibre, would contribute to the tensile strength, and the composites can improve the durability, strength, and shock absorption of such components [311]. Carbon nanotube composites are also used in ice hockey sticks for durability and controllability [310].

Carbon nanotube-reinforced rubber composite is also a kind of CNT nanocomposite, in which carbon nanotubes serve as a spring and heat conductor to improve the shape resiliency and thermal stability of the rubber component. As shown in Figure 3.54, the O-ring manufactured from the carbon nanotube-reinforced rubber composite was tested as a sealing material for the joints in pipes for drills in order to withstand severe conditions that no other sealing materials seems to be able to withstand; therefore, such composite is considered to be useful in extreme conditions such as high pressure and high temperature as in oil mining [310].

Carbon nanotubes have also been utilized to prepare metal-based composites, in which carbon nanotubes increase tensile strength and modulus. As shown in Figure 3.55, carbon nanotube-reinforced Mg composites can be processed by blending through ball (zirconia) milling and then hot pressing at 550°C in a vacuum <10 Pa, and the component can be prepared by extrusion and demonstrated 20%–30% increase in tensile strength [312a]. Commercial Al-MWNT composites have strength comparable to stainless steel at one-third the density of stainless steel [311].

Carbon nanotubes can be used in the form of coating and films to make use of their electrical properties; for example, carbon nanotube-based composite film is developed as thin-film heater for the application such as defrosting windows or sidewalks [311]. Carbon nanotubes have already been produced in the form of fibre (with diameter of around 50 μm) either through direct CVD synthesis process [37], or post-processing using fumic sulfuric acid to give dispersed SWNTs that was then extruded [313a]; this type of format can produce very long fibre; for example, using

Figure 3.54 CNT/rubber composite unloaded after 10% compression at 200°C, 70 h. Permanent set rate is shown in the image. (From Zhou, W. et al., *Adv. Mater.*, 2, 4565–4583, 2009; Hayashi, T. and Endo, M., *Compos. B Eng.*, 42, 2151–2157, 2011.)

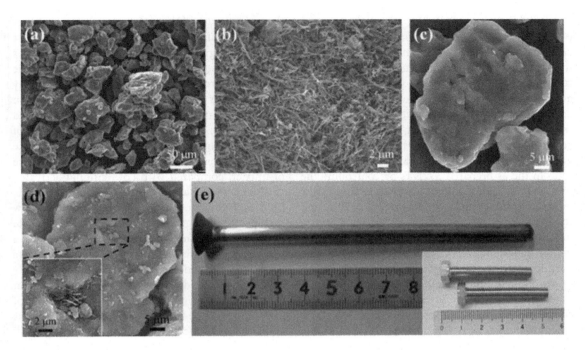

Figure 3.55 SEM images: **(a, b)** Mechanically milled AZ91D-type magnesium powders with an average diameter of 100 μm and shortened carbon nanotubes with an average length of 5 μm; **(c, d)** mechanically mixed magnesium powders containing 1% and 5% short carbon nanotubes, respectively. Note that the aggregated carbon nanotubes were observed in magnesium powders containing 5% of carbon nanotubes (inset is a magnified SEM image). **(e)** Photo of carbon nanotube-reinforced magnesium composite-based rod exhibiting clean appearance without any macro-morphological defects (Inset is a photo of clean shining two volts fabricated from extruded rods). (From Shimizu, Y. et al., *Scripta Materialia*, 58, 267–270, 2008; Bakshi, S.R. et al., *Int. Mater. Rev.*, 55, 41–64, 2010.)

a mixture of acetone and ethanol as the carbon source and ferrocene as catalyst, the resulting fibre can be in the length of several km, with tensile strength of 0.4–1.25 GPa and electrical conductivity of 5×10^5 S m^{-1}[313b]. The fibre can then be used to make yarns and sheets with electrical and mechanical properties, and they can be used as lightweight data cables and electromagnetic (EM) shielding material. In 2011, Nanocomp's nanotube-based sheet material called Emshield, was incorporated into the launched Juno spacecraft to provide protection against electrostatic discharge as the Juno spacecraft made its way through space to Jupiter [311]. In such application, the scientific and technical issues are to have good electrical conductivity and mechanical integrity, and suitable length. The process control for MWNTs is to reduce the defects and increase strength. While for SWNTs production

of monodisperse carbon nanotubes would benefit electrical properties of the materials as SWNT samples have both semiconducting and metallic carbon nanotubes. For both MWNTs and SWNTs, there is a need to increase the length so that they can be applied as reinforcements in large parts and increase the performance of fibres or sheets.

3.2.11 Clean energy applications

Carbon nanotubes have been explored for potential clean energy applications [314,315], including photovoltaics, fuel cells, and energy storage. For photovoltaics, carbon nanotubes could be used to make transparent conducting electrodes (e.g., arc-discharge SWNTs after acid treatment 70–75 Ω/sq, high transmission [>80%] at wavelengths >1200 nm) [310a], an example is for CuIn1-*x*Ga*x*Se2-based (CIGS) solar cells with the structure

SWCNT/i-ZnO/CdS/CIGS/Mo device structure (~13% efficiency) [314b]. In organic solar cells, carbon nanotubes are one of the choices to reduce the undesired carrier recombination and enhance resistance to photooxidation [314c]. The scientific and technical issues are to control the electrical conductivity of carbon nanotubes, which requires the control of the synthesis of semiconducting or conductive carbon nanotubes; there are reports to grow semiconducting carbon nanotubes in a higher ratio in the as-prepared SWNTs by controlling the process [75], but it is still difficult to selectively grow only semiconducting or conducting SWNTs during the growth process, although further purification process can give semiconducting or conducting carbon nanotube. Also, carbon nanotubes can be used for polymer electrolyte fuel cells, and the use of carbon nanotubes as catalyst supports could potentially reduced the Pt catalyst usage by 60% as compared with that of carbon black [314d]. Moreover, when N-doped vertically aligned MWNTs were used, Pt catalyst may not be required for oxygen-reduction reactions involved in fuel cells [314e].

For energy storage, MWNTs have been used to increase electrical connectivity between electrode materials and mechanical integrity of electrodes [314,315]; for example, the comparative study of carbon nanotubes and conventional carbon black in cathode with active material (e.g., $LiCoO_2$) shows that carbon nanotubes are a better conductive additive than conventional carbon black from the perspective of cycle life of

the Li-ion battery [311,314], and carbon nanotubes may even function as binder-free anodes, as shown in Figure 3.56 [315a]. Carbon nanotubes have been used in Li-ion battery for the electric vehicles to provide better performance of battery [315b]; this is a large market for the application of carbon nanotubes as there is growing need for the green society. Moreover, the forest of SWNTs can be used to realize high-performance supercapacitors [315c], and from the perspective of impact, compared with conventional activated carbon, carbon nanotubes have hollow structure, high surface area, pure carbon composition, low tortuosity, and high conductivity. These contribute to the performance of supercapacitors from the perspective of lifetime, energy, voltage (and life time), power, and current-collector, respectively [315c]. The latest developments indicate that great progress has been made in the preparation of carbon nanotube-based supercapacitors towards applications for electric vehicles [315d].

In addition, the new trend is to develop flexible energy storage for the emerging wearable technology [315e,315f]. A typical example is to prepare carbon nanotube fibres through spinning bushy-style aligned MWNTs, then use the resulting carbon fibre as the current collector and electrode to fabricate wire-shaped microsupercapacitor and microbattery. For the microsupercapacitor, the energy density was 3.47 and 1.73 mWh/cm³ while the power density was 1.08 and 0.79 w/cm³; for micro Li-ion battery, the energy densities achieved 92.84

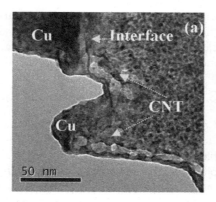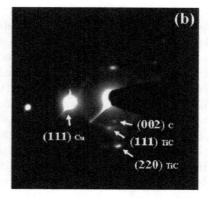

Figure 3.56 Structural analysis of the proposed anode. **(a)** HRTEM image of substrate CNT interface, showing a well-bonded interface and the presence of an interfacial phase. The selective area diffraction pattern **(b)** shows formation of TiC at the interface, which also helped in making the bonding strong and conductive.

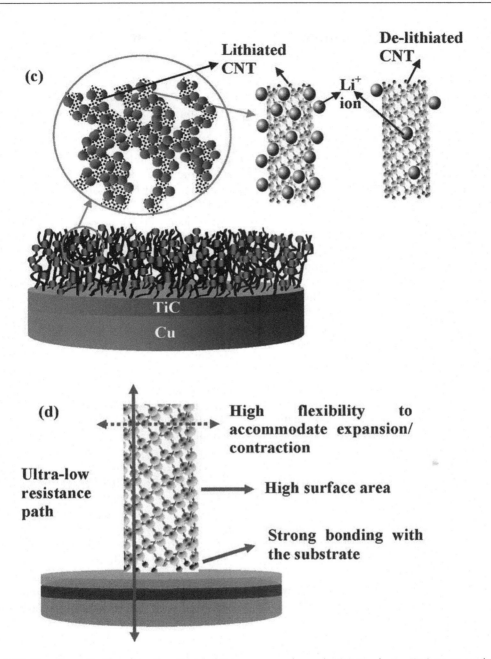

Figure 3.56 (Continued) Structural analysis of the proposed anode. **(c)** A schematic (not to scale) of the proposed lithiation delithiation mechanism, showing a huge amount of Li ion intercalated to the walls of the CNTs during lithiation and deintercalation of most of the ions in delithiation. **(d)** A schematic (not to scale) of the proposed anode structure, showing its advantageous features. (From Lahiri, I. et al., ACS *Nano*, 4, 3440–3446, 2010.)

and 35.74 mWh/cm³, and the power densities could reach 3.87 and 2.43 W/cm³ during the charge-and-discharge process [315e].

For energy storage application, the scientific and technical issues are to have good electrical conductivity and mechanical integrity; the process control is to reduce the defects (e.g., non-closed carbon ring) of carbon nanotubes, and reduce the cost of carbon nanotubes, particularly SWNTs, to allow further large-scale applications.

3.2.12 Biomedical applications

Carbon nanotube have broad biomedical applications from the diagnostic or therapeutic perspectives [316–329], such as imaging [317,318], sensors [316,320–321], therapy [319], biomedical devices, and tissue engineering, due to their size (like length of nanotubes), electronic properties (metallic or semiconducting), and structural (SWNTs or MWNTs) characteristics. For the applications of imaging, SWNTs have been used for fluorescent and photoacoustic imaging [318], because of their intrinsic emission in the near IR(NIR)-II region (950–1400 nm) and photoacoustic absorption (690 nm), respectively and large inter-band difference between the excitation and emission wavelengths for fluorescent imaging yielding high signal-to-noise ratios and increasing the detection limit [316–318], the work on mouse demonstrated the potential high resolution benefits of SWNTs in NIR-II fluorescence imaging (1–3 mm deep in the hind limbs), overcoming the challenges of conventional imaging modalities [323a]. MWNTs are good for ultrasound imaging, even better than SWNTs under similar processing conditions. The work on mice showed that ultrasound signal for carbon nanotubes was comparable to the commercial contract agents like sulphur hexafluoride [323b]. In addition, there is example to demonstrate carbon nanotube-based nano-platforms for multimodality imaging techniques combining MRI and thermoacoustic imaging (TAI) of cancer [323c]; for this type of application, carbon nanotubes (either SWNTs or MWNTs) need functionalisation to reduce their toxic effects and increase their specific binding activities. For therapy applications, carbon nanotubes can be utilized in two ways; one is to use carbon nanotubes (mainly SWNTs) to adsorb the light (like NIR laser) to generate localized heating to kill the cancer cells [319], and the other is to use carbon nanotubes as a delivery tool. For example, it is shown that chemically functionalised (with-NH_2) SWNT can be linked with peptide through covalent bond (-CO-NH_2). A bound peptide from the foot-and-mouth disease virus, corresponding to the 141–159 region of the viral envelope protein VP1, retained the structural integrity and was recognized by monoclonal and polyclonal antibodies. Moreover, this peptide-NT conjugate is immunogenic, eliciting antibody responses of the right specificity [324]. Other works include vaccine delivery tools [325,326] and gene therapy [327b].

For the applications of biosensors [316,320–321], from the perspective of sensing mechanism of target recognition and transduction, biosensors can be broadly subdivided into the following types: electrochemical biosensors, electronic transducers, immunosensors, and optical biosensors. Electrochemical biosensors are a type of widely-used biosensors due to their low-cost, ease of use, relatively fast response times, and small size. Carbon nanotubes can enhance electron transfer and suit the integration into biosensors because of their electrical and electrochemical properties. Carbon nanotubes can be in different formats, such as vertically aligned MWNTs, random MWNT film, and random SWNT films. CNT-based biosensors have been used to detect ions, metabolites, protein biomarkers, and cancers. For example, SWNT (>90% purity, 150 μm average length, and 1.4–1.5 nm diameter), after removing amorphous carbon (365°C air treatment for 90 mins), and nitric acid treatment (2.6 M, 4h to remove catalyst and functionalise with carboxy group), was deposited in 50-mm-long and 2-mm-diameter glassy carbon cylindrical rod surface as electrode (30 μm SWNT layer) through spraying a liquid mixture with 25 mg purified SWNTs, 25 mg sodium dodecyl sulphate (SDS), and 100 mL Milli-Q water at 200°C. The residual SDS was removed by heating at 280°C with air flow rate 100 cm^3 min^{-1} for 1h. RNA aptamer cl57 was then linked with SWNT film through carbodiimide chemistry to form stable amide bonds between oxidised SWNTs and -NH_2 moieties on the 3′ end of the aptamers. This potentiometric biosensor has specific recognition capacity at attomolar concentrations in real-time (concentration of 4fM in the cell using blood). This can open the door to real-time diagnostic [321b].

For electronic transducers [321c,321d], SWNT, either individual SWNT or SWNT network, can be used as electron channel between the source and the drain electrodes on a FET layout. The interaction between analyte and carbon nanotube may involve charge transfer from analyte molecules to the carbon nanotubes, or the analyte acts as a scattering potential across the carbon nanotube.

For instance, as-prepared SWNT network on 100nm SiO_2/Si substrate from CVD at 900°C using Fe nanoparticles as catalyst and CH_4/H_2 gas mixture as the process gas, were used to prepare FET devices (210 μm × 270 μm) consisting of interdigitated electrodes with 10-μm separation. For investigating interaction of single stranded DNA and SWNTs and DNA hybridisation by adsorbing ssDNA on SWNTs, the sensor can differentiate between both mutant (mut) and wild type (wt) alleles of the HFE gene, responsible for hereditary hemochromatosis (a common and easily treated disease of iron metabolism), and when DNA hybridization matched to the corresponding type alleles on the surface of SWNTs, there is a drop in conductance. The sensor can have potential to detect hereditary hemochromatosis [321d].

Immunosensors are often prepared by immobilizing recombinant antibodies or antibody fragments onto substrate and used to recognize antigens. A variety of carbon nanotube-based immunosensors have been developed for the detection of cancer biomarker and other diseases [327]. The recent example is carbon nanotube-based impedimetric immunosensors for high-performance label-free detection and quantification of anti-cholera toxin antibody, in which as-prepared commercial grade MWNTs (9.5-nm diameter, purity >95%) were deposited on glassy carbon electrode (1.6, 2.6, or 4.0 μm) and then covered by an electropolymerized layer of poly(pyrrole-NTA) generated by chronopotentiometry with the monomer 11-Pyrrol-1-yl-undecanoic acid N′,N$_\alpha$-Bis(carboxymethyl)-L-lysine amide (5 × 10^{-3} mol L^{-1}) solubilized in CH_3CN + $LiClO_4$, 0.1 mol L^{-1} and electropolymerized at a charge of 0.18 mC, then incubated for 30 min with Cholera toxin B subunit-biotin conjugated (b-CTB; lyophilized powder) (10 μL, 0.5 mg mL^{-1}). Dissolved in 1% w/v BSA/PBS to give the electrode, which can then be used for the detection of anti-cholera toxin antibody (anti-CT) through electrochemical impedance spectroscopy (EIS), the resulting impedimetric cholera sensor shows very good reproducibility, increased sensitivities, a very satisfying detection limit of 10^{-13} g mL^{-1}, and an exceptional linear range for anti-cholera detection of eight orders of magnitude (10^{-13}–10^{-5} g mL^{-1}) and a sensitivity of 24.7 ± 0.4 Ω per order of magnitude [322]. For optical biosensor, carbon nanotube-based fluorescent and photoacoustic diagnostics are mentioned earlier [320–323,327].

Another type of application is to make carbon nanotube-based biomedical parts or devices. For example, the carbon nanotube-nylon composite is used to make microcatheter, as shown in Figure 3.57. The in vivo test with dog for thrombus formation shows that the nanotube-filled nanocomposite-derived microcatheter exhibited highly reduced thrombus formation, indicating that the nanotube-filled microcatheter has lower reactivity toward blood than the pure nylon-derived one, if the blood stream is not disturbed. As shown in Figure 3.58, the nanotube-filled microcatheter shows smaller amounts of fibrils and blood cells, which means that the nanotube-filled nanocomposites-derived microcatheter exhibits anti-thrombogenicity and low blood coagulation [328]. This is a good progress in a development stage. In addition, carbon nanotubes, both MWNTs and SWNTs, have been investigated for the tissue engineering [329]. In the format of scaffold with or without polymer, in the case of carbon nanotubes without polymer, an example is to use SWNT rope-like structure (1 mm in diameter, 1.5 cm in length) to promoted the early neuronal differentiation of HCN-A94-2 neural stem cells synergistically with electrical stimulation (5 mV, 0.5 mA, 25 ms, intermittent stimulation) [329a], which shows Neurites growing along the spiral topography of the rope. The drawbacks of this type of scaffold are: (i) the structure is fragile for tissue implantation as there is no agglutinant material that adsorbs external mechanical stimuli (e.g., compression, flexion, or bending, among others); (ii) enhanced toxicity responses are caused by the direct contact of cells and tissues with CNTs from an early stage after implantation; in the case of carbon nanotubes with polymer, a broad range of polymers, including natural polymers and synthetic polymers, have been studied, a recent example is to use acid-treated MWNTs (diameters in the range of 40–90 nm and length 10–20 μm) and synthetic polymer, a highly elastic polyester, poly(octamethylene maleate (anhydride) 1,2,4-butanetricarboxylate) (124 polymer) to prepare scaffolds for cardiac tissue engineering, showing that the scaffold with carbon nanotubes can be moulded into complex shapes and crosslinked to give an elastomeric scaffold structure

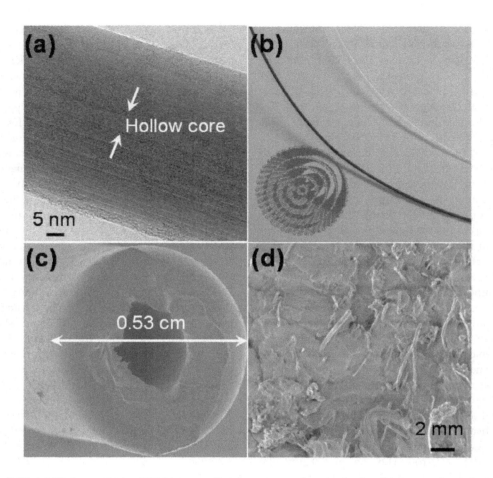

Figure 3.57 **(a)** High-resolution TEM image of carbon nanotubes (note that linear graphene layers are highly developed along the tube length), **(b)** photographs of transparent nylon-derived and opaque black-coloured nanotube-filled nanocomposite-derived microcatheters, SEM images of cross section **(c)** at low resolution and **(d)** high resolution. It is noteworthy that carbon nanotubes were dispersed homogeneously in nylon polymer. (From Endo, M. et al., *Pure Appl. Chem.*, 78, 1703–1713, 2006.)

for cardiac tissue engineering applications, and the scaffold with 0.5% CNT content (3.6 ± 0.8 V/cm) compared to materials with 0% (5.1 ± 0.8 V/cm) and 0.1% (5.0 ± 0.7 V/cm), improving greater tissue maturity [329b].

The scientific and technical issues include the potential toxicity of carbon nanotubes. The biocompatibility of carbon nanotubes depends on the geometry and surface chemistry. The process development is to control the dimension or morphology of carbon nanotubes, and subsequent development is to control the functionalization of carbon nanotubes for the device fabrication [327b], or understand potential dimension (like length) dependent effects on the behaviour of carbon nanotubes in biomedical applications [327c].

3.2.13 Environmental applications

Carbon nanotubes could be used for environmental applications through various forms and approaches. One type of applications is that carbon nanotube-based coating for anti-corrosion applications or anti-biofouling coating [330a]. When carbon nanotubes are added in the anti-corrosion coatings for metals (e.g., carbon steel), they can enhance coating stiffness and strength and provide an electric pathway for catholic protection [330b]; carbon nanotube-based anti-biofouling paint on ship hulls could discourage attachment of algae and barnacles [311].

The MWNT mats were used to electrochemically oxidize aqueous chemicals and microorganisms, and

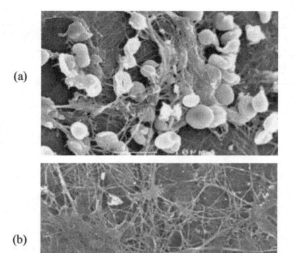

(a)

(b)

Figure 3.58 SEM images of extraluminal surface of pure nylon-derived microcatheter **(a)** and nanotube-filled nanocomposite-derived microcatheter **(b)**. (From Endo, M. et al., *Pure Appl. Chem.*, 78, 1703–1713, 2006.)

the action occurs through the surface functional groups (e.g., –OH and –COOH); therefore, the surface functional group is related to the defects of carbon nanotubes, and the density of nanotube mat is related to diameter (e.g., 17 ± 9 nm) and length (e.g., 91 ± 21 μm) of carbon nanotubes [331].

Carbon nanotubes were also used for the filtration in order to purify water [332–335]. As shown in Figure 3.59, SWNT-based filters could be used to remove dyes in the water [335]. Prototype portable water filter using a functionalised tangled carbon nanotube mesh has already been made; further development and verification is ongoing [311]. DWNTs were explored for the use in water filtration. The diameter of carbon nanotube and the way to make membrane may affect the performance of filtration [333,334]. It is worth mentioning that health and safety issues on carbon nanotubes need to be considered if they are used for applications related to drinkable water; this will be discussed later on. For the above-mentioned applications (via electrochemical oxidation), the scientific and technical issues and process control for high-performance carbon nanotubes are to control the diameter, length, defects of carbon nanotubes, and the way to process carbon

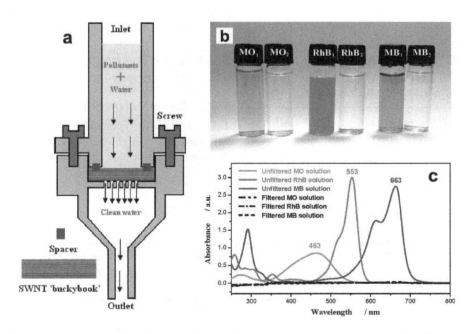

Figure 3.59 (a) Schematics of the filtration set-up used for molecular separation. **(b)** Photograph of the pristine solutions (MO_1, RhB_1 and MB_1) and the corresponding filtrates (MO_2, RhB_2 and MB_2). **(c)** Vis-a-vis spectra of the unfiltered and filtered MO, MB and RhB samples. (From Liu, Q. et al., *ACS Nano*, 3, 707–713, 2009.)

Table 3.12 Current applications for 1D nanostructures

	Research	Applied Research and Development	Demonstration	Commercial
MWNTs		Composite, biomedical, fiber		
SWNTs		Fiber, Transparent electrode, composite		R&D Sample
Si, Ge NS	PV, Sensor			
III-V NS	PV, Sensor			
II-VI NS	FET,Optoelec.,Sensor			
Oxide NS	PV, Sensor			
Carbide NS	Composite			
Nitride NS	Composite			
Elemental NS		Transparent electrode (Ag)		
Technology Readiness Level (TRL)	TRL 1 Basic principle	TRL 2 Tech. Concept	TRL 3 Proof of concept	TRL 4 Validation Lab	TRL 5 Validation Field	TRL 6 Prototype	TRL 7 Prototype real environment	TRL 8 Qualified and Tested	TRL 9 Operation

nanotubes. Based on the reviews on various type of 1D materials, Table 3.12 shows the current technology readiness level of different materials.

3.3 SCALE-UP PROTOTYPE PRODUCTION AND INDUSTRIALIZATION OF NANOWIRES AND NANOTUBES

1D nanomaterials have been prepared on a different scale including the process development in the lab, scale-up pilot production, and large-scale production on industrial scale. In this part, scale-up pilot production will be treated as prototype production, and production on large scale will be treated as industrial production.

3.3.1 Scale-up prototype production of nanowires and nanotubes

3.3.1.1 CARBON NANOTUBES

There are various types and morphology of carbon nanotubes that are now in scale-up prototype production, as highlighted in Table 3.13.

Table 3.13 Types and morphology of carbon nanotubes that are now in scale up prototype production

Morphology / Nanotube type	Random/Bulk	Bush style	Other format
MWNTs	Bulk MWNTs	Bush style MWNTs	Fibre
DWNTs	Bulk DWNTs		
SWNTs	Bulk SWNTs	Bush style SWNTs	Fibre style

3.3.1.2 MULTI-WALLED CARBON NANOTUBES

The scale-up prototype production method of MWNTs was developed with the increasing understanding of the growth mechanism of carbon nanotubes and the method depends on the structure and morphology of carbon nanotubes. The development of carbon nanofibres in the 1970–1980s was prior to the recognition of carbon nanotube by the research community in the 1990s while the industrial production of carbon nanofibres was realised in the 1980s for example, by Hyperion Carbon International. Some products could be manufactured with carbon nanotubes.

As Iijima's TEM work on carbon nanotubes was based on the sample obtained from arc discharge method, scale-up prototype production of MWNTs was via arc discharge method. During the arc discharge process, the variation of the pressure during processing, optimal pressure of 500 Torr was established, and a nanotube-to-particle yield of 25% of the starting material could be obtained. The product contained DWNTs and MWNTs. This process seems to produce low-defect DWNTs and MWNTs [25].

The CVD method is regarded as a scalable method, the processes developed include fluidized-bed process [336] and mobile-bed process [337]. Both these processes have been used for scale-up prototype production as well as on industrial scale, more detail will be discussed in the section of industrial production.

When camphor is used as the carbon source [338,339] in a simple horizontal CVD quartz tube reactor (e.g., 1 m long and 55 mm Ø quartz tube), as shown in Figures 3.60 and 3.61, 30-min process of 12 g camphor over 0.6 g Fe–Co-zeolite powder at 650°C yields 6.6 g carbon nanotubes with an as-grown purity of >91%. Since camphor is environmentally friendly, it opens door for a greener process [338]. Plasma torch-based process has been developed, and the gaseous precursor can be CH_4 [340,341], or even the use of coal [342]. There is work to use coal for arc-discharge, but this is not a conventional CVD process [343].

A schematic diagram of the arc-jet plasma reactor is shown in Figure 3.62 operating at 500 mTorr pressure [341]. The plasma is generated by the dc plasma torch (like 48 V, 300 A), pure Ar or a mixture of Ar and H_2 is used as the plasma gas. CH_4 and Ni powder are injected at the anode exit as the carbon source and metal catalyst, where CH_4 would undergo decompose at high temperature (above 5000 K) and the synthesised carbon nanotubes are collected by a stainless-steel

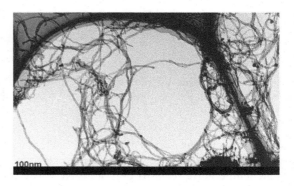

Figure 3.61 TEM image of the as-grown carbon nanotubes. (From Kumar, V. et al., *J. Phys. Chemy C*, 112, 17750–17754, 2008.

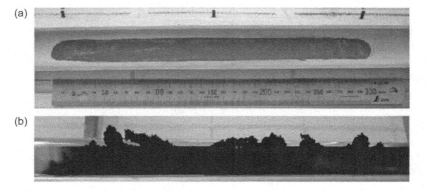

Figure 3.60 Photographs of the zeolite bed: **(a)** before CVD and **(b)** after CVD. (From Kumar, M. and Ando, Y., *Def. Sci. J.*, 58, 496–503, 2008.)

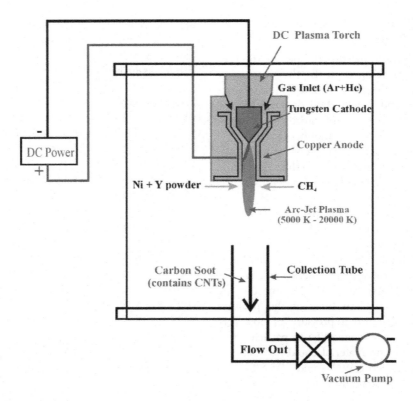

Figure 3.62 Schematic of an arc-jet plasma reactor for the synthesis of carbon nanotubes by decomposition of CH_4. (From Choi, S.I. et al., *Thin Solid Films*, 506–507, 244–249, 2006.)

collection tube under the plasma jet. A study shows that the addition of H_2 is essential to produce high temperature (e.g., to evaporate Ni powder) plasma in order to obtain carbon nanotubes, and Ni powder would not evaporate sufficiently to form nano-sized catalyst particles; therefore, the growth of carbon nanotubes is limited. However, when the content of H_2 is high (3 slm), the plasma is not stable and the growth of carbon nanotubes is also being affected negatively. Similarly, there is a limitation on the content of CH_4 in the process gas, and high content of CH_4 (4 slm) would also reduce the yield of carbon nanotubes which can be due to the lower plasma temperature to affect evaporation of Ni powder. Another factor is the plasma velocity related to the anode diameter as it seems that 7 mm diameter of anode would give a higher yield of carbon nanotubes. The as-prepared carbon nanotubes typically have a diameter of 30–60 nm and lengths of longer than 1 μm. The weight purity of carbon nanotubes can reach over 90% by optimizing the process

conditions (e.g., 2-slpm H_2, 9-mm anode diameter and 1-slpm CH_4). However, the crystallization degree of carbon nanotubes is still lower than carbon nanotubes produced by the conventional arc discharge method.

The continuous process to grow carbon nanotubes via arc jet plasma could achieve a production rate of 0.54 g/ min, and the rate is expected to increase through further process optimization [340,341]. Since the processing conditions have yet to be fully optimized, the conversion rate of CH_4 into carbon soot is below 10%. However, this conversion rate could be elevated by controlling the plasma characteristics because it might be also determined by the cooling rate of the arc-jet plasma [341]. When the coal is used as a carbon source, a continuous arc-jet method has been reported, although further optimization is required for mass production of carbon nanotubes [342]. A floating-catalyst method has been developed to produce fibres and ribbons of carbon nanotubes through direct

100 µm

Figure 3.63 Optical microscope images of different parts of a fibre showing variation in diameter. (From Motta, M. et al., *Nano Lett.*, 5, 1529–1533, 2005.)

spinning of continuous fibres from an aerogel of carbon nanotubes formed in the CVD reaction zone during the growth process, in which the mixture of thiophene and ferrocene with the hydrocarbon liquid was injected into a furnace with the temperature of 1180°C, the carrier gas was H_2 or a mixture of H_2/Ar, and the hydrocarbon source was either ethanol, ethylene glycol, or hexane. Figure 3.63 shows optical images of different parts of the resulting fibre from a process using ethanol as carbon source. The type of carbon nanotubes depends on the process conditions, as shown in Figure 3.64. Most carbon nanotubes are MWNTs [37b].

3.3.1.3 BUSH-STYLE MWNTs

Bush-style MWNTs can be used to produce carbon nanotube-based yarn [32,344], which can be further processed as filaments for the bulbs or other products. As shown in Figure 3.65 [291], the CVD process can be used to produce aligned bush-style MWNTs. With the development of the size of the substrate, the sample size could be increased. The process can be used to produce batches of aligned carbon nanotubes. The growth of super-aligned carbon nanotube arrays on

silicon wafers was successfully scaled up with wafer diameters from 1 to 4 and then to 8 inches. A low-pressure CVD (LPCVD) system was developed to achieve the growth of super-aligned CNT arrays with good uniformity to meet the industrial requirements for the batch-growth process. The heights of the super-aligned carbon nanotube arrays can be easily altered from less than 100 to 900 µm simply by tuning the growth time, and the drawable carbon nanotube arrays with a height of up to 1.5 mm have been successfully synthesized. It is found that the height of carbon nanotube array around 0.5–1.5 mm would suit the process to manufacture yarn [250]. Moreover, only the super-aligned process is good for the yarn, as shown in Figure 3.66. This realisation of the super-aligned carbon nanotube array is due to the narrower diameter distribution (e.g., tuneable around 6.2 nm [3 ~ 4 walls] to 9.2 nm [6 ~ 9 walls]) and higher density (e.g., 2.5×10^{10} tubes/cm^2 or more). It has a very clean surface, and thus very strong van der Waals interactions with neighbouring carbon nanotubes [345].

3.3.1.4 DOUBLE-WALLED CARBON NANOTUBES

DWNTs have different morphologies, but only random DWNTs have been produced on scale-up prototype production in the lab. The fixed-bed, floating catalyst and fluidized-bed-based processes have been utilized for the prototype production of DWNTs.

In the case of a fixed-bed process, Mo was added to CoMgO solid solution to act as the catalyst, when 6 g catalyst (prepared via the combustion process) was reduced in a mixture of H_2–CH_4 (e.g. 18 mol% CH_4) at a heating and cooling rates of 5°C min^{-1}, maximum temperature of 1000°C and no dwell time. The resulted dense mat of composite powder was treated with a concentrated aqueous HCl solution in order to obtain the carbon nanotubes. The HRTEM image in Figure 3.67 clearly shows DWNTs in the bundle, in the gram-scale amount of carbon nanotubes and 77% of carbon nanotubes are DWNTs, and the diameter distribution is small [346a]. When Fe-Mo-MgO was used as catalyst, and a mixture of n-hexane-Ar-H_2 was used as the precursor gases, more than 90% of the as-synthesized carbon nanotubes was DWNTs and over 900% high yield of CNTs relative to the weight of Fe-Mo in the Fe-Mo/MgO catalyst, thus, a high yield of carbon nanotubes can be obtained [346b].

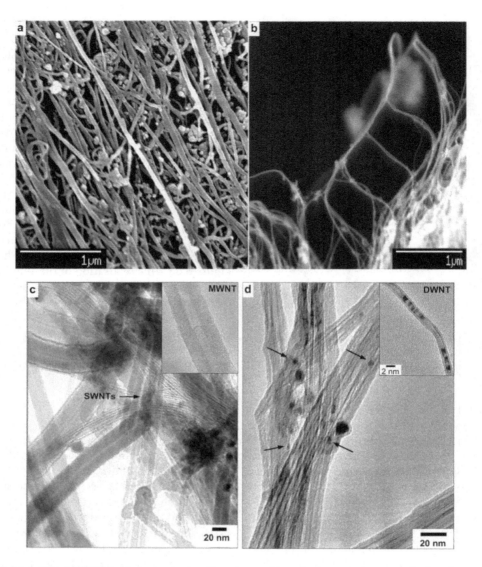

Figure 3.64 **(a)** Typical SEM micrograph showing the multi-walled nanotube strands that form the internal structure of fibres (ethanol, 500 mL/min H_2; Fe atom% 0.022). **(b)** Scanning electron micros-copy micrograph of a fracture surface of a fibre (hexane, 400 mL/min H2; Fe atom% 0.009) pointing out the characteristic branching of single-walled nanotube bundles. (The micrograph was obtained from a region close to the fibre fracture surface.) **(c)** TEM micrograph of a mixture of multi-walled and single-walled nanotubes, from fibres (ethylene glycol, 1000 mL/min [2H2/3Ar]; Fe atom% 0.016) The inset shows a multi-walled nanotube with a cross section of 30 nm, typical in this microstructure. **(d)** TEM micrograph of a fibre (hexane, 400mL/min H2; Fe atom% 0.009), predominantly composed of a mix of single- and double-walled nanotubes. Note that the average size of the Fe particles is smaller for such fibres (arrows). (From Motta, M. et al., *Nano Lett.*, 5, 1529–1533, 2005.) (A) Photograph of nanotubes being wound from the spindle (left) onto a second spindle (right). (B and C) SEM micro-graphs of a fibre that consists of well-aligned MWNTs. (D) A permanent twist introduced into a nano-tube fibre after its removal from the furnace. (From Li, Y.L. et al., *Science*, 304, 276–278, 2004; Motta, M. et al., *Nano Lett.*, 5, 1529–1533, 2005.)

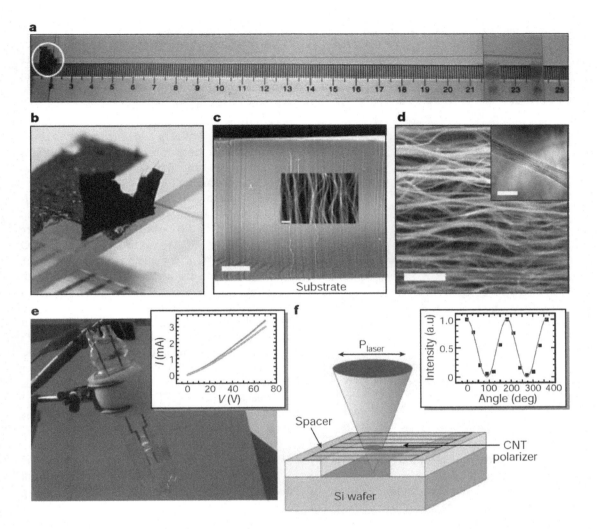

Figure 3.65 Carbon nanotube yarns. **(a, b)** A carbon nanotube yarn being continuously pulled out from a free-standing carbon nanotube array **(a)**, which is shown enlarged in **(b)** (roughly 28 magnification). **(c)** Scanning electron microscope (SEM) images of a carbon nanotube array grown on a silicon substrate, showing the superalignment of carbon nanotubes (scale bars: 100 mm; inset, 200 nm). **(d)** SEM image of the yarn in **(a)**; inset, transmission electron microscope (TEM) image of a single thread of the yarn (scale bars: 500 nm; inset, 100 nm). (e) Carbon nanotube filaments emitting incandescent light. Inset, I–V curve measured before (the lower line) and after (the upper line) light emission for 3 h at 70 V. **(f)** A carbon nanotube polarizer. Polarized laser light (Plaser) is focused on the surface of a silicon wafer and reflected back, passing the carbon nanotube polarizer twice before being collected by a CCD detector. Inset, normalized intensity (squares, ratio of the current I to I_{max}; a.u., arbitrary units) plotted against the angle between the polarization direction of the laser light and the carbon nanotube polarizer, compared with values calculated from Malus's law (the line) (From Jiang, K. et al., *Nature*, 419, 801–801, 2002.).

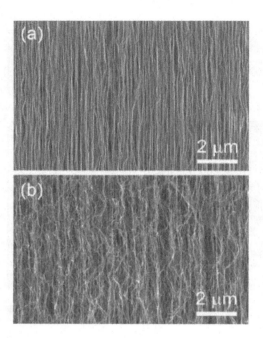

Figure 3.66 Comparison of super-aligned and ordinary arrays. **(a)** SEM side-view image of a super-aligned CNT array. **(b)** SEM side-view image of an ordinary CNT array. (From Liu, K. et al., *Nano Lett.*, 8, 700–705, 2008.)

Another approach to obtain DWNTs is via the floating catalyst method, in which ferrocene and a small amount of sulphur in the atomic Fe:S ratio of 10:1 is dissolved in xylene and forms a solution of 0.03–0.12 g/mL. This was subsequently injected by

a pump into a horizontal tube reactor, with a large H_2-Ar gas flow (2500–3500 and 500 mL/min for Ar and H_2 respectively), at a process temperature of 900°C–1180°C. DWNT filaments and films can be obtained, as shown in Figure 3.68. These long DWNTs could be produced at a rate of circa. 0.5 g/h using xylene as carbon feedstock. The inner diameter of the DWNTs varied from 1.0 to about 2.0 nm, and the interlayer distance of the DWNTs was about 0.4 nm. The growth of DWNT was affected by the argon flowing rate [346c].

DWNTs can also be grown via the fluidized-bed process. In a comparative study, when CH_4 was used as the carbon source in a mixture of 25%CH_4 + 75%Ar (4L/min) and Fe/MgO catalyst at 800°C in a packed bed reactor (PBR), the high concentration of H_2 from the high conversion of CH_4 would cause a quick reduction and sintering of the iron catalyst which inhibited the nucleation of DWNTs. Thus, the dominant carbon product changes drastically from DWNTs to MWNTs along the axial direction of the PBR. Catalyst was added in a small amount in either batch or continuous feeding mode in a fluidized-bed reactor, the DWNT product with a specific surface area of 950 m^2/g and a purity of 98% was obtained [347a].

If 2.5 wt% Fe:Mo (9:1) -Al_2O_3 catalyst at 850°C was used then the reactor was an Inconel 601 cylindrical fluidized bed with an internal diameter of 52 mm and a length of 1000 mm. 100.0 g catalyst was used in each experiment and was introduced into the reactor at room temperature

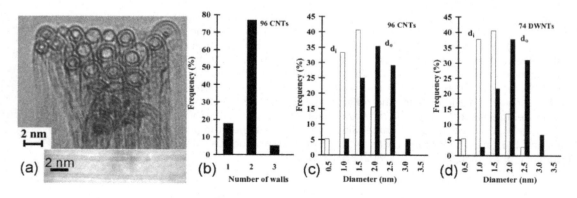

Figure 3.67 **(a)** HRTEM images of DWNTs and **(b)** distribution of the numbers of walls for the whole population (established from 96 individual carbon nanotubes). Distribution of inner (*DI*) and outer (*do*) diameter for the whole population of carbon nanotubes **(c)** and for DWNTs only **(d)**. (From Flahaut, E. et al., *Chem. Comm.*, 2003, 12: 1442–1443, 2003.)

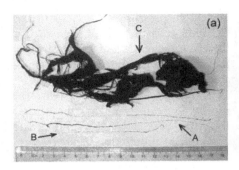

Figure 3.68 **(a)** Optical image of the long DWNT filaments and bulk DWNT films. **(b)** SEM image of the DWNT films. (From Flahaut, E. et al., *Chem. Commun.*, 12, 1442–1443, 2003; Lyu, S.C. et al., *J. Phys. Chem. B*, 108, 2192–2194, 2004; Wei, J. et al., *J. Phys. Chem. B*, 108, 8844–8847, 2004.)

under Ar with a flow rate of 3 Lmin^{-1}. A ratio of Ar:CH$_4$ of 1:1 was used and the processing time was 30 min. The catalyst and reaction conditions resulted in a yield of DWNTs exceeding 200% relative to the metal catalyst (~5 g of DWNT/batch) with circa. 75% DWNT selectivity. The synthesized DWNTs have outer diameters ranging from ~1.41 to −4.15 nm. Figure 3.69 shows the SEM images of carbon nanotubes [347b].

3.3.1.5 SINGLE-WALLED CARBON NANOTUBES

Bulk SWNTs were firstly produced using the arc-discharge process within 2 min and the total carbon mass was about 2 g, consisting of: (1) large quantities of rubbery soot condensed on the chamber walls; (2) web-like structures between the cathode and the reactor walls (no web was formed when either Y or Ni was absent); (3) a cylindrical deposit at the cathode's end; and (4) a small 'collar' (20% of the total mass) around the cathode deposit, as a black, very light and porous but free-standing material. The process is also possible to provide gram quantities of well-defined single-walled nanotubes in the form of highly crystalline bundles. In particular, (10,10) type metallic SWNTs are one of the main type of SWNTs obtained from the process [44].

Bulk SWNTs can also be produced via CVD-based process with a fixed-bed catalyst, floated catalyst, and even from the bush-style SWNTs. When 5 g Fe-Mo catalyst on the Al$_2$O$_3$-SiO$_2$ hybrid support materials was used, 1.5 g carbon nanotube materials can be obtained within 30 min, and the catalyst seems to play an important role in the synthesis using CH$_4$ as the carbon source [61].

Figure 3.69 SEM images of **(a)** Fe-Mo/Al$_2$O$_3$ catalyst particles (125–150 μm) post-synthesis and **(b)** non-purified DWNTs with a mean bundle diameter of 16 nm covering the catalyst surface. The sample was sputtered with a 5 nm layer of Au. (From Liu, Y. et al., *Chem. Eng. Technol.*, 32, 73–79, 2009; Dunens, O.M. et al., *Ind. Eng.Chem. Res.*, 49, 4031–4035, 2010.)

When ethanol is used as the carbon source, the presence of oxygen could help to increase the performance of catalyst, and the use of Fe-Co-Zeolite as a catalyst, would produce a high yield of SWNTs [59], the process is adopted by Toray Industries Inc. (Japan) with a reported capacity of 15 g/h of high purity of SWNTs [348].

In the case of floating catalyst, high-pressure CO (HiPco) process is a typical example of a scale-up prototype production. As shown in Figure 3.70, CO has been used as the carbon source with $Fe(CO)_5$ as the floating catalyst.

The high pressure (e.g., ~10 atmosphere) of CO enhanced the carbon feedstock many folds (give the value....) and significantly accelerated the disproponation of CO molecules into C atoms and thus enhanced the SWNT growth, as shown in Figure 3.71 [349a]. The scale-up prototype scale was reported with a yield of HiPco process at ~0.45 g/h. The early stage of commercial production capacity was 65 g/h by Carbon Nanotechnologies Inc. (USA) [348]. Figure 3.71 shows TEM images of the resulting carbon nanotubes, the main impurity is metal

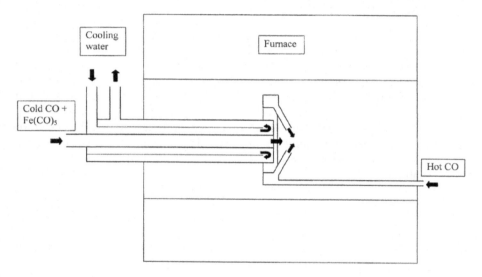

Figure 3.70 Layout of CO flow-tube reactor, showing water-cooled injector and 'showerhead' mixer. (From Nikolaev, P. et al., *Chem. Phys. Lett.*, 313, 91–97, 1999.)

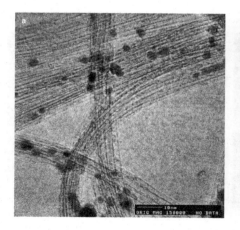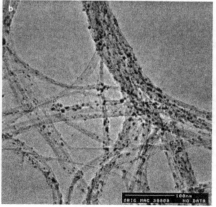

Figure 3.71 **(a)** Low-magnification TEM image of SWNT material produced using the cooled injector with showerhead mixer. Conditions: 3 atm CO with 8 ppm of $Fe(CO)5$, 1200°C. **(b)** High-magnification TEM image of SWNT material in a. (From Nikolaev, P. et al., *Chem. Phys. Lett.*, 313, 91–97, 1999.)

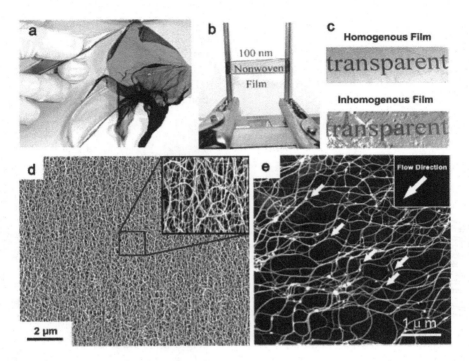

Figure 3.72 **(a)** Photo image of an as-grown 250 nm thick film. **(b)** A transparent 100 nm thick film freely stands between metallic pillars. **(c)** A 150 nm thick homogeneous (upper one) and inhomogeneous films. The importance of homogeneity is obvious. **(d)** Large-scale SEM image of a 250 nm thick film. The inset image is taken at higher magnification. **(e)** SEM image of SWNT network in a single layer. The white arrows in the image point out the Y-type junctions and the flow direction. (From Ma, W. et al., *Nano Lett.*, 7, 2307–2311, 2007.)

particles (e.g., Fe). The process gas CO can be recycled as the carbon source and the process can be operated in a continuous mode.

Since HiPco process has less control on the particle size, a floating catalyst with catalyst support process has been developed, in which the precursor of floating catalyst and precursor would be introduced to the reaction zone together. By controlling processing conditions, different diameter distributions of SWNTs could be obtained, and the process has been utilized on an industrial-scale production, more detail will be provided in the industrial section [349b].

In other cases, the floating catalyst method can be used to directly grow carbon nanotube films or strands [350]. The carbon nanotube film was grown in a horizontal tube furnace with two temperature zones, the catalyst source of ferrocene/sulphur powder was heated to 65°C–85°C and flowed into a reaction zone (over 600°C) by the mixture of 1000 sccm Ar/CH$_4$ (1–8 sccm). A film with a thickness of 100 nm was obtained in the reaction zone, as shown in Figure 3.72. The growth rates of the films were mainly determined by the sublimation rate of the catalyst. Similarly, carbon nanotube strands were produced in a vertical furnace at 1150°C. The *n*-hexane solution with a given composition of ferrocene (0.018 g/mL) and thiophene (sulphur additive, 0.4 weight %) was introduced into the reactor at a rate of 0.5 mL/min, with hydrogen as the carrier gas flowing at a rate of 250 mL/min. The process can be operated in continuous mode, and the lab scale yield was 0.5 g/h. Moreover, very long SWNT strands were produced in large yields consisting of SWNT bundle with the triangle lattice [350].

In addition to the conventional thermal CVD process, an RF plasma torch-based induction thermal plasma technology has also been used to synthesize high-quality SWNTs at large scale, but the process is not a conventional CVD process. As shown in Figure 3.73 [351], a thermal plasma jet arc discharge was used to atomize a gaseous mixture. A non-transferred thermal plasma system was operated through a direct current (DC) with a zirconium-containing W cathode and a Cu anode.

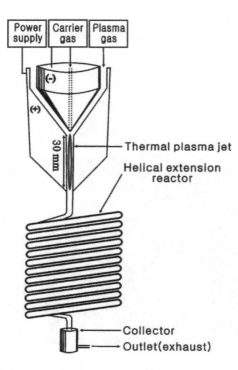

Figure 3.73 A schematic of the thermal plasma system for the continuous production of carbon nanotubes. (From Hahn, J. et al., *Carbon*, 42, 877–883, 2004.)

Thermal plasma of Ar and H_2 that was used as an auxiliary gas to enhance heat transfer between the plasma jet and reactants was generated by a high voltage of about 3 KV between the cathode and anode with DC of 100–300 A, and then flowed with near sound velocity into a Cu nozzle of 6 mm inner diameter, $Fe(CO)_5$ was bubbled with Ar as a carrier gas (3–7 ksccm) and introduced into the reaction zone; the carbon source was CO with the range of 10–50 sccm. In order to increase the yield of carbon nanotubes, a stainless steel helical-type extension reactor was coiled on the inner surface of the chamber and attached at the end of the plasma nozzle. The product soot was harvested through a collector. The temperature of the plasma jet was very high and the velocity was very fast. At a spot that was 10 mm away from the exit of the anode nozzle, the temperature was nearly 10^4K and the velocity was in the range of 100–150 m/s. Due to the high temperature, carbon nanotubes were well-graphitized. The yield of product soot depended on the process conditions: within the tested conditions, the yield was up to 11 mg/h,

the fraction of carbon nanotubes can be more than 60%. Typical carbon nanotube product was shown in SEM images with different magnification in Figure 3.74. High-resolution TEM images (HRTEM) are shown in Figures 3.75. The type of carbon nanotubes was related to process conditions.

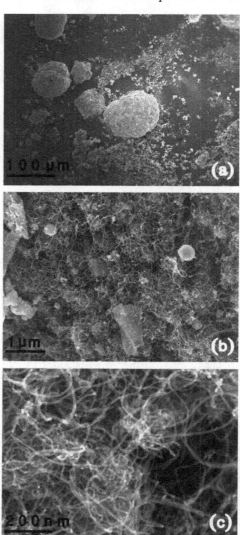

Figure 3.74 FESEM images of carbon nanotubes synthesized by the thermal plasma jet dosing with hydrogen gas. (a) A low-magnification image of the overview of the soot with cycloid shapes, (b) A medium-magnification image of cycloid shapes that are composed mostly of carbon nanotubes with minor metal particles, (c) A high-magnification image showing the existence of numerous carbon nanotubes and bundles with intricate structure. (From Hahn, J. et al., *Carbon*, 42, 877–883, 2004.)

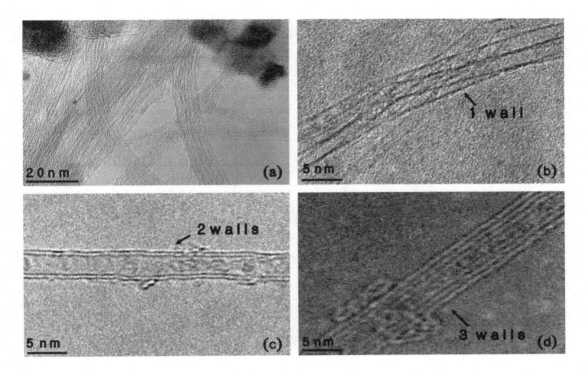

Figure 3.75 HRTEM-images of CNTs synthesized by the thermal plasma jet: **(a)** intricate and bundled, **(b)** 1 walled, **(c)** 2 walled and **(d)** 3 walled CNTs. (From Hahn, J. et al., *Carbon*, 42, 877–883, 2004.)

Without H_2 dosing, SWNTs were mostly observed, with H_2 dosing, DWNTs were mostly observed, and three-and four-walled carbon nanotubes were rarely observed. This indicated that dosing H_2 may change the type of carbon nanotubes from SWNTs to DWNTs. This type of plasma process can be operated continuously, and the reported high yield was approximately 40 wt% of SWNT that can be continuously synthesized at the high production rate of 100 g h^{-1} [351].

3.3.1.6 BUSH-STYLE SWNTs

Another approach to obtain large quantity of SWNTs is via the growth of SWNT array, then to peel off SWNTs from the substrate. It has been found that the height and density of SWNTs depended on the process conditions. In particular, maintaining the catalyst activity is important. There are various approaches, including using ethanol as the carbon source, or adding H_2O in the process gas. As shown in Figure 3.76, SWNT array was produced with ethanol as the carbon source; vertical growth of bundles occurs early and causes alignment of the SWNT film. In another case, high

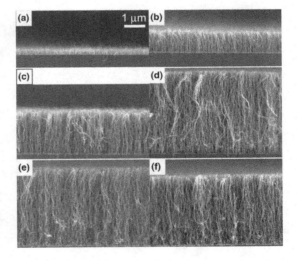

Figure 3.76 Growth of aligned SWNT films after **(a)** 15 s, **(b)** 1 min, **(c)** 3 min, **(d)** 10 min, **(e)** 30 min, and **(f)** 100 min. Vertical growth of bundles occurs early and causes alignment of the SWNT film. After 10 min the film thickness exceeds 4 lm, but the thickness decreases after longer reaction times. The scale bar applies to all images. (From Maruyama S. et al., *Chem. Phys. Lett.*, 403, 320–323, 2005.)

density of SWNT array can be obtained with the addition of H_2O in the process gas; the height is 2.5 mm in 10 min. The SWNT/catalyst weight ratio exceeds 50,000%, more than 100 times as high as that of the high-pressure carbon monoxide (HiPco) process [41]. The process can be used for the preparation of carbon nanotubes at 100 g/h [348].

3.3.1.7 OTHER 1D MATERIALS

Besides carbon nanotubes, other 1D nanomaterials have also been scaled up and produced at prototype scale. For III–V 1D nanomaterials, aerotaxy is a new method for growing semiconductor nanowires without involving a substrate, which is a scalable process for mass production of the designed nanowire materials. Aerotaxy operates by creating an aerosol of catalytic seed particles and adding precursor molecules in a tube furnace, which would encourage nanowires to grow at a very high speed of more than 1 μm/s, which is 100–1000 times faster

than the substrate-based methods. The nanowires can be collected in filters, directly onto heated substrates or dispersed in liquids [106].

Another type of process is based on the vapour-solid reaction, for example, gram quantities of Zn_3P_2 nanowires could be produced via the vapor transport of phosphorus (heated at 480°C, using 20 sccm H_2 as carrier gas in 1.5-inch-diameter quartz tube) onto heated zinc foils at 400°C under a pressure of 1 Torr [352]. In order to maximize the area of the foil over which nanowires can be obtained in one experimental run, a large 30 × 8 cm^2 foil was compacted by rolling it concentrically into a 1.5-inch diameter coil as shown in Figure 3.77. Due to the high thermal conductivity of metallic zinc, the entire foil was considered to be isothermal under the experimental conditions. Photographs of a coiled zinc foil before and after the reaction are shown in Figure 3.77a and b, respectively. SEM analysis of the deposit obtained on top of

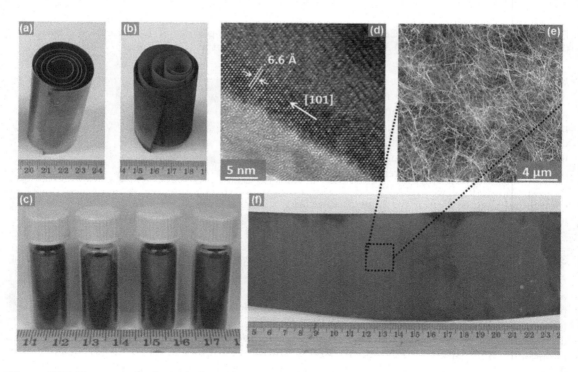

Figure 3.77 Photograph of a coiled zinc foil (a) before and (b) after the vapor transport of phosphorus. (c) Photograph of Zn_3P_2 nanowire powder obtained by brushing off the foils. (d) A transmission electron micrograph from a Zn_3P_2 nanowire indicating the growth direction of the nanowires. (e) A scanning electron micrograph of Zn_3P_2 nanowires obtained by the above-mentioned approach. (f) Photograph of the uncoiled zinc foil covered with Zn_3P_2 nanowires deposit after the vapor transport of phosphorus onto its surface. (From Brockway, L. et al., *Phys. Chem. Chem. Phys.*, 15, 6260, 2013.)

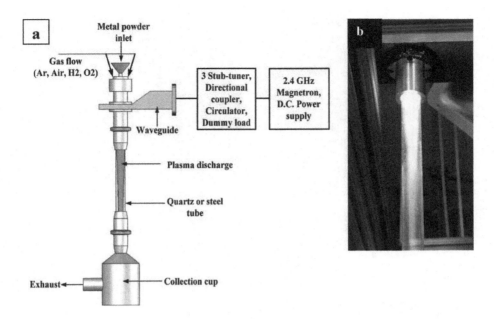

Figure 3.78 (a) Schematic of the microwave plasma jet reactor showing all the essential components. **(b)** Photograph of the high-density plasma jet discharge at 2 kW power produced in a quartz tube of 1.5 in. diameter with a plasma flame length of about 12-15 in. (From Kumar, M. and Ando, Y., *Def. Sci. J.*, 2008, 58, 496–503, 2008.)

the foils after the chemical reaction (Figure 3.77b and f) indicated that it comprised Zn_3P_2 nanowires of 30–50 nm in diameter and tens of microns long (Figure 3.77). TEM analysis (Figure 3.77d) of Zn_3P_2 nanowire revealed that the NWs were single crystalline and their growth direction was along the (101) axis. TEM-EDS analysis confirmed the Zn_3P_2 stoichiometry of the nanowires. Following the synthesis, Zn_3P_2 nanowire powder was obtained by simply brushing the foils using the back side of a razor blade (Figure 3.77c). Approximately 250 mg of nanowire powder was produced per production run.

For oxide 1D nanomaterials, ZnO nanowires are one of the most important nanomaterials for future optoelectronic nanodevices. A modified carbothermal reduction method can synthesize a large quantity of ZnO nanowires within 5 min. The synthesis was carried out in a quartz tube furnace at atmospheric pressure without using any catalysts. A mixture of nitrogen/air gas was used as a reactive and carrier gas. About 0.85 g ZnO nanowires were obtained using 1 g ZnO and 1 g graphite powder as the source materials. The prepared ZnO nanowires exhibited a hexagonal wurtzite crystal structure, with the average length

and diameter about 1.8 μm and 35 nm, respectively. A 85% yield was obtained in 5-min preparation with a production rate of 10 g/h [353a]. Another method uses a high-throughput plasma jet reactor capable of operating at atmospheric condition for the direct gas-phase oxidation of metal powders toward bulk NW production at a rate of 5 kg/day. This can be used for several metal powders such as Zn, Al, Sn, and Ti with different particle sizes and various plasma parameters, as shown in Figure 3.78 [353b]. Figure 3.79 shows low- and high-resolution SEM images of SnO_2 (a, b) and ZnO (c, d) NWs. Nanostructures of ZnO such as tripod (e), nanobrush (f), and nanocomb (g) were also observed [339].

3.3.2 Industrialization of nanotubes

Industrialization here refers to production on the scale of kg and ton, indicating that there is clear application need. For 1D materials, except previously mentioned SiC and B monofilaments that have been used for composite materials, carbon fibres have also been widely used in composite, in particular polymer matrix composites for aerospace and automotive applications, for example. These materials have diameters on the micrometre scales. Nanofibres, nanotubes, and

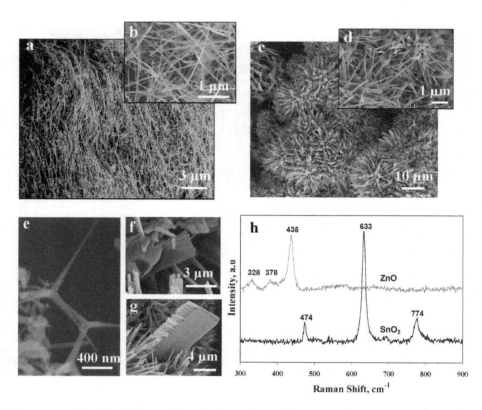

Figure 3.79 Low- and high-resolution SEM images of SnO$_2$ **(a, b)** and ZnO **(c, d)** NWs are shown. Nanostructures of ZnO such as tripod **(e)**, nanobrush **(f)**, and nanocomb **(g)** were also observed. Raman spectra of as-synthesized SnO$_2$ and ZnO NWs are shown **(h)**. (From Kumar, M. and Ando, Y., *Def. Sci. J.*, 2008, 58, 496–503, 2008.)

nanowires have diameters up to 100 nm; they can be produced through fixed-bed large-scale CVD process or a floating catalyst process. Since the TEM work of carbon nanotube in 1991 [15], 1D nanomaterials have also been produced on the industrial scale.

Besides carbon nanotubes, Ag nanowires can now be produced on an industrial scale through a solution-based method [354]. Ag nanowires are used for conductive ink that are widely in used in touch screen sensors, e-paper displays, OLEDs, and electrodes for photovoltaics like organic and CIGS solar cells. This part would focus on carbon nanotubes produced via CVD processes.

3.3.2.1 INDUSTRIALISATION OF MULTI-WALLED CARBON NANOTUBES

Many companies can provide production of MWNTs on a large scale across the world, including manufacturers in China, Japan, USA, and Europe. The production scale can reach hundreds of tons of carbon nanotubes per year in some cases.

In the case of mobile-bed process, as shown in Figure 3.80, the reactor consists of: a tubular inclined rotary kiln with a mobile bed, equipped with systems for continuous catalyst feeding and the collection of gaseous and solid products, with a mass spectrometer to analyse the exhaust gas composition, and automatic control and data acquisition. During the process, the catalyst and feeding gas, and the product collection, are all under inert gas atmosphere. The catalyst flows through the force of gravity, assisted by the rotation and inclination of the reactor, and the catalyst residence time can be adjusted through the inclination angle and rotation speed. The process temperature is up to 1050°C. The process development shows that the catalyst links with the process conditions, i.e., for specific catalyst, it has optimal process conditions that are not suitable for other catalysts. This process has a unique design for the hardware and process so that it could avoid the issues related to fixed-bed process to be scalable [355]. Figure 3.81 shows a

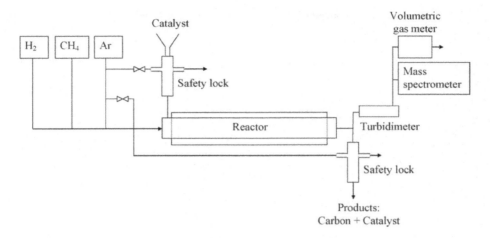

Figure 3.80 Schematic experimental set-up of the continuous reactor. (From Seashell Technology, Silver Nanowire Acquired by BASF, http://www.seashelltech.com/pressRelease_Acquisition.shtml, March 10, 2015.)

Figure 3.81 Pilot reactor for CNT production at Nanocyl SA. (From Pirard, S.L. et al., *Front. Chem. Sci. Eng.*, 11, 280–289, 2017.)

pilot reactor in the plant, and Figure 3.82 shows the images of produced SWNTs [337].

In the fluidized-bed reactor, there is a sintered porous plate used as the gas distributor at the bottom of the reactor, which also supports the weight of the solids above it before they are suspended in a fluid flow. During the process, the catalyst particles are pushed apart from one another by the up-flow of the gas at a sufficient velocity to cause mobility, and carbon nanotubes grow from the catalyst particles. Moreover, both the catalyst and carbon nanotubes were smoothly fluidized in the reactor via proper self-agglomeration [356,357].

Based on the mechanism studies of fluidized-bed process, the pilot plant was built as shown Figure 3.83. Figures 3.84 shows images of carbon nanotubes; the

Figure 3.82 SEM and TEM images of carbon nanotubes obtained with catalyst Cat-2-E using ethylene as carbon source. (From Pirard, S.L. et al., *Front. Chem. Sci. Eng.*, 11, 280–289, 2017.)

Figure 3.83 The facility for MWCTs pilot plant production. (From Wei F. et al., *Powder Technol.*, 183: 10–20, 2008.)

mass production of carbon nanotubes can reach 15 kg/h scale and MWNTs could continuously be synthesized [358]. Ultrahigh purity (99.9%) MWNTs can be obtained via the fluidized-bed process and subsequently annealed under vacuum at a temperature of 1777°C [347,356,357].

Based on the research on nano-agglomerate fluidized-bed reactor, in which when the gas velocity exceeded the apparent minimum fluidization velocity of the primary nanoparticles, nanoparticle agglomerates were formed during fluidization, i.e., nano-agglomerate fluidization, for the synthesis of carbon nanotubes, the process was called 'Nano Agglomerate Fluidized' process. Based on the growth mechanism of carbon nanotubes, the mass production of carbon nanotubes via the fluidized-bed process is proposed which would involve four scale levels from the perspective of time and length scale as follows, the schematic diagram is shown in Figure 3.85 [357]:

1. *Atomic level:* Carbon nanotubes self-assembly at the atomic scale, including the growth condition and growth mechanism of carbon nanotubes, catalyst design, and controllable synthesis of the structure of individual carbon nanotubes. The large number of carbon nanotubes are produced during the fluidized-bed process. The characteristics of carbon nanotubes, such as wall number, diameter, length, defects, centrality, crystallinity, or digitization, are determined at the atomic scale.

2. *Mesoscopic level:* As carbon nanotubes grow longer and agglomerate together, the behaviour of carbon nanotubes can be treated as at the mesoscopic level, involving their morphology, the associated

Figure 3.84 **(a)** The as-grown CNT arrays on ceramic spheres, showing the radial growth behaviour; **(b)** The released CNT arrays by simple mechanical vibration after the CO_2 oxidation. The insert shows the bare ceramic spheres after the detachment of CNT arrays, the scale bar in the insert was 1 mm. (From Huang, J.-Q. et al., *Carbon*, 48, 1441–1450, 2010.)

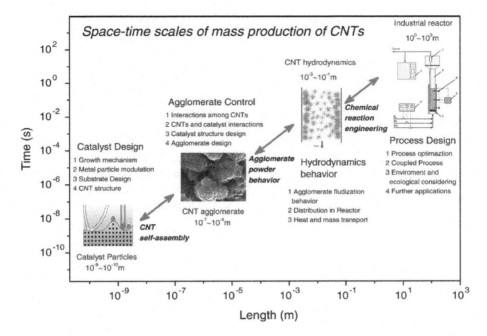

Figure 3.85 Multiscale spaceale space fsis of the mass production of CNTs. (From Zhang, Q. et al., *ChemSusChem*, 4, 864–889, 2011 and Wei, F. et al., *Powder Technol.*, 183, 10–20, 2008.)

agglomeration mechanism, and ways to control the agglomerates, as there are various types of agglomerated structures. These include carbon nanotube agglomerated particles, carbon nanotube arrays, and suspended individual carbon nanotubes. The agglomerated carbon nanotubes can be observed during the process and the control of the morphology is important for the mass production of carbon nanotubes.

3. *The flow behaviour scale:* The traditional fluidized-bed setup is required to be modified for the production of carbon nanotubes to accommodate the unique properties of carbon nanotubes and their agglomerates, and their transport properties are considered at the flow-behaviour scale.

4. *Macroscopic level:* This is macroscopic level integrating CNT process engineering, including the process reactor design and process operation, the relationship between the macroscopic operation and microscopic carbon nanotube structure, environmental and ecological considerations, and the delivery and application of carbon nanotubes, which involves traditional chemical engineering concepts like process operation.

In another case, Bayer developed its own catalyst of Co/Mn/MgO/Al$_2$O$_3$ and fluidized-bed process for the production of carbon nanotubes [359a]. Its capacity is about 200 tons per year, although due to the commercial reasons Bayer stopped the production process in 2013.

3.3.2.2 INDUSTRIALISATION OF SINGLE-WALLED CARBON NANOTUBES

In addition to MWNTs, SWNTs have also been produced on an industrial scale. SWNTs are produced in two different types of methods: floating catalyst and supported catalyst by fluidised bed. The former includes the HiPCO process, in which carbon monoxide disproponation reaction occurs at high pressure (up to 50 atm) and high temperature (800°C–1200°C). This approach was used in the early days of industrialization of SWNTs; high pressure and high temperature make this HiPCO process rather difficult to be scaled up, and commercial quantities of SWNTs are no longer made by this process. Another floating-catalyst method is to inject the catalyst into the process gas; the catalyst can be as the prepared catalyst as solid like Fe/Mo/Al$_2$O$_3$, as shown in Figure 3.86, or as the liquid like a solution of ferrocene, polyhedral oligomeric silsesquioxane, and toluene. This type of process has been utilized

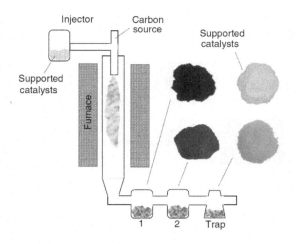

Figure 3.86 Scheme of the vertical floating technique system designed and used for continuous injection of pre-prepared alumina supported Fe/Mo catalyst powder. The final product is collected in three different containers. Images of the initial catalyst and the carbon soots collected from the different containers are also shown. (From Mora, E. et al., *Carbon*, 45, 971–977, 2007.)

commercially and can provide SWNTs to kilograms per day by a fine-chemical manufacturer (Thomas Swan) in UK with the example images shown in Figure 3.87 [359b]. Its process does not require high pressure, and the catalyst is produced *in-situ* during the growth process. As it is difficult to control the size of catalyst metal particles prior to the nucleation and growth of SWNTs, there is difficulty

in controlling diameter and chirality of SWNTs using the floating-catalyst process. Moreover, other forms of carbon as impurities exists in the product, which needs extensive purification to obtain highly pure SWNTs [360]. Instead of *in-situ* production of catalyst particles, an arc discharge-based process was developed to produce catalyst vapour, which then condenses to form nanoparticles for the catalysis of SWNTs. The resulting product contains 80% or more of high-quality single-wall carbon nanotubes [361a]. These have been marketed as TUBALL by the company OSCiAl, which built a synthesis facility named Graphetron 1.0 with a capacity of 10 tonnes in 2014 and plans to expand the capability to 60 tonnes with the addition of a second facility known as Graphetron 50 in 2017 [361b].

Another type of method is the use of supported catalyst method, in which the metal catalyst is supported in the carrier, and the process is performed via a fluidized-bed chemical process, depending on the precursor. The precursor gas can be reused to save the consumption of precursor gas. CO is used in the CoMoCAT process, in which Co and Mo catalysts are supported with silica particles, the process has been commercialized by SouthWest NanoTechnologies, Inc. (SWeNT) based on the process developed at the University of Oklahoma. The process can be performed below 900°C under the atmosphere pressure to moderate pressures and can provide control of the catalyst site, reaction temperature, carbon feed source and partial pressure, the composition

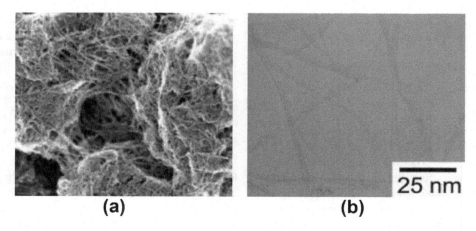

(a)   (b)

Figure 3.87 **(a)** SEM image of Elicarb® SW and **(b)** TEM image of Elicarb® SW. (From Thomas Swan, Elicarb. Carbon Nanotubes, http://www.thomasswan. co.uk/advanced-materials/ elicarb%C2%AE-carbon-nanotubes.)

of metal catalyst, and the type of catalyst support. A SWNT material, SG65i, which has an average diameter of 0.77 ± 0.02 nm, over 40% (6,5) chirality and 95%–97% semiconducting nanotubes. After removing the catalyst, the resulting material has a carbon purity greater than 95 wt.%, and relative purity typically 99%. The issue of supported catalyst CVD method is that the catalyst support is a high mass percentage of the as-prepared SWNT product. Further development to address such issue is under way to release SWNTs from supported catalyst and recycle the catalysts [361c].

In particular, CoMoCAT catalyst can be used to produce specific (6,5) chirally-pure SWNTs on a large scale (Figure 3.88).

As mentioned previously, MWNTs have been used for the production of Li-ion battery towards electric vehicle applications on the scale

of hundreds of tons, and SWNTs are used for the developing applications for transparent conductive film, conductive ink, and sensors as marketable products, such as INFLECTTM FLEX Sensor by the USA company Brewer Science [361d].

With the development of more applications, it is expected that the industrial production of carbon nanotubes will be increased. Meanwhile, further work is still needed on the controllable growth of carbon nanotubes with specific structure such as specific diameter, number of walls, and chirality.

3.3.2.3 ENVIRONMENT, HEALTH, AND SAFETY ISSUES

1D nanomaterial has similar 1D structure to asbestos fibre. The danger of asbestos fibre came to light in the 1960s, in which asbestos fibres could be inhaled into the deep lung and stick around for decades, and there are concerns that widespread use of carbon nanotubes may lead to mesothelioma, cancer of the lining of the lungs caused by exposure to asbestos [362a]. Hence, it raises the environment, health, and safety issue about 1D nanomaterials, from the perspective of manufacturing, handling, and disposal.

As mentioned previously, carbon nanofibre has already been produced commercially since the 1970–1980s. The work on 1D nanomaterials takes places in the lab and production scale particularly after Iijima's work in 1991. Currently carbon nanofibres and carbon nanotubes have now been produced on the industrial scale. An investigation on the production facility during the whole process of production of carbon nanofibres through aerosol monitoring and mobile direct-reading sampling shows that the dominant carbon nanofibre mode by particle number lies between 200 and 250 nm for both aerodynamic and mobility equivalent diameters, and the monitoring can provide information on the control of exposure limit [362b].

For the current production of carbon nanotubes, including floating catalyst HiPco process and fluidized-bed supported catalyst process, an environmental impact analysis has been conducted to assess the performance of processes. This is in terms of their environmental sustainability through a life-cycle analysis (LCA) approach, covering production in the reactor, acid treatment to remove catalyst, liquid, and solid waste disposal. The work shows that a

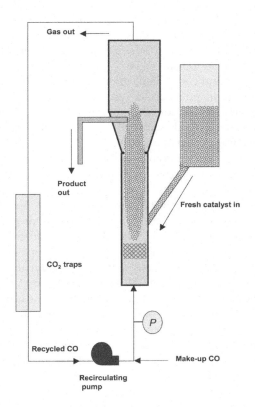

Figure 3.88 Schematic of a fluidized-bed reactor for manufacturing single-walled carbon nanotubes by the CoMoCAT process. This design exhibits extremely good temperature uniformity and is easily scalable. (From Arthur, D. et al., *MRS Bulletin*, 37, 1297–1306, 2012.)

comprehensive environmental impact assessment may produce vital data about the performance of each process in terms of various environmental impact categories [363a]. In particular, on 13 October 2016, OCSiAl announced that it is the first manufacturer of single-wall carbon nanotubes (SWNTs) to complete Registration, Evaluation, Authorization and Restriction of Chemicals (REACH) registration, indicating a technical and highly detailed analysis of SWNTs and their nature and potential risks, as well as specifications for particular methods of risk management has been performed [363b]. From the perspective of health and safety, significant effort has been made to understand the potential of exposure to 1D nanomaterials, which varies with different compositions, in particularly carbon nanotubes. Recently, biological effects of tailor-made MWNTs without functionalization were investigated *in vivo* in a two-year carcinogenicity study; the work shows a carcinogenic effect for all tested MWNTs. Moreover, aspect ratio, curvature of MWNTs seems to be an important parameter influencing the carcinogenicity of MWNTs [364]. The review of different investigations shows that MWNTs and SWNTs pose a respiratory hazard through the *in-vitro* studies in mice [365].

Although the exact adverse health effects occuring in humans after exposure to CNT and CNF are not known, the results from animal research studies indicate the need to minimize worker exposure. 1D nanomaterials can be regarded as high-aspect-ratio materials, and it is recommended to follow certain procedures and exposure limits provided by the authority. For example, UK National Institute for Occupational Safety and Health (NIOSH) has provided recommendation for the procedure of handling and exposure limit for carbon nanofibre and carbon nanotubes. It proposes a recommended exposure limit (REL) of 1 $\mu g/m^3$ elemental carbon as a respirable mass 8-hour time-weighted average (TWA) concentration, and describes strategies for controlling workplace exposures and implementing a medical surveillance program [366]. Recently, European Chemicals Agency (ECHA) has published the Infocards about MWNTS in their website about current status on various information about MWNTs [367].

For other 1D nanomaterials, they are at the early stage research in the lab. It is important to follow good lab practice, local regulations on chemicals, and guidance on high aspect nanomaterials from NIOSH before more information of data is available.

3.3.2.4 SUMMARY AND OUTLOOK

This chapter introduces historical background and the development of 1D material from the microscale to nanoscale, covering fibre, whisker, nanofibre, nanotube, and nanowire produced via CVD processes, including a brief review of various CVD processes for the preparation of carbon nanotube, Si and Ge nanowires, III–V, II–VI, oxide, carbide, nitride, and elementary 1D nanomaterials. The potential and commercial applications of 1D nanomaterials, in particular, carbon nanotubes, and the approaches for mass production of 1D nanomaterials, and industrial production process have been described.

From the perspective of specific types of 1D nanomaterial, carbon nanotubes are already produced on an industrial scale. Although great progress has been made to obtain high ratio of metallic, semiconducting, or chirality-specific carbon nanotubes, more research is needed to prepare monodisperse SWNTs or SWNTs with the desired and controllable diameter and length. For other 1D nanomaterials, research on the synthesis and new applications is essential to inspire scalable processes, which can reduce cost and further encourage wider applications.

There are a variety of 1D nanomaterials and several issues to be addressed for their further development. The first challenge is to develop processes for synthesizing the material with the desired properties. The second challenge is to explore how to integrate 1D nanomaterials into devices or complex structures, and the third challenge is to develop disruptive or indispensable applications that will demonstrate the essential advantages of 1D nanomaterials over other materials or currently available applications. The fourth challenge is to verify the environmental, health, and safety issues related to the 1D nanomaterials. Thereafter, there would be sufficient evidence and information to promote or limit the applications of 1D nanomaterials in the wider industries or the society.

In summary, 1D nanomaterials have many special properties (e.g., size confinement effects) and considerable potentials application in many fields such as photodetector and sensor. Further development is required to materialise the potentials in order to benefit the society and the civilization advancement of mankind.

REFERENCES

1. Barth, S. et al., Synthesis and applications of one-dimensional semiconductors. *Progress in Materials Science*, 2010, **55**(6): 563–627.

2. Reibold, M. et al., Carbon nanotubes in an ancient Damascus sabre. *Nature*, 2006, **444**: 286.

3. Hardy, H.K., The filamentary growth of metals. *Progress in Metal Physics*, 1956, **6**: 45–73.

4. Brenner, S.S., The growth of whiskers by the reduction of metal salts. *Acta Metallurgica*, 1956, **4**(1): 62–74.

5. Sawyer, W.E. and A. Man, Carbon for electric lights, US0229335A, June 29, 1880, p. 2.

6. Xia, Y.N. et al., One-dimensional nanostructures: Synthesis, characterization, and applications. *Advanced Materials*, 2003, **15**(5): 353–389.

7. Compton, K.G., A. Mendizza, and S.M. Arnold, Filamentary growths on metal surfaces—"Whiskers". *Corrosion*, 1951, **7**(10): 327–334.

8. Wagner, R.S. and W.C. Ellis, Vapor-liquid-solid mechanism of single crystal growth. *Applied Physics Letters*, 1964, **4**(5): 89–90.

9. Baker, R.T.K., Catalytic growth of carbon filaments. *Carbon*, 1989, **27**(3): 315–323.

10. Oberlin, A., M. Endo, and T. Koyama, Filamentous growth of carbon through benzene decomposition. *Journal of Crystal Growth*, 1976, **32**(3): 335–349.

11. Cooke, T.F., Inorganic fibers – a literature-review. *Journal of the American Ceramic Society*, 1991, **74**(12): 2959–2978.

12. Taniguchi, N., On the basic concept of "nano-technology". *Proceedings of the International Conference on Production Engineering*, Part II, Society for Precision Engineering, 1974, p. 18–23.

13. Kroto, H.W. et al., C_{60}: Buckminsterfullerene. *Nature*, 1985, **318**: 162–163.

14. Krätschmer, W. et al., Solid C_{60}: A new form of carbon. *Nature*, 1990, **347**(6291): 354–358.

15. Iijima, S., Helical microtubules of graphitic carbon. *Nature*, 1991, **354**: 56–58.

16. Holmes, J.D. et al., Control of thickness and orientation of solution-grown silicon nanowires. *Science*, 2000, **287**(5457): 1471–1473.

17. Morales, A.M. and C.M. Lieber, A laser ablation method for the synthesis of crystalline semiconductor nanowires. *Science*, 1998, **279**(5348): 208–211.

18. Hofmann, S. et al., Gold catalyzed growth of silicon nanowires by plasma enhanced chemical vapor deposition. *Journal of Applied Physics*, 2003, **94**(9): 6005–6012.

19. Monthioux, M. and V.L. Kuznetsov, Who should be given the credit for the discovery of carbon nanotubes? *Carbon*, 2006, **44**(9): 1621–1623.

20. Hofer, L.J.E., E. Sterling, and J.T. McCartney, Structure of carbon deposited from carbon monoxide on iron, cobalt and nickel. *The Journal of Physical Chemistry*, 1955, **59**(11): 1153–1155.

21. Bacon, R. and J.C. Bowman, Production and properties of graphite whiskers. *Bulletin of the American Physical Society*, 1957, **2**: 131.

22. Lieberman, M.L., C.R. Hills, and C.J. Miglionico, Growth of graphite filaments. *Carbon*, 1971, **9**(5): 633–635.

23. Iijima, S. and T. Ichihashi, Single-shell carbon nanotubes of 1 nm diameter. *Nature*, 1993, **363**: 603–605.

24. Bethune, D.S. et al., Cobalt-catalysed growth of carbon nanotubes with single-atomic-layer walls. *Nature*, 1993, **363**: 605–607.

25. Ebbesen, T.W. and P.M. Ajayan, Large-scale synthesis of carbon nanotubes. *Nature*, 1992, **358**(6383): 220–222.

26. Guo, T. et al., Self-assembly of tubular fullerenes. *The Journal of Physical Chemistry*, 1995, **99**: 10694–10697.

27. Jose-Yacaman, M. et al., Catalytic growth of carbon microtubules with fullerene structure. *Applied Physics Letters*, 1993, **62**(6): 657–659.

28. Colbert, D.T. et al., Growth and sintering of fullerene nanotubes. *Science*, 1994, **266**(5188): 1218–1222.

29. Yuan, L.M. et al., Ethylene flame synthesis of well-aligned multi-walled carbon nanotubes. *Chemical Physics Letters*, 2001, **346**: 23–28.

30. (a) Ren, Z.F. et al., Synthesis of large arrays of well-aligned carbon nanotubes on glass. *Science*, 1998, **282**(5391): 1105–1107; (b) AuBuchon, J.F., L.-H. Chen, A.I. Gapin, D.-W.

Kim, C. Daraio, and S. Jin, Multiple sharp bendings of carbon nanotubes during growth to produce zigzag morphology. *Nano Letters*, 2004, **4**: 1781–1784; (c) Qi, H.J. et al., Determination of mechanical properties of carbon nanotubes and vertically aligned carbon nanotube forests using nanoindentation. *Journal of the Mechanics Physics of Solids*, 2003, **51**: 2213–2237.

31. Bower, C. et al., Plasma-induced alignment of carbon nanotubes. *Applied Physics Letters*, 2000, **77**(6): 830–832.

32. Fan, S.S. et al., Self-oriented regular arrays of carbon nanotubes and their field emission properties. *Science*, 1999, **283**(5401): 512–514.

33. Flahaut, E. et al., Synthesis of single-walled carbon nanotube–Co–MgO composite powders and extraction of the nanotubes. *Journal of Materials Chemistry*, 2000, **10**(2): 249–252.

34. Zhang, R., Zhang, Y., Zhang, Q., Xie, H., Qian, W., and Wei, F., Growth of half-meter long carbon nanotubes based on Schulz–Flory distribution. *ACS Nano*, 2013, **7**(7): 6156–6161.

35. Yamada, T. et al., Size-selective growth of double-walled carbon nanotube forests from engineered iron catalysts. *Nature Nanotechnology*, 2006, **1**(2): 131–136.

36. Thess, A. et al., Crystalline ropes of metallic carbon nanotubes. *Science*, 1996, **273**(5274): 483–487.

37. (a) Li, Y.L., I.A. Kinloch, and A.H. Windle, Direct spinning of carbon nanotube fibers from chemical vapor deposition synthesis. *Science*, 2004, **304**(5668): 276–278; (b) Motta, M., Y.-L. Li, I. Kinloch, and A. Windle, Mechanical properties of continuously spun fibers of carbon nanotubes. *Nano Letters*, 2005, **5**: 1529–1533.

38. Vander Wal, R.L., T.M. Ticich, and V.E. Curtis, Diffusion flame synthesis of single-walled carbon nanotubes. *Chemical Physics Letters*, 2000, **323**(3–4): 217–223.

39. Eres, G. et al., Molecular beam-controlled nucleation and growth of vertically aligned single-wall carbon nanotube arrays. *The Journal of Physical Chemistry B*, 2005, **109**(35): 16684–16694.

40. Huang, S., X. Cai, and J. Liu, Growth of millimeter-long and horizontally aligned single-walled carbon nanotubes on flat substrates. *Journal of the American Chemical Society*, 2003, **125**(19): 5636–5637.

41. (a) Hata, K. et al., Water-assisted highly efficient synthesis of impurity-free single-walled carbon nanotubes. *Science*, 2004, **306**(5700): 1362–1364; (b) Einarsson, E., Y. Murakami, M. Kadowaki, and S. Maruyama, Growth dynamics of vertically aligned single-walled carbon nanotubes from in situ measurements. *Carbon*, 2008, **462008**: 923–930.

42. Zhang, G. et al., Ultra-high-yield growth of vertical single-walled carbon nanotubes: Hidden roles of hydrogen and oxygen. *Proceedings of the National Academy of Sciences*, 2005, **102**(45): 16141–16145.

43. Li, W.Z. et al., Large-scale synthesis of aligned carbon nanotubes. *Science*, 1996, **274**(5293): 1701–1703.

44. Journet, C. et al., Large-scale production of single-walled carbon nanotubes by the electric-arc technique. *Nature*, 1997, **388**(6644): 756–758.

45. (a) Zhou, W. et al., Copper catalyzing growth of single-walled carbon nanotubes on substrates. *Nano Letters*, 2006, **6**: 2987–2990; (b) Ding, L., D. Yuan, and J. Liu, Growth of high-density parallel arrays of long single-walled carbon nanotubes on quartz substrates. *Journal of the American Chemical Society*, 2008, **130**(16): 5428–5429.

46. Guo, T. et al., Self-assembly of tubular fullerenes. *Journal of Physical Chemistry*, 1995, **99**(27): 10694–10697.

47. Baker, R.T.K. et al., Nucleation and growth of carbon deposits from the nickel catalyzed decomposition of acetylene. *Journal of Catalysis*, 1972, **26**(1): 51–62.

48. Baker, R.T.K. et al., Formation of filamentous carbon from iron, cobalt and chromium catalyzed decomposition of acetylene. *Journal of Catalysis*, 1973, **30**: 86–95.

49. Baker, R.T.K. et al., Effect of the surface state of iron on filamentous carbon formation. *Journal of Catalysis*, 1982, **77**(1): 74–84.

50. Rostrup-Nielsen, J. and D.L. Trimm, Mechanisms of carbon formation on nickel-containing catalysts. *Journal of Catalysis*, 1977, **48**(1–3): 155–165.

51. Baird, T., J.R. Fryer, and B. Grant, Structure of fibrous carbon. *Nature*, 1971, **233**: 329–330.

52. Evans, E.L. et al., Growth of filamentary carbon on metallic surfaces during the pyrolysis of methane and acetone. *Carbon*, 1973, **11**: 441–445.

53. Tessonnier, J.-P. and D.S. Su, Recent progress on the growth mechanism of carbon nanotubes: A review. *ChemSusChem*, 2011, **4**(7): 824–847.

54. Bedewy, M. et al., Collective mechanism for the evolution and self-termination of vertically aligned carbon nanotube growth. *The Journal of Physical Chemistry C*, 2009, **113**(48): 20576–20582.

55. AuBuchon, J.F. et al., Multiple sharp bendings of carbon nanotubes during growth to produce zigzag morphology. *Nano Letters*, 2004, **4**(9): 1781–1784.

56. Teo, K.B.K. et al., Uniform patterned growth of carbon nanotubes without surface carbon. *Applied Physics Letters*, 2001, **79**(10): 1534–1536.

57. Ding, F., K. Bolton, and A. Rosén, Nucleation and growth of single-walled carbon nanotubes: A molecular dynamics study. *Journal of Physical Chemistry B*, 2004, **108**: 17369–17377.

58. (a) Cheng, H.M. et al., Large-scale and low-cost synthesis of single-walled carbon nanotubes by the catalytic pyrolysis of hydrocarbons, *Applied Physics Letters*, 1998, **72**: 3282–3284; (b) Dai H. et al., Single-wall nanotubes produced by metal-catalyzed disproportionation of carbon monoxide. *Chemical Physics Letters*, 1996, **260**(3–4): 471–475.

59. Maruyama, S. et al., Low-temperature synthesis of high-purity single-walled carbon nanotubes from alcohol. *Chemical Physics Letters*, 2002, **360**: 229–234.

60. Li, Y. et al., How catalysts affect the growth of single-walled carbon nanotubes on substrates. *Advanced Materials*, 2010, **22**(13): 1508–1515.

61. Kocabas, C. et al., Improved synthesis of aligned arrays of single-walled carbon nanotubes and their implementation in thin film type transistors. *The Journal of Physical Chemistry C*, 2007, **111**(48): 17879–17886.

62. Cassell, A.M. et al., Directed growth of free-standing single-walled carbon nanotubes. *Journal of the American Chemical Society*, 1999, **121**(34): 7975–7976.

63. Huang, L. et al., Cobalt ultrathin film catalyzed ethanol chemical vapor deposition of single-walled carbon nanotubes. *The Journal of Physical Chemistry B*, 2006, **110**(23): 11103–11109.

64. Pisana, S. et al., Flying and crawling modes during surface-bound single wall carbon nanotube growth. *The Journal of Physical Chemistry C*, 2007, **111**(46): 17249–17253.

65. Yao, Y. et al., Crinkling ultralong carbon nanotubes into serpentines by a controlled landing process. *Advanced Materials*, 2009, **21**(41): 4158–4162.

66. Yuan, D. et al., Horizontally aligned single-walled carbon nanotube on quartz from a large variety of metal catalysts. *Nano Letters*, 2008, **8**(8): 2576–2579.

67. (a) Takagi, D. et al., Carbon nanotube growth from semiconductor nanoparticles. *Nano Letters*, 2007, **7**(8): 2272–2275; (b) Zhou, W. et al., Copper catalyzing growth of single-walled carbon nanotubes on substrates. *Nano Letters*, 2006, **6**(12): 2987–2990; (c) Takagi, D. et al., Single-walled carbon nanotube growth from highly activated metal nanoparticles. *Nano Letters*, 2006, **6**(12): 2642–2645.

68. Liu, B. et al., Metal-catalyst-free growth of single-walled carbon nanotubes. *Journal of the American Chemical Society*, 2009, **131**(6): 2082–2083.

69. Kang, L. et al., Growth of close-packed semiconducting single-walled carbon nanotube arrays using oxygen-deficient TiO_2 nanoparticles as catalysts. *Nano Letters*, 2015, **15**(1): 403–409.

70. Zhou, W. et al., Copper catalyzing growth of single-walled carbon nanotubes on substrates. *Nano Letters*, 2006, **6**(12): 2987–2990.

71. Uchino, T. et al., Catalyst free low temperature direct growth of carbon nanotubes. 2005, IEEE.

72. Huang, S. et al., Metal-catalyst-free growth of single-walled carbon nanotubes on substrates. *Journal of the American Chemical Society*, 2009, **131**(6): 2094–2095.

73. Yang, F. et al., Chirality-specific growth of single-walled carbon nanotubes on solid alloy catalysts. *Nature*, 2014, **510**(7506): 522–524.

74. Ritschel, M. et al., Rhenium-catalyzed growth carbon nanotubes. *The Journal of Physical Chemistry C*, 2007, **111**(24): 8414–8417.

75. Li, Y. et al., Preferential growth of semiconducting single-walled carbon nanotubes by a plasma enhanced CVD method. *Nano Letters*, 2004, **4**(2): 317–321.

76. Qu, L., F. Du, and L. Dai, Preferential syntheses of semiconducting vertically aligned single-walled carbon nanotubes for direct use in FETs. *Nano Letters*, 2008, **8**(9): 2682–2687.

77. Ding, L. et al., Selective growth of well-aligned semiconducting single-walled carbon nanotubes. *Nano Letters*, 2009, **9**(2): 800–805.

78. Harutyunyan, A.R. et al., Preferential growth of single-walled carbon nanotubes with metallic conductivity. *Science*, 2009, **326**(5949): 116–120.

79. An, L., L.E. McNeil, and J. Liu, Synthesis of nearly uniform single-walled carbon nanotubes using identical metal-containing molecular nanoclusters as catalysts. *Journal of the American Chemical Society*, 2002, **124**(46): 13688–13689.

80. Zhang, S. et al., Arrays of horizontal carbon nanotubes of controlled chirality grown using designed catalysts. *Nature*, 2017, **543**(7644): 234–238.

81. Smalley, R.E. et al., Single wall carbon nanotube amplification? En route to a type-specific growth mechanism. *Journal of the American Chemical Society*, 2006, **128**(49): 15824–15829.

82. Yao, Y. et al., "Cloning" of single-walled carbon nanotubes via open-end growth mechanism. *Nano Letters*, 2009, **9**(4): 1673–1677.

83. Liu, J. et al., Chirality-controlled synthesis of single-wall carbon nanotubes using vapour-phase epitaxy. *Nature Communications*, 2012, **3**: 1199.

84. Yu, X. et al., Cap formation engineering: From opened C_{60} to single-walled carbon nanotubes. *Nano Letters*, 2010, **10**(9): 3343–3349.

85. Omachi, H. et al., Initiation of carbon nanotube growth by well-defined carbon nanorings. *Nature Chemistry*, 2013, **5**(7): 572–576.

86. Sanchez-Valencia, J.R. et al., Controlled synthesis of single-chirality carbon nanotubes. *Nature*, 2014, **512**(7512): 61–64.

87. Liu, B. et al., Nearly exclusive growth of small diameter semiconducting single-wall carbon nanotubes from organic chemistry synthetic end-cap molecules. *Nano Letters*, 2015, **15**(1): 586–595.

88. (a) Liu, B. et al., Chirality-controlled synthesis and applications of single-wall carbon nanotubes. *ACS Nano*, 2017, **11**(1): 31–53; (b) Zhou, W., X. Bai, E. Wang, and S. Xie, Synthesis, structure, and properties of single-walled carbon nanotubes, *Advanced Materials*, 2009, **21**: 4565–4583.

89. Schmidt, V. et al., Silicon nanowires: A review on aspects of their growth and their electrical properties. *Advanced Materials*, 2009, **21**(25–26): 2681–2702.

90. Fuhrmann, B. et al., Ordered arrays of silicon nanowires produced by nanosphere lithography and molecular beam epitaxy. *Nano Letters*, 2005, **5**(12): 2524–2527.

91. Wang, Y. et al., Epitaxial growth of silicon nanowires using an aluminium catalyst. *Nature Nanotechnology*, 2006, **1**(3): 186–189.

92. Islam, M.S. et al., Ultrahigh-density silicon nanobridges formed between two vertical silicon surfaces. *Nanotechnology*, 2004, **15**: L5–L8.

93. Wang, N. et al., Transmission electron microscopy evidence of the defect structure in Si nanowires synthesized by laser ablation. *Chemical Physics Letters*, 1998, **283**: 368–372.

94. Johnson, E.R. and J.A. Amick, Formation of single crystal silicon fibers. *Journal of Applied Physics*, 1954, **25**(9): 1204–1205.

95. Sunkara, M.K. et al., Bulk synthesis of silicon nanowires using a low-temperature vapor–liquid–solid method. *Applied Physics Letters*, 2001, **79**(10): 1546–1548.

96. Miyamoto, Y. and M. Hirata, Role of agents in filamentary growth of amorphous silicon. Japanese *Journal of Applied Physics*, 1976, **15**(6): 1159–1160.

97. Yamada, I. et al., Silicon nanowire growth on Si and SiO$_2$ substrates by rf magnetron sputtering in Ar/H$_2$. *Applied Physics Express*, 2015, **8**(6): 066201.

98. Saif Islam, M. et al., A novel interconnection technique for manufacturing nanowire devices. *Applied Physics A*, 2005, **80**(6): 1133–1140.

99. Li, F.J., S. Zhang, and J.-W. Lee, Rethinking of the silicon nanowire growth mechanism during thermal evaporation of Si-containing powders. *Thin Solid Films*, 2014, **558**: 75–85.

100. Liu, Y. et al., Arrays of one-dimensional germanium cone-like nanostructures: Preparation and application as fluorescent pH sensor. *The Journal of Physical Chemistry C*, 2011, **115**(44): 21599–21603.

101. Wu, Y. and P. Yang, Germanium nanowire growth via simple vapor transport. *Chemistry of Materials*, 2000, **12**(3): 605–607.

102. Kamins, T.I. et al., Growth and structure of chemically vapor deposited Ge nanowires on Si substrates. *Nano Letters*, 2004, **4**(3): 503–506.

103. Dick, K.A., A review of nanowire growth promoted by alloys and non-alloying elements with emphasis on Au-assisted III? V nanowires. *Progress in Crystal Growth and Characterization of Materials*, 2008, **54**(3–4): 138–173.

104. Kuykendall, T. et al., Metalorganic chemical vapor deposition route to GaN nanowires with triangular cross sections. *Nano Letters*, 2003, **3**(8): 1063–1066.

105. Jensen, L.E. et al., Role of surface diffusion in chemical beam epitaxy of in as nanowires. *Nano Letters*, 2004, **4**(10): 1961–1964.

106. Heurlin, M. et al., Continuous gas-phase synthesis of nanowires with tunable properties. *Nature*, 2012, **492**(7427): 90–94.

107. Nickl, J.J. and W. Just, Das wachstum von galliumarsenid-kristallen nach dem VLS-mechanismus. *Journal of Crystal Growth*, 1971, **11**(1): 11–20.

108. Duan, X.F. and C.M. Lieber, General synthesis of compound semiconductor nanowires. *Advanced Materials*, 2000, **12**(4): 298–302.

109. Shimada, T. et al., Size, position and direction control on GaAs and InAs nanowhisker growth. *Superlattices and microstructures*, 1998, **24**(6): 453–458.

110. Cirlin, G.E. et al., Self-catalyzed, pure zincblende GaAs nanowires grown on Si(111) by molecular beam epitaxy. *Physical Review B*, 2010, **82**(3): 035302.

111. Persson, A.I. et al., Solid-phase diffusion mechanism for GaAs nanowire growth. *Nature Materials*, 2004, **3**(10): 677–681.

112. He, M. et al., Growth of GaN nanowires by direct reaction of Ga with NH$_3$. *Journal of Crystal Growth*, 2001, **231**(3): 357–365.

113. Han, W. et al., Synthesis of GaN–carbon composite nanotubes and GaN nanorods by arc discharge in nitrogen atmosphere. *Applied Physics Letters*, 2000, **76**(5): 652–654.

114. (a) Duan, X.F. and C.M. Lieber, Laser-assisted catalytic growth of single crystal GaN nanowires. *Journal of the American Chemical Society*, 2000, **122**(1): 188–189; (b) Kuykendall, T. et al., Metalorganic chemical vapor deposition route to GaN nanowires with triangular cross sections. *Nano Letters*, 2003, **3**(8): 1063–1066.

115. Calarco, R. et al., Size-dependent Photoconductivity in MBE-Grown GaN–Nanowires. *Nano Letters*, 2005, **5**(5): 981–984.

116. Borgstrom, M.T. et al., Synergetic nanowire growth. *Nature Nanotechnology*, 2007, **2**(9): 541–544.

117. (a) Boulanger J. et al., Polytype formation in GaAs/GaP axial nanowire heterostructures. *Journal of Crystal Growth*, 2011, **332**: 21–26; (b) Husanu, E. et al., Growth of defect-free GaP nanowires. *Nanotechnology*, 2014, **25**(20): 205601.

118. Vaddiraju, S. et al., Synthesis of group III antimonide nanowires. *The Journal of Physical Chemistry C*, 2007, **111**(20): 7339–7347.

119. Guo, Y.N. et al., Structural characteristics of GaSb/GaAs nanowire heterostructures grown by metal-organic chemical vapor deposition. *Applied Physics Letters*, 2006, **89**(23): 231917.

120. Lugstein, A. et al., Synthesis of nanowires in room temperature ambient: A focused ion beam approach. *Applied Physics Letters*, 2006, **88**(16): 63114.

121. Conesa-Boj, S. et al., Gold-free ternary III–V antimonide nanowire arrays on silicon: Twin-free down to the first bilayer. *Nano Letters*, 2014, **14**(1): 326–332.

122. Ihn, S.-G. and J.-I. Song, InAs nanowires on Si substrates grown by solid source molecular beam epitaxy. *Nanotechnology*, 2007, **18**(35): 355603.

123. Cai, X.M. et al., A systematic study of chemical vapor deposition growth of InN. *Applied Surface Science*, 2008, **255**(5): 2153–2158.

124. Zhang, J. et al., Vapor–solid growth route to single-crystalline indium nitride nanowires. *Journal of Materials Chemistry*, 2002, **12**(4): 802–804.

125. Chang, Y.-K. and F.C.-N. Hong, Synthesis and characterization of indium nitride nanowires by plasma-assisted chemical vapor deposition. *Materials Letters*, 2009, **63**(21): 1855–1858.

126. Stoica, T. et al., Photoluminescence and intrinsic properties of MBE-grown InN nanowires. *Nano Letters*, 2006, **6**(7): 1541–1547.

127. Chao, C.K. et al., Catalyst-free growth of indium nitride nanorods by chemical-beam epitaxy. *Applied Physics Letters*, 2006, **88**(23): 233111.

128. (a) Trentler, T., Solution- liquid- solid growth of indium phosphide fibers from organometallic precursors: Elucidation of molecular and nonmolecular components of the pathway. *Journal of the American Chemical Society*, 1997, **119**: 2172; (b) Bao, J. et al., Optical properties of rotationally twinned InP nanowire heterostructures. *Nano Letters*, 2008, **8**(3): 836–841.

129. Radhakrishnan, G., A. Freundlich, and B. Fuhrmann, Chemical beam epitaxy of highly ordered network of tilted InP nanowires on silicon. *Journal of Crystal Growth*, 2009, **311**(7): 1855–1858.

130. (a) Yang, Y. et al., Fabrication and optical property of single-crystalline InSb nanowire arrays. *Journal of Materials Sciences*, 2007, **42**: 2753; (b) Zhou, J. F. et al., Anisotropic growth of indium antimonide nanostructures. *European Physical Journal D*, 2007, **43**: 283–286; (c) Mandl, B. et al., Crystal structure control in Au-free self-seeded InSb wire growth. *Nanotechnology*, 2011, **22**(14): 145603.

131. Thelander, C. et al., Electrical properties of InAs 1–x Sb x and InSb nanowires grown by molecular beam epitaxy. *Applied Physics Letters*, 2012, **100**(23): 232105.

132. Vogel, A.T. et al., Fabrication of high-quality InSb nanowire arrays by chemical beam epitaxy. *Crystal Growth & Design*, 2011, **11**(5): 1896–1900.

133. Cesare Soci, X.-Y.B.D.P.R.A. and W. Deli, A systematic study on the growth of GaAs nanowires by metal organic chemical vapor deposition. *Nano Letters*, 2008, **8**(12): 4275–4282.

134. Plante, M.C. and R.R. LaPierre, Growth mechanisms of GaAs nanowires by gas source molecular beam epitaxy. *Journal of Crystal Growth*, 2006, **286**(2): 394–399.

135. Zhao, S. et al., III-Nitride nanowire optoelectronics. *Progress in Quantum Electronics*, 2015, **44**: 14–68.

136. Utama, M.I.B. et al., Synthesis and optical properties of II–VI 1D nanostructures. *Nanoscale*, 2012, **4**(5): 1422.

137. Jie, J.S. et al., One-dimensional II–VI nanostructures: Synthesis, properties and optoelectronic applications. *Nano Today*, 2010, **5**(4): 313–336.

138. Hao, Y. et al., Periodically twinned nanowires and polytypic nanobelts of ZnS? The role of mass diffusion in vapor? Liquid? Solid Growth. *Nano Letters*, 2006, **6**(8): 1650–1655.

139. Mondal, S.P. et al., Characteristics of CdS nanowires grown in a porous alumina template using a two-cell method. *Nanotechnology*, 2007, **18**(9): 095606.

140. Xu, D. et al., Preparation of CdS single-crystal nanowires by electrochemically induced deposition. Advanced Materials, 2000, **12**: 520–522.

141. Yan, P. et al., A cluster growth route to quantum-confined CdS nanowires. *Chemical Communications*, 1999, (14): 1293–1294.

142. Ma, R.M. et al., Synthesis of high quality n-type CdS nanobelts and their applications in nanodevices. *Applied Physics Letters*, 2006, **89**(20): 203120.

143. Duan, X.F. et al., High-performance thin-film transistors using semiconductor nanowires and nanoribbons. *Nature*, 2003, **425**(6955): 274–278.

144. (a) Shen, G.Z. et al., Heteroepitaxial growth of orientation-ordered ZnS nanowire arrays. *The Journal of Physical Chemistry C*, 2008, **112**: 12299; (b) Barrelet, C.J. et al., Synthesis of CdS and ZnS nanowires using single-source molecular precursors. *Journal of the American Chemical Society*, 2003, **125**(38): 11498–11499.

145. Utama, M.I.B. et al., Vertically aligned cadmium chalcogenide nanowire arrays on muscovite mica: A demonstration of epitaxial growth strategy. *Nano Letters*, 2011, **11**(8): 3051–3057.

146. Schierhorn, M. et al., Photoelectrochemical performance of CdSe nanorod arrays grown on a transparent conducting substrate. *Nano Letters*, 2009, **9**(9): 3262–3267.

147. Ma, C. et al., Single-crystal CdSe nanosaws. *Journal of the American Chemical Society*, 2004, **126**(3): 708–709.

148. Shan, C.X., Z. Liu, and S.K. Hark, highly oriented zinc blende CdSe nanoneedles. *Applied Physics Letters*, 2005, **87**(16): 163108.

149. (a) Tang, Z. et al., Spontaneous organization of single CdTe nanoparticles into luminescent nanowires. *Science*, 2002, **297**: 237; (b) Neretina, S. et al., Vertically aligned wurtzite CdTe nanowires derived from a catalytically driven growth mode. *Nanotechnology*, 2007, **18**(27): 275301.

150. Park, W.I. et al., Transformation of ZnTe nanowires to CdTe nanowires through the formation of ZnCdTe–CdTe core–shell structure by vapor transport. *Journal of Materials Chemistry*, 2008, **18**(8): 875.

151. Meng, X.M. et al., Structure- and size-controlled ultrafine ZnS nanowires. *Chemical Physics Letters*, 2003, **382**(3–4): 434–438.

152. Chan, S.K. et al., MBE-grown cubic ZnS nanowires. *Journal of Electronic Materials*, 2008, **37**(9): 1433–1437.

153. Shi, L., Y. Xu, and Q. Li, Controlled fabrication of ZnSe arrays of well-aligned nanorods, nanowires, and nanobelts with a facile template-free route. *The Journal of Physical Chemistry C*, 2009, **113**(5): 1795–1799.

154. Zhu, Y.-C. and Y. Bando, Preparation and photoluminescence of single-crystal zinc selenide nanowires. *Chemical Physics Letters*, 2003, **377**(3–4): 367–370.

155. Zhang, X. et al., Routes to grow well-aligned arrays of ZnSe nanowires and nanorods. *Advanced Materials*, 2005, **17**(11): 1405–1410.

156. Solanki, R. et al., Atomic layer deposition of ZnSe/CdSe superlattice nanowires. *Applied Physics Letters*, 2002, **81**(20): 3864–3866.

157. Chan, Y.F. et al., ZnSe nanowires epitaxially grown on GaP(111) substrates by molecular-beam epitaxy. *Applied Physics Letters*, 2003, **83**(13): 2665–2667.

158. (a) Li, Y.D. et al., A novel chemical route to ZnTe semiconductor nanorods. *Advanced Materials*, 1999, **11**(10): 847–850; (b) Li, L. et al., Fabrication and characterization of single-crystalline ZnTe nanowire arrays. *The Journal of Physical Chemistry B*, 2005, **109**(25): 12394–12398.

159. Meng, Q., C. Jiang, and S.X. Mao, Temperature-dependent growth of zinc-blende-structured ZnTe nanostructures. *Journal of Crystal Growth*, 2008, **310**(20): 4481–4486.

160. Janik, E. et al., ZnTe nanowires grown on GaAs(100) substrates by molecular beam epitaxy. *Applied Physics Letters*, 2006, **89**(13): 133114.

161. Lu, J.G., P. Chang, and Z. Fan, Quasi-one-dimensional metal oxide materials—Synthesis, properties and applications. *Materials Science and Engineering: R: Reports*, 2006, **52**(1–3): 49–91.

162. Pung, S.-Y., K.-L. Choy, and X. Hou, Tip-growth mode and base-growth mode of Au-catalyzed zinc oxide nanowires using chemical vapor deposition technique. *Journal of Crystal Growth*, 2010, **312**(14): 2049–2055.

163. (a) Dai, L. et al., Fabrication and characterization of In₂O₃ nanowires. *Applied Physics A: Materials Science & Processing*, 2002, **75**(6): 687–689; (b) Li, C., Diameter-controlled growth of single-crystalline In₂O₃ nanowires and their electronic properties. *Advance Materials*, 2003, **15**: 143–146.

164. (a) Kim, S., S. Fujita, and S. Fujita, ZnO nanowires with high aspect ratios grown by metalorganic chemical vapor deposition using gold nanoparticles, *Applied Physics Letters*, 2005, **86**: 153119; (b) Heo, Y.W. et al., UV photoresponse of single ZnO nanowires. *Applied Physics A*, 2004, **80**(3): 497–499.

165. Wang, Z.L., Zinc oxide nanostructures: Growth, properties and applications. *Journal of Physics: Condensed Matter*, 2004, **16**(25): R829–R858.

166. Huang, M.H. et al., Room-temperature ultraviolet nanowire nanolasers. *Science*, 2001, **292**(5523): 1897–1899.

167. Huang, M.H. et al., Catalytic growth of zinc oxide nanowires by vapor transport. *Advanced Materials*, 2001, **13**(2): 113–116.

168. Sun, Y., G.M. Fuge, and M.N.R. Ashfold, Growth of aligned ZnO nanorod arrays by catalyst-free pulsed laser deposition methods. *Chemical Physics Letters*, 2004, **396**(1–3): 21–26.

169. Pung, S.-Y. et al., Preferential growth of ZnO thin films by the atomic layer deposition technique. *Nanotechnology*, 2008, **19**(43): 435609.

170. (a) Pan, Z.W., Z.R. Dai, and Z.L. Wang, Nanobelts of semiconducting oxides. *Science*, 2001, **291**(5510): 1947–1949; (b) Kong, X.Y. and Z.L. Wang, Spontaneous polarization-induced nanohelixes, nanosprings, and nanorings of piezoelectric nanobelts. *Nano Letters*, 2003, **3**: 1625–1631.

171. Liu, Z. et al., Laser ablation synthesis and electron transport studies of tin oxide nanowires. *Advanced Materials*, 2003, **15**(20): 1754–1757.

172. Miao, J.-J. et al., Ultrasonic-induced synthesis of CeO₂ nanotubes. *Journal of Crystal Growth*, 2005, **281**(2–4): 525–529.

173. Liu, X.M. and Y.C. Zhou, Electrochemical deposition and characterization of Cu₂O nanowires. *Applied Physics A*, 2005, **81**(4): 685–689.

174. Vantomme, A. et al., Surfactant-assisted large-scale preparation of crystalline CeO₂ nanorods. *Langmuir*, 2005, **21**(3): 1132–1135.

175. (a) Jiang, X., T. Herricks, and Y. Xia, CuO nanowires can be synthesized by heating copper substrates in air. *Nano Letters*, 2002, **2**(12): 1333–1338; (b) Cao, M. et al., A simple route towards CuO nanowires and nanorods. *Journal of Nanoscience and Nanotechnology*, 2004, **4**(7): 824–828.

176. Pol, V.G. et al., Synthesis of europium oxide nanorods by ultrasound irradiation. *The Journal of Physical Chemistry B*, 2002, **106**(38): 9737–9743.

177. Shi, K. et al., Assembling porous Fe₂O₃ nanowire arrays by electrochemical deposition in mesoporous silica SBA-16 films. *Acta Chimica Sinica-Chinese Edition*, 2004, **62**(19): 1859–1861.

178. (a) Zhang, J. et al., A simple method to synthesize β-Ga₂O₃ nanorods and their photoluminescence properties. *Journal of Crystal Growth*, 2005, **280**(1–2): 99–106; (b) Chang, P., Z. Fan, W. Tseng, A. Rajagopal, and J. Lu, β-Ga₂O₃ nanowires: Synthesis, characterization, and p-channel field-effect transistor. *Applied Physics Letters*, 2005, **87**: 222102.

179. (a) Park, G.-S. et al., Structural investigation of gallium oxide (β-Ga$_2$O$_3$) nanowires grown by arc-discharge. *Journal of Crystal Growth*, 2000, **220**(4): 494–500; (b) H. Kim and N. Kim, Synthesis of βGa2O3 nanowires by an MOCVD approach. *Applied Physics A*, 2005, **81**(4): 763–765.

180. Chen, Y. et al., The effect of Mg vapor source on the formation of MgO whiskers and sheets. *Journal of Crystal Growth*, 2002, **245**(1–2): 163–170.

181. Wang, X. et al., Sol-gel template synthesis of highly ordered MnO$_2$ nanowire arrays. *Journal of Power Sources*, 2005, **140**(1): 211–215.

182. Zheng, D. et al., One-step preparation of single-crystalline β-MnO$_2$ nanotubes. *The Journal of Physical Chemistry B*, 2005, **109**(34): 16439–16443.

183. Xu, C. et al., Large-scale synthesis of rutile SnO$_2$ nanorods. *Solid State Communications*, 2003, **125**(6): 301–304.

184. Cao, Y.-L. et al., Rapid synthesis of lead oxide nanorods by one-step solid-state chemical reaction at room temperature. *Chinese Journal of Chemistry*, 2004, **22**(11): 1288–1290.

185. Wu, Y. et al., Superconducting MgB$_2$ nanowires. *Advanced Materials*, 2001, **13**(19): 1487.

186. Xu, H. et al., Fabrication and characterization of highly ordered zirconia nanowire arrays by sol-gel template method. *Materials Chemistry and Physics*, 2003, **80**(2): 524–528.

187. Wu, J.-M. et al., Thermal evaporation growth and the luminescence property of TiO$_2$ nanowires. *Journal of Crystal Growth*, 2005, **281**(2–4): 384–390.

188. Zhang, M., Y. Bando, and K. Wada, Sol-gel template preparation of TiO$_2$ nanotubes and nanorods. *Journal of Materials Science Letters*, 2001, **20**(2): 167–170.

189. Xu, C. et al., A simple and novel route for the preparation of ZnO nanorods. *Solid State Communications*, 2002, **122**: 175–179.

190. Zheng, M.J. et al., Fabrication and optical properties of large-scale uniform zinc oxide nanowire arrays by one-step electrochemical deposition technique. *Chemical Physics Letters*, 2002, **363**(1–2): 123–128.

191. Bae, S.Y., H.W. Seo, and J. Park, Vertically aligned sulfur-doped ZnO nanowires synthesized via chemical vapor deposition. *The Journal of Physical Chemistry B*, 2004, **108**(17): 5206–5210.

192. (a) Huang, M.H., Y. Wu, H. Feick, N. Tran, E. Weber, and P. Yang, Catalytic growth of zinc oxide nanowires by vapor transport. *Advanced Materials*, 2001, **13**(2): 113–116; (b) Park, W.I. et al., Metalorganic vapor-phase epitaxial growth of vertically well-aligned ZnO nanorods. *Applied Physics Letters*, 2002, **80**: 4232–4234.

193. (a) Liang, J. et al., Chromate nanorods/nanobelts: General synthesis, characterization, and properties. *Inorganic Chemistry*, 2005, **44**(25): 9405–9415; (b) Zhu, D., H. Zhu, and Y. Zhang, Microstructure and magnetization of single-crystal perovskite manganites nanowires prepared by hydrothermal method. *Journal of Crystal Growth*, 2003, **249**: 172–175.

194. (a) Dai, H.J. et al., Synthesis and characterisation of carbide nanorods. *Nature*, 1995, **375**: 769–772; (b) Yuan, Y.W. and J.S. Pan, The effect of vapor phase on the growth of TiC whiskers prepared by chemical vapor deposition. *Journal of Crystal Growth*, 1998, **193**: 585–591.

195. (a) Zhang, H.Z. et al., Boron carbide nanowires with uniform CNx coatings. *New Journal of Physics*, 2007, **9**: 13; (b) Zhao, P. et al., Carbon nitride nanowire bundle and tubes: Solid-state synthesis, characterization and photoluminescent properties. *Chemical Research in Chinese Universities*, 2009, **25**(4): 412–416.

196. He, C.N. et al., Fabrication of aluminum carbide nanowires by a nano-template reaction. *Carbon*, 2010, **48**(4): 931–938.

197. Pan, J., R. Cao, and Y. Yuan, A new approach to the mass production of titanium carbide, nitride and carbonitride

whiskers by Spouted Bed Chemical Vapor Deposition. *Materials Letters*, 2006, **60**(5): 626–629.

198. Singh, P., T. Nautiyal, and S. Auluck, Electronic and vibrational properties of vanadium-carbide nanowires. *Journal of Applied Physics*, 2012, **112**(6): 063502.

199. Yang, R. et al., Effect of growth promoters on chemistry synthesis of Cr$_3$C$_2$ nanowhiskers from water-soluble precursors. *Journal of Crystal Growth*, 2017, **458**: 133–139.

200. Liu, Q. et al., Chemical vapour deposition of zirconium carbide and silicon carbide hybrid whiskers. *Materials Letters*, 2010, **64**(4): 552–554.

201. Johnsson, M., Synthesis of boride, carbide, and carbonitride whiskers. *Solid State Ionics*, 2004, **172**(1–4, SI): 365–368.

202. Lin, H. et al., Cobalt-doping in molybdenum-carbide nanowires toward efficient electrocatalytic hydrogen evolution. *Advanced Functional Materials*, 2016, **26**(31): 5590–5598.

203. Futamoto, M., I. Yuito, and U. Kawabe, Hafnium carbide and nitride whisker growth by chemical vapor-deposition. *Journal of Crystal Growth*, 1983, **61**(1): 69–74.

204. (a) Johnsson, M. and M. Nygren, Carbothermal synthesis of TaC whiskers via a vapor-liquid-solid growth mechanism. *Journal of Materials Research*, 1997, **12**(9): 2419–2427; (b) Zhao, G. et al., A study on in situ growth of TaC whiskers in Al$_2$O$_3$ matrix powder for ceramic cutting tools. *Materials Research Bulletin*, 2012, **47**(8): 2027–2031.

205. Wang, S.J. et al., Growth and characterization of tungsten carbide nanowires by thermal annealing of sputter-deposited WCx films. *Applied Physics Letters*, 2004, **85**(12): 2358–2360.

206. Terrones, M. et al., Pure and doped boron nitride nanotubes. *Materials Today*, 2007, **10**(5): 30–38.

207. Arenal, R. et al., Root-growth mechanism for single-walled boron nitride nanotubes in laser vaporization technique. *Journal of the American Chemical Society*, 2007, **129**(51): 16183–16189.

208. Cumings, J. and A. Zettl, Mass-production of boron nitride double-wall nanotubes and nanococoons. *Chemical Physics Letters*, 2000, **316**(3–4): 211–216.

209. Lourie, O.R. et al., CVD growth of boron nitride nanotubes. *Chemistry of Materials*, 2000, **12**(7): 1808–1810.

210. Wang, J. et al., Low temperature growth of boron nitride nanotubes on substrates. *Nano Letters*, 2005, **5**(12): 2528–2532.

211. Han, W.Q. et al., Synthesis of boron nitride nanotubes from carbon nanotubes by a substitution reaction. *Applied Physics Letters*, 1998, **73**(21): 3085–3087.

212. (a) Golberg, D. et al., Boron nitride nanotubes and nanosheets. *ACS Nano*, 2010, **4**: 2979–2993; (b) Smith, M.W. et al., Very long single- and few-walled boron nitride nanotubes via the pressurized vapor/condenser method. *Nanotechnology*, 2009, **20**(50): 505604; (c) Zhi, C., Y. Bando, C. Tan, and D. Golberg, Effective precursor for high yield synthesis of pure BN nanotubes. *Solid State Communication*, 2005, **135**: 67–70.

213. Zhi, C. et al., Effective precursor for high yield synthesis of pure BN nanotubes. *Solid State Communications*, 2005, **135**(1–2): 67–70.

214. Lee, C.H. et al., Effective growth of boron nitride nanotubes by thermal chemical vapor deposition. *Nanotechnology*, 2008, **19**(45): 455605.

215. Lee, C.H. et al., Patterned growth of boron nitride nanotubes by catalytic chemical vapor deposition. *Chemistry of Materials*, 2010, **22**(5): 1782–1787.

216. Liu, B. et al., 352 nm ultraviolet emission from high-quality crystalline AlN whiskers. *Nanotechnology*, 2010, **21**: 075708.

217. (a) Chen, L.Y. et al., A low-temperature route to nanocrystalline boron nitride whiskers and flakes. *Materials Letters*, 2004, **58**(27–28): 3634–3636; (b) Wang, J., C.H. Lee, and Y.K. Yap, Recent advancements in boron nitride nanotubes. *Nanoscale*, 2010, **2**: 2028–2034.

218. Sakamoto, Y. and M. Takaya, Growth of carbon nitride using microwave plasma CVD. *Thin Solid Films*, 2005, **475**(1–2): 198–201.

219. Li, F. et al., Additive-assisted nitridation to synthesize Si_3N_4 nanomaterials at a low temperature. *Journal of the American Ceramic Society*, 2009, **92**(2): 517–519.

220. Pan, J.S., R.X. Cao, and Y.W. Yuan, A new approach to the mass production of titanium carbide, nitride and carbonitride whiskers by Spouted Bed Chemical Vapor Deposition. *Materials Letters*, 2006, **60**(5): 626–629.

221. Zhang, Y. et al., Rapid synthesis of cobalt nitride nanowires: highly efficient and low-cost catalysts for oxygen evolution. *Angewandte Chemie-International Edition*, 2016, **55**(30): 8670–8674.

222. Wang, Z. et al., Copper-nitride nanowires array: An efficient dual-functional catalyst electrode for sensitive and selective non-enzymatic glucose and hydrogen peroxide sensing. *Chemistry-A European Journal*, 2017, **23**(21): 4986–4989.

223. Zong, F.J. et al., Structural properties and photoluminescence of zinc nitride nanowires. *Applied Physics Letters*, 2005, **87**(23): 233104.

224. Ning, J.Q. et al., 418 cm(-1) Raman scattering from gallium nitride nanowires: Is it a vibration mode of N-rich Ga-N bond configuration? *Applied Physics Letters*, 2007, **91**(10): 103117.

225. Takahashi, T., H. Itoh, and S. Noguchi, Formation of zirconium nitride films and whiskers by chemical vapor-deposition technique. *Nippon Kagaku Kaishi*, 1975, (4): 627–631.

226. Takasu, T., T. Watari, and A. Kato, Growth of niobium nitride whiskers by chemical vapor-deposition and superconducting properties. *Nippon Kagaku Kaishi*, 1985, (7): 1477–1482.

227. (a) Li, C.Z., H. Sha, and N.J. Tao, Adsorbate effect on conductance quantization in metallic nanowires. *Physical Review B*, 1998, **58**: 6775–6778; (b) Karim, S. et al., Synthesis of gold nanowires with controlled crystallographic characteristics. *Applied Physics A*, 2006, **84**(4): 403–407.

228. Sears, G.W., Mercury whiskers. *Acta Metallurgica*, 1953, **1**(4): 457–459.

229. Vayssieres, L., L. Rabenberg, and A. Manthiram, Aqueous chemical route to ferromagnetic 3-d arrays of iron nanorods. *Nano Letters*, 2002, **2**(12): 1393–1395.

230. Sadki, E.S., S. Ooi, and K. Hirata, Focused-ion-beam-induced deposition of superconducting nanowires. *Applied Physics Letters*, 2004, **85**(25): 6206–6208.

231. Luxmoore, I.J. et al., Low temperature electrical characterisation of tungsten nano-wires fabricated by electron and ion beam induced chemical vapour deposition. *Thin Solid Films*, 2007, **515**(17): 6791–6797.

232. Choi, H. and S. Park, Seedless growth of free-standing copper nanowires by chemical vapor deposition. *Journal of the American Chemical Society*, 2004, **126**: 6248–6249.

233. Richter, G. et al., Ultrahigh strength single crystalline nanowhiskers grown by physical vapor deposition. *Nano Letters*, 2009, **9**(8): 3048–3052; (b) Ding J. et al., One-step synthesis of pure Cu nanowire/carbon nanotube coaxial nanocables with different structures by arc discharge. *Journal of Physics and Chemistry of Solids*, 2011, **72**: 1519–1523.

234. Lee, H.-B.-R. et al., Growth of Pt nanowires by atomic layer deposition on highly ordered pyrolytic graphite. *Nano Letters*, 2013, **13**(2): 457–463.

235. Kim, C. et al., Copper nanowires with a five-twinned structure grown by chemical vapor deposition. *Advanced Materials*, 2008, **20**(10): 1859–1863.

236. Benson, J. et al., Chemical vapor deposition of aluminum nanowires on metal substrates for electrical energy storage applications. *ACS Nano*, 2012, **6**(1): 118–125.

237. Morelock, C.R., Sub-micron whiskers by vapor deposition. *Acta Metallurgica*, 1962, **10**(2): 161–167.

238. Chan, K.T. et al., Oriented growth of single-crystal Ni nanowires onto amorphous SiO_2. *Nano Letters*, 2010, **10**(12): 5070–5075.

239. Sears, G.W., A mechanism of whisker growth. *Acta Metallurgica*, 1955, **3**(4): 367–369.

240. Zhou, J. et al., Growth of large-area aligned molybdenum nanowires by high temperature chemical vapor deposition: Synthesis, growth mechanism, and device application. *The Journal of Physical Chemistry B*, 2006, **110**(21): 10296–10302.

241. Li, J.-L. et al., Spontaneous formation of ordered indium nanowire array on Si(001). *Applied Physics Letters*, 2001, **79**(17): 2826–2828.

242. Mohanty, P. et al., Synthesis of single crystalline tellurium nanotubes with triangular and hexagonal cross sections. *The Journal of Physical Chemistry B*, 2006, **110**(2): 791–795.

243. Vaddiraju, S., H. Chandrasekaran, and M.K. Sunkara, Vapor phase synthesis of tungsten nanowires. *Journal of the American Chemical Society*, 2003, **125**(36): 10792–10793.

244. Wang, Y. et al., Synthesis of one-dimensional TiO_2/V_2O_5 branched heterostructures and their visible light photocatalytic activity towards Rhodamine B. *Nanotechnology*, 2011, **22**(22): 225702.

245. Lo Nigro, R. et al., Template-free and seedless growth of pt nanocolumns: imaging and probing their nanoelectrical properties. *ACS Nano*, 2007, **1**(3): 183–190.

246. Compton, K.G. et al., Filamentary growths on metal surfaces–"whiskers". *Corrosion*, 1951, **7**: 327–334.

247. Thompson, W.A., Lead whisker growth from the metallic vapor phase. *The Journal of Chemical Physics*, 1978, **68**(4): 1854–1856.

248. Yao, Z. et al., Carbon nanotube intramolecular junctions. *Nature*, 1999, **402**(6759): 273–276.

249. Ouyang, M. et al., Atomically resolved single-walled carbon nanotube intramolecular junctions. *Science*, 2001, **291**(5501): 97–100.

250. Li Y. et al., Dopant-free GaN/AlN/AlGaN radial nanowire heterostructures as high electron mobility transistors. *Nano Letters*, 2006, **6**(7): 1468–1473.

251. Algra, R.E. et al., Twinning superlattices in indium phosphide nanowires. *Nature*, 2008, **456**(7220): 369–372.

252. Mohan, P., J. Motohisa, and T. Fukui, Fabrication of InP/InAs/InP core-multishell heterostructure nanowires by selective area metalorganic vapor phase epitaxy. *Applied Physics Letters*, 2006, **88**(13): 133105.

253. Hu, J.T. et al., Controlled growth and electrical properties of heterojunctions of carbon nanotubes and silicon nanowires. *Nature*, 1999, **399**(6731): 48–51.

254. Yang, C. et al., Single p-Type/Intrinsic/n-Type silicon nanowires as nanoscale avalanche photodetectors. *Nano Letters*, 2006, **6**(12): 2929–2934.

255. Lauhon, L.J. et al., Epitaxial core-shell and core-multishell nanowire heterostructures. *Nature*, 2002, **420**(6911): 57–61.

256. Wu, Y. et al., Single-crystal metallic nanowires and metal/semiconductor nanowire heterostructures. *Nature*, 2004, **430**(6995): 61–65.

257. Li, Q. and C. Wang, One-step fabrication of uniform Si-Core/CdSe-sheath nanocables. *Journal of the American Chemical Society*, 2003, **125**(33): 9892–9893.

258. Zhang, Y.F. et al., Growth, optical, and electrical properties of single-crystalline Si–CdSe biaxial p–n heterostructure nanowires. *The Journal of Physical Chemistry C*, 2007, **111**(39): 14343–14347.

259. Wang, C.R. et al., ZnSe-Si Bi-coaxial nanowire heterostructures. *Advanced Functional Materials*, 2005, **15**(9): 1471–1477.

260. Hu, J.Q. et al., Epitaxial heterostructures: Side-to-side Si-ZnS, Si-ZnSe biaxial nanowires, and sandwichlike ZnS-Si-ZnS triaxial nanowires. *Journal of the American Chemical Society*, 2003, **125**(37): 11306–11313.

261. Dick, K.A. et al., The morphology of axial and branched nanowire heterostructures. *Nano Letters*, 2007, **7**(6): 1817–1822.

262. Dayeh, S.A. et al., Growth, defect formation, and morphology control of germanium-silicon semiconductor nanowire heterostructures. *Nano Letters*, 2011, **11**(10): 4200–4206.

263. Carnevale, S.D. et al., Three-dimensional GaN/AlN nanowire heterostructures by separating nucleation and growth processes. *Nano Letters*, 2011, **11**(2): 866–871.

264. Lehmann, S. et al., High crystal quality wurtzite-zinc blende heterostructures in metal-organic vapor phase epitaxy-grown GaAs nanowires. *Nano Research*, 2012, **5**(7): 470–476.

265. Fontcuberta i Morral, A. et al., Prismatic quantum heterostructures synthesized on molecular-beam epitaxy GaAs nanowires. *Small*, 2008, **4**(7): 899–903.

266. Qian, F. et al., Core/multishell nanowire heterostructures as multicolor, high-efficiency light-emitting diodes. *Nano Letters*, 2005, **5**(11): 2287–2291.

267. Qian, F. et al., Gallium nitride-based nanowire radial heterostructures for nanophotonics. *Nano Letters*, 2004, **4**(10): 1975–1979.

268. Li, Q.W. et al., Sustained growth of ultralong carbon nanotube arrays for fiber spinning. *Advanced Materials*, 2006, **18**(23): 3160–3163.

269. Ohlsson, B.J. et al., Growth and characterization of GaAs and InAs nano-whiskers and InAs/GaAs heterostructures. *Physica E: Low-Dimensional Systems and Nanostructures*, 2002, **13**(2–4): 1126–1130.

270. Kanungo, D.P. et al., Selective area growth of III–V nanowires and their heterostructures on silicon in a nanotube template: Towards monolithic integration of nano-devices. *Nanotechnology*, 2013, **24**(22): 225304.

271. Jiang, X.C. et al., InAs/InP radial nanowire heterostructures as high electron mobility devices. *Nano Letters*, 2007, **7**(10): 3214–3218.

272. Lind, E. et al., Improved subthreshold slope in an InAs nanowire heterostructure field-effect transistor. *Nano Letters*, 2006, **6**(9): 1842–1846.

273. Caroff, P. et al., High-quality InAs/InSb nanowire heterostructures grown by metal–organic vapor-phase epitaxy. *Small*, 2008, **4**(7): 878–882.

274. Ghalamestani, S.G., M. Ek, and K.A. Dick, Realization of single and double axial InSb-GaSb heterostructure nanowires: Realization of single and double axial InSb-GaSb heterostructure nanowires. *Physica Status Solidi (RRL) – Rapid Research Letters*, 2014, **8**(3): 269–273.

275. Chen, Z. et al., Protein microarrays with carbon nanotubes as multicolor Raman labels. *Nature Biotechnology*, 2008, **26**(11): 1285–1292.

276. Chang, P.-C. et al., β-Ga$_2$O$_3$ nanowires: Synthesis, characterization, and p-channel field-effect transistor. *Applied Physics Letters*, 2005, **87**(22): 222102.

277. Lai, C.W. et al., In situ synthesis and phase transformation of In$_2$O$_3$/Sb core-shell nanostructures. *Journal of Crystal Growth*, 2005, **282**(3–4): 383–388.

278. Han, S. et al., Transition metal oxide core–shell nanowires: Generic synthesis and transport studies. *Nano Letters*, 2004, **4**(7): 1241–1246.

279. He, R.R. et al., Functional bimorph composite nanotapes. *Nano Letters*, 2002, **2**(10): 1109–1112.

280. Yan, J. et al., Structure and cathodoluminescence of individual ZnS/ZnO biaxial nanobelt heterostructures. *Nano Letters*, 2008, **8**(9): 2794–2799.

281. An, S.J. et al., Heteroepitaxial fabrication and structural characterizations of ultrafine GaN/ZnO coaxial nanorod heterostructures. *Applied Physics Letters*, 2004, **84**(18): 3612–3614.

282. Park, W.I. et al., Quantum confinement observed in ZnO/ZnMgO nanorod heterostructures. *Advanced Materials*, 2003, **15**(6): 526–529.

283. Wong, Y.-h. and Q. Li, Study of the crystallinity of ZnO in the Zn/ZnO nanocable heterostructures, Electronic supplementary information (ESI) available: SEM and XRD. See http://www.rsc.org/suppdata/jm/b4/b400646a. *Journal of Materials Chemistry*, 2004, **14**(9): 1413.

284. Wu, J.-J. et al., Heterostructures of ZnO–Zn coaxial nanocables and ZnO nanotubes. *Applied Physics Letters*, 2002, **81**(7): 1312–1314.

285. (a) Gothard, N. et al., Controlled growth of Y-junction nanotubes using Ti-doped vapor catalyst. *Nano Letters*, 2004, **4**(2): 213–217; (b) Wang D. et al., Rational Growth of Branched and Hyperbranched Nanowire Structures. *Nano Letters*, 2004, **4**(5): 871–874.

286. Du, F. et al., Preparation of tunable 3D pillared carbon nanotube-graphene networks for high-performance capacitance. *Chemistry of Materials*, 2011, **23**(21): 4810–4816.

287. (a) Zhai, T. et al., Design and fabrication of rocketlike tetrapodal CdS nanorods by seed-epitaxial metal–organic chemical vapor deposition. *Crystal Growth & Design*, 2007, **7**: 488–491; (b) Zhu, Y.C., Y. Bando, D.F. Xue, and D. Golberg, Nanocable-aligned ZnS tetrapod nanocrystals. *Journal of the American Chemical Society*, 2003, **125**: 16196–16197; (c) Zhai, T. et al., Synthesis of single-crystal ZnS nanoawls via two-step pressure-controlled vapor-phase deposition and their optical properties. *Crystal Growth & Design*, 2007, **7**(8): 1388–1392.

288. Gautam, U.K. et al., Synthesis, structure, and multiply enhanced field-emission properties of branched ZnS nanotube – In nanowire core-shell heterostructures. *Acs Nano*, 2008, **2**(5): 1015–1021.

289. Yang, R. et al., Single-crystalline branched zinc phosphide nanostructures: synthesis, properties, and optoelectronic devices. *Nano Letters*, 2007, **7**(2): 269–275.

290. Wang, B. et al., Self-assembled and Pd decorated Zn_2SnO_4/ZnO wire-sheet shape nano-heterostructures networks hydrogen gas sensors. *Sensors and Actuators B: Chemical*, 2014, **195**: 549–561.

291. (a) Jiang, K., Q. Li, and S. Fan, Spinning continuous carbon nanotube yarns. *Nature*, 2002, **419**: 801–801; (b) Jiang, K.L. et al., Superaligned carbon nanotube arrays, films, and yarns: a road to applications. *Advanced Materials*, 2011, **23**(9): 1154–1161.

292. Hsu, Y.-J. and S.-Y. Lu, One-step preparation of coaxial CdS–ZnS nanowires. *Chemical Communications*, 2004, **2004**(18): 2102–2103.

293. Fu, X.L., L.H. Li, and W.H. Tang, Preparation and characterization of CdS/Si coaxial nanowires. *Solid State Communications*, 2006, **138**(3): 139–142.

294. Sun, X.H. et al., One-dimensional silicon–cadmium selenide heterostructures. *The Journal of Physical Chemistry C*, 2007, **111**(24): 8475–8482.

295. Dai, G. et al., Ordered CdS micro/nanostructures on CdSe nanostructures. *Nanotechnology*, 2009, **20**(12): 125601.

296. Zhai, T. et al., Design and fabrication of rocketlike tetrapodal CdS nanorods by seed-epitaxial metal–organic chemical vapor deposition. *Crystal Growth & Design*, 2007, **7**(3): 488–491.

297. Fan, X. et al., Formation and photoelectric properties of periodically twinned ZnSe/SiO_2 nanocables. *The Journal of Physical Chemistry C*, 2009, **113**(3): 834–838.

298. Wang, K. et al., Direct growth of highly mismatched Type II ZnO/ZnSe core/shell nanowire arrays on transparent conducting oxide substrates for solar cell applications. *Advanced Materials*, 2008, **20**(17): 3248–3253.

299. Liu, Z. et al., Supramolecular chemistry on water-soluble carbon nanotubes for drug loading and delivery. *ACS Nano*, 2007, **1**(1): 50–56.

300. Tian, W. et al., Flexible ultraviolet photodetectors with broad photoresponse based on branched ZnS-ZnO heterostructure nanofilms. *Advanced Materials*, 2014, **26**(19): 3088–3093.

301. Jung, Y., D.-K. Ko, and R. Agarwal, synthesis and structural characterization of single-crystalline branched nanowire heterostructures. *Nano Letters*, 2007, **7**(2): 264–268.

302. Li, Y. et al., Dopant-free GaN/AlN/AlGaN radial nanowire heterostructures as high electron mobility transistors. *Nano Letters*, 2006, **6**(7): 1468–1473.

303. Murphy, M.W. et al., Optical emission of biaxial ZnO–ZnS nanoribbon heterostructures. *The Journal of Chemical Physics*, 2009, **130**(8): 084707.

304. Sun, T. et al., Synthesis and characterization of Zn_3P_2/ZnS core/shell nanowires. *Physics Letters A*, 2011, **375**(21): 2118–2121.

305. Shen, G. et al., Self-organized hierarchical ZnS/SiO$_2$ nanowire heterostructures. *The Journal of Physical Chemistry B*, 2006, **110**(14): 7199–7202.

306. Kirmse, H. et al., TEM analysis of the container effect of Au-based catalyst droplets during vapour-liquid-solid growth of axial ZnTe/CdTe nanowires. *Crystal Research and Technology*, 2009, **44**(10): 1047–1053.

307. Cao, Y.L. et al., Coaxial nanocables of p-type zinc telluride nanowires sheathed with silicon oxide: Synthesis, characterization and properties. *Nanotechnology*, 2009, **20**(45): 455702.

308. Zhang, Q. et al., The road for nanomaterials industry: A review of carbon nanotube production, post-treatment, and bulk applications for composites and energy storage. *Small*, 2013, **9**(8): 1237–1265.

309. Wang, Z.L., Zinc oxide nanostructures: Growth, properties and applications. *Journal of Physics-Condensed Matter*, 2004, **16**(25): R829–R858.

310. (a) Zhou W., X. Bai, E. Wang, and S. Xie, Synthesis, structure, and properties of single-walled carbon nanotubes. *Advanced Materials*, 2009, **2**: 4565–4583; (b) Hayashi, T. and M. Endo, Carbon nanotubes as structural material and their application in composites. *Composites Part B: Engineering*, 2011, **42**(8): 2151–2157.

311. De Volder, M.F.L. et al., Carbon nanotubes: Present and future commercial applications. *Science*, 2013, **339**(6119): 535–539.

312. (a) Shimizu, Y. et al., Multi-walled carbon nanotube-reinforced magnesium alloy composites. *Scripta Materialia*, 2008, **58**(4): 267–270; (b) Bakshi, S.R., D. Lahiri, and A. Agarwal, Carbon nanotube reinforced metal matrix composites – a review. *International Materials Reviews*, 2010, 55(1): 41–64.

313. (a) Ericson, L.M. et al., Macroscopic, neat, single-walled carbon nanotube fibers. *Science*, 2004, **305**(5689): 1447–1450; (b) Zhong, X. et al., Continuous multilayered carbon nanotube yarns. *Advanced Materials*, 2010, **22**: 692–696.

314. (a) Dai, L. et al., Carbon nanomaterials for advanced energy conversion and storage. *Small*, 2012, **8**(8): 1130–1166; (b) Contreras, M.A. et al., Replacement of transparent conductive oxides by single-wall carbon nanotubes in Cu(In, Ga)Se2-based solar cells. *The Journal of Physical Chemistry C*, 2007, **111**(38): 14045–14048; (c) Lee, J., J. Park, S. Lee, H. Kim, S. Yoo, and S. Kim, Selective electron- or hole-transport enhancement in bulk-heterojunction organic solar cells with N- or B-doped carbon nanotubes. *Advanced Materials*, 2011, **23**: 629–633; (d) Matsumoto, T., T. Komatsu, K. Arai, T. Yamazaki, M. Kijima, H. Shimizu, Y. Takasawa, and J. Nakamura, Reduction of Pt usage in fuel cell electrocatalysts with carbon nanotube electrodes. *Chemical Communications*, 2004, **2004**: 840–841; (e) Gong K., F. Du, Z. Xia., M. Durstock, and L. Dai, Nitrogen-doped carbon nanotube arrays with high electrocatalytic activity for oxygen reduction. *Science*, 2009, **323**: 760–764.

315. (a) Lahiri, I. et al., High capacity and excellent stability of lithium ion battery anode using interface-controlled binder-free multiwall carbon nanotubes grown on copper. *ACS Nano*, 2010, **4**: 3440–3446; (b) Technews, http://technews.cn/2016/05/10/cnanotechnology/, 2016; (c) Izadi-Najafabadi, A. et al., Extracting the full potential of single-walled carbon nanotubes as durable supercapacitor electrodes operable at 4 V with high power and energy density. *Advanced Materials*, 2010, **22**: E235–E241; (d) Hqew, http://tech.hqew.com/news_1277555, 2016; (e) Ren, J. et al., Twisting carbon nanotube fibers for both wire-shaped micro-supercapacitor and micro-battery. *Advanced Materials*, 2013, **25**: 1155–1159; (f) Zhang, Y., W. Bai, J. Ren, W. Weng, H. Lin, Z. Zhang, and H. Peng, Super-stretchy lithium-ion battery based on carbon nanotube fiber. *Journal of Materials Chemistry A*, 2014, **2**: 11054–11059.

316. Heller, D.A. et al., Single-walled carbon nanotube spectroscopy in live cells: Towards long-term labels and optical sensors. *Advanced Materials*, 2005, **17**(23): 2793–2799.

317. De La Zerda, A. et al., Carbon nanotubes as photoacoustic molecular imaging agents in living mice. *Nature Nanotechnology*, 2008, **3**(9): 557–562.

318. Hong, S.Y. et al., Filled and glycosylated carbon nanotubes for in vivo radioemitter localization and imaging. *Nature Materials*, 2010, **9**(6): 485–490.

319. (a) Kam, N.W.S. et al., Carbon nanotubes as multifunctional biological transporters and near-infrared agents for selective cancer cell destruction. *Proceedings of the National Academy of Sciences of the United States of America*, 2005, **102**(33): 11600–11605; (b) Liang, X. et al., Dye-conjugated single-walled carbon nanotubes induce photothermal therapy under the guidance of near-infrared imaging. *Cancer Letters*, 2016, **383**: 243–249.

320. (a) Lin, Y., F. Lu, Y. Tu, and Z. Ren, Glucose biosensors based on carbon, nanotube nanoelectrode ensembles. *Nano Letters*, 2004, **4**(2): 191–195; (b) Kurkina, T. et al., Label-free detection of few copies of DNA with carbon nanotube impedance biosensors. *Angewandte Chemie International Edition*, 2011, **50**(16): 3710–3714.

321. (a) Heller, D.A. et al., Multimodal optical sensing and analyte specificity using single-walled carbon nanotubes. *Nature Nanotechnology*, 2009, **4**(2): 114–120; (b) Zelada-Guillén, G.A., A. Tweed-Kent, M. Niemann, H.U. Göringer, J. Riu, and F.X. Rius, Ultrasensitive and real-time detection of proteins in blood using a potentiometric carbon-nanotube aptasensor. *Biosensors and Bioelectronics*, 2013, **41**: 366–371; (c) Allen, B., P. Kichambare and A. Star, Carbon nanotube field-effect-transistor-based biosensors. *Advanced Materials*, 2007, **19**: 1439–1451; (d) Star, A., E. Tu, J. Niemann, J. Gabriel, C. Joiner, and C. Valcke, Label-free detection of DNA hybridization using carbon nanotube network field-effect transistors. *Proceedings of the National Academy of Sciences of the United States of America*, 2006, **103**: 921–926.

322. Palomar, Q., C. Gondran, M. Holzinger, Marks R., and S. Cosnier, Controlled carbon nanotube layers for impedimetric immunosensors: High performance label free detection and quantification of anti-cholera toxin antibody. *Biosensors and Bioelectronics*, 2017, **97**: 177–183.

323. (a) Hong, G. et al., Multifunctional in vivo vascular imaging using near-infrared II fluorescence. *Nature Medicine*, 2012, **18**: 1841–1846; (b) Delogu, L.G. et al., Functionalized multiwalled carbon nanotubes as ultrasound contrast agents. *Proceedings of the National Academy of Sciences of the United States of America*, 2012, **109**: 16612–16617; (c) Ding, W. et al., Targeted Fe-filled carbon nanotube as a multifunctional contrast agent for thermoacoustic and magnetic resonance imaging of tumor in living mice, nanomed. *Nanotechnology Biology Medicine*, 2016, **12**: 235–244.

324. Pantarotto, D., C.D. Partidos, and R. Graff, Synthesis, structural characterization, and immunological properties of carbon nanotubes functionalized with peptides. *Journal of the American Chemical Society*, 2003, **125**(20): 6160–6164.

325. Bianco, A., K. Kostarelos, and M. Prato, Making carbon nanotubes biocompatible and biodegradable. *Chemical Communications*, 2011, **47**(37): 10182.

326. Salvador-Morales, C. et al., Complement activation and protein adsorption by carbon nanotubes. *Molecular Immunology*, 2006, **43**(3): 193–201.

327. (a) Sireesha, M., V. Babu, and S. Ramakrishna, Functionalized carbon nanotubes in bio-world: Applications, limitations and future directions. *Materials Science and Engineering B*, 2017, **223**: 43–63; (b) Yang, W. et al., Carbon nanotubes for biological and biomedical applications. *Nanotechnology*, 2007, **18**(41): 412001; (c) Liu Z., S. Tabakman, K. Welsher, and H. Dai, Carbon nanotubes in biology and medicine: in vitro and in vivo detection, imaging and drug delivery. *Nano Research*, 2009, **2**: 85–120; (d) Tîlmaciu C. and M. Morris, Carbon nanotube biosensors. *Frontiers in Chemistry*, 2015, **3**: 59.

328. Endo, M., T. Hayashi, and Y.-A. Kim, Large-scale production of carbon nanotubes and their applications. *Pure and Applied Chemistry*, 2006, **78**(9): 1703–1713.

329. (a) Huang, Y.J., H.C. Wu, N.H. Tai, and T.W. Wang, Carbon nanotube rope with electrical stimulation promotes the differentiation and maturity of neural stem cells. *Small*, 2012, **8**: 869–877; (b) Ahadian, S., L. Huyer, M. Estili, B. Yee, N. Smith, Z. Xu, Y. Sun, and M. Radisic, Moldable elastomeric polyester-carbon nanotube scaffolds for cardiac tissue engineering. *Acta Biomaterialia*, 2017, 52: 81–91.

330. (a) Beigbeder, A., P. Degee, S.L. Conlan, R.J. Mutton, A.S. Clare, M.E. Pettitt, M.E. Callow, J.A. Callow, and P. Dubois, Preparation and characterisation of silicone-based coatings filled with carbon nanotubes and natural sepiolite and their application as marine fouling-release coatings. *Biofouling*, 2008, **24**: 291–302; (b) Park, S. and M, Shon, Effects of multi-walled carbon nano tubes on corrosion protection of zinc rich epoxy resin coating. Journal of Industrial and Engineering Chemistry, 2015, 21: 1258–1264.

331. Gao, G. and C.D. Vecitis, Electrochemical carbon nanotube filter oxidative performance as a function of surface chemistry. *Environmental Science & Technology*, 2011, **45**(22): 9726–9734.

332. Rahaman, M.S., C.D. Vecitis, and M. Elimelech, Electrochemical carbon-nanotube filter performance toward virus removal and inactivation in the presence of natural organic matter. *Environmental Science & Technology*, 2012, **46**(3): 1556–1564.

333. Holt, J.K. et al., Fast mass transport through sub-2-nanometer carbon nanotubes. *Science*, 2006, **312**(5776): 1034–1037.

334. Corry, B., Designing carbon nanotube membranes for efficient water desalination. *The Journal of Physical Chemistry B*, 2008, **112**(5): 1427–1434.

335. Liu, Q. et al., In situ assembly of multi-sheeted buckybooks from single-walled carbon nanotubes. *ACS Nano*, 2009, **3**(3): 707–713.

336. Wang, Y. et al., The large-scale production of carbon nanotubes in a nano-agglomerate fluidized-bed reactor. *Chemical Physics Letters*, 2002, **364**(5): 568–572.

337. Pirard, S.L., S. Douven, and J.-P. Pirard, Large-scale industrial manufacturing of carbon nanotubes in a continuous inclined mobile-bed rotating reactor via the catalytic chemical vapor deposition process. *Frontiers of Chemical Science and Engineering*, 2017, **11**(2): 280–289.

338. Kumar, V. et al., Gas-phase, bulk production of metal oxide nanowires and nanoparticles using a microwave plasma jet reactor. *The Journal of Physical Chemistry C*, 2008, **112**(46): 17750–17754.

339. Kumar, M. and Y. Ando, Gigas growth of carbon nanotubes. *Defence Science Journal*, 2008, **58**(4): 496–503.

340. Hahn, J. et al., New continuous gas-phase synthesis of high purity carbon nanotubes by a thermal plasma jet. *Carbon*, 2004, **42**(4): 877–883.

341. Choi, S.I. et al., Continuous process of carbon nanotubes synthesis by decomposition of methane using an arc-jet plasma. *Thin Solid Films*, 2006, **506–507**: 244–249.

342. Tian, Y.J. et al., Effect of catalysis on coal to nanotube in thermal plasma. *Catalysis Today*, 2004, **89**(1–2): 233–236.

343. (a) Pang, L.S.K. and M.A. Wilson, Nanotubes from coal. *Energy & Fuels*, 1993, **7**(3): 436–437; (b) Li, Y.F. et al., Bamboo-shaped carbon tubes from coal. *Chemical Physics Letters*, 2002, **366**(5–6): 544–550.

344. Jiang, K., Q. Li and S. Fan, Nanotechnology: Spinning continuous carbon nanotube yarns. *Nature*, 2002, 419: 801.

345. Liu, K. et al., Controlled growth of super-aligned carbon nanotube arrays for spinning continuous unidirectional sheets with tunable physical properties. *Nano Letters*, 2008, **8**(2): 700–705.

346. (a) Flahaut, E., R. Bacsa, A. Peigney, and C. Laurent, Gram-scale CCVD synthesis of double-walled carbon nanotubes. *Chemistry Communications*, 2003, **12**: 1442–1443; (b) Lyu, S.C. et al., Large-scale synthesis of high-quality double-walled carbon nanotubes by catalytic decomposition of n-hexane. *The Journal of Physical Chemistry B*, 2004, **108**: 2192–2194; (c) Wei, J. et al., Large-scale synthesis of long double-walled carbon nanotubes. *The Journal of Physical Chemistry B*, 2004, **108**(26): 8844–8847.

347. (a) Liu Y., W. Qian, Q. Zhang, G. Ning, G. Luo, Y. Wang, D. Wang, and F. Wei, Synthesis of high-quality, double-walled

carbon nanotubes in a fluidized bed reactor. *Chemical Engineering & Technology*, 2009, **32**(1): 73–79; (b) Dunens, O.M., K.J. MacKenzie, and A.T. Harris, Large-scale synthesis of double-walled carbon nanotubes in fluidized beds. *Industrial & Engineering Chemistry Research*, 2010, **49**(9): 4031–4035.

348. Kumar, M. and Y. Ando, ChemInform abstract: Chemical vapor deposition of carbon nanotubes: A review on growth mechanism and mass production. *Journal of Nanoscience and Nanotechnology*, 2010, **10**(6): 3739–3758.

349. (a) Nikolaev, P., M.J. Bronikowski, K.R. Bradley, F. Rohmund, Colbert D.T., K.A. Smith, and R.E. Smalley, Gas-phase catalytic growth of single-walled carbon nanotubes from carbon monoxide. *Chemistry Physics Letters*, 1999, **313**(1–2): 91–97.; (b) Shaffer, M.S.P., I. Kinloch, and A.H. Windle, Synthesis of nanoscaled carbon materials, US 7135159B2; (c) Mora, E., T. Tokune, and A.R. Harutyunyan, Continuous production of single-walled carbon nanotubes using a supported floating catalyst, *Carbon*, 2007, **45**: 971–977.

350. (a) Ma, W. et al., Directly synthesized strong, highly conducting, transparent single-walled carbon nanotube films, *Nano Letters*, 2007, **7**: 2307–2311; (b) Zhu, H.W. et al., Direct synthesis of long single-walled carbon nanotube strands. *Science*, 2002, **296**(5569): 884–886.

351. (a) Hahn, J., J.H. Han, J.E. Yoo, H.Y. Jung, and J.S. Suh, New continuous gas-phase synthesis of high purity carbon nanotubes by a thermal plasma jet. *Carbon*, 2004, **42**: 877–883; (b) Kim, K.S. et al., Large-scale production of single-walled carbon nanotubes by induction thermal plasma. *Journal of Physics D: Applied Physics*, 2007, **40**(8): 2375–2387; (c) Maruyama, S., E. Einarsson, Y. Murakami, and T. Edamura, Growth process of vertically aligned single-walled carbon nanotubes. *Chemical Physics Letters*, 2005, **403**: 320–323.

352. Brockway, L. et al., Large-scale synthesis and in situ functionalization of Zn_3P_2 and Zn_4Sb_3 nanowire powders. *Physical Chemistry Chemical Physics*, 2013, **15**(17): 6260.

353. (a) Zhou, Z. et al., Rapid mass production of ZnO nanowires by a modified carbothermal reduction method. *Materials Letters*, 2011, **65**(5): 832–835; (b) Kumar, V., J.H. Kim, C. Pendyala, B. Chernomordik, and M.K. Sunkara, Gas-phase, bulk production of metal oxide nanowires and nanoparticles using a microwave plasma jet reactor. *Journal of Physical Chemistry C*, 2008, **112**(46): 17750–17754.

354. Seashell Technology, Silver Nanowire Acquired by BASF, http://www.seashelltech.com/pressRelease_Acquisition.shtml, March 10, 2015.

355. Douven, S. et al., Large-scale synthesis of multi-walled carbon nanotubes in a continuous inclined mobile-bed rotating reactor by the catalytic chemical vapour deposition process using methane as carbon source. *Chemical Engineering Journal*, 2012, **188**: 113–125.

356. Huang, J.Q. et al., Process intensification by co2 for high quality carbon nanotube forest growth: double-walled carbon nanotube convexity or single-walled carbon nanotube bowls? *Nano Research*, 2009, **2**(11): 872–881.

357. (a) Zhang, Q. et al., Carbon nanotube mass production: Principles and processes. *ChemSusChem*, 2011, **4**(7): 864–889; (b) Wei, F. et al., The mass production of carbon nanotubes using a nano-agglomerate fluidized bed reactor: A multiscale space/time analysis. *Powder Technology*, 2008, **183**: 10–20.

358. Huang, J.-Q. et al., The release of free standing vertically-aligned carbon nanotube arrays from a substrate using CO_2 oxidation. *Carbon*, 2010, **48**(5): 1441–1450.

359. (a) Buchholz, S. et al., Process for the production of carbon nanotubes in a fluidized bed, US 9050572 B2; (b) Thomas Swan, Elicarb® Carbon Nanotubes, http://www.thomas-swan.co.uk/advanced-materials/elicarb%C2%AE-carbon-nanotubes.

360. Moisala, A., A.G. Nasibulin, and E.I. Kauppinen, The role of metal nanoparticles in the catalytic production of single-walled carbon nanotubes—A review. *Journal of Physics: Condensed Matter*, 2003, **15**(42): S3011–S3035.

361. (a) Predtechensky, M.R., O.M. Tukhto, and I.Y. Koval, System and method for producing carbon nanotubes, US8551413B2, 2013; (b) Issa, D., TUBALL single wall carbon nanotubes, https://www.nanotechmag.com/high-yield-technology-synthesis-high-purity-tuball-single-wall-carbon-nanotubes-nano-tech-2017-japan/, 2017; (c) Arthur, D. et al., Carbon nanomaterial commercialization: Lessons for graphene from carbon nanotubes. *MRS Bulletin*, 2012, **37**(12): 1297–1306; (d) Brewer, Science, InFlect Flex Sensors, https://www.brewerscience.com/products/inflect-flex-sensor/, 2017.

362. (a) Poland, C.A. et al., Carbon nanotubes introduced into the abdominal cavity of mice show asbestoslike pathogenicity in a pilot study. *Nature Nanotechnology*, 2008, **3**: 423–428; (b) Evans, D.E. et al., Aerosol monitoring during carbon nanofiber production: mobile direct-reading sampling. *The Annals of Occupational Hygiene*, 2010, **54**(5): 514–531.

363. (a) Singh, A. et al., Environmental impact assessment for potential continuous processes for the production of carbon nanotubes. *American Journal of Environmental Sciences*, 2008, **4**(5): 522–534; (b) Bergeson, L., Company Announces First REACH Registration of Single Wall Carbon Nanotubes, http://www.nanotech-now.com/columns/?article=1125, 2016.

364. Rittinghausen, S. et al., The carcinogenic effect of various multi-walled carbon nanotubes (MWCNTs) after intraperitoneal injection in rats. *Particle and Fibre Toxicology*, 2014, **11**: 59.

365. Lam, C.-W. et al., A review of carbon nanotube toxicity and assessment of potential occupational and environmental health risks. *Critical Reviews in Toxicology*, 2006, **36**(3): 189–217.

366. Niosh, Occupational Exposure to Carbon Nanotubes and Nanofibers. https://www.cdc.gov/niosh/docs/2013-145/pdfs/2013-145.pdf, 2013.

367. European Chemicals Agency, https://echa.europa.eu/substance-information/-/substanceinfo/100.217.898, 2017.

CVD of flat monolayer of 2D atomics honeycomb structure and their applications

MANOJ KUMAR SINGH, DHANANJAY K. SHARMA, GONZALO OTERO-IRURUETA, AND MARÍA J. HORTIGÜELA

4.1 INTRODUCTION

The field of research in graphene—a single atomic layer of graphitic carbon and other two-dimensional materials has been enjoying extraordinary growth during the past decade because of its extraordinary properties that make it a suitable material for a wide range of technological applications [1–4]. Their excellent charge-carrier mobility quickly grabbed the attention of the scientific community, bringing a hope that one day it will compete with silicon to be the material for the next generations for certain applications in the electronic industries. Furthermore, many extraordinary properties, such as its 2.3% absorption in the white light spectrum, high surface area, high Young's modulus, and excellent thermal conductivity, have all been reported. Because of its remarkable properties, applications using graphene in a wide range of areas, including high-speed electronic and optical devices, energy generation and storage, hybrid materials, chemical sensors, and even DNA sequencing, have all been explored. Beyond this, very recently, graphene-based field-effect transistors (GFETs) were used, which shows its exciting and bright prospects for sensing applications due to its greatly higher sensitivity and stronger selectivity [5–7]. GFETs play a role for

the preparation of ultra-sensitive nanoelectronic biosensors because of its perfect characteristics, including high carrier mobility, strong and flexible structure, and so on. In their work, Seon Joo Park and co-workers were successfully able to make an ultrasensitive and flexible FET olfactory system, which they named 'Bioelectronic nose' [8]. There are numerous examples in which graphene was used for biological applications, for example for the detection of living cell and bacteria sensors [9], GFETs functionalized with E.coli antibodies [10], and FETs based on graphene for the detection of malaria-infected red blood cells [11].

A variety of proof-of-concept devices have also been demonstrated. Materials science had a major scientific breakthrough in 2004, when Novoselov and Geim isolated the first single-layer two-dimensional (2D) material, graphene, through the Scotch tape exfoliation of graphite, and received the Nobel Prize in 2010. However, the obtained samples were suitable only for fundamental studies. Moreover, pristine graphene itself is unlikely to be used for the fabrication of logical circuits operated at room temperature with low standby power dissipation because graphene has no band gap. The result is a small current on/off ratio in graphene field-effect transistors. The prerequisite for such applications is the mass production of graphene in a controlled manner because the number of graphene layers as well as the defects in these graphene layers significantly influences the subsequent transport properties. In this direction, some other 2D materials were also discovered like TMDCs structures to overcome these problems. Very recently one such 2D lattice, silicene, germanene, an sp^2-hybridized equivalent of graphene, offers better compatibility with silicon processing and may provide solutions for some of the problems of graphene associated with the lack of an energy gap. Methods such as mechanical exfoliation, liquid-phase exfoliation, chemical vapour deposition (CVD), and molecular beam epitaxy have been developed in order to make single layers and a few layers of 2D lattices. Despite these efforts, the fine control of the number and structure of graphene and other 2D lattices over an entire substrate remains a major challenge. More realistic samples, from an industry point of view, were obtained by CVDs and related techniques on metallic substrates.

In this chapter, we discussed in detail the structure, electronic properties, deposition mechanism of graphene, and cousins (silicene and germanene). In addition to this, we will also present an overview of the large-scale production and towards industrialisation of these novel 2D materials for the real state-of-the-art applications in the field of electronics, clean energy, biomedical, and environmental field [12–14].

4.2 GRAPHENE

Graphene, a two-dimensional, single-layer sheet of sp^2-hybridized carbon atoms, has attracted world-wide attention and research interest, owing to its exceptional physical properties, such as high electronic conductivity, good thermal stability, and excellent mechanical strength. Interestingly, graphene is a basic building block for graphitic materials of all other dimensionalities. It can be wrapped up into 0D fullerenes, rolled into 1D nanotubes or stacked into 3D graphite.

The story of graphene is quite old. In fact, we had been using it for a long time when Petroski in 1564 invented pencils using graphite (which is composed of stack graphene) but we were not aware that we were making graphene in a simple way (by pressing pencil against a sheet of paper). Moreover, theoretically, graphene has been studied for 60 years and was widely used for describing properties of various carbon-based nanostructured materials. After 40 years, it was realized that graphene holds

promising electronic and physical properties. In the year 2004, A.K. Geim and his colleague K.S. Novoselov were able to successfully isolate a single layer of graphene from highly oriented pyrolytic graphene (HOPG) with simple scotch tape method by mechanical exfoliation. For this achievement the group was awarded the Noble Prize in 2010. Up to now, the best quality graphene, in terms of structural integrity, has been obtained by this method. Consequently, only flat graphene sheets (several micrometres in size) can be produced, and the number of exfoliated layers is not easily controlled. However, graphene of large area and high quality with low structural defects is needed for practical applications such as microelectronics, optoelectronics (solar cells, touch screens, liquid crystal displays), biomedical, sensors, nanofluids, graphene-based batteries, super-capacitors, and energy applications. In this direction, a 'bottom-up' synthesis route using CVD holds promise for the large-area graphene deposition.

In following section, we will present an overview of large-area synthesis and electronic properties of graphene by CVD by using state-of-the-art techniques.

4.2.1 Electronic structure of single-layer graphene

Graphene is made out of carbon atoms arranged in hexagonal structure, as represented in Figure 4.1. The structure can be seen as a triangular lattice with a basis of two atoms per unit cell. The lattice vectors can be written as

$$a_1 = \frac{a}{2}\left(3, \sqrt{3}\right), a_2 = \frac{a}{2}\left(3, -\sqrt{3}\right) \qquad (4.1)$$

Where a $\approx$ 1.42 Å is the carbon-carbon distance. The reciprocal-lattice vectors are given by

$$b_1 = \frac{2\pi}{3a}\left(1, \sqrt{3}\right), b_2 = \frac{2\pi}{3a}\left(1, -\sqrt{3}\right) \qquad (4.2)$$

In the physics of graphene, the two points K and K' at the corners of the graphene Brillouin zone (BZ) are represented as Dirac points. Their positions in momentum space are given below

$$K = \left(\frac{2\pi}{3a}, \frac{2\pi}{3\sqrt{3}a}\right), K' = \left(\frac{2\pi}{3a}, -\frac{2\pi}{3\sqrt{3}a}\right), \qquad (4.3)$$

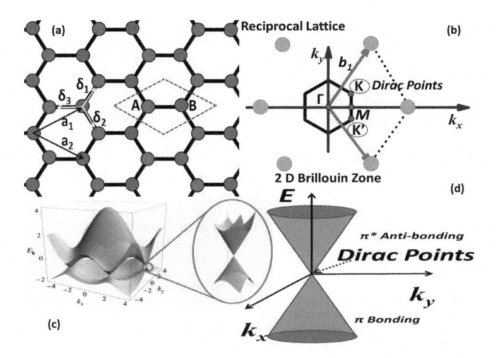

Figure 4.1 Lattice structure of graphene **(a)** Honeycomb lattice, **(b)** its Brillouin zone, electronic dispersion in the honeycomb lattice **(c)** Energy spectrum (in units of t) for finite values of t and t′ for t = 2.7 eV and t′ = −0.2t with zoomed energy bands close to one of the Dirac points, **(d)** The energy bands of two-dimensional graphene are smooth-sided cones.

The three nearest-neighbour vectors in real space are given by

$$\delta_1 = \frac{a}{2}(1,\sqrt{3}),\ \delta_2 = \frac{a}{2}(1,-\sqrt{3}),\ \delta_3 = -a\,(1,0)\ (4.4)$$

While the six second-nearest neighbours are located at $\delta'_1 = \pm a_1$, $\delta'_2 = \pm a_2$, $\delta'_3 = \pm(a_2 - a_1)$.

The tight-binding Hamiltonian for electrons in graphene considering that electrons can hop to both nearest and next nearest neighbour atom, has the form (we use units such that $\hbar = 1$)

$$H = -t \sum_{\langle i,j\rangle,\sigma}\left(a^\dagger_{\sigma,i}b_{\sigma,j}+H.c\right)$$

$$-t'\sum_{\langle\langle i,j\rangle\rangle,\sigma}\left(a^\dagger_{\sigma,i}a_{\sigma,j}+b^\dagger_{\sigma,i}b_{\sigma,j}+H.c.\right)$$

(4.5)

Where $a_{i,\sigma}(a^\dagger_{i,\sigma})$ annihilates (creates) an electron with spin $\sigma(\sigma=\uparrow,\downarrow)$ on site R_i on sublattice A (an equivalent definition is used for sublattice A (an equivalent definition is used for sublattice B),

$t(\approx 2.8\text{eV})$ is the nearest-neighbour hopping energy difference between different sub lattices), and t' is the next nearest-neighbour hopping energy (hopping in the same lattice). The energy bands derived from this Hamiltonian have the form (P. R. Wallace, *Physical Review*, 1947, 71, 622–634)

$$E_\pm(k) = \pm t\sqrt{3+f(k)}-t'f(k),$$

$$f(k)=2\cos\left(\sqrt{3}k_y a\right)+4\cos\left(\frac{\sqrt{3}}{2}k_y a\right)\cos\left(\frac{3}{2}k_x a\right),\ (4.6)$$

Where the plus sign applies to the upper (π^*) and the minus sign the lower (π) band. It is clear from Equation (4.6) that the spectrum is symmetric around zero energy if $t'=0$. For finite values of t', the electron-hole symmetry is broken and the π and π^* band becomes asymmetric. In Figure 4.1, we show the full-band structure of graphene with both t and t'. In the same figure, we also show the zoom-in of the band structure close to one of the Dirac points (at the K or K' point in the BZ) (Figure 4.1). This dispersion can be obtained by expanding the full-band structure, Equation (4.6), close to the K (or K') vector, Equation (4.3), as $k = K + q$, with $|q| \ll |K|$ [15].

$$E_{\pm}(q) \approx v_F |q| + O\left[(q/K)^2\right], \qquad (4.7)$$

Where q is the momentum measured relatively to the Dirac points and v_F is the Fermi velocity, given by $v_F = 3ta/2$, with a value $v_F \cong 1 \times 10^6$ m/s. This result was first performed by Wallace (1947) [15].

4.2.2 Graphene properties and synthesis

Graphene exhibits more remarkable electronic, mechanical, optical, and thermal properties than those obtained in any other material, justifying its nickname of a 'miracle material'. Many graphene characteristics measured in experiments have reached theoretically predicted limits such as: room-temperature electron mobility of 2.5×10^5 cm^{2}V^{-1} s^{-1} [16] (theoretical limit ~2×10^5 cm^{2}V^{-1}s^{-1}) [17]; a Young's modulus of 1 TPa and intrinsic strength of 130 GPa [18], (very close to that predicted by theory [19]), very high thermal conductivity (above 3,000 W mK^{-1}; [20]), optical absorption of exactly $\pi\alpha \approx <2.3\%$ (in the infrared limit, where α is the fine structure constant) [21], complete impermeability to any gases [22], and ability to sustain extremely high densities of electric current (a million times higher than copper) [23]. In addition to this, graphene can be readily chemically functionalized by various heteroatoms and functional molecules for exotic applications in the field of nanotechnology [24].

These many superior properties have been achieved only with highest-quality samples (mechanically exfoliated graphene [25]) and for graphene deposited on special substrates like hexagonal boron nitride substrates. However, equivalent characteristics have not been observed on graphene prepared by other techniques. There is a lot of research underway to stabilize the condition for the high-quality mass production of graphene for real state-of-the-art device applications at industry level. Currently, there are many methods being used and developed to prepare graphene of various dimensions, shapes, and quality. Here, we will concentrate on only scalable methods such as: (i) liquid phase exfoliation, (ii) synthesis on SiC, and (iii) CVD.

4.2.2.1 GRAPHENE ON SINGLE CRYSTALS GROWN UNDER UHV CONDITIONS

Graphene has been synthesized on different metallic single crystals like Pt(111), Ir(111), or Ru(0001).

Figure 4.2 shows STM images (a and b) and a LEED pattern (Figure 4.2c) obtained after growing single-layer graphene on Pt(111) under UHV conditions.

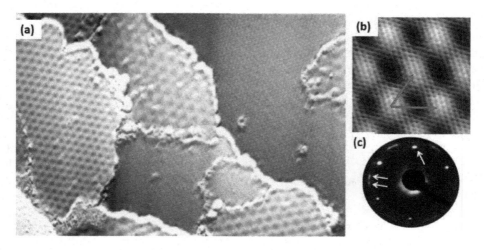

Figure 4.2 STM images **(a, b)** and **(c)** LEED pattern obtained after growing single-layer graphene on Pt(111) under UHV conditions.

STM reveals a rich variety of Moire-like superstructures covering the whole surface, while characteristic LEED pattern shows bright spots (white arrow in Figure 4.2b) related with the Pt(111) surface together with a low-intensity ring that can be related with the graphene sheet. A high-resolution STM image of one of these Moire-like superstructures (Figure 4.2b) shows the honeycomb structure of the carbon atoms in the background while the aspect of the STM image is modulated by the Moire. The red arrows indicate the vectors of the unit cell, which, in the example presented here, are around 1.5 nm large.

The analysis of the STM images combined with that one obtained by LEED indicates that the graphene sheet on Pt(111) is formed by small domains that follow different angular orientation with respect to the main crystallographic directions of the metallic surface. Thus, when the molecular precursors are introduced in the experimental facility, graphene starts growing in several points that are spatially separated. Then, as the coverage increases, the islands slowly grow covering the surface until the borders of neighbouring domains come in contact. Interestingly, the intensity of the graphene ring detected by LEED is modulated as a function of the angle (see yellow arrows in Figure 4.2c) rather than uniform. This modulated intensity indicates that the structural angular orientation of the graphene domains follows specifics angles with respect to the Pt(111) main crystallographic directions. A phenomenological model, based on the strain of the graphene domains following different on-surface angular orientation, predicts up to 22 different possibilities [26].

Similar Moire-like superstructures have been observed on other metallic substrates like Ru(0001) [27], Rh(111) [28], and Ir(111) [29].

Interestingly, the observed Moire patterns, in particular their orientation, corrugation, and periodicity, depend strongly on the particular substrate. That indicates a non-neglected correlation between the metallic atoms of the last layer of the substrate and the graphene sheet. The XPS characterization of graphene shows that both the shapes of the carbon spectra and their binding energies strongly depend on the particular metallic substrate. The reference sample, HOPG, presents a sharp peak that can be fitted by only one peak centred at a binding energy of 284.23 eV and 0.35 eV of FWHM. Also, the spectra corresponding to graphene on Pt(111) and Ir(111) are very sharp and can be

fitted by only one component. The BE and FWHM obtained for the first case are 283.97 eV and 0.34 eV, while for the second case are 284.16 eV and 0.4 eV, respectively. On the other hand, the Rh(111) and Ru(0001) cases present several differences. The XPS spectra show that at least two peak components are necessary for fitting the spectra. The BEs obtained for the Rh(111) case are 284.94 eV and 284.41 eV, while their respective FWHM are 0.56 eV and 0.46 eV. In the Ru(0001) case values of 285.12 and 284.52 eV are obtained for the BEs and 0.61 eV and 0.48 eV for the FWHM, respectively. This splitting of the C 1s core level peak with respect to the HOPG sample indicates that the carbon atoms of the graphene sheets are under different chemical environments. Thus, in the last two cases, the modulation of the Moire structures observed in the topographic STM images have a counterpart in the XPS, indicating an alternating sequence of high and low interaction between carbon atoms and the metallic substrate.

4.2.2.2 LIQUID PHASE EXFOLIATION

In order to develop the commercial applications from the laboratory, it is quite necessary to have industrially scalable methods for the large production of defect-free graphene; liquid exfoliation technique can be used to have large production [12]. In this technique, exfoliation of graphite occurs in a solvent with a surface tension that favours an increase in the total area of graphite crystallites. Moreover, this liquid phase exfoliation is not a new technique. Liao et al. used this method of probe-tip sonication for the dispersion of carbon nano-tubes in epoxy resin nanocomposites [30].

Manoj Singh et al. reported a two-step method for obtaining a homogenous colloidal suspension of single or FLG sheets up to 0.15 mg mL(-1) in N, N-dimethylformamide (DMF) solution. Their method includes high-temperature heat treatment (~2273 K) of commercially available pyrolytic graphite powders in a vacuum (having the range of 1.3×10^{-5} mbar) for 3h followed by probe-tip sonication for 2h of as-obtained heat-treated graphite in DMF solution. The pyrolytic graphite has weakly bonded graphene layers by interlayer interaction forces along the c-axis, which can easily slide against each other and can also be peeled off. The heat treatment at high temperature under vacuum and tip-probe sonication makes liquid-phase exfoliation easier with DMF solutions. The author was successfully able to obtain single or FLG (few-layer graphene, two to six layers) dispersed in

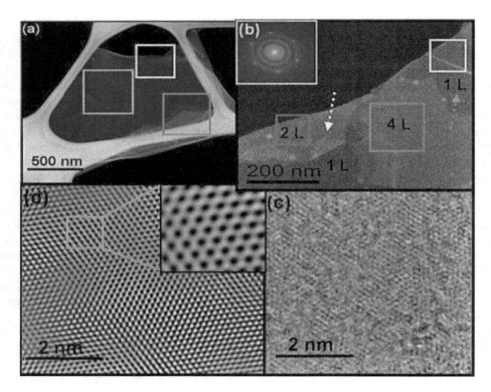

Figure 4.3 HRTEM images of a freely suspended graphene membrane. **(a)** Bright-field TEM image of a suspended graphene membrane. **(b)** Magnified view of the region denoted by a box in **(a)**; the inset shows 2D FFT performed in the region indicated with a box. **(c)** HRTEM image of single layer graphene acquired from the region indicated with a dotted arrow in **(b)**. **(d)** Reconstructed image after filtering in the frequency domain to remove unwanted noise, for clarity. The inset shows the hexagonal graphene network.

DMF solution. Although the percentage of single-layer graphene was lower than FLG, they were able to explain the rotational stacking faults with various rotation angles in two-, three-, and six-layer graphene sheets (Figure 4.3). They gave the quite informative knowledge about the introduction of Moire pattern along with rotation stacking faults in AB Bernal stacked graphene bilayers, which changes the dispersion relationship close to the k-point from parabolic (AB) to linear band behaviour (rotation disorder) and leads to some monolayer graphene properties being observed in two-layer and FLG films.

In another work by Paton et al., large production of graphene has been highlighted by shear exfoliation in liquids [12]. The work can be described in the following ways, by generating high shear using (Silverson model L5M) with closely spaced (~100 µm) rotor/stator combination (Figure 4.4). The graphite along with solvent N-methly-2-pyrrolidone (NMP)

mixed with aqueous surfactant solutions (sodium cholate, NaC) creates large volume suspensions. After centrifugation of the obtained aqueous solution, large quantities of high-quality graphene nanosheets, including some monolayers, were obtained. More experimental details can be found in the literature [12]. To confirm the quality of graphene, the author did Raman spectroscopy, transmission electron microscopy, x-ray photoelectron spectroscopy, and atomic force microscopy. All these characterizations confirm the good quality of graphene even by using liquid exfoliation techniques.

4.2.2.3 GRAPHENE ON SILICON CARBIDE

Silicon carbide (SiC) is a compound present in nature that can crystallize following different symmetries. In this section we will focus on the graphene layer obtained on a 6H-SiC(0001) surface. A standard procedure for obtaining high-quality graphene on this substrate consists in heating the samples at

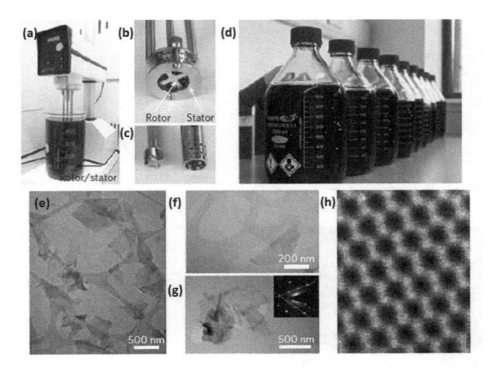

Figure 4.4 Large-scale production of graphene by shearing mixing process; **(a)** A complete Silverson model L5M high-shear mixer with mixing head with rotor in a 5 L beaker of graphene dispersion, **(b)** Close-up view of mixing head with rotor and stator, **(c)** mixing head separated from stator, **(d)** dispersed graphene by shear exfoliation kept in beaker, **(e)** wide-field TEM image, **(f)** TEM images of individual nanosheets, **(g)** monolayer image electron diffraction pattern (inset), **(h)** high-resolution TEM image.

1950 K in a controlled atmosphere [31]. The preparation method includes a first step of H-etching of the surface followed by an annealing at 1950 K into an atmosphere of 900 mbar of argon. The samples prepared by following this method normally show very large high-quality single-layer graphene terraces.

Figure 4.5 shows a standard structural and electronic characterization of the samples performed by different experimental techniques. Figure 4.5a shows a diffraction pattern performed by low-energy electron diffraction (LEED) technique. The diffraction patterns, consisting of sharp spots, show a high level or order in the graphene samples and also their orientation (white arrows) with respect to the substrate main crystallographic direction (grey arrows). Thus, the basal plane unit vectors of graphene are rotated 30° with respect to that one corresponding to the SiC(0001) surface. In Figure 4.5b the C 1s core level peak obtained by XPS shows a rich variety of components related with carbon atoms under different chemical

environments. At lower binding energies appear the component related with the carbon atoms of the SiC substrate (indicated as SiC in the Figure 4.5). Then, a main sharp peak dominates the spectrum and is related with the carbon atoms of the graphene sheet (indicated as G). Finally, two components, indicated as S1 and S2, come from the $(6\sqrt{3} \times 6\sqrt{3})$ R30° interface layer (buffer layer) of this sample. Interestingly, the electronic properties of the graphene sheet can be probed by angle-resolved photoelectron spectra near the K point Figure 4.5c. It shows the characteristic band structure that correspond with high-quality single-layer graphene samples [32]. The position of the Dirac energy (grey arrow in Figure 4.5c) at 0.45 eV below the Fermi energy indicated an electro doping of the graphene sheet from the substrate. Finally, the Raman spectra presented in Figure 4.5d compares the quality of the graphene samples obtained by the methodology above (bottom spectra) described with respect to other procedures (top spectra). The characteristic

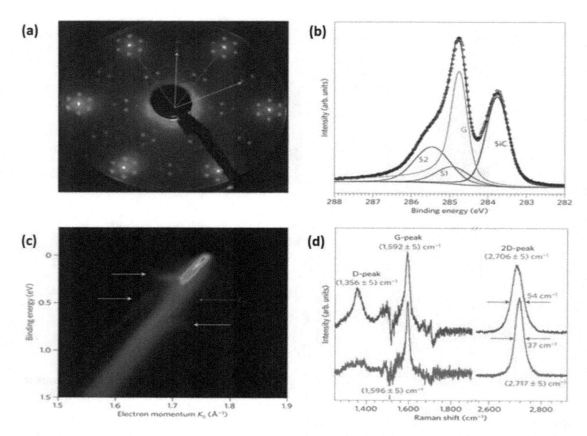

Figure 4.5 Atomic and electronic structure of ex-situ-grown monolayer graphene, **(a)** showing LEED pattern at 74 eV having diffraction spots due to SiC(0001) substrate and graphene lattice, **(b)** XPS peak of C1s core-level spectrum measured at a photon energy of 700 eV **(c)** ARPES image of π bands in the vicinity of the K-point of the hexagonal Brillouin zone measured along the Γ K-direction, **(d)** Comparison of Raman spectra of Ar-grown (bottom spectra) and UHV-grown (top spectra) epitaxial graphene on 6H–SiC(0001). (From Emtsev, K.V. et al., *Nat. Mater.*, 8, 203–207, 2009.)

G and 2D lines are clearly observed. The narrowest shape of the 2D line of the red spectra together with the low intensity of the D peak indicates that the quality of the sample obtained by the above preparation procedure is better than other techniques. Importantly, the blue-shift (38 cm⁻¹) detected in the 2D peak with respect to exfoliated graphene [33] is due to a compressive strain of the graphene sheet produced by the substrate.

4.2.2.4 GRAPHENE BY STATE-OF-THE-ART TECHNIQUE CHEMICAL VAPOUR DEPOSITION

4.2.2.4.1 Graphene on metals

Metals are one of the most-used substrates for synthesizing graphene on surfaces by CVD and related techniques like plasma-enhanced vapour deposition (PECVD). Comparing graphene grown using CVD and PECVD, CVD is much more preferred due to certain drawbacks which are required for the commercialization. In PECVD, a distance must be maintained between the plasma initial stage and the deposition stage to allow the plasma to diffuse to the substrate. The deposition is therefore not taking place in the direct plasma but downstream, which does not provide the heat treatment of sample (generally required for the re-crystallizing of the substrate). Another disadvantage of using the PECVD is that it can only be used at very low pressure [34]. Also, it would be interesting to mention that in plasma systems the gases collide with the surface and very easily could damage the films. Therefore, the devices may deteriorate during the process. These problems are not found in the CVD process. Furthermore, researchers are more focused on the CVD process than the PVD.

The standard CVD protocol for growing a graphene sample on these kinds of substrates consists of introduction of a molecular precursor rich in carbon atoms while the sample is at high temperature (>1300 K) under a controlled reductive atmosphere. Among others, molecular precursors like methane or ethylene are widely used around the world while polycrystalline copper foils are standard substrates. Copper foils present several advantages with respect to other substrates, such as low cost and, importantly, the low solubility of the carbon species inside its bulk. Thus, the carbon atoms remain on the surface of the substrate, facilitating the formation of the graphene sheet.

In the following, we will discuss in detail the experimental methodology employed for growing graphene on metals; we will also show the most relevant results obtained until now.

4.2.2.4.2 Growth protocols

Figure 4.6a shows a photograph of a standard CVD system. It is formed by a vacuum circuit with a vacuum pump at one corner, an introduction of gases step at the opposite, and an alumina tube in the centre surrounded by a cylindrical furnace. The vacuum pump is used for removing the atmosphere inside the alumina tube in order to replace it by high-quality gases at the desired pressure. Thus, after *in-situ* cleaning of the substrate by H_2/Ar gases at high temperature, the C-rich molecular precursor is added to the atmosphere. These precursors react with the copper surface and decompose on it. Furthermore, the carbon atoms spontaneously reorganize on the surface forming the graphene sheet.

Alternatively, several studies were performed on single-crystal metals like Pt(111), Ir(111), or Ru(0001) under UHV conditions in order to deeply characterize the structural and electronic properties of the samples, avoiding any kind of contamination.

As in the cases of standard CVD experiments, in the case of the samples grown at UHV a carbon-rich molecule is introduced in the system while the sample is heated at high temperature. One of the main advantages of the last methodology is that it allows the *in-situ* characterization of the samples with different experimental techniques like STM, LEED, or one based on synchrotron radiation facilities.

4.2.2.4.3 Graphene on copper foils

For the growth of graphene on copper (Good Fellow 99.999%), the process can be described in four stages as shown in the figure. During the first stage the Cu substrate is subjected to hydrogen annealing (A_r:H_2 = 80:20 standard cubic centimetres per minute [sccm]) at 1073 K to remove traces of oxide(s) for 20 min at ~26 mbar. An initial span of 15 minutes is required to achieve a constant temperature of 1073 K. After this the system is vacuumised back to 1.3×10^{-2} mbar accompanied by drop in temperature of the system as shown in Figure 4.7. CH_4/H_2 (10:50 sccm) is filled in to return the pressure to ~26 mbar and Ta filament is put-on to gain a substrate temperature of ~873 K. The charge of CH_4/H_2 current let-in undergoes precracking cleavage on the Ta filament (~2073 K), forming an active flux of highly charged carbon and hydrogen radicals in the third stage. The radical formation was allowed to proceed for 10 min to permit the formation of sufficient flux of carbon radicals. In the final step, we simply put-off the filament and again introduce Ar/H_2 (80:20 sccm) in the system without change in pressure until the substrate temperature drops to 473 K.

Consequently, the H_2 feed was cutoff and Ar flow was continued till the system reached room temperature (R.T.). Further detail can be found in the described journal. The schematic diagram of the process is shown in Figures 4.7 and 4.8.

The properties/qualities of graphene grown by CVD techniques are much more enhanced as compared to liquid exfoliation (graphene oxides [GOs] synthesized by various methods). The comparative studies have been illustrated in Table 4.1.

4.2.3 From the laboratory to the industry

Conventional transparent electrodes (indium tin oxide, ITO) are commonly electrodes which are used in various applications including solar cells, touch sensors, and flat panel displays. Although due to its remarkable properties it offers sheet resistance of less than 100 Ω Υ^{-1}, optical transparency of ~90%, and unlimited scalability, the sheet resistance offered by graphene grown on ~280 Ω Υ^{-1} [35,43,44] on nickel substrates by CVD is only on the scale

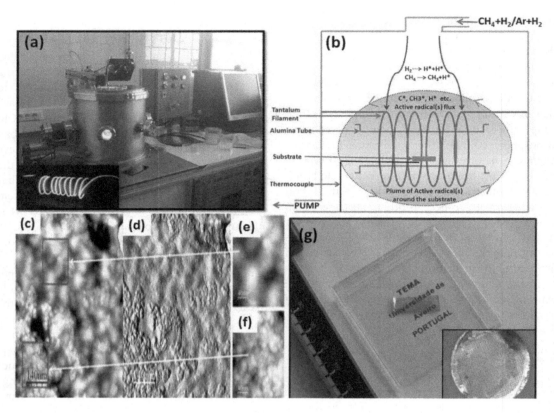

Figure 4.6 HFTCVD setup with schematic diagram of graphene growth, (a) homemade hot-filament thermal CVD set-up for large-area graphene film deposition (inset shows Ta filament (2073 K) wound around alumina tube, (b) schematic of graphene growth deposition and formation of active flux, of highly charged carbon and hydrogen radicals by catalytic reaction of gaseous precursors with the filament, (c) AFM topography of graphene covered area and uncovered region simultaneously, (d) AFM amplitude image of graphene on Au/glass, (e, f) shows highly magnified AFM topography indicated by square region, (g) large-area as-deposited graphene film deposited on Cu substrate (inset shows optical image bilayer graphene transferred on Cu TEM grid).

of a few centimetres. Due to outstanding [25], mechanical [18,22], and chemical [24,45] properties of graphene, the industrial application of graphene on flexible electronics is one of the most demanding tasks, now and in the future.

Li et al. discovered an efficient way for production of graphene but only at the centimetre level on copper substrates [46]; however, they created possibilities for large-scale production of high-quality graphene for real practical application. Moreover, Bae et al. use the same technique with modified roll-type substrate fitting inside a tubular furnace to maximizing the scale and homogeneity of the graphene. 30-inch graphene films were grown using CVD onto flexible copper substrates (Figure 4.9). The process states that the roll of copper foil is inserted into an 8-inch-wide

tubular quartz reactor and heated to 1273K with the flowing of 8 sccm of H_2 at 1.2×10^{-1} mbar. After reaching 1273 K, the annealing of sample was done in order to remove the traces of oxides and to increase the grain size of copper without changing flow rate or pressure. After annealing the mixture of CH_4 and H_2, the sample was supplied with rates of 24 and 8 sccm at 6.10×10^{-1} mbar for 30 mins. Finally, the sample is rapidly cooled to room temperature (~283 K s^{-1}) with flowing H_2 under a pressure of 1.2×10^{-1} mbar [47].

Very recently, large-area graphene was synthesized by CVD (showing the scale production without losing the homogeneity of the films) by Bae et al. Using an 8″ wide, double tubular, quartz reactor in CVD, they showed the way for the production of a monolayer graphene film to be

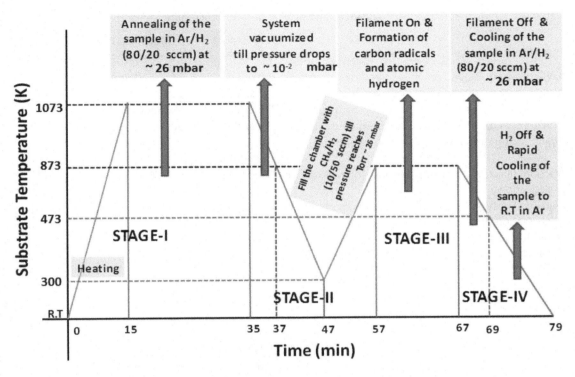

Figure 4.7 Experimental process diagram for graphene growth on Cu substrate by HFTCVD.

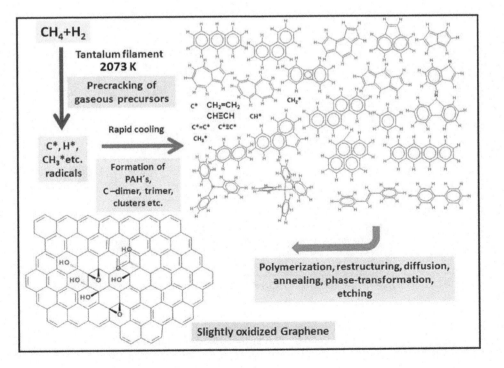

Figure 4.8 Probable mechanism of graphene growth by HFTCVD.

Table 4.1 Comparison of sheet resistances and optical transmittance between graphene and GO

Graphene	Resistance/ conductance	Transmittance wavelength	References
CVD grown graphene on Ni followed by transfer	280 $\Omega\ \Upsilon^{-1}$	76% (550 nm)	[35]
CVD grown graphene on Ni followed by transfer	200 $\Omega\ \Upsilon^{-1}$	85% (550 nm)	[36]
CVD grown graphene on Ni followed by transfer	770 $\Omega\ \Upsilon^{-1}$	90% (500–1000 nm)	[37]
CVD grown graphene on Cu followed by transfer	350 $\Omega\ \Upsilon^{-1}$	90% (550 nm)	[38]
Spin coating of reduced GO –SiO_2 composite	0.45 Scm^{-1} (28 nm thick)	95% (550 nm)	[39]
Vacuum filtration of GO suspension, followed by reduction	$4.3 \times 10^4\ \Omega\ \Upsilon^{-1}$	73% (550 nm)	[40]
Vacuum filtration of graphene platelets made by sonication of graphite in n-methyl-2-pyrrolidone (NMP)	$3 \times 10^3\ \Omega\ \Upsilon^{-1}$	75% (550 nm)	[41]
Spray deposition of graphene platelets made by sonicating graphite in dimethylformamide (DMF)	$5 \times 10^3\ \Omega\ \Upsilon^{-1}$	90%	[42]

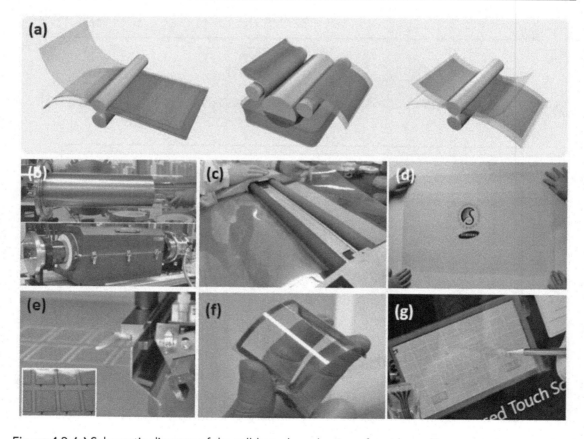

Figure 4.9 **(a)** Schematic diagram of the roll-based production of graphene films grown on a copper foil, **(b)** 39 inch width and 8-inch-thick quartz reactor in which copper foil wrapping of an approximately 7.5-inch quartz tube is to be inserted, **(c)** Using thermal release tape to a PET film at 393 K for support and adhesion, **(d)** large-area graphene film transferred on a 35-inch PET sheet, **(e)** silver electroding on graphene/PET film, **(f)** flexible assembled graphene/PET touch panel, **(g)** graphene-based touch-screen panel connected to a computer.

synthesized on a roll of copper foil with dimensions as large as 30″ in the diagonal direction. It is worth mentioning that the graphene obtained showed excellent optical properties. In another new and innovative approach to CVD synthesis, ultra large-area graphene based on selective Joule heating was presented by Kobayashi et al. The author describes a roll-to-roll CVD system comprising a stainless steel vacuum chamber with a pair of winders, along with a pair of current-feeding electrode rollers (separated by a distance of 0.4 m). Copper foil (230 mm wide, >100 m, 36 μm thick, >99.9% pure) was rolled between two rollers and heated to a temperature of ~1000°C by Joule heating under vacuum chamber with the pressure maintained at 1000 Pa with the flow of CH_4 (450 sccm) and H_2 (50 sccm) [48]. It would be interesting to mention that Yamada et al. have also reported large-scale and high-throughput roll-to-roll CVD synthesis of graphene on 297 mm wide copper foil at a relatively low temperature ~400°C by assisting the dissociation of methane using microwave plasma, but the obtained graphene lagged behind for use in electronic applications due to high sheet resistance of >10 kΩ/sq [49]. From the literature, it can be concluded that the quality of graphene improves with increasing growth temperature; also, the growth rate of graphene is an important factor to be considered for efficient graphene film production.

Liquid exfoliation can yield the mass-production graphene using various techniques but, due to low electrical properties (shown in Table 4.1) compared to graphene by CVD, it does not get more attention for industrial applications. The GO is mainly synthesized by the Brodie, Staudenmaier, or Hummers' method or modification of these methods. These methods involve oxidation of graphite to various levels. Potassium chlorate ($KClO_3$) with Nitric acid (HNO_3) is used to oxidize graphite in Brodie and Staudenmaier whereas Hummers' method involves strong treatment of potassium permanganate ($KMnO_4$) and sulfuric acid (H_2SO_4). The treatment of graphene leads to polar oxygen functional groups, rendering it hydrophilic in nature. Furthermore, GO can be more reduced using several reducing agents such as hydrazine [50], hydroquinone [51], sodium borohydride [$NaBH_4$] [52,53], and ascorbic acid [54], which affects the properties of graphene.

4.2.4 Quality comparison

The first report on CVD synthesis were reported in 2006 by Somani et al., and since then the growth and development of high-quality, large-area on catalytic metal substrate were taken into account [55].

The production of the large-scale graphene films was so far on polycrystalline, but the research effort are more focused to control the domain size, the number of graphene layers, the density of grain boundaries, the defects, etc. Using CVD technique, one can easily produce graphene on polycrystalline metals with size up to micrometre- to millimetre-size domains. However, the graphene films grown on Ni foils do not yield uniform monolayer graphene; in most cases it resembles a mixture of monolayer and few layers (polygraphene). On the other hand, experiments proved that Cu is an excellent candidate for making large-area of uniform thickness (95%) of single-layer graphene. This could be due to the low solubility of C in the Cu. Previous results show that graphene growth on Cu is somehow surface-mediated and self-limiting [46]. It will be worth mentioning that in Cu-catalysed, CVD-produced graphene growth, the hexagonal lattice of Cu(111) favoured the high quality of the as-grown graphene. Furthermore, during growth in CVD on Cu(111), the orientation of graphene nuclei became well controlled with domain boundaries of Cu.

The production of graphene by graphite oxide (the most popular wet chemical method) using Hummer's method is based on the exfoliation mechanism of the oxidative intercalation and production of oxygen-containing functional groups on the graphene layers, which helps the dispersion and stabilization of GO sheet in the water [56]. The story of Hummers' method started in 1859, when Brodie oxidized graphite in the presence of potassium chlorate and fuming nitric acid and beyond that it slowly evolved to the widely used Hummers' method that involves the combination of sodium nitrate, potassium permanganate, and sulfuric acid [57]. Since then, efforts have been made to improve on the Hummers' method by eliminating the use of sodium nitrate (which produces toxic nitrous gas) [58]. It will be worth mentioning that GO is lagging behind due to its chemical inhomogeneity, batch-to-batch reproducibility, and the inevitable creation of irreparable hole defects on the graphene sheets during the oxidation, which affects the conductivity of GO. However, Eigler et al. recently showed the production of GO with larger region of pristine network at low-temperature oxidation [59].

For the large-scale production of graphene there could be possibilities of using SiC by performing the thermal decompositions. When the SiC substrates are annealed at high temperatures, Si atoms selectively desorb from the surface and the C atoms left behind naturally form FLG. Because SiC is a wide-band-gap semiconductor, FLG on SiC can serve as a graphene substrate for electronics applications. However, two fundamental problems must be solved before the thermal decomposition method can be used at wide scale. First, the FLG obtained does not meet the uniform thickness distribution, as the electronic properties of FLG strongly depends on its thickness. Secondly, hitherto the mechanism behind the factors affecting the physical properties of FLGs is not fully understood.

Considering the main factors for industrialization of graphene production (namely the production cost, scalability, reproducibility, processability, and performance of the graphene products), the current promising methods of production (by using CVD, GO, and SiC) each have their own advantages and limitations. In the case of GO (using Hummers' method), we can obtain graphene (GO) with the advantage of high yielding and high dispersibility. However, it suffers from the use of a potentially explosive process and also it could have structural inhomogeneity. The SiC could be a candidate for commercialization at wide scale for future application, but due to the problem discussed above it may lose its importance. Finally, graphene grown by CVD technique is acquiring the potential interest of the researcher and more and more R&D is focused in this field due to low-cost producibility and processability; in addition, the performance of the graphene is good and can be implemented in the semi-conductor industries.

4.2.5 Doping of graphene

Doping in graphene-related systems has attracted a lot of research activity in the past years. In fact, the mechanism of the doping in the carbon materials is not new; the first reported doping in carbon material was done in carbon nanotubes using by poly(ethyleneimine) (PEI)—an electron-donating polymer. A similar concept was adopted to dope graphene, which results in n-type behaviour [60,61]. It was observed that the concentration of holes and electrons was imbalanced,

which enhanced the electron conduction and reduced the hole conduction (similar phenomena are observed when graphene is used as electrodes beneath the metal electrodes like the diazonium salts).

Doping of the graphene can be done in two distinct ways: (i) electrically by changing gate voltage (known as surface transfer doping) and (ii) chemically by introducing the dopants (known as substitutional doping) in honeycomb structure, which shifts the Dirac point relative to the Fermi level. The involvement of foreign atoms such as nitrogen, boron, etc. would disturb the sp^2 hybridization of carbon atoms, known in turn as n- or p-type doping.

P-type doping drives the Dirac points of graphene above the Fermi level, and n-type doping drives the Dirac points below the Fermi level. By the addition of atoms with fewer valence electrons than carbon, like boron, etc., one can achieve p-type doping; n-type doping is generally achieved by adding atoms with more valence electrons than carbon, like nitrogen. Normally, the Dirac point of the pristine undoped graphene is at the zero and if the same shifts up—it becomes n type and if it goes down—it converts to p type. If the band gap opens, graphene-based transistors will have higher on/off ratio [62].

There are various methods that can be successfully applied to dope graphene in both n and p type. Literature shows that ammonia is best suitable for incorporating free electrons, which turns graphene into n-type graphene [45,63], whereas boron is responsible for p-type doping [64]. Moreover, it can be concluded that researchers are more interested to dope graphene for n-type behaviour than p-type.

It would be interesting to mention that Arun et al. used p-toluenesulfonic acid (PTSA) for converting pristine graphene into n-type graphene. They first deposited CVD-manufactured graphene on Si/SiO_2 using PMMA coating for reducing the breakage and creation of defects. Afterwards, they soaked the as-transferred graphene in PTSA solution (0.1 M) for the certain period of time. Upon drying, the graphene changed its behaviour and they were successfully able to make it into n-type [65]. In another work, Ito et al. explained the role of precursors and gas phase in carbon films, especially graphene. Using numerous precursors, namely methane, methylamine, nitromethane, ethylamine, ethanolamine, nitro-ethane, acetonitrile, benzene, pyride, aniline, and nitrobenzene, the author studied the effect and observed the interesting results in Raman spectroscopy (D, G, D′, 2D, I_D/I_G, $I_{D'}/I_G$, and $I_{2D}/I_{D'}$).

Figure 4.10 shows the schematic overview of the possible nitrogen function present in n-doped graphene. It also shows the study of different aliphatic nitrogen compounds namely methylamine, ethylamine, ethanol amine, nitromethane, nitroethane, acetonitrile, pyridine, aniline, nitrobenzene, and benzene [66].

Being a two-dimensional material with large surface areas and ultra-high conductivity, which changes rapidly when external molecules (gas/other chemicals) adsorb on the surface of it, graphene becomes an excellent candidate for sensors with high sensitivity that can detect even individual molecules [67]. Literature shows that graphene as sensor can be used to detect bacteria [9], glucose [68], pH and proteins [69,70], etc. Furthermore, GO can be also be used for the detection for DNA [71,72] and for DNA sequencing [73–78].

4.3 SILICENE

Silicene is a two-dimensional (2D) honeycomb sheet. From a structural point of view, it remembers graphene but each carbon atom is replaced by a silicon atom. It is proposed as an ideal low-dimensional material because of their amazing structural, electronic, and magnetic properties. The exploration of free-standing silicene sheets with theoretical tools predicted that this 2D systems is stable, opening the doors for its experimental synthesis [79,80]. Interestingly, the theoretical works suggested that free-standing silicene should present a linear energy-momentum dispersion with Dirac cones at the corners of the Brillouin zone.

Thus, today there is a sharp competition to fabricate silicene in wafer scale in order to supply a wide range of industries. As in the case of graphene,

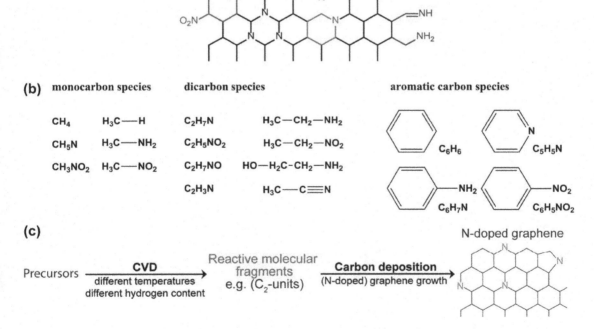

Figure 4.10 (a) Possible nitrogen incorporation in N-doped graphene, (b) CVD precursors with different functional groups as used by Ito et al. (namely methane, methylamine, ethylamine, ethanol amine, nitromethane, nitroethane, acetonitrile, pyridine, aniline, nitrobenzene), (c) schematic formation of nitrogen-containing graphene by CVD. (Reprinted with permission from Ito, Y. et al., *ACS Nano*, 8, 3337–3346, 2014. Copyright 2014 American Chemical Society.)

growth protocols based on CVD techniques could play a key role in their large-scale synthesis.

In the following sections, we will discuss the properties of freestanding and on-surfaces silicene, their synthesis, and applications. Finally, we will show as an example how these materials can be implemented in a transistor for further electronic nano-devices.

4.3.1 Electronic structure of silicene

Free-standing silicene, showed in Figure 4.11a, is a stable material [79] but its synthesis in wafer scales in order to introduce it in the industries will require hard work during the next years. However, its properties can be predicted and explored by theoretical tools. Thus, its honeycomb structure resembles that of graphene but presents an important difference. Calculations found two stable structures, low- and high-buckled ones, indicating that the silicon atoms of the silicene sheet are vertically displaced with respect to the basal plane (down panel of Figure 4.11a). The origin of this buckling is in the mixture of sp^2 and sp^3 electronic hybridization between the silicon atoms in a honeycomb configuration [79,81]. From an energetic point of view the low-buckled configuration is the most stable, yielding a lattice constant of 0.384 nm and an angle of 102° for the bonding between neighbouring silicon atoms [82].

Probably the most relevant feature that we can find in silicene is its electronic properties. Figure 4.11b shows the calculated electronic structure of freestanding silicene for the low-buckling

solution. Interestingly, the band structure has linearly dispersing π and $\pi*$ bands crossing at the K points and forming Dirac cones. As in the case of graphene, this feature implies that the charge carriers behave as massless relativistic particles, introducing an exciting scenario in which several nano-electronic applications could be developed based on this new material or by combining it with others like graphene.

4.3.2 Synthesis of silicene on surfaces

The synthesis of silicene is still quite difficult. As an example, and by comparing with the case of graphene, it is not possible to obtain silicene by the simplest method of exfoliation of bulk silicon. Most of the silicene samples have been fabricated on silver substrates by following the ideas of CVD techniques but under ultra-high vacuum (UHV) conditions. The UHV system allows gaining control on the environment, avoiding strange molecular species that could affect the silicene growth.

This section shows the methodology developed for synthesising silicene on silver, the most-used metallic substrate, and briefly explores the advances performed on other metallic substrates like gold and iridium.

4.3.2.1 FIRST SYNTHESIS OF SILICENE ON SILVER SUBSTRATES

The first relevant experiments for the rational synthesis of silicene and related structures were

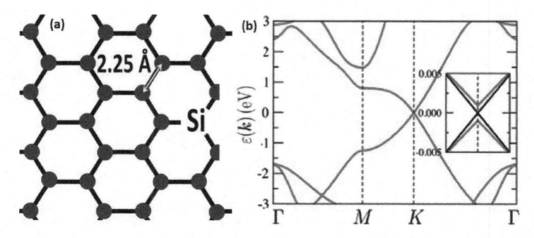

Figure 4.11 Lattice structure of Silicene **(a)** and its band structures **(b)**, inset showing the electronic bandgap around the Dirac point. (From Lars, M. et al., *J. Phys. Condens. Matter*, 25, 395305, 2013.)

performed by depositing silicon atoms on metallic substrates under UHV conditions [84]. In these experiments, Si atoms were sublimated by heating up a piece of a Si wafer located just in front of the substrate. The observed superstructures were eventually characterized and no direct relation was found with silicene. However, that pioneering work opened the door for the subsequent experiments. Thus, by intensively repeating the above experimental protocol but by changing the experimental conditions as well as the substrate, silicene nanoribbons were finally synthesised on Ag(110) [85,86]. Silicene nanoribbons were less than 2 nm in width and several nanometres in length and their characterization showed that their electronic structure resembles that of graphene [87].

After the synthesis of silicene nanoribbons the next goal was their expansion in order to obtain a full 2D sheet on a surface. In 2010, by carefully controlling the experimental conditions during the growth, with special care on the substrate temperature, quantity of silicon atoms and deposition rate, Lalmi et al. claimed the synthesis of one single sheet of epitaxial silicene on Ag(111) [88]. In these experiments the substrate temperature was 523 K and silicene accommodates on the Ag(111) substrate following a commensurate $(2\sqrt{3} \times 2\sqrt{3})$ R30° superstructure.

Figure 4.12 shows an atomically resolved STM image of silicene on Ag(111) in which the honeycomb structure is clearly observed. By carefully analysing the STM images it is possible to discriminate the six silicon atoms that form each hexagon in the silicene sheet. Importantly, three of the Si atoms appear brighter than the other three. Thus, the silicene atomic structure is formed by two sub-lattices with different contrast in the STM images. Among other electronic contributions to the image, this corrugation was related with a small buckling in the silicene lattice as predicted by theoretical works, i.e., the silicon atoms are displaced perpendicular to the basal plane [79].

Further experiments [90–95] showed that silicene forms a rich variety of superstructures on the Ag(111) surface depending on the experimental conditions of the growth [89]. Figure 4.12b shows a set of LEED patterns obtained for samples prepared by slightly changing the substrate temperature and deposition times. Silicene accommodates on Ag(111) in 4 × 4; $\sqrt{13} \times \sqrt{13}$ R13.9°; $\sqrt{19} \times \sqrt{19}$ R23.4°; $\sqrt{3} \times \sqrt{3}$ R30°; 3.5 × 3.5; $4/\sqrt{3} \times 4/\sqrt{3}$ superstructures.

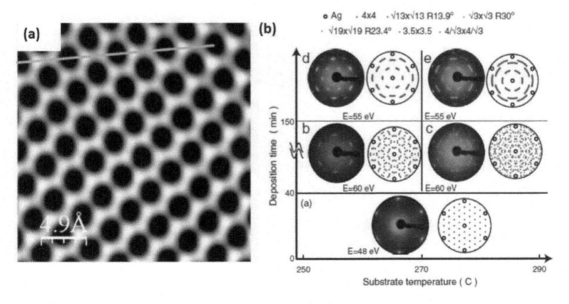

Figure 4.12 **(a)** Atomically resolved STM image showing honeycomb structure, revealing two sublattices, **(b)** phase evolution of Silicene on Ag (111). (From Lalmi, B. et al., *Appl. Phys. Lett.*, 97, 223109, 2010; Arafune, R. et al., *Surf. Sci.*, 608, 297–300, 2013.)

4.3.2.2 SILICENE ON OTHER SUBSTRATES

Ag(111) is the most-used substrate for growing silicene but the above experimental protocol has been extrapolated to other metals. In this section we show some of the most representative advances obtained until now.

Following the ideas explored on Ag(110), flat Si nanoribbons were synthesised on the (2 × 1) reconstructed Au(110) surface kept at 673 K after depositing 0.3 ML of silicon atoms [95,96]. Figure 4.13a shows an STM image of the obtained structures. Importantly, the structure showed in Figure 4.13a drastically depends on the silicon coverage. For lower coverages (0.2 layers) the observed superstructure is a 2D surface alloy with a stoichiometry close to Au_3Si; the nanoribbons only grow after increasing the coverage at 0.3 ML.

Also, due to its catalytic properties, Iridium is another metal explored in order to grow silicene. By slightly changing the growing protocol, buckled ($\sqrt{7} \times \sqrt{7}$)R19.1° silicene was obtained by depositing Si on an Ir(111) surface kept at 643 K (Figure 4.13c) [97]. The relative position of the silicon atoms with respect to the underneath iridium atoms (on top, hcp hollow or fcc hollow sites) produce the observed buckling. That indicates, at a first glance, a non-neglected silicene-substrate interaction. However, theoretical calculation shows that the Si-Si bonding is stronger than the interaction between Si atoms and the iridium surface, providing an indication of the silicene formation on this surface.

4.3.3 Controversy – discrepancies

Due to the high level of complexity, the proposed models for silicene on surfaces are not free of discrepancies. Thus, on one hand the measured electronic properties showed a linear band dispersion in the silicene on Ag(111) system, indicating the existence of Dirac electrons [98–100]. On the other, however, this linear dispersion was attributed to the strong hybridization between silicon and silver atoms [79]. With this in mind and in contrast to previous works, recent studies draw a different scenario for the silicon deposition on Ag(111) and Pt(111) [101–103].

On Ag(111), an intermixing between silicon and silver atoms could occur followed by a surface precipitation of crystalline (sp³-bonded silicon) as the coverage is increased [101].

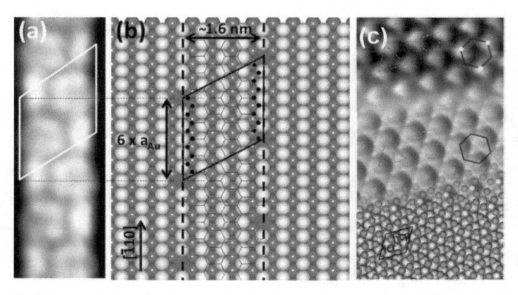

Figure 4.13 **(a)** Shows a representation of the nanoribbons on the Au(110)-(2 × 1) surface, **(b)** proposed model based on the STM image and the LEED pattern, **(c)** zoomed-in STM image of the silicon layer (top), below showing simulated STM image, correlating the experimental results in the same triangles and hexagons, (bottom) Top view of the relaxed atomic model of the ($\sqrt{3} \times \sqrt{3}$)silicene/($\sqrt{7} \times \sqrt{7}$) Ir(111) configuration. (From Rachid Tchalala, M. et al., *Appl. Phys. Lett.*, 102, 083107, 2013; Meng, L. et al., *Nano Lett.*, 13, 685–690, 2013.)

In a similar way, a recent study discussed the $(\sqrt{19} \times \sqrt{19})R23.4°$ superstructure obtained upon depositing silicon on a Pt(111) surface kept at 753 K. After carefully characterising the atomic and electronic structure by a set of experimental and theoretical tools, they concluded that a silicide (Si_3Pt) is formed on this surface instead of a silicene layer [102]. This work is especially important because the authors expand the formation of alloys to other systems like Si_3Ir instead of silicene on Ir(111).

These works point out that today more experiments have to be performed in order to elucidate the exact nature of the low-dimensional silicon on metals systems from an atomistic point of view and the origin of their electronic properties. However, the synthesis of silicene is a growing field in which significant advances appear day by day.

4.3.4 From the laboratory to the industry

Today, most of the experimental works focus on the synthesis and characterization of silicene on metals. The lack of a protocol for the separation of the silicene sheet from the substrates is one of the main drawbacks in order to use silicene in realistic applications. Much effort has to be dedicated in the next years in order to overcome this goal. However, if their isolation and transfer to an arbitrary substrate becomes a routine process during the next years the applications will be further expanded.

With this in mind, a new methodology, denoted silicene-encapsulated delamination with native electrodes (SEDNE), has been recently developed (Figure 4.14) [103]. It consists of a series of steps performed under well-controlled conditions. First, an Ag(111) thin film is growth on a mica template, followed by the synthesis of silicene as discussed before under UHV conditions. Then, the silicene layer is protected by growing a thin film of Al_2O_3 on it, encapsulating the silicene film in a sandwich. At this point silicene is protected on one face by the Al_2O_3 thin film and on the other by the Ag(111)/mica system allowing for removal of the sample from the UHV facility. Finally, the Ag(111) thin film was separated from the mica template with the help of a blade. Tao et al. [103] developed and used this methodology in order to fabricate a transistor based on silicene. In their work, the encapsulated silicene was placed on a device substrate (SiO_2/Si) with the Al_2O_3 face in contact with the SiO_2 film. On the other face of the device, the silver electrodes (source and drain) were patterned by electron beam lithography followed by etching to produce backgate transistor devices.

4.4 GERMANENE

In 2004, the graphene success diverted the interest of researchers to thinking more about single-atom-thick sheets (two-dimensional materials) from crystalline solids. Due to the fact that 2D materials have fundamentally different electronic

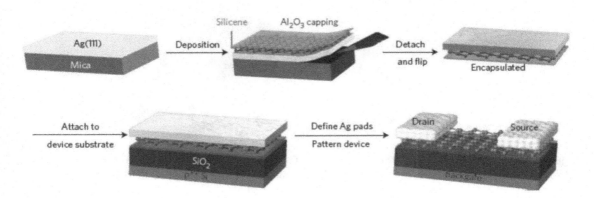

Figure 4.14 Showing schematics diagram of silicene and its synthesis–transfer–fabrication process. (From Tao, L. et al., *Nat Nano.*, 10, 227, 2015.)

structure and properties from their parent materials [104], now the time has come to play the structural and electronic properties of the materials. Researchers started innovating some new materials, leading them to ponder the other 2D crystalline structures like Germanene, Silicene, etc. Germanene (Figure 4.15) is the name given to the material which is made up of a single layer (monolayer, ML) of germanium with honeycomb-like structure. In the following section, we will be discussing more about the Germanene, its structural and electronic properties, followed by growth mechanisms on different substrates using different techniques.

Le Lay et al. from Aix-Marseille University in 2014 was able to confirm the growth of Germanene on Au(111) using molecular beam epitaxy (MBE). The author named it as cousin of graphene. 'We have provided compelling evidence of the birth of nearly flat germanene—a novel, synthetic germanium allotrope which does not exist in nature. It is a new cousin of graphene' [105].

4.4.1 Electronic structure of germanene

Germanene has some of the same characteristics as Graphene and Silicene, namely high mobility of the carriers and potential optical applications (a single layer could lead to optical transparency). The only thing which makes it different from graphene is that it has certain band-gap (Figure 4.15b) as compared to graphene and in case of bulk structure, germanene does not form van der Waal's layered structures. The atomic arrangement is similar to graphene, i.e., honeycomb structure with buckles in their stable state, with sp²-sp³ hybridization (Figure 4.15a).

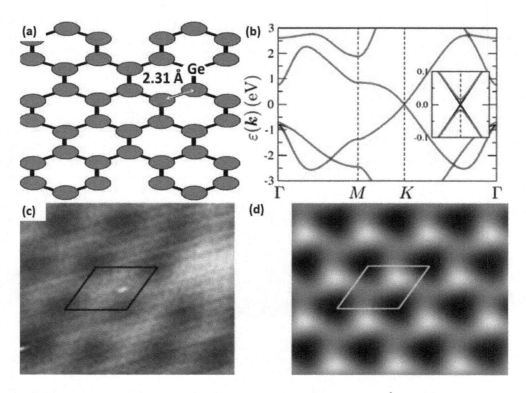

Figure 4.15 (a) Showing the electronic structure of Ge with d_{Ge-Ge} as 2.31 Å, (b) band gap of Ge as 23.9 meV (low-buckled honeycomb structure, for planar it is 4.0 meV) [109], (c) STM image of ($\sqrt{7} \times \sqrt{7}$) superstructure of Ge on Au(111), (d) simulated STM image of ($\sqrt{3} \times \sqrt{3}$) structure. (From Dávila, M.E. et al., New J. Phys., 16, 95002, 2014.)

The buckling (vertical distance separating the two atomic planes in the structure) nature of germanene is due to stability in its lower state, which can be in the form of chair (most stable form), boat, or wash board (unstable and flat form) [106]. Balendhran et al. summarize the structural and electronic parameters of germanene. It is worth noting that the vertical distance separating the two atomic planes in these structures is known as buckling parameter (δ) and the variation in the buckling parameter creates the changes in the electronic properties which are explained by density of electronic states.

The Raman spectra (calculated using Gaussian broadening) of Ge shown by Scalise et al., shows a strong peak at around 300 cm^{-1}, in which the peak is more pronounced as compared to Si because of the buckling [107]. Also, the peaks are hardly distinguishable due to low intensity of the materials.

4.4.2 First synthesis of germanene on gold

In 1994, M. Gothelid et al. was successfully able to study the Ge (111)-Au($\sqrt{3} \times \sqrt{3}$) R30°. They used sputtering technique over Ga-doped Ge wafer from the range of 0.004 – 0.01-Ω cm. The pressure maintained was about 1.3×1^{-9} mbar, in which Au was evaporated from Tungsten-filament evaporator, with the sample kept at either room temperature (RT) or 673 K. Earlier gold deposition of 1/10 of Å, which later changed to 1/4 of Å, formed the structure of c(2 × 8) with weak $\sqrt{3}$ and mixture of $\sqrt{3}$ with the split of (2 × 2). The same phenomena were observed in Ge(111)-Al surface [108]. Later in 2014, Dávila et al. deposited Ge using molecular beam epitaxy on Au(111). Using synchrotron radiation core-level spectroscopy measurement and density functional theory calculation, they were able to identify the structure, i.e., ($\sqrt{3} \times \sqrt{3}$) R30° along with ($\sqrt{7} \times \sqrt{7}$) R(19.1°) [105] (Figure 4.15c and d). However, the author also discussed one of the phases obtained at ~473 K growth temperature at about 1 ML (monolayer) coverage from 32% attenuation of the Au $4f_{7/2}$ core level intensity. The as-grown Ge was either $\sqrt{19} \times \sqrt{19}$ R(23.4°) supercell with reference to Au(111) 1 × 1, a 5 × 5 one or a ($\sqrt{7} \times \sqrt{7}$) R(19.1°), which corresponds to 2 × 2 germanene-reconstructed epitaxial sheet (projected in-plane Ge-Ge distance d_{Ge-Ge} = 0.221 nm) or to a $\sqrt{3} \times \sqrt{3}$ R30° (with

Ge-Ge distance d_{Ge-Ge} = 0.255 nm) and freestanding germanene have d_{Ge-Ge} = 0.238 nm. It is interesting to note that $\sqrt{19} \times \sqrt{19}$ R(23.4°) and 5 × 5 corresponds to 3 × 3 (projected in-plane Ge-Ge distance d_{Ge-Ge} = 0.242 nm) and $\sqrt{13} \times \sqrt{13}$ R(13.9°) with d_{Ge-Ge} = 0.231 nm [105].

4.4.3 Germanene on platinum

In 2014, under UHV condition with the base pressure of about 2×10^{-10} mbar using electron-beam evaporator, Li et al. grew germanium on Pt(111), which later after annealing formed ($\sqrt{19} \times \sqrt{19}$) superstructure with respect to the substrate. The result was confirmed by low-energy electron diffraction (LEED) and scanning tunnelling microscopy (STM). Moreover, the author used density functional theory (DFT)-based on ab initio calculation, which supported their results revealing a honeycomb structure of germanium with 2D continuity on the Pt(111) with buckled conformation.

The author used LEED to confirm the growth, superstructure of $\sqrt{19} \times \sqrt{19}$ germanium layer. Figure 4.16 shows the six outer symmetric bright spots of pristine Pt(111) with additional distinct spots belonging to germanium superstructures. For the sake of better understanding the author sketched a map of the diffraction spots of the superstructures in both reciprocal space and real space (Figure 4.16). The arrows in Figure 4.16c reveals the reciprocal vector whose diffraction pattern are shown in the real space.

Furthermore, to study more about the germanium adlayer in detail, STM measurement was carried out as shown in Figure 4.16, in which black rhombus resembles one of the supercells with the orientation of about 23° relative to Pt(1-10) [bar on 1]. The periodicity of the brightest protrusions in STM image is about 1.2 nm is confirmed using line profile which is equal to the dimension of ($\sqrt{19} \times \sqrt{19}$) superlattice of the Pt(111) (as the lattice constant of Pt(111) is 0.277 nm; and $\sqrt{19} \times 0.277$ nm = 1.21 nm). It thanks to the line that the author found that different germanium atoms in the adlayer comprise different heights (0.60 Å) with respect to underlying Pt lattice; this was further confirmed using DFT calculations. The corrugation value of buckled height of freely standing germanene in vacuum is

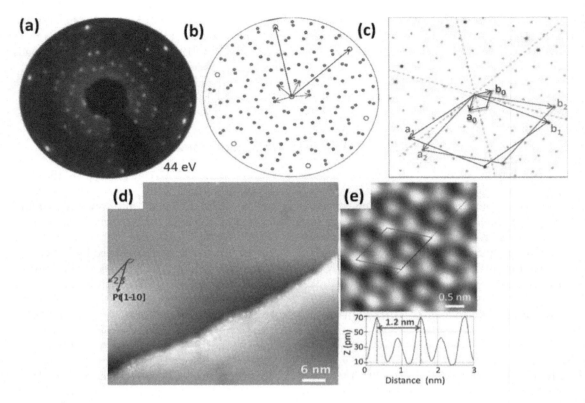

Figure 4.16 (a) Showing the LEED patterns of ($\sqrt{19} \times \sqrt{19}$) (brightest six outermost spots from Pt(111)), (b) reciprocal space showing the various vectors as indicated in arrows, (c) corresponding diagram in real space, (d) STM image of ($\sqrt{19} \times \sqrt{19}$) superstructure of germanium adlayer formed on the Pt (111) surface; the angle between the two arrows is about 23°, (e) Zoomed - in STM image of the germanium adlayer and the line profile revealing the periodicity of the germanium superstructure (1.2 nm).

0.64 Å, which implies Ge-Pt interaction is similar to the moderate formation of silicene on Ag(111) in Si-Ag interaction [110].

4.5 CONCLUSIONS

In this nascent field of 2D materials, the salient differences between bulk and few or single layers are just starting to be understood. Initially, the R&Ds were focussed on graphene and other 2D materials and were concerned about the various strategies to prepare single-layer and multilayer assemblies in solution, on substrates, and on the wafer scale. There are numerous exciting opportunities in developing the growth of high-quality large-area materials with controllable layer thicknesses. This would not only expand our understanding of the underlying physics associated in nanoscience and nanotechnology but potentially lead to the discovery of unanticipated phenomena and applications of 2D materials.

Although the large-scale production of graphene materials has been realized, many issues need to be addressed to advance their industrial applications. As any other product, the cost/performance ratio is the greatest concern for companies when determining whether graphene can be used in their products. For commercialization, the controllability for mass production in order to have uniform graphene sheets, realizing direct growth of large-area, uniform, defect-free, few-layer graphene films on arbitrary substrates at low temperatures, and the efficient, intact, and clean transfer from metal substrates without sacrificing the metals so that they can be re-used are still challenges that need to be overcome [111]. A very good review on basic synthesis method, product, production capacity along with main application products of several big graphene manufacturers in worldwide has been illustrated nicely by Wencai Ren and Hui-Ming Cheng [112].

Due to the same predicted linear electronic dispersion properties as found in graphene, Silicene can be touted as the next graphene. Additionally, it is easy to induce a bandgap opening due to the reduced crystal symmetry. Moreover, it is also compatible with current silicon microelectronics.

Furthermore, due to a desired quantum-spin Hall effect in buckled germanene and high-T_c superconductivity found in doped germanene, tremendous progress can be found in research that motivated us to investigate analogous 2D crystalline systems. It can be easily anticipated that researchers are developing methods for germanene growth and now thinking to migrate from graphene.

This Chapter highlights three examples of CVD of 2D materials such as graphene, silicene and germanene. It is worth noting that there are other 2D materials with increasing research interests. These include hexagonal boron nitride, transition metal di-chalcogenides and MXenes, for examples that can also be fabricated using CVD.

REFERENCES

1. Van Noorden, R., Production: Beyond sticky tape. *Nature*, 2012. **483**(7389): S32–S33.
2. Segal, M., Material history: Learning from silicon. *Nature*, 2012. **483**(7389): S43–S44.
3. Schmidt, C., Bioelectronics: The bionic material. *Nature*, 2012. **483**(7389): S37–S37.
4. Brody, H., Graphene. *Nature*, 2012. **483**(7389): S29–S29.
5. Dong, X., et al., Electrical detection of femtomolar DNA via gold-nanoparticle enhancement in carbon-nanotube-network field-effect transistors. *Advanced Materials*, 2008. **20**(12): 2389–2393.
6. Fu, D., et al., Differentiation of gas molecules using flexible and all-carbon nanotube devices. *The Journal of Physical Chemistry C*, 2008. **112**(3): 650–653.
7. Dong, X., et al., Label-free electronic detection of DNA using simple double-walled carbon nanotube resistors. *The Journal of Physical Chemistry C*, 2008. **112**(26): 9891–9895.
8. Park, S.J., et al., Ultrasensitive flexible graphene based field-effect transistor (FET)-type bioelectronic nose. *Nano Letters*, 2012. **12**(10): 5082–5090.
9. Mohanty, N. and V. Berry, Graphene-based single-bacterium resolution biodevice and DNA transistor: Interfacing graphene derivatives with nanoscale and microscale biocomponents. *Nano Letters*, 2008. **8**(12): 4469–4476.
10. Huang, Y., et al., Graphene-based biosensors for detection of bacteria and their metabolic activities. *Journal of Materials Chemistry*, 2011. **21**(33): 12358–12362.
11. Ang, P.K., et al., Flow sensing of single cell by graphene transistor in a microfluidic channel. *Nano Letters*, 2011. **11**(12): 5240–5246.
12. Paton, K.R., et al., Scalable production of large quantities of defect-free few-layer graphene by shear exfoliation in liquids. *Nature Materials*, 2014. **13**(6): 624–630.
13. Fei, Z., et al., Gate-tuning of graphene plasmons revealed by infrared nano-imaging. *Nature*, 2012. **487**(7405): 82–85.
14. Chen, J., et al., Optical nano-imaging of gate-tunable graphene plasmons. *Nature*, 2012. **487**(7405): 77–81.
15. Wallace, P.R., The band theory of graphite. *Physical Review*, 1947. **71**(9): 622–634.
16. Mayorov, A.S., et al., Micrometer-scale ballistic transport in encapsulated graphene at room temperature. *Nano Letters*, 2011. **11**(6): 2396–2399.
17. Morozov, S.V., et al., Giant intrinsic carrier mobilities in graphene and its bilayer. *Physical Review Letters*, 2008. **100**(1): 016602.
18. Lee, C., et al., Measurement of the elastic properties and intrinsic strength of mono-layer graphene. *Science*, 2008. **321**(5887): 385–388.
19. Liu, F., P. Ming, and J. Li, *Ab initio* calculation of ideal strength and phonon instability of graphene under tension. *Physical Review B*, 2007. **76**(6): 064120.
20. Balandin, A.A., Thermal properties of graphene and nanostructured carbon materials. *Nature Materials*, 2011. **10**(8): 569–581.
21. Nair, R.R., et al., Fine structure constant defines visual transparency of graphene. *Science*, 2008. **320**(5881): 1308.

22. Bunch, J.S., et al., Impermeable atomic membranes from graphene sheets. *Nano Letters*, 2008. **8**(8): 2458–2462.

23. Moser, J., A. Barreiro, and A. Bachtold, Current-induced cleaning of graphene. *Applied Physics Letters*, 2007. **91**(16): 163513.

24. Elias, D.C., et al., Control of graphene's properties by reversible hydrogenation: Evidence for graphane. *Science*, 2009. **323**(5914): 610–613.

25. Geim, A.K. and K.S. Novoselov, The rise of graphene. *Nature Materials*, 2007. **6**(3): 183–191.

26. Merino, P., et al., Strain-driven Moiré super-structures of epitaxial graphene on transition metal surfaces. *ACS Nano*, 2011. **5**(7): 5627–5634.

27. Marchini, S., S. Günther, and J. Wintterlin, Scanning tunneling microscopy of graphene on Ru(0001). *Physical Review B*, 2007. **76**(7): 075429.

28. Wang, B., et al., Coupling epitaxy, chemical bonding, and work function at the local scale in transition metal-supported graphene. *ACS Nano*, 2010. **4**(10): 5773–5782.

29. Sun, Z., et al., Topographic and electronic contrast of the graphene moiré on Ir(111) probed by scanning tunneling microscopy and noncontact atomic force microscopy. *Physical Review B*, 2011. **83**(8): 081415.

30. Liao, Y.-H., et al., Investigation of the dispersion process of SWNTs/SC-15 epoxy resin nanocomposites. *Materials Science and Engineering: A*, 2004. **385**(1–2): 175–181.

31. Emtsev, K.V., et al., Towards wafer-size graphene layers by atmospheric pressure graphitization of silicon carbide. *Nature Materials*, 2009. **8**(3): 203–207.

32. Bostwick, A., et al., Quasiparticle dynamics in graphene. *Nature Physics*, 2007. **3**(1): 36–40.

33. Ferrari, A.C., et al., Raman spectrum of graphene and graphene layers. *Physical Review Letters*, 2006. **97**(18): 187401.

34. Woehrl, N., et al., Plasma-enhanced chemical vapor deposition of graphene on copper substrates. *AIP Advances*, 2014. **4**(4): 047128.

35. Kim, K.S., et al., Large-scale pattern growth of graphene films for stretchable transparent electrodes. *Nature*, 2009. **457**(7230): 706–710.

36. Chen, J.-H., et al., Intrinsic and extrinsic performance limits of graphene devices on SiO$_2$. *Nat Nano*, 2008. **3**(4): 206–209.

37. Reina, A., et al., Layer area, few-layer graphene films on arbitrary substrates by chemical vapor deposition. *Nano Letters*, 2009. **9**(8): 3087–3087.

38. Li, X., et al., Transfer of large-area graphene films for high-performance transparent conductive electrodes. *Nano Letters*, 2009. **9**(12): 4359–4363.

39. Watcharotone, S., et al., Graphene–silica composite thin films as transparent conductors. *Nano Letters*, 2007. **7**(7): 1888–1892.

40. Eda, G., G. Fanchini, and M. Chhowalla, Large-area ultrathin films of reduced graphene oxide as a transparent and flexible electronic material. *Nature Nanotechnology*, 2008. **3**(5): 270–274.

41. De, S., et al., Flexible, transparent, conducting films of randomly stacked graphene from surfactant-stabilized, oxide-free graphene dispersions. *Small*, 2010. **6**(3): 458–464.

42. Blake, P., et al., Graphene-based liquid crystal device. *Nano Letters*, 2008. **8**(6): 1704–1708.

43. Reina, A., et al., Large area, few-layer graphene films on arbitrary substrates by chemical vapor deposition. *Nano Letters*, 2009. **9**(1): 30–35.

44. Cai, W., et al., Large area few-layer graphene/graphite films as transparent thin conducting electrodes. *Applied Physics Letters*, 2009. **95**(12): 123115.

45. Wang, X., et al., N-doping of graphene through electrothermal reactions with ammonia. *Science*, 2009. **324**(5928): 768–771.

46. Li, X., et al., Large-area synthesis of high-quality and uniform graphene films on copper foils. *Science*, 2009. **324**(5932): 1312–1314.

47. Bae, S., et al., Roll-to-roll production of 30-inch graphene films for transparent electrodes. *Nature Nanotechnology*, 2010. **5**(8): 574–578.

48. Kobayashi, T., et al., Production of a 100-m-long high-quality graphene transparent conductive film by roll-to-roll chemical vapor deposition and transfer process. *Applied Physics Letters*, 2013. **102**(2): 023112.

49. Yamada, T., et al., A roll-to-roll microwave plasma chemical vapor deposition process for the production of 294 mm width graphene films at low temperature. *Carbon*, 2012. **50**(7): 2615–2619.

50. Stankovich, S., et al., Synthesis of graphene-based nanosheets via chemical reduction of exfoliated graphite oxide. *Carbon*, 2007. **45**(7): 1558–1565.

51. Wang, G., et al., Facile synthesis and characterization of graphene nanosheets. *The Journal of Physical Chemistry C*, 2008. **112**(22): 8192–8195.

52. Si, Y. and E.T. Samulski, Synthesis of water soluble graphene. *Nano Letters*, 2008. **8**(6): 1679–1682.

53. Shin, H.-J., et al., Efficient reduction of graphite oxide by sodium borohydride and its effect on electrical conductance. *Advanced Functional Materials*, 2009. **19**(12): 1987–1992.

54. Dua, V., et al., All-organic vapor sensor using inkjet-printed reduced graphene oxide. *Angewandte Chemie International Edition*, 2010. **49**(12): 2154–2157.

55. Somani, P.R., S.P. Somani, and M. Umeno, Planer nano-graphenes from camphor by CVD. *Chemical Physics Letters*, 2006. **430**(1–3): 56–59.

56. Stankovich, S., et al., Graphene-based composite materials. *Nature*, 2006. **442**(7100): 282–286.

57. Brodie, B.C., On the atomic weight of graphite. *Philosophical Transactions of the Royal Society of London*, 1859. **149**: 249–259.

58. Marcano, D.C., et al., Improved synthesis of graphene oxide. *ACS Nano*, 2010. **4**(8): 4806–4814.

59. Eigler, S., et al., Wet chemical synthesis of graphene. *Advanced Materials*, 2013. **25**(26): 3583–3587.

60. Shim, M., et al., Polymer functionalization for air-stable n-type carbon nanotube field-effect transistors. *Journal of the American Chemical Society*, 2001. **123**(46): 11512–11513.

61. Farmer, D.B., et al., Chemical doping and electron–hole conduction asymmetry in graphene devices. *Nano Letters*, 2009. **9**(1): 388–392.

62. Liu, H., Y. Liu, and D. Zhu, Chemical doping of graphene. *Journal of Materials Chemistry*, 2011. **21**(10): 3335–3345.

63. Zhao, L., et al., Visualizing individual nitrogen dopants in monolayer graphene. *Science*, 2011. **333**(6045): 999–1003.

64. Kim, Y.A., et al., Raman spectroscopy of boron-doped single-layer graphene. *ACS Nano*, 2012. **6**(7): 6293–6300.

65. Singh, A.K., et al., Molecular n-doping of chemical vapor deposition grown graphene. *Journal of Materials Chemistry*, 2012. **22**(30): 15168–15174.

66. Ito, Y., et al., Chemical vapor deposition of n-doped graphene and carbon films: The role of precursors and gas phase. *ACS Nano*, 2014. **8**(4): 3337–3346.

67. Schedin, F., et al., Detection of individual gas molecules adsorbed on graphene. *Nature Materials*, 2007. **6**(9): 652–655.

68. Wang, Y., et al., Nitrogen-doped graphene and its application in electrochemical biosensing. *ACS Nano*, 2010. **4**(4): 1790–1798.

69. Ang, P.K., et al., Solution-gated epitaxial graphene as pH sensor. *Journal of the American Chemical Society*, 2008. **130**(44): 14392–14393.

70. Ohno, Y., et al., Electrolyte-gated graphene field-effect transistors for detecting pH and protein adsorption. *Nano Letters*, 2009. **9**(9): 3318–3322.

71. He, S., et al., A graphene nanoprobe for rapid, sensitive, and multicolor fluorescent DNA analysis. *Advanced Functional Materials*, 2010. **20**(3): 453–459.

72. Lu, C.-H., et al., A graphene platform for sensing biomolecules. *Angewandte Chemie International Edition*, 2009. **48**(26): 4785–4787.

73. Xu, M., D. Fujita, and N. Hanagata, Perspectives and challenges of emerging single-molecule DNA sequencing technologies. *Small*, 2009. **5**(23): 2638–2649.

74. Postma, H.W.C., Rapid sequencing of individual DNA molecules in graphene nanogaps. *Nano Letters*, 2010. **10**(2): 420–425.

75. Schneider, G.F., et al., DNA translocation through graphene nanopores. *Nano Letters*, 2010. **10**(8): 3163–3167.

76. Merchant, C.A., et al., DNA translocation through graphene nanopores. *Nano Letters*, 2010. **10**(8): 2915–2921.

77. Garaj, S., et al., Graphene as a subnanometre trans-electrode membrane. *Nature*, 2010. **467**(7312): 190–193.

78. Wehling, T.O., et al., Molecular doping of graphene. *Nano Letters*, 2008. **8**(1): 173–177.

79. Cahangirov, S., et al., Two- and one-dimensional honeycomb structures of silicon and germanium. *Physical Review Letters*, 2009. **102**(23): 236804.

80. Guzmán-Verri, G.G. and L.C. Lew Yan Voon, Electronic structure of silicon-based nanostructures. *Physical Review B*, 2007. **76**(7): 075131.

81. Liu, L., et al., Spin-torque ferromagnetic resonance induced by the spin hall effect. *Physical Review Letters*, 2011. **106**(3): 036601.

82. Takagi, N., et al., Silicene on Ag (111): Geometric and electronic structures of a new honeycomb material of Si. *Progress in Surface Science*, 2015. **90**(1): 1–20.

83. Lars, M., P. Olivia, and B. Friedhelm, Massive Dirac quasiparticles in the optical absorbance of graphene, silicene, germanene, and tinene. *Journal of Physics: Condensed Matter*, 2013. **25**(39): 395305.

84. Léandri, C., et al., Growth of Si nanostructures on Ag(0 0 1). *Surface Science*, 2007. **601**(1): 262–267.

85. Le Lay, G., C. Léandri, H. Oughaddou, J.-P. Biberian, P. De Padova, M.E. Dávila, B. Ealet, A. Kara, Physics and chemistry of silicene nano-ribbons. *Applied Surface Science*, 2009. **256**(2): 524529.

86. Kara, A., et al., Physics of silicene stripes. *Journal of Superconductivity and Novel Magnetism*, 2009. **22**(3): 259–263.

87. De Padova, P., et al., Evidence of graphene-like electronic signature in silicene nanoribbons. *Applied Physics Letters*, 2010. **96**(26): 261905.

88. Lalmi, B., et al., Epitaxial growth of a silicene sheet. *Applied Physics Letters*, 2010. **97**(22): 223109.

89. Arafune, R., et al., Structural transition of silicene on Ag(111). *Surface Science*, 2013. **608**(0): 297–300.

90. De Padova, P., et al., Evidence of Dirac fermions in multilayer silicene. *Applied Physics Letters*, 2013. **102**(16): 163106.

91. Chen, L., et al., Spontaneous symmetry breaking and dynamic phase transition in monolayer silicene. *Physical Review Letters*, 2013. **110**(8): 085504.

92. Chen, L., et al., Evidence for Dirac fermions in a honeycomb lattice based on silicon. *Physical Review Letters*, 2012. **109**(5): 056804.

93. Resta, A., et al., Atomic structures of silicene layers grown on Ag (111): Scanning tunneling microscopy and noncontact atomic force microscopy observations. *Scientific Reports*, 2013. **3**: 2399.

94. Feng, B., et al., Evidence of silicene in honeycomb structures of silicon on Ag(111). *Nano Letters*, 2012. **12**(7): 3507–3511.

95. Lin, C.-L., et al., Substrate-induced symmetry breaking in silicene. *Physical Review Letters*, 2013. **110**(7): 076801.

96. Rachid Tchalala, M., et al., Formation of one-dimensional self-assembled silicon nanoribbons on Au(110)-(2 × 1). *Applied Physics Letters*, 2013. **102**(8): 083107.

97. Meng, L., et al., Buckled silicene formation on Ir(111). *Nano Letters*, 2013. **13**(2): 685–690.

98. Vogt, P., et al., Silicene: Compelling experimental evidence for graphenelike two-dimensional silicon. *Physical Review Letters*, 2012. **108**(15): 155501.

99. Tsoutsou, D., et al., Evidence for hybrid surface metallic band in (4 × 4) silicene on Ag(111). *Applied Physics Letters*, 2013. **103**(23): 231604.

100. Avila, J., et al., Presence of gapped silicene-derived band in the prototypical (3 × 3) silicene phase on silver (111) surfaces. *Journal of Physics: Condensed Matter*, 2013. **25**(26): 262001.

101. Mannix, A.J., et al., Silicon growth at the two-dimensional limit on Ag(111). *ACS Nano*, 2014. **8**(7): 7538–7547.

102. Švec, M., et al., Silicene versus two-dimensional ordered silicide: Atomic and electronic structure of Si-($\sqrt{19}\times\sqrt{19}$)R23.4°/Pt(111). *Physical Review B*, 2014. **89**(20): 201412.

103. Tao, L., et al., Silicene field-effect transistors operating at room temperature. *Nature Nanotechnology*, 2015. **10**: 227.

104. Bianco, E., et al., Stability and exfoliation of germanane: A germanium graphane analogue. *ACS Nano*, 2013. **7**(5): 4414–4421.

105. Dávila, M.E., S. Cahangirov, A. Rubio and G. Le Lay, Germanene: A novel two-dimensional germanium allotrope akin to graphene and silicene. *New Journal of Physics*, 2014. **16**(9): 95002.

106. Balendhran, S., et al., Elemental analogues of graphene: Silicene, germanene, stanene, and phosphorene. *Small*, 2015. **11**(6): 633.

107. Scalise, E., et al., Vibrational properties of silicene and germanene. *Nano Research*, 2013. **6**(1): 19–28.

108. Yang, W.S. and F. Jona, Atomic structure of Ge{111} and reactions with Al. *Solid State Communications*, 1982. **42**(1): 49–53.

109. Liu, C.-C., W. Feng, and Y. Yao, Quantum spin hall effect in silicene and two-dimensional germanium. *Physical Review Letters*, 2011. **107**(7): 076802.

110. Gao, J. and J. Zhao, Initial geometries, interaction mechanism and high stability of silicene on Ag(111) surface. *Scientific Reports*, 2012. **2**.

111. Zurutuza, A. and C. Marinelli, Challenges and opportunities in graphene commercialization. *Nature Nanotechology*, 2014. **9**(10): 730–734.

112. Ren, W. and H.-M. Cheng, The global growth of graphene. *Nature Nanotechnology*, 2014. **9**(10): 726–730.

5

CVD of superlattice films and their applications

GUILLAUME SAVELLI

This book chapter is dedicated to the recent advances of superlattice (SL) films made by chemical vapour deposition (CVD), from their growth to their applications. The CVD processing steps and deposition mechanisms are explained, and the different SL structures are defined. Recent achievement of quantum well (QW) and quantum dots superlattices are also presented. Otherwise, the growth of such structures allows the materials to obtain specific properties, mainly in electrical, optical, and thermal fields, as will be detailed in this chapter. Finally, the applications and future prospects of superlattices are presented.

5.1 INTRODUCTION

Superlattices were introduced in 1970 by Esaki and Tsu with their pioneer research on synthesized semiconductor quantum nanostructures.[1]

Superlattices can offer unique and superior properties in many fields, such as their mechanical, optical, electrical, thermal properties, etc., as will be shown in Section 5.3.

In this section, definitions of QW and quantum dots superlattices are given as well as the different deposition mechanisms allowing their growth.

5.1.1 Definitions

Basically, superlattices can be defined as periodic structures alternating layers of two or more materials. Quantum dots superlattices can be defined as superlattices where one of the continuous layers is replaced by an array of dots. These dots can be regularly stacked or disordered in the matrix.

The thickness of one layer (and/or dots) is typically several nanometres.

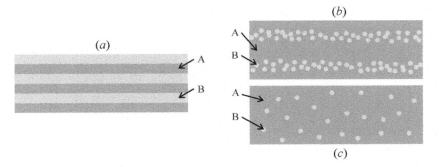

Figure 5.1 Basic schematic definitions of a SL **(a)**, a regularly stacked, **(b)** and a disordered, **(c)** quantum dots SL, alternating two materials A and B.

Figure 5.1 represents schematically such structures.

More precisely, superlattices count among the quantum-confined structures. A quantum-confined structure is one in which the movement of the electrons or holes is restricted in one or more directions.[2] An isolated QW, in which the particle is confined in only one direction, has two free directions (2-D system). Further confinement is obtained for a nanowire with two confined directions and one free direction (1-D system) and for a quantum box (0-D system) where the particle is confined along the three directions (no free direction).

Quantum well SLs (QWSL) have to be differentiated from multiple QW (MQW) structures. Indeed, in QWSL, the barriers separating neighbouring wells are very thin so the wave functions of adjacent wells overlap. The electrons are thus delocalized as they can easily tunnel out. At the opposite, for MQW structures, the number of wells is smaller or the barriers are thicker such that the wave functions of adjacent QWs do not overlap.

By further confining the material's charge carriers over the three dimensions, a quantum dot is obtained. Confining the electrons in a quantum dot will cause the apparition of discrete possible energy values for the charge carriers. By the same mechanism as that for QWs, it is possible to tailor up to a certain point the position of the permitted energy levels of the material by controlling the size of the quantum dot. Quantum dots are generally employed in the form of a dispersion of the particles in a solvent or in a similar way of QWSLs. In this case, the material is called quantum dot SL (QDSL) and is obtained by successively stacking quantum dots inside a matrix. As has been shown

in Figure 5.1 and will be detailed later, QDSL can be grown as disordered or regularly stacked structures.

As nanowires and nanotubes, QWSLs and QDSLs are different ways of thin-film nanostructuration. They allow materials to have specific properties, making them very widespread in some applications such as optoelectronics or thermoelectrics, for example. More details will be given later in this chapter.

5.1.2 Superlattice band structures

The superlattice miniband structures depend on the type of heterostructure (type I, type II, or type III) defined as below:

- In type I, as illustrated in Figure 5.2a, the bottom of the conduction band and the top of the valence subband are formed in the same semiconductor layer. For example, this is the case of GaAs-(Ga,Al)As superlattices, but also (Ga,In)As-InP or (Cd,Hg)Te-CdTe systems. It is worth noting that the band offsets can vary between these different systems.
- In type II, the electron wells occur in one material and the hole wells occur in the other. As shown in Figure 5.2b, in this case, the bands are uncrossed, i.e., the bottom of the electron wells is at higher energy than the top of the hole wells. For example, this is the case of Si-Ge and (Ga,In)As-GaAs systems.
- The type III is similar to the type II, but with crossed gap. Thus, here, the electron wells are at a lower energy than the hole wells. As shown in Figure 5.2c, the bandgap of material B lies entirely outside the bandgap of material A. For example, this is the case of InAs-(Ga,In)Sb superlattices.

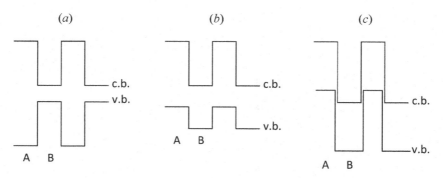

Figure 5.2 Schematic band structures (c.b. and v.b. corresponding for conduction band and valence band respectively) of type I **(a)**, type II **(b)** and type III **(c)** SLs made of materials A and B.

5.1.3 Superlattices deposition techniques

By definition, QWSL and QDSL consist of a stack of very thin (continuous or not) layers of which the thickness needs to be controlled at the nanometre scale. To grow such structures, some deposition techniques are available: physical vapour deposition (PVD), molecular beam epitaxy (MBE), sputtering, pulsed laser deposition (PLD) and CVD.[3]

Sputtering is a process where atoms are ejected from a solid target material due to bombardment of the target by energetic particles like atoms or ions. Here, a potential difference between the target and the reactor wall is applied to create plasma made from electrons, ions, and photons. Thanks to the electric field, positive species of plasma collide with the cathode (target), leading to the sputtering of atoms.

Sputtering is a well-adapted method to produce metal/ceramic multilayers, for example Metal/Al_2O_3, either by alternately sputtering metallic (Co, Pt, W, Fe, Al) and ceramic (as Al_2O_3) targets.[4,5] Sputtering is also one of the best methods to produce multilayers with amorphous metal or amorphous semiconductor layers.

MBE is an advanced ultra-high-vacuum facility to make compound semiconductor materials with great precision (less than 0.01 nm) and purity (more than 99.99999%). As shown in Figure 5.3,

an MBE system is composed of an evaporator where the deposition occurs under ultra-high vacuum (UHV) conditions (10^{-8} Pa). Ultra-pure elements are heated in separate effusion cells until they begin to slowly sublime; the gaseous elements then condense on the wafer where they may react with each other. The absence of carrier gases as well as the ultra-high vacuum environment results in the highest achievable purity of the grown films.

MBE is often employed to produce monocrystalline superlattices. Superlattices are obtained by alternating the use of one or more suitable effusion cells, as a function of the desired materials. Various characterization tools can also be integrated, such as reflection high-energy electron diffraction (RHEED) and Auger electron spectroscopy (AES), and are available for *in-situ* characterization. This system is used to produce very high-quality nanostructures as it allows to control precisely the thickness of each layer, and so to grow very thin layers. On the other hand, the deposition rate is low (typically less than 3 μm per hour).

PLD, or laser ablation, consists in vaporizing material from the surface of a bulk target, that is then collected onto a substrate, by using a short-pulse laser (some tens of nanoseconds). The main advantage of this process is the possibility to deposit a multicomponent film with the same composition as the target. PLD is thus an adapted method to deposit stoichiometric compounds and multilayered materials from multiple targets.

Another way to grow superlattices structures is the CVD.

5.2 CVD PROCESSING, DEPOSITION MECHANISMS, AND STRUCTURES

In contrast to PVD processes (as MBE or sputtering for example), the growth of crystals by CVD is made by chemical reaction and not physical deposition.

Chemical processes are used to produce high-purity and high-performance solid materials to the point where they are often used in the semiconductor industry to produce thin films and nanostructured materials (as superlattices, nanowires, nanotubes, etc.). That is why CVD is a perfectly adapted method to grow semiconductor

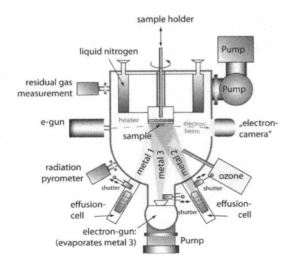

Figure 5.3 Example of schematic drawing of an MBE tool. (From Bauer, E., *Zeitschrift für Kristallographie*, 110, 395–431, 1958.)

superlattices such as Si/SiGe, GaAs/AlGaAs, or Bi_2Te_3/Sb_2Te_3.

A variety of reaction types (for example pyrolysis, reduction, and oxidation) is used to produce semiconductors and metals, as well as oxides and nitrides.

Thanks to the diversity of materials that can be deposited, metal-organic CVD (MOCVD) is one of the main techniques to deposit semiconductor films in general, and semiconductor superlattices in particular. Indeed, by precisely controlling the temperature, pressure, and gas composition, high-quality superlattices can be produced.

Unlike the CVD process where only gaseous precursors are used, MOCVD can use solid, liquid, and gaseous precursors. Thus, unlike traditional silicon semiconductors, these materials may contain combinations of group III and group V; group II and group VI; group IV; or group IV, V, and VI elements.

The CVD or MOCVD deposition mechanisms are schematically given in Figure 5.4 for QWSL (a-b-c) and QDSL (d-e-f) nanostructures. The example is given for a QWSL alternating Si and Ge layers and a QDSL with Ge dots inside a Si matrix.

After the transportation of material A reactants (SiH_4 in this example) by forced convection in the gaseous flow area (convection made with a carrier gas, not illustrated in Figure 5.4), a diffusion of the reactive species and surface reactions occur (chemical separation of reactive species, diffusion and adsorption on growth sites, etc.), leading to the material A deposition (Si in the example). Then, after the product's desorption of the chemical reactions and the evacuation of the gaseous flow, the same steps are repeated for the material B (GeH_4 precursor for a Ge layer deposition). More details on the processing steps will be presented in the next section, notably the process flow difference between the QWSL and QDSL deposition.

MOCVD and more generally thermal CVD are the preferred manufacturing process for the superlattice-based compound semiconductor electronic devices. Note that the atomic layer deposition (ALD) is also used for high-k oxide insulators for silicon memory and micro-processor ICs. Most process temperatures are in the 500°C–1100°C range, but lower temperatures (lower than 250°C) can be achieved for oxide processes.

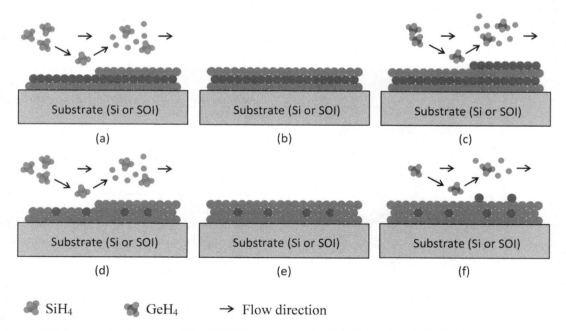

SiH$_4$ GeH$_4$ → Flow direction

Figure 5.4 Schematic drawing of the (MO)CVD process for QWSL **(a–c)** and QDSL **(d–f)** growths. Here, examples are given for a QWSL alternating Si and Ge layers and a QDSL with Ge dots inside a Si matrix.

Moreover, if MOCVD and more generally thermal CVD present the same process parameters and superlattice quality (main difference consists in the choice of materials to deposit), the main process difference with MBE technique, which is also largely used for the growth of QWSL and QDSL, consists in the pressure range; as presented previously, MBE deposition of the required atoms occurs in an ultra-high vacuum ambient, atomic layer by atomic layer obtained from atomic or molecular beam sources. Thus, if MBE has the lowest growth rates of all processes, it also is used for the manufacturing of superlattice-based commercial electronic circuits, and well adapted for QWSL and QDSL researches. Typical deposition temperatures range from 400°C to 700°C.

5.2.1 Superlattices processing steps

The SL processing steps can depend on the employed CVD type, as a function of the operating pressure (atmospheric pressure, low pressure, or ultra-high vacuum), the assisted reactions (as plasma, aerosol, or UV), and the nature of grown materials (usual CVD or MOCVD). Nevertheless, the main deposition steps of QWSL and QDSL growth can be summarized in Figure 5.5.

The alternation of the two materials A and B forming the superlattices is made by alternating gas parameters and potentially by changing the deposition chamber parameters for the two materials. It should be noted that a composite material can be composed from several precursors inserted simultaneously in the chamber.

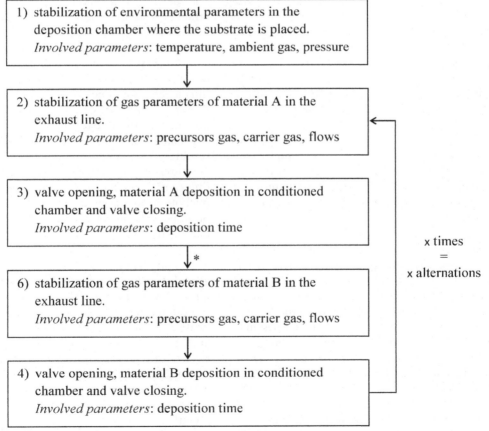

environmental parameters in the deposition chamber can be changed between materials A and B.

Figure 5.5 Schematic diagram of QWSL and QDSL deposition steps.

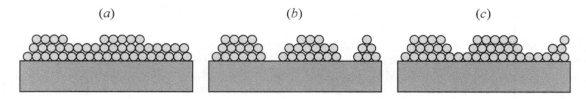

(a)   *(b)*   *(c)*

Figure 5.6 The three primary modes of epitaxial growth: Frank–van der Merwe **(a)**, Volmer-Weber **(b)** and Stranski–Krastanov **(c)**.

The other main distinction between the growth of QWSL and QDSL consists in the parameters used for the gas: generally, for the QDSL growth, the precursors are more diluted in the carrier gas, so their partial pressure are lower, and the time deposition is shorter than for a continuous layer, allowing the control of the dot size and density.

CVD depositions allow the obtaining of diversified QWSL proving the large choice of materials that can be grown thanks to the different existing CVD types. For example, a 1 nm/5 nm Bi_2Te_3/Sb_2Te_3 semiconductor QWSL from the family of V–VI can been grown by MOCVD at 225°C,[7] a 8 nm/8 nm $Si/Si_{85}Ge_{15}$ semiconductor QWSL from the family of IV can been grown by reduced-pressure CVD at 1000°C,[8] and a 5 nm/11 nm SiN_x/SiO_x nitride-oxide QWSL can be obtained by plasma-enhanced CVD at 350°C.[9]

These examples show also that very low thicknesses can be obtained by CVD techniques while keeping very high structure quality (interfaces, crystallinity) and that the deposition temperature can vary in a large range according to the used CVD type.

5.2.2 QWSL and QDSL deposition mechanisms

The role of temperature is of great importance for the diffusion of atoms. The higher is the temperature, the higher is the mobility of the species on the surface. Thus, the deposition process can be limited by the diffusion of gaseous species through the boundary layer (for the higher temperatures) or by the surface diffusion and reactions of the adsorbed species (for the lower temperatures).

Another important parameter that controls the solid phase growth is the interaction of the deposited species with the substrate atoms. Assuming that sufficient temperature is given to the system to assure

surface diffusion, species presenting a high interaction (for example, the growth of Si on Si), will grow layer by layer. This is described by the Frank–van der Merwe model[10] and is shown in Figure 5.6a. In the opposite situation, where the interaction between the deposited atoms is higher than the interaction between the growing species and the substrate, the growth will be characterized by an island growth, also called the Volmer–Weber growth (Figure 5.6b). The third growth mechanism, also called the Stranski–Krastanov model, is characterized by a layer-by-layer growth until a critical thickness is reached and then by an island growth (Figure 5.6c).

In the case of the Volmer–Weber growth, if the deposition continues up to the formation of a thin film, a polycrystalline SL will be produced.

For a Frank–van der Merwe growth type, the growth can occur epitaxially, i.e., the grown film will follow the substrate's crystalline orientation resulting in a monocrystalline SL. If the temperatures employed are low (resulting in a small diffusion of surface atoms and larger number of nucleation points) or if the substrate has defects or impurities, a polycrystalline SL will be produced. The same is true for the Stranski–Krastanov growth type.

It is important to notice that the Volmer-Weber growth type, if well controlled, can result in the formation of quantum dots. This mechanism is typical for metallic growth on insulating substrate and was observed in recent research, as Ti and Mo-based nano-islands grown onto SiGe substrates.[11] By controlling the deposition temperature and duration, the islands remained isolated without forming a continuous layer and acted as quantum dots precursors, used for the QDSL growth.

An example of such growth is given in Figure 5.7, representing a TEM view of a silicide QDSL grown by reduced-pressure CVD.

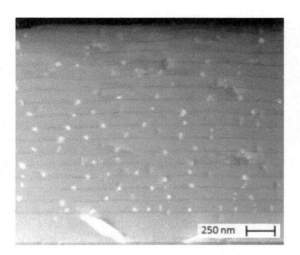

Figure 5.7 Cross-section TEM image of a TiSi$_2$/SiGe monocrystalline QDSL grown by CVD using the Volmer-Weber growth type. (From Silveira Stein, S., Croissance et caractérisation de superréseaux de boites quantiques à base de siliciures métalliques et SiGe pour des applications thermoélectriques, PhD dissertation, Université Grenoble 1, 2014.)

Figure 5.8 Cross-section TEM image of an 8 nm/8 nm Si/SiGe monocrystalline QWSL grown by CVD illustrating the layers wave typical of a modified Stranski–Krastanov growth. (From Savelli, G., Etude et développement de composants thermoélectriques à base de couches minces, PhD dissertation, Université Grenoble 1, 2007.)

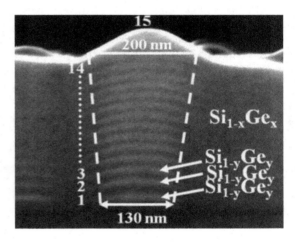

Figure 5.9 Cross-section SEM image of a Si$_{1-y}$Ge$_y$/Si$_{1-x}$Ge$_x$ monocrystalline QDSL focusing on a Si$_{1-y}$Ge$_y$ "dot column" where vertical alignment took place during the CVD growth (dots thickness is about 15 nm), typical behavior of a modified Stranski–Krastanov growth. (From Hauser, D. et al., *Thin Solid Films*, 520, 4259, 2012.)

The Stranski–Krastanov type occurs when the grown material and the substrate have a similar crystalline structure but a different lattice parameter. A basic example is the growth of germanium layer on silicon substrate. In this case, the growth is bidimensional for the first layers as the germanium wet the silicon surface well. When the thickness increases, tridimensional islands appear, allowing the stress to be minimized in the germanium layer. But for the silicon-germanium-based QWSL and QDSL, the mechanism is slightly different and called modified Stranski–Krastanov growth. Indeed, here the growth mode is first bidimensional then the accumulated stress in the layer becomes too important and the growth continues with a wavy form. It allows the stress to be relaxed elastically without generating crystalline defect (as dislocation), which is so different from the classical Stranski–Krastanov growth. Such wavy Si/SiGe QWSL is illustrated in Figure 5.8.

Figure 5.9 presents a SEM image of a Si$_{1-y}$Ge$_y$/Si$_{1-x}$Ge$_x$ monocrystalline QDSL.[12] A strong vertical ordering between the dots and progressive enlargement of base width can be clearly observed. This is a typical behaviour for Ge/Si QDSL identified as a modified Stranski–Krastanov growth.[13]

Surface dots nucleate preferentially above buried Si$_{1-y}$Ge$_y$ dots due to the induced local strain field.

As explained, the stress induced in the structure during the growth plays a critical role. Indeed, by definition, QWSL and QDSL are composed of several different thin film materials with different lattice parameters. The stacking of thin layers can induce high stress, which can lead to the formation of dislocations. These dislocations can be localized or can be present along the entire structure.

The dislocation formation is due to three main mechanisms: a homogeneous nucleation, a grain boundary initiation, and interfaces between the lattice and the surface, precipitates, etc.

For example, take again the example of a Si/SiGe thin-film QWSL epitaxially grown onto a silicon wafer. In this case, the difference on the lattice parameter of the two materials can disturb the growth. In order to accommodate the atoms according to the silicon substrate's lattice, the lattice parameter of the SiGe is reduced along the horizontal direction and increased along the vertical direction. The higher the Ge content of the SiGe, the more this effect is noticed. This phenomenon will result in an increase of the internal elastic energy of the material. The thicker the film is, the higher is the stored energy. A further increase in the elastic energy of the system will lead to the creation of dislocations and relaxing of the structure. In this case, the lattice parameter of the SiGe film will no longer match the Si film, but it will be rather the one of a bulk material. A further growth of the film will then result in the formation of dome-like structures, resulting in a Stranski-Krastanov-like growth, as explained before. This feature can be employed to create quantum dots/nano-islands structures based on SiGe/Si.[14,15] The shape of the obtained nano-islands can thus be further tailored by annealing the obtained material.[16] The thickness where the system will relax and form dislocations is called the critical thickness and is dependent on the Ge content of the SiGe film.[17]

Figure 5.10 presents two examples of mono-crystalline QWSL (a) and QDSL (b) where misfit dislocations and threading dislocations are highlighting.

5.2.3 QWSL and QDSL structures

One of the greatest advantages of CVD compared to other deposition types is the possibility to grow quickly QWSL and QDSL with different crystallographic structures: monocrystalline, polycrystalline, or amorphous. The two first are usually obtained using the usual thermal CVD (UHVCVD, LPCVD, RPCVD, MOCVD) techniques, while amorphous structures are generally grown by PECVD.

The differentiation between mono- and polycrystalline structures is mainly obtained by changing growth temperature and pressure combinations. Thus, for example, chemically identical QWSL can be obtained in the form of mono- or polycrystalline structures only by changing the growth temperature.

Recent monocrystalline QWSL have been successfully grown with a very high crystallinity structure and clearly defined interfaces, such as a 12 nm/12 nm GaAs/AlAs QWSL made by MOCVD,[18] or a 6 nm/10 nm Si/SiGe QWSL grown by UHVCVD.[19]

(a)

(b)

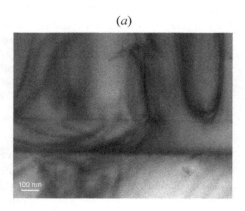
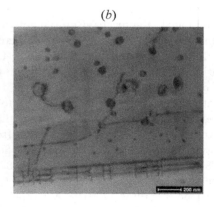

Figure 5.10 Cross-section TEM images of a Si/SiGe monocrystalline QWSL. (From Bauer, E., *Zeitschrift für Kristallographie*, 110, 395–431, 1958.) **(a)** and a TiSi$_2$/SiGe QDSL. (From Bauer, E., *Zeitschrift für Kristallographie*, 110, 395–431, 1958.) **(b)** Grown by CVD showing dislocations and interfacial stress.

Other recent amorphous QWSL with very thin layers has been obtained by PECVD such a 2 nm/2 nm a-Si:H/a-SiGe:H.[20,21]

Otherwise, the polycrystalline structures can propose typical columnar growths, as much for QWSL as for QDSL. Thus, the grain structure in most polysilicon films deposited at temperatures between 600°C and 700°C is columnar with a grain size typically increasing from the bottom of the layer to the top.[22] Figure 5.11 gives an example of such polycrystalline QWSL (a) and QDSL (b) with a columnar growth and increasing grain size.

Note that for polycrystalline QDSL, recent works have shown that dots tend to agglomerate at the grain boundaries (see inset of Figure 5.11b), generating an inhomogeneity of QD inside the material.

The probable reason for the preferential growth near the grain boundaries is that these highly disordered sites lower the energy for the nucleation of the dots.

Moreover, for QDSL structures, independently of their crystallinity, the quantum dots can be disordered in the whole matrix and so placed completely randomly, or they can be aligned on same lines and so regularly stacked (see Figure 5.12). This alignment can be in the vertical direction (see example in Figure 5.9)[12,13,23,24] or horizontal direction (see Figure 5.12b).[25]

New recent amorphous QDSL have also been grown by PECVD showing, for example, a high density of nanocrystalline Si quantum dots (~2 nm diameter) integrated in a SiC_x:H matrix (2–3 nm thickness).[26]

(*a*) (*b*)

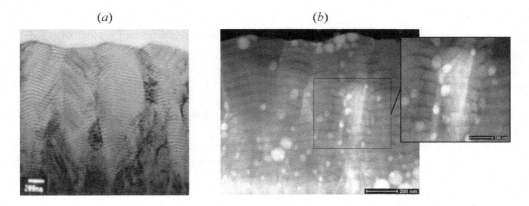

Figure 5.11 Cross-section TEM images of a polycrystalline Si/SiGe QWSL. (From Bauer, E., *Zeitschrift für Kristallographie*, 110, 395-431, 1958.) **(a)** and a polycrystalline silicide QDSL. (From Bauer, E., *Zeitschrift für Kristallographie*, 110, 395–431, 1958.) **(b)** Grown by CVD showing the columnar growth of the nanostructure.

(*a*) (*b*)

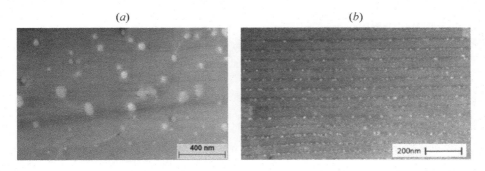

Figure 5.12 Cross-section TEM images of a randomly organized silicide Ti-based QDSL **(a)** and a regularly stacked silicide Mo-based QDSL **(b)** grown by CVD. (From Bauer, E., *Zeitschrift für Kristallographie*, 110, 395–431, 1958.)

5.3 MAIN QWSL AND QDSL PROPERTIES

In this section, the influence of superlattices on their main properties is explained, especially on their mechanical, electrical, optical, and thermal properties.

5.3.1 Mechanical properties

In 1970, J. Koehler theoretically predicted that by using superlattices made of nanolayers of materials alternating high and low elastic constants, shearing resistance could be improved by up to 100 times, mainly as the dislocations caused by the Frank-Read source cannot operate in the nanolayers.[27]

This increase of mechanical hardness of such superlattice materials was later confirmed, for example with polycrystalline CrN/TiN superlattices deposited by magnetron sputtering system.[28] This superlattice system overcomes the primary limitation of CrN coatings, low hardness, while retaining their excellent adhesion and oxidation resistance.

In CVD system, similar results have been obtained with TiN/Ti-B-N multilayers deposited by plasma-assisted CVD (PACVD).[29] It has been shown that for these superlattices, illustrated in Figure 5.13, increased hardness and residual stresses, as well as new structural and functional properties, could be obtained, notably by combining low-friction TiN layers and nanocomposite Ti-B-N layers. Moreover, multilayer coatings

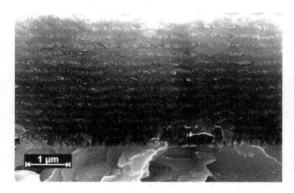

Figure 5.13 SEM fracture cross-section of a TiN/Ti-B-N multilayer containing 32 individual layers deposited by PACVD. (From Stoiber, M. et al., *Surf. Coat. Tech.*, 177-178, 348, 2004.)

with increased resistance to plastic deformation (i.e., a high H^3/E^2 ratio) could be designed, where an increase in "fracture toughness" by a factor of 2 was obtained compared to TiN and Ti-B-N single layers.

5.3.2 Electrical properties

One of the key physical quantities needed to predict the electrical properties of QWSL and QDSL structures is the density of states. The density of states characterizes the number of states per unit volume given a small interval of energies. It can be expressed as:

$$dN = D(E)d(E) \qquad (5.1)$$

where N is the number of states per unit volume and $D(E)$ is the density of states.

A general definition for the density of states is:

$$D(E) = \frac{1}{\Omega}\sum_{v}\delta(E - E_v) \qquad (5.2)$$

where Ω is the sample volume and E_v is the energy associated with the state.

For two-dimensional structures like QWSL and QDSL, the $E(k)$ dispersion relation can be written as:

$$E = E_q + \frac{\overline{h}k_x^2}{2m^*} + \frac{\overline{h}k_y^2}{2m^*} \qquad (5.3)$$

where E_q is the confinement energy in the z direction. Thus, Eq. (5.2) becomes:

$$D_{2D}(E) = \frac{2}{S}\sum_{q,k_x,k_y}\delta\left(E - E_q - \frac{\overline{h}k_x^2}{2m^*} - \frac{\overline{h}k_y^2}{2m^*}\right) \qquad (5.4)$$

where S is the area of the QW surface. This expression can finally be written:

$$D_{2D}(E) = \frac{m^*}{\pi\overline{h}^2}\sum_{q}\Theta(E - E_q) \qquad (5.5)$$

where Θ is the Heaviside step-function. Thus, Figure 5.14 shows the density of states for a two-dimensional structure. $D_{2D}(E)$ is also a discrete staircase-like function where each discrete value is associated with a different energy state E_q.

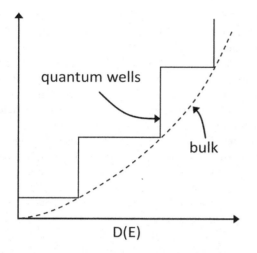

Figure 5.14 Density of state D(E) of a two-dimensional structure (QW) compared to a bulk crystal (three-dimensional system).

The most important feature from an application point of view for semiconductor QWs is the possibility of new electronic transitions between the confined discrete levels (see Figure 5.15). Formed by repeating the QW structure successively, a QWSL is more suitable to take advantage of these features.

The strength of these transitions is determined by the overlap of wavefunctions. This property combined with the dependence of E_g as a function of well parameters can be used to engineer devices for the desired application. Moreover, the quantum confinement effect allows the density of states to shift to higher energies.

5.3.3 Optical properties

One of the most interesting optical properties is the absorption which takes part, particularly for the design of photodetectors.

An approximate expression for the absorption coefficient is[30]:

$$\alpha(\omega) \approx \frac{\pi e^2}{c\varepsilon_0 n_r m_0^2 \omega} \left|p_{cv}(0)\right|^2$$

$$\frac{2}{\Omega}\sum_k \delta\left(E_c(k) - E_v(k) - \bar{h}\omega\right)$$

(5.6)

where ω is the frequency, e the electronic charge, c the vacuum light velocity, ε_0 the vacuum permittivity, n_r the real part of the complex refractive index, and $p_{cv}(0)$ is the matrix element involving Bloch functions in conduction and valence bands. The summation in Eq. (5.6) over the Bloch wave vector k defines a density of states but, in contrast to Eq. (5.2), the energy of two bands is affected. This term is called optical joint density of states.

The optical transitions taking place in a symmetric QW are also shown in the previous Figure 5.15. The first type is the intersubband transition corresponding to the transitions between energy levels within the same conduction or valence bands (5.15b). The second type is the interband transition corresponding to the transitions between energy levels in the conduction and valence bands (5.15c). Note that Equation (5.6) has been obtained for interband transitions. For intersubband transitions, the

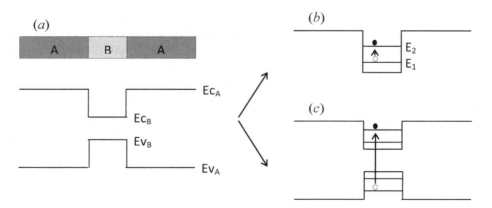

Figure 5.15 Schematic representation of a QW made of materials A and B showing the resulting band diagram **(a)**, an intersubband transition **(b)** and an interband transition **(c)**.

effect of occupation through the Fermi factors has to be considered.

The absorption edge is more abrupt for interband transitions in QW than in bulk material because of the optical joint density of states. Note that the absorption depends on the number of QW and that a large number of periods is necessary to achieve a similar active volume than in bulk materials. Thus, QWSL-based devices allow the contrast in photodetection near the band gap to be improved.

Moreover, the intersubband transitions produce spectrally selective absorption. This feature is advantageous in many applications where the wavelength performs. The intersubband transitions are so suitable for infrared photodetectors while interband transitions are more suitable at higher energies in the visible and ultraviolet spectra.

5.3.4 Thermal properties

The nanostructuration of materials impacts greatly their thermal properties. In some fields, like thermoelectric applications, materials need to have the lowest thermal conductivity, but keeping a high electrical conductivity.

Heat transport in semiconductors has contributions of both of the two modes, i.e., electrons and phonons, as described by the Eq. (5.7).

$$\lambda = \lambda_e + \lambda_l \qquad (5.7)$$

where λ_e is the electronic contribution and λ_l the lattice contribution (phonons) to the thermal conductivity. The electronic contribution can be approximately expressed by the Wiedemann–Franz law, considering no inelastic collisions between electrons and phonons.

$$\lambda_e = L_0 \times \sigma \times T \qquad (5.8)$$

where L_0 is the Lorentz number, σ the electrical conductivity, and T the temperature. The lattice contribution λ_l can be expressed as:

$$\lambda_l = \frac{C_v \times v \times \tau}{3} \qquad (5.9)$$

where C_v is the phonon's specific heat, v is the phonon's average speed, and τ is the phonon's relaxation time. This equation is obtained from the kinetic gas theory applied to phonons and is valuable at low temperatures, where the phonon's dispersion relation is negligible.

It can be seen from Eq. 5.2 that the electronic contribution for the heat transport λ_e depends on the material's electrical conductivity. This is easily understood, since electrons are the particles responsible for heat transport. By increasing the number of charge carriers by further doping the material, the electrical conductivity is increased but λ_e is increased as well; that is not a suitable solution to decrease the total thermal conductivity.

Equation 5.3 shows the contribution of the phonon's relaxation time τ to the lattice thermal conductivity λ_l. The relaxation time depends on the collision mechanisms, which scatter the phonons responsible for the heat transport.

The phonon relaxation time can be divided in different parts, corresponding to the different scattering sources present in a material, as described by Matthiesen law:

$$\tau^{-1} = \tau_{ph}^{-1} + \tau_{imp}^{-1} + \tau_{int}^{-1} + \tau_{gb}^{-1} + \tau_{np}^{-1} \qquad (5.10)$$

where τ_{ph} is the phonon-phonon scattering contribution, τ_{imp} refers to collisions with structural defects present in crystalline lattices and ionized impurities linked to the doping, τ_{gb} is the contribution due to grain boundaries, and τ_{np} is the contribution from nanoparticles inside the matrix.

Thus, growing QWSL and QDSL makes possible the materials to change their thermal conductivity by creating new phonon-scattering mechanisms inside the matrix.

It should be noticed that the influence of the different scattering mechanisms depends on the phonon wavelength. Indeed, long-wavelength phonons will mostly interact with grain boundaries and nanometric inclusions and short-wavelength phonons will interact mostly with atomic defects such as alloying and dopant atoms.

Most phonon-scattering mechanisms can be represented in the case of a polycrystalline QDSL (see Figure 5.16). The schematic drawing of this structure is similar to the columnar structure illustrated in Figure 5.11b where nanodots are mainly localized at the grain boundaries. In this kind of structure, short-wavelength phonons are effectively scattered by atomic defects and dopant atoms **(1)** while most of the long-wavelength phonons are scattered by grain boundaries **(2)**,

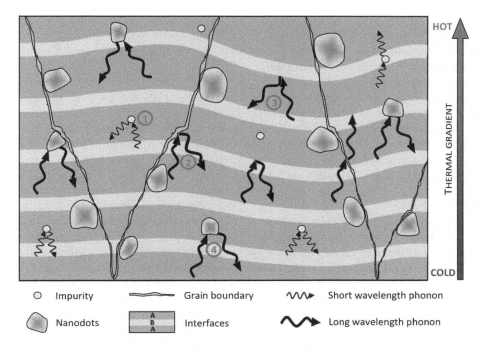

○	Impurity	⌇⌇⌇⌇►	Short wavelength phonon
Nanodots	Interfaces	Long wavelength phonon

Figure 5.16 Schematic diagram illustrating various phonon-scattering mechanisms present in a poly-crystalline QDSL. (Modified from Vineis, C.J. et al., *Adv. Mater.*, 22, 3970, 2010.)

layer interfaces (**3**), and nanoparticles (**4**). The size and shape of nanodots can greatly differ amongst themselves.

Note that the grain boundary acts as a natural scattering site for phonons but also for electrons. This explains why polycrystalline materials have typically a smaller thermal and electrical conductivity than a monocrystalline solid with the same doping level and stoichiometry. That is why the influence of nanostructuration via QWSL or QDSL on thermal conductivity is more highlighted for monocrystalline structure than for polycrystalline.

Thus, very low thermal conductivity values have been measured for such structures grown by CVD. For example, Bi_2Te_3/Sb_2Te_3 QWSL grown by MOCVD have shown a 50% reduction in the lattice thermal conductivity for different periods (1 nm/5 nm, 3 nm/3 nm, and 2 nm/4 nm).[31] More recent examples have also shown a thermal conductivity reduction: for SiGe-based QDSL reduction until 50% has been obtained for mono-crystalline structures,[32] for silicides QDSL grown by RPCVD, reduction until 27% has been obtained for monocrystalline structures.[11]

5.4 APPLICATIONS

Recent QWSL and QDSL grown by CVD techniques have been used in different and varied applications going from electroluminescence[9] to memory devices,[34] including photonics, optoelectronics, and thermoelectrics. The three last ones that may be considered as the main application are detailed below.

5.4.1 Photonics

Photodetectors are devices used for the detection of light, in most cases of optical powers. There are many types of photodetectors: photodiodes, phototransistors, photoresistors, photomultipliers, pyroelectric photodetectors, thermal detectors, solar cells, and others.

Depending on the application, a photodetector has to fulfil various requirements: responsivity, noise, speed, contrast, selectivity, linearity, and detectivity.

The specific properties of QWSL, particularly the optical and electrical properties, have facilitated the development of advanced concepts and applications related to photodetectors, leading to extended functionalities (like selective absorption) and improved performances (like photocurrent gain).

Photodetectors for the visible and near-infrared spectral regions are generally made from bulk silicon or III–V alloys such as GaInAs. Since these devices work very well, the main application for QWSL-based photodetectors is in the infrared spectral region and for especially demanding applications such as avalanche photodiodes and solar cells.[2]

5.4.1.1 SOLAR CELLS

The power generated by a solar cell is given by the product of the photocurrent and the voltage across the diode. These two parameters are determined by the band gap of the semiconductor. QWSL-based devices can give better performance than the bulk devices because the optimization of the current and voltage can be separated.[34] This is because the voltage is primarily determined by the band gap of the barrier regions, while the absorption edge is determined by the band gap of the QWs.

Recently, hydrogenated amorphous Si/SiGe QWSL have been grown by PECVD for solar cells applications. Compared to III–V materials, hydrogenated amorphous Si/SiGe QWSL have the advantages of large area production, lower prices, and lower temperature deposition. Better conversion efficiency has been measured for the QWSL compared to an a-Si:H-based solar cell.[20,21]

Moreover, a superlattice solar cell with InGaAs well and GaAsP barrier was also realized recently for the first time.[35] This 3.7 nm/5.45 nm InGaAs/GaAsP QWSL was grown by MOCVD. An increase in short-circuit current by 2 mA/cm^2 was achieved with 60 stacks of the 3.7-nm-thick wells as compared with a GaAs p-i-n reference cell. Combining a minimal drop in open-circuit voltage V_{oc} with an apparent enhancement in short-circuit current I_{sc}, the SL solar cell exhibited an overall efficiency of approximately 18%, which is slightly higher than the value for the control cell.

Up to now, the record of solar cell efficiency integrating QWSL is obtained from InGaAs/GaAsP structure grown by MOCVD.[36] It consists in alternating 50 strain-balanced 95 Å/196 Å $In_{0.1}Ga_{0.9}As$/ $GaAs_{0.911}P_{0.089}$. With the increase in I_{sc} and V_{oc} due to the use of distributed Bragg reflectors (DBR), an efficiency of 27% has been measured at 328 suns.

To date, this InGaAs/GaAsP strain-balanced QW solar cell (SB-QWSC) grown by MOCVD exhibits the highest QW solar cell efficiency.

5.4.1.2 AVALANCHE PHOTODIODES

Avalanche photodiodes (APDs) are detectors adapted for many applications in telecommunications and single-photon counting. The avalanche multiplication mechanism plays a critical role in determining the photodetection gain, the noise, and the gain-bandwidth product. Commercially available III–V semiconductor APDs are typically engineered with different bandgap materials in the absorption and multiplication regions.[37] The absorption layer has a relatively narrow band gap to allow larger absorption, whereas the multiplication region has a wider band gap to reduce the dark current at the high electric fields required. It has been demonstrated that the dark current can be reduced by incorporating adequately designed superlattices into the multiplication layer to form a SL avalanche photodiode (SL-APD).[38] Figure 5.17 presents a schematic diagram of a typical InGaAs/InP/InGaAsP/InAlAs SL-APD. Light is absorbed in the bulk InGaAs layer and the resulting photocurrent is multiplied by the avalanche process in the InGaAsP/InAlAs SL region. The spatial periodicity of the SL reduces the dark current.

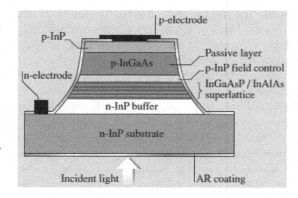

Figure 5.17 Schematic diagram of a typical InGaAs/InP/InGaAsP/InAlAs SL avalanche photodiode (SL-APD). (From Fox, M. and Ispasoiu, R., Quantum wells, superlattices, and band-gap engineering, in *Handbook of Electronic and Photonic Materials*, S. Kasap and P. Capper, Eds. Springer, Boston, MA, pp. 1021–1040, 2007.)

5.4.1.3 INTER-SUB-BAND DETECTORS

The engineered band structure of QWs leads to the possibility of inter-sub-band (ISB) transitions, which take place between confined states within the conduction or valence bands (see Figure 5.15b). The transitions typically occur in the infrared spectral region.

Infrared detectors are required for applications in defence, night vision, astronomy, thermal mapping, gas-sensing, etc. Quantum-well ISB photodetectors (QWIPs) are designed so that the energy separation of the confined levels is matched to the chosen wavelength. A major advantage of QWIPs over the conventional approach employing narrow-gap semiconductors is the use of mature GaAs-based technologies.

GaN/AlN SLs grown by MOCVD have also been used for QWIP detectors.[39] This 150 periods 1 nm/1.5 nm GaN/AlN QWSL, illustrated in Figure 5.18a, presents in the absorption spectrum a clearly resolved absorption peak at around 0.64 eV, corresponding to a wavelength of 1.9 mm, which is not present for the absorption spectrum measured for the sample without the SL structure (see Figure 5.18b).

This 1.9 mm absorption can be mainly induced by the transition from the first confined electron sub-band to the second confined electron sub-band in the GaN QW. Such results are very encouraging as they open new routes for the fabrication of 2D photodetector arrays based on ISBTs working in the optical fibre communication wavelength region.

5.4.1.4 VCSEL

Vertical-cavity surface-emitting lasers (VCSEL) are semiconductor lasers, more specifically laser diodes with a monolithic laser resonator, where the emitted light leaves the device in a direction perpendicular to the surface of the chip. The resonator (cavity) is made with two semiconductor reflectors and, between those, there is an active region (gain structure). Both Bragg reflectors and active regions can be usually made of QWSL, such as AlGaAs/AlAs, GaInAs/InP or InP/InAlGaAs SLs grown by MOCVD.[40–42]

SLs simultaneously offer high reflectivity and low series resistance. The low cost and high yield associated with the semiconductor SL DBR VCSEL and the improved confinement offered with proton bombardment make the device an attractive candidate for use in optical communications and in other optoelectronic applications. A typical GaAs-based VCSEL schematic diagram is given in Figure 5.19. As shown, the QWs that comprise the gain medium are placed at the centre of the cavity formed between two DBR mirrors.

The conventional VCSEL structures grown on GaAs substrates operate in the wavelength range 700–1100 nm. Much work is currently focused on extending the range of operation to the telecommunication wavelengths of 1300 nm and 1550 nm.

VCSELs have many applications. They are used as transmitters for optical fibre communications, gas sensing, miniature optical clocks, thresholdless laser, etc. An application area which was developed later, but has acquired a large market volume,

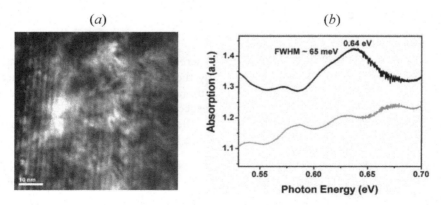

Figure 5.18 Cross-sectional TEM image of the AlN/GaN QWSL grown by MOCVD **(a)** and the infrared absorption spectrum (black line = QWSL, grey line = without QWSL) measured at room temperature under the normal incidence condition **(b)**. (From Zhou, H. et al., *Jpn. J. Appl. Phys.*, 46, 5128, 2007.)

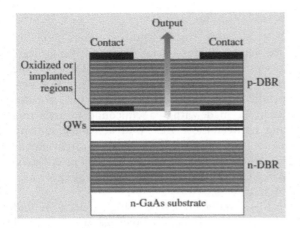

Figure 5.19 Schematic diagram of a typical GaAs-based vertical-cavity surface-emitting laser (VCSEL). (From Fox, M. and Ispasoiu, R., *Quantum wells, superlattices, and band-gap engineering*, in *Handbook of Electronic and Photonic Materials*, S. Kasap and P. Capper, Eds. Springer, Boston, MA, pp. 1021-1040, 2007.)

is that of computer mice. A laser mouse with a VCSEL as light source can have high tracking precision combined with low electricity consumption, which is important for battery-powered devices.

Thanks to the high gain coefficients to the QW structures, their small size and very low threshold currents, the development of VCSEL has been very rapid, and many local-area fibre networks operating around 850 nm currently employ VCSEL devices.

5.4.2 Optoelectronics

In optoelectronics, quantum cascade lasers (QCL) are one of the most important sources of coherent light in the infrared region. As an electron moves through the QWSL, it undergoes ISB transitions and tunnels through the next thin film, which acts a barrier region. Emerging from the barrier into another QW, it undergoes another identical ISB transition, creating thus a cascade of coherent photons.[43]

InGaAs/InAlAs QWSL grown by MOCVD are usually used for such devices.[44,45]

Moreover, it has been recently suggested that by using a QDSL instead of QWSL, an increase of the device properties could be obtained by eliminating the movement of charge carriers along the QW plane.[46]

QCL are used in different applications, going from medical to military fields.[43]

Recently, a novel QCL active-region design employing the use of metamorphic buffer layer (MBL) grown on GaAs substrates holds the potential for achieving high continuous wave (CW) performance devices. It corresponds to a low-temperature sensitivity of the device electro-optical characteristics and low device thermal resistance at short emission wavelength ($\lambda \approx 3.6~\mu m$).[47] MBL are virtual substrates commonly used to reduce threading-dislocation densities for heteroepitaxial-mismatched films. But MBL can significantly degrade the interfacial structure and compositional uniformity of the heterostructures grown atop, hence, the difficulty to grow high-quality QDSL-based active materials. In this study, the novel active region consists of an $In_xGa_{1-x}As/Al_yIn_{1-y}As$ QWSL SL structure grown by MOCVD. As illustrated in Figure 5.20a, the QDSL presents high quality with absence of dislocations. Moreover, Figure 5.20b shows that the measured transition wavelength is 4.72 μm and the simulated transition wavelength is 4.13 μm. This is the first observation of ISB absorption from SL structures grown on an MBL.

5.4.3 Thermoelectrics

Thanks to their optimized electrical and thermal properties, QWSL and QDSL are excellent materials for thermoelectric applications. Thermoelectric materials can be classified with their figure-of-merit ZT defined by:

$$ZT_m = \frac{\sigma S^2}{\lambda} T_m \qquad (5.11)$$

where σ, S, λ, and T_m are the electrical conductivity, the Seebeck coefficient, and the mean temperature (average of hot T_h and cold T_c temperatures), respectively.

The maximum conversion efficiency Φ_{max} is given by:

$$\Phi_{max} = \frac{T_h - T_c}{T_h} \times \frac{\sqrt{1 + ZT_m} - 1}{\sqrt{1 + ZT_m} + T_c/T_h} \qquad (5.12)$$

In Eq. (5.12), the first term relates to the Carnot efficiency and the second term is the thermoelectric efficiency, which is function of the figure-of-merit.

(a)

(b)

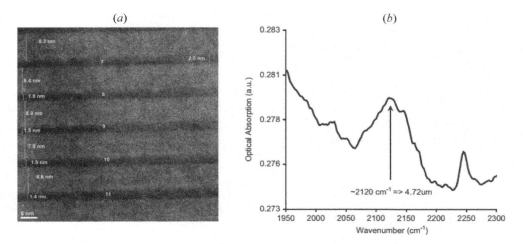

Figure 5.20 **(a)** High-resolution (HR) TEM image of an $In_xGa_{1-x}As/Al_yIn_{1-y}As$ QWSL grown atop AlGaInAs MBL by MOCVD. The dark layers correspond to the InGaAs (wells) and the lighter layers to AlInAs (barriers); **(b)** Measured intersubband absorption of the SL. (From Mawst, L.J. et al., *J. Cryst. Growth*, 370, 230, 2013.)

Figure 5.21 represents the evolution of the maximum conversion efficiency Φ_{max} as a function of figure-of-merit ZT and the temperature difference ΔT. As shown, the higher ZT is, the higher the conversion efficiency is. Thus, Eq. (5.11) shows that to increase ZT, there are two means: increase the power factor σS^2 and/or decrease the thermal conductivity λ. As explained previously, nanostructures as QWSL and QDSL allow thermal conductivities to be lower than their equivalent non-structured thin films thanks to an addition to phonon diffusion mechanisms.[48] Some recent works have also shown that QDSL can increase the TE materials properties. For example, Ti-based silicide SiGe QDSL grown by CVD have shown a simultaneously increase of their power factor and decrease of their thermal conductivity.[49]

Currently, the highest figure-of-merit measured in literature for all materials taken together (bulk and thin films) was obtained for a Bi_2Te_3/Sb_2Te_3 QWSL grown by MOCVD (shown in Figure 5.22), with a ZT value of 2.4 at 300 K.[50]

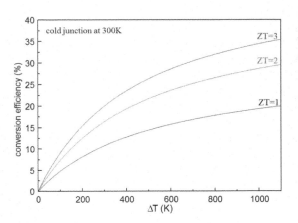

Figure 5.21 Evolution of maximum conversion efficiency Φ_{max} as a function of figure-of-merit ZT and the temperature difference ΔT.

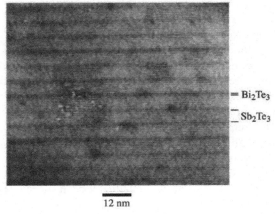

Figure 5.22 Cross-section TEM images of a 1 nm/5 nm Bi_2Te_3/Sb_2Te_3 QWSL grown by MOCVD, exhibiting a record ZT of 2.4 at 300 K. (From Venkatasubramanian, R. et al., *Appl. Phys. Lett.*, 75, 1104, 1999.)

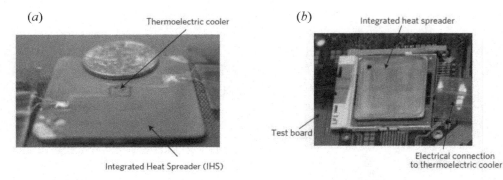

Figure 5.23 Photographs of the thermoelectric device attached to the underside of the heat spreader **(a)** and of the state-of-the-art electronic package with the thermoelectric cooler beneath the heat spreader **(b)**. (From Chowdhury, I. et al., *Nature Nanotechnol.*, 4, 235, 2009.)

One of the main applications of thermoelectrics is cooling, based on the Peltier effect.

Recently, the Bi_2Te_3/Sb_2Te_3 QWSL with ZT of 2.4 has been integrated by Intel Corporation in a thin film thermoelectric device to cool electronic chips, as shown in Figure 5.23.[51] It combines an active and passive cooling. The passive cooling is obtained thanks to the integrated heat spreader and the active one with the use of the TE cooler which integrates the QWSL. Thus, the active (7.3°C) plus passive (7.6°C) hot-spot cooling provides a total cooling superior to 14°C at a heat flux ~1250 W/cm². This on-demand cooling is very useful to prevent chips from overheating and thermal damages.

This is the first practical application of nanostructured thermoelectrics to solve a real-world chip problem.

Moreover, Laird Technologies, a developer of components and solutions for electromagnetic interference protection and thermal protection of electronics, has recently acquired U.S.-based manufacturer of thin-film thermoelectric technologies Nextreme Thermal Solutions. They currently commercialize TE products (generator and cooler, as illustrated in Figure 5.24) integrating the high ZT Bi_2Te_3/Sb_2Te_3 QWSL.

In generator mode, such devices can supply up to 24 mW for a temperature difference of 50°C and in cooling mode, they can generate a temperature difference of 50°C for a current of 1.1 A, making them the most competitive TE thin films devices.

Figure 5.24 Illustration of a thermoelectric power generator integrating Bi_2Te_3/Sb_2Te_3 QWSL commercialized by Laird technologies. (From http://www.lairdtech.com/nextreme/#.U330X_kwtaQ.)

5.4.4 Future prospects

SLs were introduced 40 years ago to increase the materials range for electronic and optoelectronic applications. The development of QWs and quantum dots with more and more decreasing dimensions was dictated by the technological advancements.

New opportunities for QWSL and QDSL may appear in fields where nanostructures-based components could increase.

For example, advances in photodetection have taken advantage of the properties offered by quantum systems. But new optimizations in nanostructures will lead to increased performances. A better control of technology should allow an improvement of the photodetection capabilities. New promising horizons in this field exist, especially the research on photodetectors based on the piezoelectric effect. Indeed, the use of piezoelectric materials in nanostructures could enhance many structures such as SL avalanche photodetectors and high-gain photodetectors. Presently, however, the use of piezoelectric nanostructured materials is technically immature and the current objective is to extend the photodetection range to shorter or longer wavelengths and increase sensitivity.[30]

The discovery of new quantum phenomena will also able to lead to new uses of the nanostructures. Otherwise, the development of a more and more sophisticated MOCVD deposition tool allows the growth of nanostructures, which was usually developed by MBE. With more and more equivalent materials quality compared to MBE technique, MOCVD is the preferred growth technique in manufacturing due to its higher growth rate and multi-wafer capability. The recent growth of InAs/GaSb SLs is a typical example of a SL which was normally grown by MBE, and is now obtained by MOCVD.[52]

Nevertheless, industrialization of such films is not yet automatic. From on part, the relation gain in performances compared to the time (and so money) necessary to grow the film has to be very advantageous to convince manufacturers to integrate SL into their devices.

On the other hand, one of the challenges for the SL is their reliability under thermal stress and their faculties to still have their nanostructuration. Indeed, for example, dots in QDSL can move inside the matrix and agglomerate together, changing thus their size and density, and so modifying their properties.

Thus, the topic of semiconductor nanostructures moves very rapidly and it is difficult to imagine how it will change in the future. Some fields have progressed very quickly from research into the commercial sector, such as VCSEL and thermoelectrics, while others, such as QCLs, have taken many years to come to maturity. Nanostructures have enhanced the performance of conventional devices (such as thermoelectric devices, edge-emitting lasers), and in others, they have led to radically new devices (such as VCSEL, QCLs). If, at present, the main commercial use for SLs is in

thermoelectrics, LED, and laser diodes, it remains to be seen whether some of the other devices described here (QW solar cells, SLAPDs, QWIPs) will come to commercial applications.

REFERENCES

1. L. Ezaki and R. Tsu, Superlattice and negative differential conductivity in semiconductors, *IBM J Res Dev* 14, 61–65, 1970.
2. M. Fox and R. Ispasoiu, Quantum wells, superlattices, and band-gap engineering, in *Handbook of Electronic and Photonic Materials*, S. Kasap and P. Capper (Eds.), Springer, Boston, MA, pp. 1021–1040, 2007.
3. A.S. Edelstein and R.C. Cammaratra, *Nanomaterials: Synthesis, Properties and Applications*, 2nd edition, CRC Press, Boca Raton, FL, 1998.
4. C. Morawe and H. Zabel, Metal/Al$_2$O$_3$ multilayers as high-temperature x-ray mirrors, *Mater Res Soc Symp Proc* 382, 363–368, 1995.
5. O. Lenoble, P. Bauer, J.F. Bobo, H. Fischer, M.F. Ravet and M. Piecuch, Thermal behaviour and magnetic properties of Fe/Al$_2$O$_3$ multilayers, *J Phys Cond Mat* 6, 3337–3346, 1994.
6. F. Baiutti, G. Christiani and G. Logvenov, Towards precise defect control in layered oxide structures by using oxide molecular beam epitaxy, *Beilstein J Nanotechnol* 5, 596–602, 2014.
7. R. Venkatasubramanian, T. Colpitts and B. O'Quinn, Low-temperature organometallic epitaxy and its application to superlattice structures in thermoelectrics, *Appl Phys Lett* 75, 1104, 1999.
8. G. Savelli, "Etude et développement de composants thermoélectriques à base de couches minces", PhD dissertation, Université Grenoble 1, 2007.
9. H.-Y. Tai, Y.-H. Lin and G.-R. Lin, Wavelength-shifted yellow electroluminescence of Si quantum-dot embedded 20 pair SiNx/SiOx superlattice by Ostwald ripening effect, *IEEE Photonics J* 5, 6600110, 2013.
10. E. Bauer, Phänomenologische Theorie der Kristallabscheidung an Oberflächen. II, *Zeitschrift für Kristallographie* 110, 395–431, 1958.

11. S. Silveira Stein, Croissance et caractérisation de super-réseaux de boites quantiques
à base de siliciures métalliques et SiGe pour
des applications thermoélectriques, PhD
dissertation, Université Grenoble 1, 2014.

12. Hauser D., G. Savelli, M. Plissonnier, and
J. Simon, Growth of heavily doped monocrystalline and polycrystalline SiGe-based
quantum dot superlattices, *Thin Solid Films*
520, 4259–4263, 2012

13. K. Brunner, Si/Ge nanostructures, *Rep Prog
Phys* 65, 2002.

14. D. Hauser, Elaboration de super-réseaux
de boîtes quantiques à base de SiGe et
développement de dispositifs pour l'étude
de leurs propriétés thermoélectriques, PhD
diss., Université Grenoble 1, 2011.

15. I. Berbezier and A. Ronda, "SiGe nanostructures", *Surf Sci Rep* 64, 47–98, 2009.

16. J.J. Zhang, A. Rastelli, H. Groiss, J. Tersoff,
F. Schäffler, O.G. Schmidt, and G. Bauer,
Shaping site-controlled uniform arrays of
SiGe/Si(001) islands by in situ annealing,
Appl Phys Lett 95, 183102, 2009.

17. D.J. Paul, Si/SiGe heterostructures: From
material and physics to devices and circuits,
Semicond Sci Tech 19, R75, 2004.

18. M.N. Luckyanova, Detecting coherent
phonon wave effects in superlattices using
time-domain thermoreflectance, PhD
dissertation, Massachusetts Institute of
Technology, Cambridge, MA, 2012.

19. M.-J. Wu, H.-C. Wen, S.-C. Wu et al.,
Nanomechanical characteristics of annealed
Si/SiGe superlattices, *Appl Surf Sci* 257,
8887–8893, 2011.

20. W.H. Son, S.K. Lee, Y.S. Moon, T.Y. Lee and
S.Y. Choi, Characteristics of amorphous
silicon thin-film solar cells of a-Si:H/a-SiGe:H
superlattices in different thickness for barrier and well layers, *Mol Cryst Liq Cryst* 586,
61–68, 2013.

21. J.E. Kim, J.W. Shin, W.H. Son and S.Y. Choi,
The optical properties of a-Si:H/a-SiGe$_x$:H
superlattice structure to apply intrinsic layer
in solar cell, *Mol Cryst Liq Cryst* 532, 15–20,
2010.

22. A.D. McConnell, S. Uma, and K.E.
Goodson, Thermal conductivity of doped
polysilicon layers, *J Microelect Sys* 10,
360–369, 2001.

23. G. Capellini, M. De Seta, L. Di Gaspare
and F. Evangelisti, Evolution of Ge/Si (001)
islands during Si capping at high temperature, *J Appl Phys* 98, 124901, 2005.

24. V. Le Thanh, P. Boucaud, D. Débarre, Y. Zheng,
D. Bouchier and J.-M. Lourtioz, Nucleation
and growth of self-assembled Ge/Si (001)
quantum dots, *Phys Rev B* 58, 13115, 1998.

25. S. Silveira Stein, G. Savelli, P. Faucherand,
G. Bernard-Granger and L. Montes, Growth
and characterization of QDSL (Quantum Dots
Superlattices) of metal silicides in an n-doped
SiGe matrix for thermoelectric applications,
*Proceedings of the 14th IEEE International
Conference on Nanotechnology*, Toronto,
Canada, pp. 64–69, 2014.

26. J. Ma, J. Ni, J.J. Zhang, Q. Liu, XD. Zhang
and Y. Zhao, Quantum-confined photoluminescence from size-controlled boron doped
nanocrystalline-Si:H/a-SiCx:H superlattice,
Thin Solid Films 573, 191–195, 2014.

27. J. Koehler, Attempt to design a strong solid,
Phys Rev B 2, 547–551, 1970.

28. P. Yashar, S.A. Barnett, J. Rechner and
W.D. Sproul, Structure and mechanical properties of polycrystalline CrN/TiN
superlattices, *J Vac Sci Technol A* 16,
2913–2918, 1998.

29. M. Stoiber, S. Perlot, C. Mitterer, M.
Beschliesser, C. Lugmair and R. Kullmer,
PACVD TiN/Ti–B–N multilayers: From
micro- to nano-scale, *Surf Coat Tech*
177–178, 348, 2004.

30. C. Rivera, Photodetectors based on quantum-well structures: Theory, properties and
novel concepts, in *Quantum Wells: Theory,
Fabrication and Applications*, A. Ruyter
and H. O'Mahoney (Eds.), Nova Science
Publishers, New York, pp. 1–46, 2009.

31. R. Venkatasubramanian, Lattice thermal
conductivity reduction and phonon localiza-
tionlike behavior in superlattice structures,
Phys Rev B 61, 3091, 2000.

32. G. Savelli, D. Hauser, H. Michel and
J. Simon, Growth, electrical and thermal
properties of doped mono and polycrystalline SiGe-based quantum dot superlattices,
AIP Conf Proc 1449, 275–278, 2012.

33. K. Mallick and M.J. Witcomb, Application
of quantum dots in organic memory
devices: A brief overview, in *Quantum Dots:*

Research, Technology and Applications, R.W. Knoss (Ed.), Nova Science Publishers, New York, pp. 651–668, 2008.

34. K. Barnham, I. Ballard, J. Barnes et al., Quantum well solar cells, *Appl Surf Sci* 113/114, 722–733, 1997.

35. Y. Wang, Y. Wen, H. Sodabanlu, K. Watanabe, M. Sugiyama and Y. Nakano, A superlattice solar cell with enhanced short-circuit current and minimized drop in open-circuit voltage, *IEEE J Photovolt* 2, 387–392, 2012.

36. D.C. Johnson, I.M. Ballard, K.W.J. Barnham et al., Optimisation of photon recycling effects in strain-balanced quantum well solar cells, *Proceedings of the 4th World Conference on Photovoltaic Energy Conversion*, Waikoloa, HI, pp. 26–31, 2006.

37. J. Wei, J.C. Dries, H. Wang, M.L. Lange, G.H. Olsen and S.R. Forrest, Optimization of 10-Gb/s long-wavelength floating guard ring InGaAs–InP avalanche photodiodes, *IEEE Photon Technol Lett* 14, 977–979, 2002.

38. A. Suzuki, A. Yamada, T. Yokotsuka, K. Idota and Y. Ohiki, Dark current reduction of avalanche photodiode using optimized InGaAsP/InAlAs superlattice structure, *Jpn J Appl Phys* 41, 1182–1185, 2002.

39. H. Zhou, W. Liu and J. Chua, Normal incidence intersubband absorption in GaN/AlN superlattices grown on facet-controlled epitaxial lateral overgrown GaN/Sapphire templates, *Jpn J Appl Phys* 46, 5128–5130, 2007.

40. D. Schmitz, M. Deschler, F. Schulte and H. Juergensen, State-of-the-art control of growth of superlattices and quantum wells, *Mat Sci Eng B* 35, 102–108, 1995.

41. J.-H. Kim, B.-S. Yoo, J.-H. Shin et al., Fabrication method of densely spaced 1.55 μm multiple-wavelength vertical-cavity surface-emitting laser array structure for the application of dense wavelength division multiplexing, *Jpn J Appl Phys* 43, 137–139, 2004.

42. S. Ishida, T. Miyamoto and F. Koyama, Short-period GaInAs/InP superlattice for distributed Bragg reflector, *Jpn J Appl Phys* 45, L723–L725, 2006.

43. B. Saleh and M. Teich, *Fundamental of Photonics*, Wiley, Hoboken, NJ, 2007.

44. M. D'Souza, J.-C. Shin, D. Xu et al., Deep-well 4.8 μm-emitting quantum-cascade lasers grown by MOCVD, *Conference Digest–IEEE International Semiconductor Laser Conference*, 4636001, 454–456, 2008.

45. C. Wang, A. Goyal, R. Huang et al. Strain-compensated GaInAs/AlInAs/InP quantum cascade laser materials, *J Cryst Growth* 312, 1157–1164, 2010.

46. R.A. Suris and I.A. Dmitriev, Quantum-dot based quantum cascade lasers: Arguments in favor, *IEEE Winter Topicals*, WTM 5730037, 43–44, 2011.

47. L.J. Mawst, J.D. Kirch, C.-C. Chang et al., InGaAs/AlInAs strain-compensated super-lattices grown on metamorphic buffer layers for low-strain, 3.6 μm-emitting quantum-cascade-laser active regions, *J Cryst Growth* 370, 230–235, 2013.

48. C.J. Vineis, A. Shakouri, A. Majumdar and M.G. Kanatzidis, Nanostructured thermoelectric: Big efficiency gains from small features, *Adv Mater* 22, 3970–3980, 2010.

49. G. Savelli, S. Silveira Stein, G. Bernard-Granger et al., Titanium-based silicide quantum dot superlattices for thermoelectrics applications, *Nanotechnology* 26, 275605, 2015.

50. R. Venkatasubramanian, E. Siivola, T. Colpitts and B. O'Quinn, Thin film thermoelectric devices with high room temperature figures of merit, *Nature* 413, 597, 2001.

51. I. Chowdhury, R. Prasher, K. Lofgreen et al., On-chip cooling by superlattice-based thin-film thermoelectrics, *Nat Nano* 4, 235, 2009.

52. L.-G. Li, S.-M. Liu, S. Luo et al., Metalorganic chemical vapor deposition growth of InAs/GaSb type II superlattices with controllable AsxSb1-x interfaces, *Nanoscale Res Lett* 7:160, 2012.

CVD coatings

KWANG LEONG CHOY

Chemical vapour deposition (CVD) and its variants have been widely used in the production of films and coatings for a variety of structural, functional, and biomedical applications. The background of CVD of films and coatings has been given in Chapter 1. The use of CVD for large-area on-line deposition of transparent conducting oxides (e.g., fluorine-doped tin oxides), low-E coatings during float glass production for the building industry or solar cell applications, Si_3N_4 as antireflective-layer, Si-based thin-film solar cells, III–V thin films for LED, for examples, have been highlighted in Chapter 1. The CVD of graphene films has also been described in detail in Chapter 2. This chapter highlights further the CVD of films and coatings, especially thicker films with special attention to their industrial applications. These include high performance protective coatings for cutting tools, thermal barrier coatings and aluminised coatings for aeroengine components, thick metal coatings on graphite and composites for fusion applications, thick silicon coatings for the semiconductor industry and high precision optical mirrors for space and terrestrial applications, fibre coatings for reinforced metal matrix composites, optical fibres for telecommunications, and parylene coatings for industrial applications. CVD free standing synthetic diamonds for optics, acoustic, high-power electronics, environmental, and sensor beyond tribological applications will also be highlighted. In addition, CVD of thick film deposition onto 3D objects for high-temperature commercial applications (e.g., crucibles for crystal growth, thrusters and chambers for propulsion) will also be presented.

6.1 ADVANCED PROTECTIVE COATINGS FOR CUTTING TOOLS

CVD has been exploited commercially for coating tools against wear, abrasion, corrosion oxidation, thermal shock, and metal fatigue, as well as to reduce friction. Allowing tools to operate at higher temperatures, results in faster cutting speeds, extended tool life, improves quality, and reliability, and increased material removal rate—thus, leading to greater productivity. CVD tends to produce thicker coatings (i.e., 5–20 microns) as compared to physical vapour deposited (PVD) coatings (i.e., 0.5–5 microns) and are very wear-resistant. This makes them suitable for cutting steels and cast irons, whereas PVD coatings are more suited for workpiece materials that are tough or work-harden easily (e.g., stainless steel, high-temperature alloys, and nonferrous alloys). CVD can be performed in atmosphere or vacuum; however, processing temperatures tend to be high (>1000°C). Therefore, the use of CVD tends to be limited to substrate materials that can withstand high temperatures such as cemented carbide tools (made from cobalt-bonded tungsten carbide) and is less suited for tool steels and high-speed steels (HSS) where they would need to be heat treated after the CVD coating process. The high processing temperatures produce hard and wear resistant coatings with strong chemical and metallurgical bonds to the substrates, as well as higher bond strength compared to those deposited by PVD at lower deposition temperatures (400°C–600°C). Such enhanced coating adhesion is desirable to protect tools from sliding friction which causes rapid wear due to the severe shearing stresses generated during heavy metal-forming applications. Therefore, one of the most common applications of CVD is the deposition of protective coatings for metal-forming tools including dies, punches, forging and stamping tools, forming and seaming rolls etc [1,2].

The CVD of TiC coatings on cemented carbide tools was introduced by Sandvik in 1969, which has led to a significant increase in tool lifetime. Figure 6.1 shows the timeline of selected key developments in CVD coatings on hardmetals and cemented carbide tools.

Other hard coatings such as TiN and Ti(C, N) as well as thermal barrier coatings (TBCs) of Al_2O_3 have also been developed and applied as wear-protective coatings. These coatings are typically deposited using metal halide precursors that react with nitrogen or carbon-based gaseous precursors in a reducing environment according to the chemical reactions below:

$$TiCl_4(g) + N_2(g) + H_2(g) \xrightarrow{1000°C} TiN(s)$$
$$+ 4\,HCl(g) + H_2(g)$$

$$TiCl_4(g) + CH_4(g) + H_2(g) \xrightarrow{1030°C} TiC(s)$$
$$+ 4HCl(g) + H_2(g)$$

$$4TiCl_4(g) + 2CH_3CN(g) + 9H_2(g) \xrightarrow{1000°C}$$
$$4Ti(C,N)(s) + 2CH_4(g) + 16HCl(g)$$

$$2AlCl_3 + 3/2CO_2 + H_2 \xrightarrow{1000°C}$$
$$Al_2O_3(s) + HCl(g)$$

Each type of coating has unique hardness, chemical, and physical characteristics that make it appropriate for different material classes. The coating type is chosen based on a number of variables, but the most important factors are the workpiece's material properties and applications. The relative resistance of CVD coatings to the

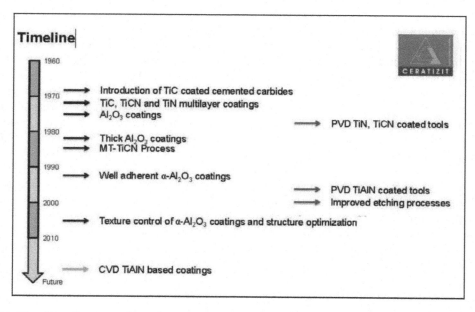

Figure 6.1 Timeline of selected key developments in CVD coatings on hard metals. (From Brookes, K.J., *Metal Powder Report*, March–April 2014, p. 22–27.)

Table 6.1 Relative resistance of CVD coatings to the main types of cemented carbide cutting tool wear

Coating	Adhesion wear	Abrasion wear	Diffusion wear	Oxidation
TiC	✓ ✓✓	✓ ✓	✓ ✓	✓
TiN	✓	✓	✓ ✓ ✓	✓ ✓ ✓
Ti(C, N)	✓ ✓	✓ ✓	✓ ✓	✓ ✓
Al$_2$O$_3$	✓ ✓	✓ ✓ ✓	✓ ✓ ✓	✓ ✓ ✓

✓: indicating little resistance; ✓✓: good resistance; ✓✓✓: excellent resistance.

main types of cemented carbide cutting tool wear is summarised in Table 6.1.

Plansee/CERATIZIT introduced the world's first CVD of TiC/TiCN/TiN multi-layer coating (called GoldMaster) in 1973 [3]. This was an exciting development. Subsequently, a range of multilayer or compositionally graded coating systems containing a combination of TiC, TiN, Ti(C, N) and Al$_2$O$_3$ have been developed in different sequences to optimise cutting speeds and extend tool life [4].

The CVD deposition of these protective coatings tend to occur at high temperatures >1000°C, which has led to decarburisation of tungsten carbide at the matrix surface and W, C, and Co diffusion into the coatings, causing the formation of a brittle layer (η-layer: Co$_3$W$_3$C) at the substrate-coating interface [5].

In order to overcome such problems, in the 1980s and 1990s various CVD coating companies focused on the development of moderate-temperature (MT) CVD of TiCN which would allow coating deposition to occur at a lower temperature (i.e., by 100°C or more) as well as the development of thick and thermally stable α-type Al$_2$O$_3$ coatings via H$_2$S doping during CVD deposition (between 800°C and 1000°C) [5,6]. These have greatly reduced the formation of the undesirable brittle η-layer, as well as the carbon level in the carbide, leading to better adhesion and resistance to thermal cracking, which is desirable for high-speed and high-efficiency cutting.

The current trends in the tooling market development include [7]:

1. Higher cutting speeds for increased productivity
2. Dry machining and/or use of minimal quantity of cooling lubricants to reduce cost and the environmental impact

3. Difficult-to-machine materials, e.g., high strength materials, for lighter components and constructions

The introduction of new regulations to minimise environment impact of the cooling lubricants and the increase in the cost of cooling lubricants have prompted the demand for more temperature-resistant coated cemented carbide. Hence, Al$_2$O$_3$, has been developed as a promising coating material for high-speed metal cutting to meet the above demand for the tooling market due to its high thermal and chemical stability, and wear resistance [7,8]. Moreover, CVD still seems to be the only deposition technique that can economically deposit high purity Al$_2$O$_3$ coatings with a well-controlled structure and composition.

The study by M'Saoubi et al. [6,9] of Selco Tools has shown that the performance of modern CVD alumina coatings can be enhanced substantially by controlling the microstructure and crystal orientation. A clearly enhanced ability of (0 0 0 1) textured layer to undergo uniform plastic deformation was confirmed. Furthermore, the textured coating has a smoother surface, thus reducing friction. Moreover, the textured coatings are more heat-resistant, and at least 10% tougher and harder than conventional coatings. Walter has developed a commercial CVD-Al$_2$O$_3$-grade coatings that can counter thermal cracking at the cutting edge and minimise tribochemical wear on the flank face. Such coatings together with new surface treatments have led to a 50%–100% improvement in cutting performance in milling, turning, and drilling applications [10].

An example of a modern commercial multi-layered CVD coating is that of Ti(C, N)/Al$_2$O$_3$/TiN on a cemented carbide tool for enhanced coating adhesion, toughness and wear properties through microstructural and process optimisations and post-treatments [11]. The coating system shows the use of: (a) CVD-deposited hard Ti(C, N) layer with fine columnar grained structure to provide abrasive wear resistance in order to reduce the flank wear; (b) CVD of α-Al$_2$O$_3$ layer which is chemically inert with low-thermal-conductivity, high-oxidation-resistant, high-strength and thermally stable as a thermal barrier and resistant to crater wear and to improve plastic deformation resistance. The controlled CVD growth of Al$_2$O$_3$ layer is used to direct the strongest crystal facets towards the cutting zone to greatly increase wear resistance, and the top surface is smooth and does not stick to the chip. (c) Finally, a bright-gold-coloured TiN-coating is deposited onto the insert flank to impart lubricity and wear resistance. Subsequently, the multilayer-coated engineering parts are heat treated to improve edge toughness in interrupted cuts and reduce smearing tendencies. Table 6.2 shows examples of a range of multilayer CVD hard coatings that have been developed commercially with tailored thicknesses and combinations of coating materials to achieve hardness, oxidation resistance, and low friction coefficients for specific cutting, milling, and turning functions.

In order to meet increasing demands of the metal cutting industry for faster and harder cutting tools with improved cutting performance, as well as milling of stainless austenitic steels and titanium alloys, where toughness is required, alternative cutting materials and very hard, wear-resistant coatings are needed. For the best cutting performance for specific applications and work piece materials, the correct combination of carbide grade and coating material, is crucial. Holzschuh has explored CVD of ternary borides in the Ti–B–N and Zr–B–N systems which are known to have metallic character, small grain size, and a combination of high hardness and toughness, making them ideal for cutting applications [13].

TiBN and ZrBN coatings were deposited onto standard cemented carbide tool inserts by LPCVD using TiCl$_4$, ZrCl$_4$, BCl$_3$ precursors and N$_2$ and H$_2$ for balance at deposition temperatures of 850°C–1050°C. Small amounts of BCl$_3$ added to the gas phase (0.06 vol.%) resulted in grain refinement. A further increase of BCl$_3$ content in the gas phase from 0.06 to 2.1 vol.%, decreased grain size further (Figure 6.2). Increasing B-content increased hardness from HV 3475–4800 (34–47 GPa) in TiBN. Such high hardness as well as their residual compressive stresses are unique for CVD coatings on carbide tools. Due to the moderate deposition temperature, coatings did not show any decarburization effects. The multilayer coating of TiB$_2$/TiN exhibited good cutting performance [13,14]. For extreme cutting conditions, nanocomposite coatings and nano-scale multilayered composite coatings for cutting tools have been developed and reviewed in Chapter 7 under CVD of nanocomposite coatings.

CVD has evolved greatly over the years with improved accuracy, efficiency and adaptation to hard-to-cut materials. This has been achieved by working on the development of novel ceramic coating materials and/or fine microstructure and

Table 6.2 Multilayer CVD hard coatings, their properties, and properties of the coated tools

Type	Composition/color	Thickness (microns)	Micro-hardness	Coefficient of friction	Oxidation temp.	Max. working temp.	Process temp
CVD	TiN/TiCN/TiC/TiN Gold	4–12	2400–3200	0.35	440°C	600°C	1050°C
CVD	TiC Silver/Gray	4–12	3200–3400	0.35	440°C	600°C	1050°C
CVD	TiC/(Mo, W)S$_2$ Silver/Gray	4–12	3200–3400	0.15	440°C	600°C	1050°C
CVD	TiN/TiCN/Al$_2$O$_3$/TiN Gold	4–9	2200–2400	0.35	N/A	880°C	1050°C
CVD	TiN/TiCN/Al$_2$O$_3$ Black	4–9	2200–2400	0.35	N/A	880°C	1050°C

Source: Richter Precision Inc. Coating Technical Data.

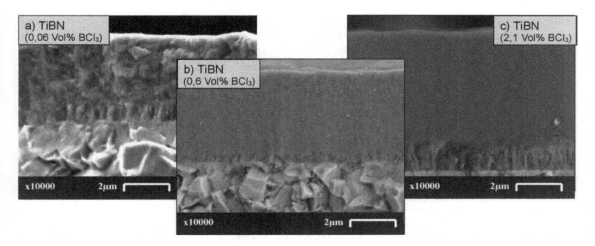

Figure 6.2 Influence of the microstructure of TiBN coatings on the BCl_3 content in the gas phase: **(a)** 0.06 vol.% BCl_3, **(b)** 0.6 vol.% BCl_3 and **(c)** 2.1 vol.% BCl_3. (From Holzschuh, H., *Thin Solid Films*, 469, 92–98, 2004.)

coating processes by leading tool companies such as Sandvik, Walter, Ceratizit, Richter Precision, Ionbond, Seco Tools, Mitsubishi, Sumitomo, Kyocera, etc. Some details of processing conditions and dopants remain proprietary.

Other protective coatings for tools include diamond and diamond like carbon films. Plasma-assisted CVD has been used to deposit diamond that enables super-hard diamond-coated carbide cutting tools, inserts, drills, and wear parts. Diamond-coated tools are suitable for machining abrasive non-metallic materials, non-ferrous metals, and abrasive non-ferrous metals. These include graphite, green ceramic, fibre-reinforced plastics, carbon fibre composites, metal matrix composites, and aluminium die castings with high silicon content. There are several companies active in the CVD production of diamond-coated tools. These include sp3 Diamond Technologies, Element Six, Sandvik, Mitsubishi, Sumitomo, and Kennametal.

Sp3 Diamond Technologies, for example, has developed thin-film CVD diamond coatings (typically 6–20 μm or up to 50 μm for more demanding applications) on cemented carbide tools using a patented hot-filament CVD deposition process that maximises the number of tools in each reactor run for economical operation and low power consumption as shown in Figure 6.3. The diamond coating is grown in a vacuum chamber using feed gases of hydrogen and methane in a 50:1 ratio where the gaseous precursors are decomposed, with the methane broken down into carbon and

Figure 6.3 Industrial CVD coater for super-hard diamond-coated carbide cutting tools, inserts, drills, and wear parts. (From *What makes a CVD diamond coated cutting tool "best of breed"?* http://www. sp3diamondtech.com/products.php.)

hydrogen by the hot filament (2200°C). The carbon atoms nucleate and grow tiny crystals of diamond at a rate of 0.5–1.0 μm/hour where in time these tiny diamond crystallites will grow into a continuous diamond film on the tool substrate heated to around 850°C. HSS would not survive such high temperatures. A typical reactor run to grow a 40-micron film on cutting inserts can take about 2 days; therefore, the reactor is packed with as many tools as possible (200–300 inserts at a time, or 65–100 round tools) to be coated in each run to make production more economical. The inserts or tools need to have a compatible carbide grade (with Co content <10%, ideally <5%) to ensure the growth of diamond coating with strong bonding to the substrate and long tool life. In addition, the carbide surface also need to be pre-treated and roughened so that the diamond can grow into the surface and attach itself through mechanical interlocking of the diamond coating and carbide substrate. Diamond-coated tools have extremely low friction coefficients and consistently sharp edges, allowing the use of cutting speeds two to three times higher than carbide with significant longer wear life (10–50 times the life of carbide) [15].

Diamond-like carbon (DLC) consists of a mixture of sp^2 (graphite) and sp^3 (diamond) phases. It is amorphous, chemically inert, hard, wear resistant and has a low coefficient of friction. Such properties of DLC make it suitable for tool applications such as the machining of cast and wrought aluminium under dry or low lubricating conditions. Commercial DLC films are deposited by vacuum-plasma-assisted CVD (PACVD) and PVD methods (e.g., sputtering). There are different types of DLC, with physical properties including hardness and coefficient of friction dependant on processing methods, sp^3 content, dopant, and doping level, as shown in Table 6.3. The sp^3 content in DLC can be tailored accordingly for specific applications. Metal and hydrogen-containing DLC (Me-DLC or a-C:H:Me) exhibits hardness within the range 800–2200 HV with 35% sp^3, metal-free DLC (a-C:H, a-C) typically 1500–4000 HV and up to 75% sp^3, whereas tetrahedral amorphous carbon (ta-C) can be 4000–9000 HV with 80–85% sp^3 [16,17].

6.2 THERMAL BARRIER COATINGS (TBCS)

Ceramic thermal barrier coatings (typically 150–250 μm) act as thermal insulating layers to protect Ni-based superalloy-based gas turbine components, extending engine life and enabling them to operate at higher temperatures (>1200°C), thereby increasing efficiency and reducing fuel consumption. TBCs have been widely adopted by the gas turbine industry for aviation, where these coatings are predominantly deposited by plasma spraying

Table 6.3 DLC coatings and their properties

	Carbon-based coatings				
	Metal-doped DLC Me-DLC (WC-C.H)	Amorphous hydrogenated DLC (a-C:H)	Silicon-doped DLC (a-C:H-Si)	Sputtered DLC (a-C)	Hydrogen-free DLC (ta-C)
Method	PVD/PACVD	PACVD	PACVD	PVD	PVD
Hardness (HV 0.05)	800–2200	1500–3500	1500–2500	2000–4000	3000–7000
Coefficient of friction	0.1–0.2	0.05–0.15	0.05–0.1	0.05–0.1	0.02–0.1
Internal stress (Gpa/μm)	0.1–1.5	1–3	1–3	2–6	1–3
Thickness (μm)	1–10	1–10	1–10	1–3	1–3
Industrial use	+++	+++	+++	+++	+++
Mass production	+++	+++	+++	+++	+++

Source: Plasma technology for you, http://www.hauzertechnocoating.com/en/plasma-coating-explained/dlc-coating/.

and electron beam physical vapour deposition (EBPVD), for stationary and rotating gas turbine parts respectively. Plasma sprayed coatings tend to produce splat-like porous coatings with low thermal conductivity (0.5–1.0 W m^{-1} K^{-1}), whereas EBPVD deposits dense stress-strain tolerant columnar coatings with higher thermal conductivity (e.g., 1.5–2.0 W m^{-1} K^{-1}) [18]. However, both techniques tend to be line-of sight. Hence, they are not well suited to uniformly coating complex turbine components, such as turbine nozzle guide vanes, which can lead to hot spots in the coated component.

The non-line-of-sight capability of CVD, plus the ability to deposit stress-strain tolerant columnar microstructures with high throwing power makes it a promising alternative to TBC deposition with good conformal coverage. TBCs have been deposited via conventional thermally assisted CVD with use of metal organic precursor to lower the deposition temperatures and enhance deposition rates. The rate of deposition can also be enhanced using auxiliary energy sources such as microwave plasma or laser heating to assist CVD deposition and growth of thick coatings. Different variants of CVD of TBCs have been explored and highlighted in the following sections.

6.2.1 Conventional thermally assisted CVD

For practical TBC applications, a high deposition rate, similar or higher than that of EBPVD (~250 µm/hr) is required to produce thick coatings economically. However, most conventional CVD methods tend to have low deposition rates (<20 µm/hr) [19,20] and use halide precursors which require high deposition temperatures (>900°C) and have corrosive HCl by-products which cause high-temperature degradation of the Ni-superalloy substrates.

Alternative chemical precursors with lower processing temperatures, such as metal organic-based precursors (e.g., Zr(thd)$_4$ and Y(thd)$_3$) were explored by Wahl et al. [21] and delivered by Ar carrier gas into a hot-wall reactor which reacted with oxygen at a lower deposition temperature

(as low as circa. 627°C). Thick films of tetragonal and cubic YSZ with a columnar structure were deposited at a reasonably high deposition rate of ~50 µm/hr. Higher deposition rates of 100 µm/hr can be achieved by increasing the evaporation temperature of precursors in a gas phase, however, this leads to coatings with poor adhesion to the substrates and fewer columnar character.

6.2.2 Plasma-assisted CVD

Plasma is an auxiliary energy source that has been used to enhance the deposition rate of CVD of thick ZrO$_2$ and YSZ coatings [22,23]. Thick, columnar YSZ coatings (circa 65–200 µm) (see Figure 6.4) with the desired tetragonal t′ phase have been deposited using microwave plasma-enhanced CVD using metal halide precursors with deposition rates upto 250 µm/hr [23].

6.2.3 Laser-assisted CVD

Goto's team has developed a high-power laser-assisted CVD process (see Figure 6.5) to deposit YSZ coatings with the desired columnar structure at a high deposition rate (~660 µm/h). Chemical vapour precursors of Zr(dpm)$_4$ and Y(dpm)$_3$,

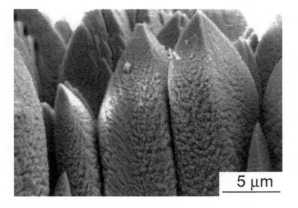

5 µm

Figure 6.4 Microwave plasma-enhanced CVD deposited YSZ with tetragonal t′ phase and strong (200) orientation. (From Préauchat, B., and Drawin, S., *Surf. Coat. Technol.*, 142, 835–842, 2001.)

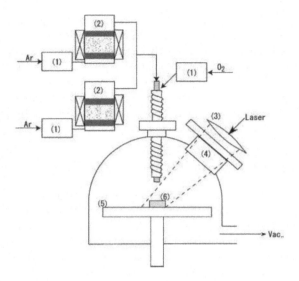

Figure 6.5 Schematic diagram of laser CVD apparatus: (1) mass flow controller, (2) precursor evaporator, (3) optical lens, (4) quarts window, (5) substrate folder, (6) substrate. (From Kimura, T., and Goto, T., *Mater. Trans.*, 44, 421–424, 2003.)

were generated and held at 220°C and 170°C, respectively. The precursors were delivered by Ar gas to the reactor via a double tube nozzle perpendicular to the heated substrate (temperature controlled from room temperature to 750°C) and O_2 gas was separately introduced into the chamber 15 mm above the substrate [24,25]. A laser beam of Nd:YAG (wavelength 1063 nm, power 260 W, laser spot 15 mm in diameter) was emitted through a quartz window onto the substrate for uniform

coating of substrates (13 mm × 13 mm × 2 mm) at 0.93 kPa.

The deposition rate increased significantly from 1–3 μm/h to 300 μm/h when the laser power increased from ≤60 W to ≥100 W. A high deposition rate of 660 μm/h can be achieved with a further increase of the precursor concentration in the gas phase. Such a high deposition rate is caused by plasma formation during the deposition which allows the dissociation of precursors to occur and produced reactive clusters [26]. The substrate temperature also has a strong effect on the crystalline structure and surface morphology of YSZ coatings. When the stage temperature increased from 100°C to 750°C K, a well-grown columnar coating was observed, similar to those columnar coatings deposited by EBPVD.

Garcia and Goto [24] have compared laser-assisted CVD with thermal and plasma-assisted CVD processes as shown in Table 6.4, with particular focus on the crystal phases formed. The Monoclinic phase is undesirable for TBC applications.

Since laser CVD seems to be a promising method for the deposition of TBCs, Goto's team has used this process to coat gas turbine blades with YSZ films as shown in Figure 6.6 [27]. This could potentially offer lower costs of metal organic chemical precursors, process scale-up and thermal shock resistance, and improved thermal cycling properties of the TBC blades. In addition to coating process development, efforts have been made to develop functionally graded TBCs via CVD to enhance TBC properties by minimising thermal expansion mismatch between the ceramic YSZ top coat and the bond coat [28].

Table 6.4 CVD deposition conditions and several properties of YSZ intended for TBCs

Process	Precursors	Deposition temperature (K)	Phase	Y_2O_3 content (mol%)	Deposition rate (μm h^{-1})	Coating thickness (μm)
Thermal CVD	$Zr(thd)_4$, $Y(thd)_3$	900–1300	Cubic	1.4–19	50	
Thermal CVD	$Zr(thd)_4$, $Y(thd)_3$	723–1123	Cubic + monoclinic	5–30	0.12–42	1
Thermal CVD	$Zr(dpm)_4$, $Y(dpm)_3$	873–1173	Tetragonal	2.7	100	25
Plasma CVD	$ZrCl_4$, $Y(thd)_3$	973–1173	t′	3–4.4	100–252	65–200
Plasma CVD	$Zr(dpm)_4$, $Y(dpm)_3$				>300	100

Source: Vargas Garcia, J.R., and Goto, T., *Sci. Technol. Adv. Mater.*, 4, 397–402, 2003.
Key: t′, non-transformable tetragonal phase; thd, tetramethyl-heptanedionate; dpm, dipivaloylmethanato.

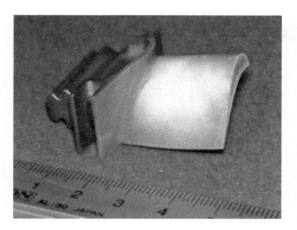

Figure 6.6 Laser CVD of YSZ coated Ni-superalloy blade. (From Goto, T., *J. Wuhan Univ. Technol.-Mater. Sci. Ed.*, 31, 1–5, 2016.)

6.2.4 Electrostatic spray-assisted vapour deposition (ESAVD)

Choy's team has developed a cost-effective non-vacuum ESAVD method [29,30] and applied this non-line-of-sight method for the deposition of thick ceramic thermal barrier coatings on 3-D components such as turbine aerofoils as shown in Figure 6.7 [14]. The details of the process have been described in Chapter 1. A precursor mixture of alkoxide of zirconium and yttrium was used

to deposit adherent 8wt% Y_2O_3-ZrO_2 thick coating (150–400 µm) with the desired stress-strain tolerance and a columnar-like microstructure (see Figure 6.8 [15]) at a moderate temperature of ~550°C and a reasonably high deposition rate (>30 µm/hr). The cross-section SEM of the polished sample of ESAVD deposited ceramic thermal barrier coating in Figure 6.9 reveals the unique columnar-like structure with fine microcracks that can help to lower the thermal conductivity of the coating. Therefore, the nonvacuum ESAVD method has the potential to greatly reduce the cost of TBC production as well as to extend the range of turbine components and objects with a complex shape to be coated. Furthermore, the ease of the formulation of chemical precursors and the ability to control the coating structure and composition at molecular level enables the ESAVD process to be used as a flexible and cost-effective tool for the development of new ceramic TBC without the need to fabricate dedicated powders/targets, and without the difficulties of controlling coating stoichiometry associated with the EBPVD and plasma-spraying processes. Initial thermal cycle test results indicate that ESAVD TBCs compare favourably with standard air-plasma-spray (APS) and EBPVD coatings.

Figure 6.8 Cross-section SEM of the ESAVD deposited thick columnar-like 8wt% Y_2O_3-ZrO_2 coating onto Ni-superalloy. (From Vyas, J.D., and Choy, K.L., *Mater. Sci. Eng. A*, 277, 206–212, 2000.)

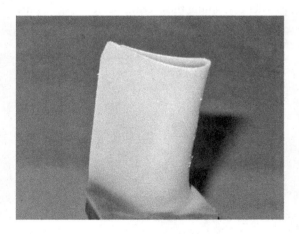

Figure 6.7 An example of ESAVD coated turbine aerofoil. (From Choy, K.L., *Prog. Mater. Sci.*, 48, 57–170, 2003.)

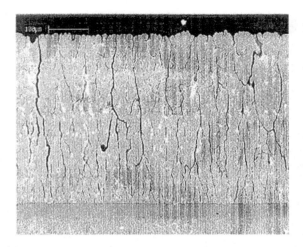

Figure 6.9 Cross-section SEM of the polished sample of ESAVD deposited 8wt% Y_2O_3-ZrO_2 coating.

6.3 DIFFUSION COATINGS

CVD has also been used to deposit metal coatings, which subsequently form alloy coatings via the inter-diffusion of one or more coating elements and component base material at elevated temperatures to form diffusion coatings. The elements may be in the solid, liquid or vapour state.

Variants of CVD such as pack-cementation and vapour-phase-diffusion processes are high-temperature coating processes used commercially for producing oxidation, corrosion and/or wear resistance coatings via aluminising (900°C–1100°C), chromising (900°C–1100°C), boriding (800°C–1000°C), and siliconizing (900°C–940°C) for a wide range of aero and industrial gas turbine blades and vanes, engine parts in chemical and petroleum processes, pump shafts, valves, cylinder, liner, nuts, bolts, etc [31, 32]. Examples of coated parts are shown in Figure 6.10.

Aluminising is a commonly used industrial processes for producing diffusion bond coats (e.g., aluminides) by aluminising the outer surface of aero and industrial turbine blades. This helps not only to improve adhesion of the ceramic top coat to the Ni-superalloy substrate by minimising their thermal expansion mismatch but also to provide the desired resistance to high-temperature oxidation and corrosion in extreme environments.

During the pack-cementation process, an engineering component (e.g., turbine blade) is packed in powder containing the coating element (e.g., a pure metal, alloy or compound), an activator (or mixture of activators) and an inert material that acts as a diluent to distribute the pack constituents, in order

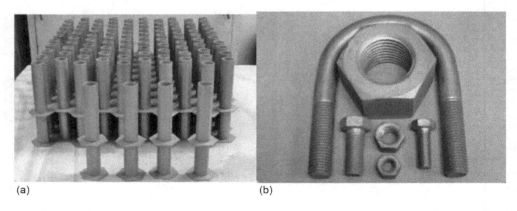

(a)   (b)

Figure 6.10 Diffusion coated parts via aluminising and pack-cementation of **(a)** ferrules **(b)** fasteners. (From Diffusion Alloys, *Aluminising*, http://www.diffusion-alloys.com/content/aluminising.)

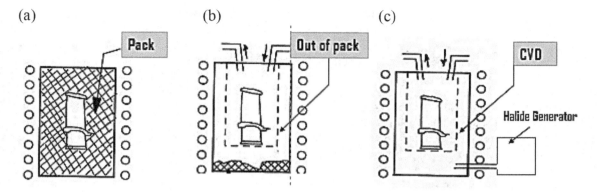

Figure 6.11 Schematic diagrams illustrate the pack-cementation and vapour-phase-diffusion processes: **(a)** closed reactor; **(b)** semi-open reactor; **(c)** open reactor.

to prevent sintering and support the component(s) as shown in Figure 6.11a. This is a hot-wall process and it is typically performed in a closed reactor and the gaseous reactants are generated within the packed bed, unlike the usual vapour generation routes in CVD (e.g., Figure 6.11c). The theory and use of pack-cementation coatings for superalloys has been reviewed [33,34]. This process has been first exploited by the Chromalloy gas turbine corporation to improve the performance and lifetime of gas turbine parts in the 1950s [35]. The process has been developed further by introducing Pt and Pd into the coatings system via electroplating prior to aluminising to increase the life of the component further [36–38]. The drawbacks of the pack-cementation method include: disposal of a large amount of wasted starting powder; ease of incorporating contaminants; long coating cycle times; and laborious cleaning of the coated parts after pack-cementation, prior to subsequent heat treatment for the formation of the diffusion coating [29]. This has prompted the development of pack-cementation process into an "out-of-pack" process to address some of the drawbacks. In the "out-of-pack" process, the coating element source and halides are contained in a

tray beneath and out of contact with the component. The coating vapour is piped through internal passages using a carrier gas into the reactor. This is also a hot-wall process and is carried out in a semi-open reactor as shown in Figure 6.11b. Such equipment configurations create the possibility of coating internal cooling passages. In the vapour diffusion process, an external generator is used to generate the vapour precursor and this is piped continuously into the substrate holder containing the component. This hot-wall process is performed in an open reactor as shown in Figure 6.11c.

All these are variants of CVD based on thermochemical processes that work on the basis that the activator is decomposed and releases a volatile halogen, which reacts with the metal to form a volatile metal halide. The metal halide then decomposes in the vicinity of the heated surface of engineering part, from where it diffuses into the base metal to form an oxidation and/or wear-resistant coating. The majority of commercial processes use halide compounds as the activator, selected on the basis of their high vapour pressure at temperature and low cost. The processing steps, assuming isothermal conditions, can be summarised in the equation below:

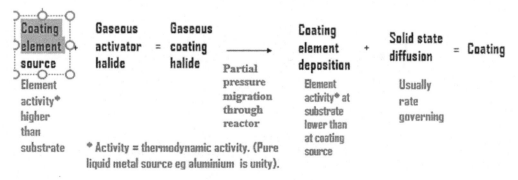

The vapour-phase-diffusion process (as described in Figure 6.11c), or so-called chemical vapour aluminising (CVA), is performed under low pressure and is widely used commercially to deposit aluminium which subsequently diffuses into the substrate to form intermetallic compounds such as nickel aluminides as shown in Figure 6.11b. This non-line-of-sight method produces high-purity coatings with well-controlled doping levels, and well-controlled thickness, uniformity and conformity compared to pack-cementation-based methods. Furthermore, non-line-of-sight of CVA also enables the deposition of coatings on the inner surfaces of turbine aerofoils. CVA has been used commercially for more than 20 years to protect blades and vanes in the hot section of gas turbines against oxidation and hot corrosion, as well as producing bond coats for industrial and aerospace gas turbines. Aluminium coatings can be also doped with such performance-enhancing elements such as Cr, Si, Hf, Zr, etc. The properties of these types of CVD coatings are highly dependent on the chemistry of the base materials, especially for high-activity coatings.

Figure 6.12 shows an example of an industrial CVA coating system to produce diffusion coatings for protection against hot corrosion and high-temperature oxidation [39]. Advanced co-deposition CVA processes with the addition of metallic elements to the aluminide coatings are being developed for producing modified coating properties and structures of multiple metal diffusion coatings with elements such as Al, Cr, Si, Co, and Y [40].

The processing temperature has a strong influence on the microstructure of diffusion coatings. At high activity and low temperature (750°C–950°C), the pack process uses pure aluminium powder with halide activators and alumina diluent. The coating grows by the inward diffusion of aluminium and incorporates base material elements, e.g., chromium, into the nickel aluminide layer [41]. The aluminium content is usually ~40 wt% and borders on the Ni_2Al_3 phase field. If required, heat treatment may be applied to improve the coating ductility by lowering the aluminium content into the β-NiAl phase field. The chemistry of the base substrate material would strongly influence the properties of the resultant diffusion coating properties.

For a low-activity, high-temperature (1025°C–1150°C) pack, out-of-pack or vapour-phase-diffusion processes use a lower activity aluminium alloy source with halide activator and alumina diluent (when required). The coating grows by outward diffusion of nickel and therefore contains very few base metal elements. The aluminium content is usually 20–30 wt% and the coating consists entirely of β-NiAl phase.

The high-active and low-active regimes of nickel aluminisation can be accessed through variation of the HCl/H_2 flow ratio, as confirmed by Wierzba et al. [42]. Wierzba used modelling to simulate diffusion during aluminisation of nickel and its super-alloys, MAR-M200+Hf and CMSX-4, and compared the predictions with the results of experiments performed by the authors.

6.4 THICK SILICON COATINGS

Thick, dense, adherent, smooth (surface roughness, R_a ~ 0.2–0.4 μm), and ultra-high purity (UHP, 99.999%) amorphous silicon coatings with

Figure 6.12 An example of a commercial CVA coating system for vacuum aluminising for the turbine industry. (From Ionbond UK Ltd, *CVD technology*, http://www.engineeringarena.co.uk/companies/ionbond-uk-ltd/products/bernex-aluvap-cva-coating-system.)

low residual stress are deposited using conformal plasma enhanced CVD process (PECVD) at a low-deposition temperature ($<150°C$) for applications including:

- Coating of the silicon wafer handling and processing chamber (often made of 6061 Al alloy) in order to greatly reduce the levels of contamination of Fe, Cu, and Cr in silicon wafers
- High-precision optical mirrors for space and terrestrial applications

Fully dense, strongly adherent Si coatings provide excellent porosity-free surfaces with no post-deposition surface finish. Hence, they do not trap or generate particulate matter and are easy-to-clean. Such UHP Si coatings are compatible with a wide variety of substrates, including ceramics, metals, alloys, graphite, and polymers in thin (~10 nm) and thick ($<100\,\mu m$). Table 6.5 summarises the properties

Table 6.5 Properties of UHP silicon coating deposited by PECVD

Substrates	Metals, ceramics (AlN, Al$_2$O$_3$, Quartz, YSZ, Graphite, etc.), and polymers
Structure	Amorphous, contains hydrogen
Deposition Temperature	<150°C
Use Temperature	−50°C to 600°C
Electrical Resistivity	$5 * 10^8$ Ω-cm; (lower values are also possible)
Hardness	600 DPHN
Wear/abrasion resistance	Good
Corrosion Resistance	Excellent resistance in aqueous, acidic and alkaline environments
Thickness	Uniform and conformal coating, ranging in thickness from a few nm to 150 μm (6 mils)
Substrate size and geometry	Any shape including complex ones, up to a size of 36″ dia. can be coated

Source: Gunda, N. et al., *Strongly adherent thick coatings of ultra high purity silicon*, http://www.surmet.com/docs/(2)Article_UHPSi.pdf.

of PECVD deposited-dense, hard and conformal UHP Si coatings, which have low residual stress.

The non-line-of-sight CVD and its variant processes enable the deposition of think amorphous silicon coatings onto complex shapes.

6.5 THICK METAL COATINGS

Tungsten has the highest melting temperature of all refractory metals (3422°C). It tends to be difficult to form due to its high hardness, and the most common manufacturing technique is powder metallurgy. However, W can be easily deposited with CVD using the common gaseous precursor tungsten hexafluoride, WF$_6$. CVD-deposited W coatings have high purity, density, and thermal conductivity with a fine, well controlled structure and thickness (μm to mm) required to meet the challenging coating requirements for fusion applications such as abrasion resistance to particle bombardment, high thermal load, high operating temperatures and high thermal conductivity. Other applications of pure W include electronic switches, filaments of incandescent light bulb, and elements to produce X-rays. For X-ray anodes manufacturing, tungsten is mixed with rhenium to improve alloy's ductility, high temperature resistance, corrosion resistance and tensile strength. Such X-ray anodes are used in medical imaging devices and they are integrated into angiography, mammography, computed tomography, and cardiology equipment, as well as for other non-medical applications such as non-destructive inspection and security checks. In addition, thick metal coatings such as W and Re have been also deposited by CVD onto graphite or composites for fusion applications. Figure 6.13a shows a CVD reactor for the deposition of W-Re coatings and Figure 6.13b shows CVD deposited W layers with Re interlayers to reduce the thermal stress and for crack resistance [44]. Tungsten-rhenium deposits can range from 1 μm to 1 mm. In order to deposit a tungsten-rhenium alloy, two precursors are used: tungsten hexafluoride (WF$_6$) and rhenium hexafluoride (ReF$_6$) with dihydrogen being the carrier gas. The chemical reactions which occur on the substrate surface are as follows:

$$WF_6 + 3H_2 \rightarrow W + 6HF$$

$$ReF_6 + 3H_2 \rightarrow Re + 6HF$$

(a)

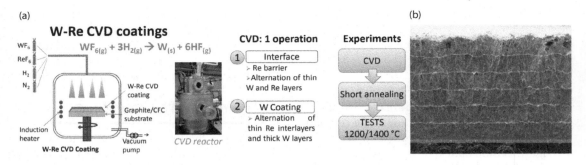

(b)

Figure 6.13 **(a)** CVD reactor for the deposition of W-Re coatings; **(b)** cross-section SEM of W layers with Re interlayers. (From Acerde, *Chemical vapor deposition (CVD)*, http://www.acerde.com/process/technology.)

6.6 POLYMERIC COATINGS

In addition to the deposition of metal, ceramic, and semiconductor materials, CVD is also an established method used in industry for the deposition of polymers (see Chapter 1, Section 1.4). One industrial example is CVD of Parylene, an inert linear chained polymer with a long, high molecular weight (~500,000) chain of carbon, hydrogen and sometimes chlorine, fluorine and other compounds attached to an aromatic carbon ring structure. Parylene coatings were first developed in 1947 by Michael Szwarc. He discovered that the pyrolysis of xylenes under vacuum formed thin polymer films that exhibit interesting properties. Parylenes cannot be produced by conventional methods (e.g., moulding or extrusion) due to their high molecular weight, high melting temperatures and crystallinity (see Table 6.6). In addition, Parylene has low solubility in organic and other media, except at temperatures >175°C, so it cannot be formed by casting. William Gorham at Union Carbide devised an efficient method of depositing these films based on CVD in the late 1960s, which opened the commercial applications of CVD of Parylene coatings.CVD is an established technique used in industry to deposit Parylene via a three-stage vapor deposition process which can be summarised as follows [45–47]:

Stage 1: *Chemical vapour precursor generation:*
The solid Parylene dimer starting material in powder form is vaporized into a dimer gas at ~150°C inside a vaporiser.
Stage 2: *Pyrolysis of precursor:* The dimer gas subsequently flows into a heater chamber where it

undergoes pyrolysis at 650°C–700°C to cleave the dimer into a monomer vapor.
Stage 3: *Coating deposition:* The monomer vapor is drawn into the coating chamber (maintained at room-temperature) via a vacuum pump, where the monomers polymerise and are deposited as a clear poly-para-xylyene polymer superthin/thin-film (from angstroms, to microns).

The deposition rate is typically 1–2 μm per hour. Conformal coating deposition can be achieved irrespective of the complexity of the substrate/component geometry.

Types: There are various types of Parylene coatings which have unique properties. Table 6.6 gives a summary of the common types of Parylene used in industry.

Tables 6.7 through 6.9 compare the physical and mechanical, thermal, and electrical properties of different types of Parylene with other conformal polymeric coating materials, respectively. In general, Parylenes have higher elasticity and low coefficient of friction than other polymers.

Parylenes are thermally stable up to ~220°C in oxygen-free, vacuum, or inert atmospheres. The density and crystallinity of Parylene increases with annealing which can increase hardness, and improve the abrasion resistance of Parylene. In general, dielectric breakdown strengths of Parylenes are better than other conformal polymeric coatings (e.g., epoxides, slicones, and urethanes).

An alternate route to Parylene AF-4 has been developed as shown in Table 6.6 above. The advantage to this process is the low cost of synthesis for

Table 6.6 Commercial types of parylene coatings

Type of parylene	Characteristics and properties
Parylene N • A completely linear, highly crystalline material where the polymer is manufactured from di-p-xylylene (paracyclophane), a dimer synthesized from p-xylylene.	• Higher dielectric strength than Parylene C and the dielectric constant does not vary with changes in frequency. • Able to penetrate crevices more effectively than type C because of the higher level of molecular activity that occurs during deposition. • Normally used high frequency applications because of its low dissipation and dielectric constant values.
Parylene C • Most common Parylene and it is chemically different from Type N, having a chlorine atom on the benzene ring.	• Can be deposited at a faster rate than Parylene N, and has a lower throw capability. • An associated reduction in crevice penetration activity. • A very low permeability to moisture and corrosive gases. • A useful combination of electrical and Physical properties. • The dielectric breakdown strength of Parylene C is better than N, especially for films below 5 μm (see Table 6.9).
Parylene D • This polymer is produced from the same raw material as Parylene N, and the two aromatic hydrogens are being substituted by the chlorine atoms.	• Similar in properties to Parylene C with the added ability to withstand slightly higher use temperatures. • Physical strength and electrical properties are maintained at higher temperatures. • Halogen atoms increase the density of the coating, making the coating a better diffusion barrier. • Due to its high molecular weight, Parylene D suffers from poor conformality and uniformity across-the-chamber at room temperature (135°C threshold temperature). It is thus less used than Parylene C.

Source: http://www.paryleneinc.com; http://www.paryleneengineering.com/basics_of_parylene.htm; https://scscoatings.com/what-is-parylene/parylene-properties/.

the liquid precursors and catalysts can be used to lower the processing temperature resulting in higher-quality polymer thin film with less char in the pyrolysis zone [52]. Besides pure Parylene, copolymers and nanocomposites Parylene have also been deposited at near-room temperature via thermal chemical vapor deposition of poly(chloro-*p*-xylylene)/SiO$_2$ nanocomposites [53].

6.6.1 Applications

Parylene films deposited by CVD are chemically inert, hydrophobic, optically transparent, pin-hole free, abrasion resistant, thermally stable up to 220°C,

mechanically stable from −200°C to +150°C, oxidation resistant up to 1350°C (Parylene AF-4, HT, SF), have extremely high dielectric strength (7 kV/mil), high elasticity, biocompatible, non-toxic, relatively stress free, and salt resistant, as well as resistant to fungus and bacteria. They also have a low coefficient of friction (AF-4, HT, SF) and can act as dry-film lubricants removing the requirement for liquid release agents [45–47].

Parylene films deposited by CVD have numerous commercial applications as shown in Figure 6.14. These include environmentally protective coatings for metals, glass, ceramics, electronics, elastomers, and other polymers against

Table 6.7 Physical and mechanical properties of parylene as compared with other conformal coating materials

Properties	Method	Parylene N	Parylene C	Parylene D
Secant (Young's) Modulus (psi)	1	350,000	400,000	380,000
Tensile Strength (psi)	2	6,000–11,000	10,000	11,000
Yield Strength (psi)	2	6,100	8,000	9,000
Elongation to Break (%)	2	20–250	200	10
Yield Elongation (%)	2	2.5	2.9	3.0
Density (g/cm³)	3	1.10–1.12	1.289	1.418
Index of Refraction ($n_D{}^{23}$)	4	1.661	1.639	1.669
Water Absorption (% after 24 hours)	5	Less than 0.1		
Rockwell Hardness	6	R85	R80	R80
Coefficient of Friction Static	7			
Dynamic		0.25	0.29	0.33
		0.25	0.29	0.31

Source: http://www.scscookson.com/parylene/properties.cfm.

Table 6.8 Thermal properties of parylene

Properties	Method	Parylene N	Parylene C	Parylene D
Melting Point (°C)	1	420	290	380
T5 Point (°C) (Modulus = (10^5 psi))	1	160	125	125
T4 Point (°C) (Modulus = (10^4 psi))	1	>300	240	240
Linear Coef. of Expan. at 25°C ($*10^5$, (°C)$^{-1}$)	–	6.9	3.5	3.8
Thermal Conductivity at 25°C ($*10^{-4}$ cal/ (cm·s·°C))	2	3.0	2.0	–
Specific Heat at 20°C (cal/g·°C)	–	0.20	0.17	–

Source: http://www.scscookson.com/parylene/properties.cfm.

Table 6.9 Electrical properties of parylene

Properties	Parylene N	Parylene C	Parylene D
Dielectric Strength, dc volts/mil short time, 1 mil times	7,000	5,600	5,500
Corrected to 1/8 in	630	500	490
Volume Resistivity. Ohm·cm, 23°C, 50% RH	1.4×10^{17}	8.8×10^{16}	1.2×10^{17}
Surface Resistivity. Ohms, 23°C, 50% RH	10^{15}	10^{14}	10^{16}
Dielectric Constant			
60 Hz	2.65	3.15	2.84
1°KHz	2.65	3.10	2.82
1°MHz	2.65	2.95	2.80
Dissipation Factor			
60 Hz	0.0002	0.020	0.004
1°KHz	0.0002	0.019	0.003
1°MHz	0.0006	0.013	0.002

Source: http://www.scscookson.com/parylene/properties.cfm.

(a)   **(b)**

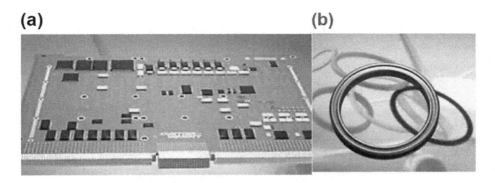

Figure 6.14 **(a** and **b)** Applications of CVD deposited Parylene films for various engineering and biomedical Components. (From http://vsiparylene.com/parylene; http://www.parylene.co.uk/about-parylene.)

chemicals (e.g., organic solvents, inorganic reagents, acids), moisture, gases, temperatures, corrosive bodily fluids, and fungus for a variety of key industries include components for aerospace, defence, automotive, marine, electronic, and medical (e.g., catheters) applications, especially those operating in the harshest of environments. Figure 6.14a shows a typical application of Parylene on critical electronics and devices [47]. Due to the hydrophobic properties of Parylenes, they are also deposited on elastomers (e.g., O-rings, gaskets, and seals) as shown in Figure 6.14b [47] to remove the tackiness of silicone and rubber components and reduce friction in plastic and metal sub-assemblies. Parylene coatings have also been used to protect and lubricate medical devices, and diagnostics for over 25 years. These include implants, pacemakers, leads, PCBs, surgical tools, surgical entry devices, trocars, speculums, needles and probes, catheters, and medical electronics and diagnostics. Parylene is FDA approved and safe to use within the body. Parylene coatings have been tested to the relevant ISO10993 standard for long-term implantable use.

6.7 FIBRE COATINGS AND CERAMIC MONOFILAMENT FIBRE PRODUCTION

CVD has been established as the preferred commercial method for the manufacturing of large-diameter, high-strength, and high-modulus ceramic monofilament fibres (e.g., B and SiC), which consist of CVD deposited thick (85–130 µm) ceramic coatings of B or SiC onto C or W fibre cores (diameter of circa. 10–33 µm).

6.7.1 Boron fibres

Ceramic monofilament fibres are produced in single-filament CVD reactors as shown in Figure 6.15a, which also shows the production facility in Figure 6.15c [54]. Prior to chemical vapour deposition of B, the W-core filament is pre-heated to 1200°C in hydrogen to remove any oxide and impurities from the surface. For the C-core fibre, it is pre-coated with 1–2 µm pyrolytic graphite. Boron monofilament fibres are produced using CVD of BCl_3 in H_2 at about 1300°C according to the chemical reaction: $2BCl_3(g) + 3H_2(g) \rightarrow 2B(s) + 6HCl(g)$, in a cold-wall CVD reactor where the B is deposited onto W or C filaments continuously drawn through the long reactor (~4 m) via mercury seals at both ends at a high speed (>13 m/min). Mercury is used as a gas seal and contact electrode to heat up the filament resistively. Deposition tends to be performed at atmospheric pressure avoiding the use of expensive vacuum systems, thus minimising the cost of production. Moreover, high processing temperatures and pressure with low precursor flow rate favour the mass transport rate limiting step, leading to a high deposition rate. The key process parameters during fibre production are temperature of the core filament of W or C, reactant gas input ratio with the reducing gas hydrogen, precursor flow rates, and spooling speed (i.e., fibre residence time). The filament temperature can be controlled by adjusting the voltage applied to the mercury electrode as well as varying composition of the precursors. Significant progress has been made inceramic monofilament fibre production process and many processing details remain proprietary [29,55].

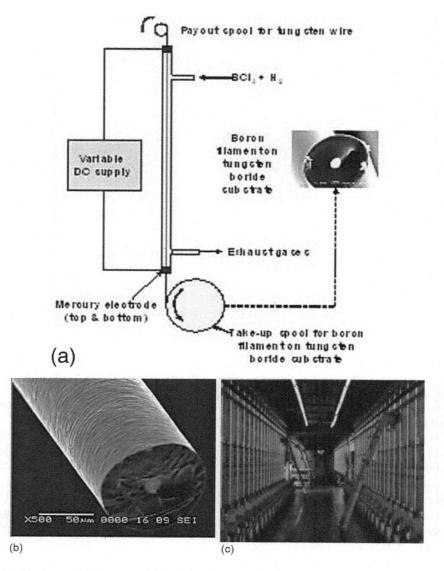

Figure 6.15 **(a)** A schematic diagram of the CVD of B monofilament fibre; **(b)** SEM micrograph of CVD fabricated B fibre; **(c)** CVD production facility for B monofilament fibres consisting of multiple reactors. (From Specialty Materials, Inc., *Boron fiber*, http://www.specmaterials.com/boronfiber.htm.)

The as-produced fibres may be subjected to post processing by chemical and thermal treatments, and/or further coatings to improve their oxidation resistance/chemical compatibility with matrix materials (when used in a composite). Figure 6.15b shows the commercial boron monofilament which consists of amorphous boron deposited onto a tungsten wire substrate with a fully boride coated tungsten core, and the resulted diameters of the monofilament fibres of 102 μm and 142 μm by Specialty Materials. The properties of the boron fibre are described in Table 6.10 [54]. The fracture behaviour of boron fibres has been studied [56].

CVD boron filaments were first introduced in the 1960s and explored for reinforced metal matrix composites (MMCs) such as Ti-MMCs and Al-MMCs. However, the interfacial reactions between B filaments and the metal matrices leads to the formation of brittle reaction products and degraded the strength of fibres during composition manufacturing which have hindered the applications of B fibres in MMCs. Nowadays the

Table 6.10 Properties of B monofilament fibre

Property	Units	4-mil boron (typical)	5.6-mil boron (typical)	3-mil boron (typical)	8-mil boron (typical)	11-mil boron (typical)
Diameter	μm	102	142	76	203	279
	in. × 10⁻³	4	5.6	3	8	11
Cross-Section				Round		
Aspect Ration				Continuous		
Density	g/cc	2.61	2.48	2.82	2.44	2.41
	lbs/in³	0.094	0.089	0.102	0.088	0.087
Thermal	PPM/°C	4.5	4.5	4.5	4.5	4.5
Expansion	PPM/°F	2.5	2.5	2.5	2.5	2.5
Tensile	MPa	3,600	4,000	–	–	–
Strength	ksi	520	580	–	–	–
Tensile	GPa	400	400	–	–	–
Modulus	msi	58	58	–	–	–
Compression	MPa	>6000	–	–	–	–
Strength (est.ª)	ksi	>900	–	–	–	–
Hardness	Knoop	3,200	3,200	3,200	3,200	3,200

Source: Specialty Materials, Inc., *Boron fiber*, http://www.specmaterials.com/boronfiber.htm.
ª Filament compression strength calculated from composite lamina property and back-out factor.

B monofilament fibres are predominantly used in resin-matrix composites, especially in the construction of high tensile strength and lightweight tapes for aerospace structures [54]. There are also limited applications for high-end sporting products (e.g., golf clubs, tennis racket, and fishing rods [54,57]), while high costs hinder widespread applications. The B fibre surface is textured to provide a good interface in resin-matrix composites and eliminating any need for sizing treatments. The high-strength, high-modulus B fibres are used in both tensile and compressive applications. It is available commercially as a unidirectional prepreg tape (see Figure 6.16) consisting of 50 vol% boron fibres with 121°C and 176°C cure epoxies. In compression, the unidirectional boron composite exhibits a modulus of 30 million psi and stress-strain curves are linear-to-failure (400,000 psi failing stress) [54].

6.7.2 SiC fibres

CVD is the established method for manufacturing SiC monofilament fibres (ranging from 100 to 140 μm) by companies such as Specialty Materials (US) and Tisics (UK). The deposition rig to

Figure 6.16 Boron prepreg tapes. (From Specialty Materials, Inc., *Boron fiber*, http://www.specmaterials.com/boronfiber.htm.)

produce SiC monofilament fibre is similar to the system used for B fibre production in Figure 6.15a. SiC monofilament is produced using gaseous precursors of the silane compound (e.g., CH_3SiCl_3, CH_3ClSiH) and H_2/Ar at a reaction temperature of 1000°C–1400°C [29]. If CH_3SiCl_3 is used as the reactant, the chemical reaction is proceeds as $CH_3SiCl_3 \rightarrow SiC(s) + 3HCl(g)$. Depending on the deposition conditions, stoichiometric, Si-rich or C-rich SiC can be deposited, allowing the microstructure, physical and mechanical properties of the SiC fibres to be varied. The deposition of

(a)   (b)

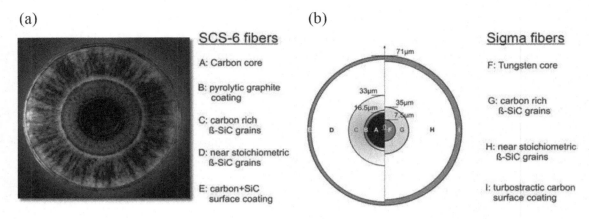

SCS-6 fibers

A: Carbon core

B: pyrolytic graphite coating

C: carbon rich ß-SiC grains

D: near stoichiometric ß-SiC grains

E: carbon+SiC surface coating

Sigma fibers

F: Tungsten core

G: carbon rich ß-SiC grains

H: near stoichiometric ß-SiC grains

I: turbostractic carbon surface coating

Figure 6.17 Cross-section SEM of a **(a)** SCS SiC fibre; **(b)** schematic diagram (not to scale) comparing the cross-sections of SiC monofilament fibres of SCS-6 (specialty materials) and sigma (Tisics). (Adapted from Specialty Materials, Inc., *Boron fiber*, http://www.specmaterials.com/boronfiber.htm; Flores, O. et al., *Adv. Eng. Mater.*, 16, 621–636, 2014; Kanouté, P. et al., *Arch. Comput. Method Eng.*, 16, 31–75, 2009.)

stoichiometric SiC requires a high ratio of hydrogen to silane whereas C-rich SiC requires a low ratio of hydrogen to silane or argon-silane, while a very large excess of H leads to Si-rich SiC [58]. Figure 6.17 shows a schematic diagram comparing the cross-sections of the structures of SCS-6 and Sigma fibres produced by Specialty Materials and Tisics, respectively. The tensile strength of SiC monofilament fibres is very sensitive to surface defects and abrasion, hence they are coated with an outer layer of a 1.5–3.0 μm protective carbon–silicon coating, consisting mostly of carbonaceous material with embedded SiC particles. Such an outerlayer not only helps to preserve but also double the strength of the SiC fibre without the outer layer [59,60]. Similarly, the Sigma fibre has an outer

~5 μm thick turbostratic carbon coating; however, without nano-scale SiC particles [60]. The microstructure and composition of these SiC fibres has been reviewed in detail [55,61]. The properties and costs of different commercial available SCS series and Sigma by CVD are summarised and compared in Table 6.11.

6.7.3 Applications

These SiC monofilament fibres are used as reinforcements in polymer, ceramic, and metal matrix composites. However, high fibre production costs have hindered the commercialisation and widespread use of SiC fibres for advanced composites, hence applications are limited to

Table 6.11 Properties and cost details of commercially available CVD-derived ceramic SiC fibres

Trade name (manufacture)	Structure	Diameter (μm)	Density (g cm⁻³)	Tensile strength (GPa)	Elastic modulus (GPa)	Approx. price (2013)
SCS-6 (Specialty Materials)	SiC on 33 μm C core	142	3.1	3.9	380	€6000/kg
SCS-9A (Specialty Materials)	SiC on 33 μm C core	78	2.8	≈3.5	≈310	Not comm. available at present
SCS-Ultra (Specialty Materials)	SiC on 33 μm C core	142	3.1	5.9	415	€8300/kg
Sigma (TISICS)	SiC on W	100	3.4	3.3	380	€7600/kg
		140	3.4	4.0	400	

Source: Flores, O. et al., *Adv. Eng. Mater.*, 16, 621–636, 2014.

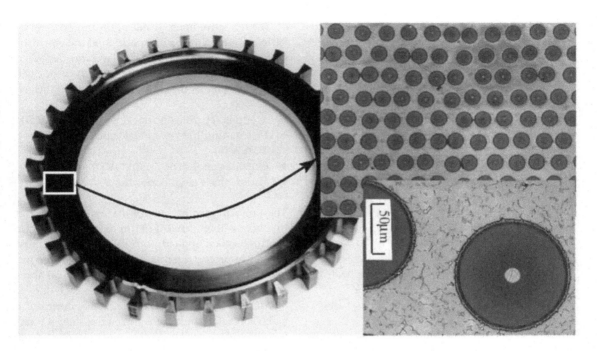

Figure 6.18 ASiC-Ti metal matrix composite bling which is being considered for future engine design. Insets shows the cross-sectional SEM of the SiC fibres reinforced Ti-MMCs at low and high magnifications. (From Kanouté, P. et al., *Arch. Comput. Method Eng.*, 16, 31–75, 2009.)

high performance and high value components for aerospace and space applications. In these applications, SiC fibres are reinforced in Ti-MMCs, Al-MMCs, intermetallic matrix composites, and CMCs for the production of high-strength, high-stiffness, low-density composites that can maintain their properties at high temperatures [54,62]. These include potential structural materials for aerojet components and compressor blades [29]. Figure 6.18 shows the SiC monofilament fibre reinforced Ti-composite of a simplified version of a "bling" (or "bladed ring") for possible replacement of compressor or turbine discs in aeronautical turboengines. Finite element modelling has been performed on the mechanical and thermomechanical responses of this composite component [62].

SCS-6 SiC monofilaments have also been used in glass-ceramic matrices, for example, barium aluminosilicate [63] and strontium aluminosilicate [64], which show chemical compatibility and phase stability upto 1600°C; they are being explored for high-temperature applications by the *Nasa Lewis Research Center*. Figure 6.19 shows a

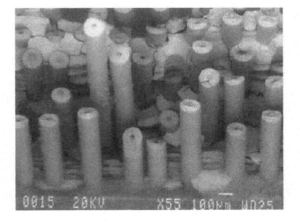

Figure 6.19 SEM micrograph of fracture surfaces of unidirectional CVD SiC fibre reinforced strontium aluminosilicate composite. (From Bansal, N.P., *Mater. Sci. Eng.*, A231, 117–127, 1997.)

SEM micrograph of the fracture surface of SCS SiC fibre-strontium aluminosilicate composite after a three-point bend test, where long fibre pull-out lengths are observed indicating a weak fibre-matrix interface and a tough composite [64].

6.8 OPTICAL FIBRES

Optical fibres are widely used in telecommunications, fibre optic sensors, and fibre lasers. Commercial optical fibres are fabricated by direct melting or from a glass rod (preform). However, most optical fibres are fabricated from the preform using a variant of CVD, whereby silica layers are deposited to make a preform using a mixture of gaseous reactants (e.g., $SiCl_4/GeC_4/POCl_3/BCl_3/O_2$) that undergo the following chemical reactions:

$$SiCl_4(g) + O_2(g) \rightarrow SiO_2(s) + 2\ Cl_2\ (g)$$

$$GeCl_4\ (g) + O_2(g) \rightarrow GeO_2(s) + + 2\ Cl_2(g)$$

$$4\ POCl_3\ (g) + 3\ O_2(g) \rightarrow 2\ P_2O_5\ (s) + + 6\ Cl_2(g)$$

$$4\ BCl_3(g) + 3\ O_2(g) \rightarrow 2\ B_2O_3\ (s) + + 6\ Cl_2(g)$$

The silica layers can be doped during the deposition process in order to control the refractive index. The use of dopants such as GeO_2 and P_2O_5 increases the refractive index of glass, whereas B_2O_3 has the opposite effect [65]. There is demand for high performance rare earth (RE) doped, high-power (in the multi-kW range) silica fibre lasers with high efficiency, reliability, and beam quality [66]. The solubility of RE (e.g., Yb) ions in silica can be improved and thus the RE content increased without phase separation and crystallisation by adding co-dopants such as Al and/or phosphorus during deposition [67].

The refractive index profile of the preform can be varied by adjusting the composition of the precursor mixture during deposition. Silica layers are deposited either on the outside of a mould or on the inside of a fused silica tube by the various CVD methods below:

- Modified chemical vapor deposition (**MCVD**)
- Plasma-modified chemical vapor deposition (**PMCVD**)
- Outside vapor deposition (**OVD**)
- Vapor-phase axial deposition (**AVD**)

The preform is subsequently drawn to a rod and finally to a fibre of 100–125 μm diameter. A plastic coating is then applied onto the fibre's outermost surface to protect against surface damage. The use of CVD variants for the production of optical fibres has been reviewed in references [29,65,68,69].

MCVD was developed by Bell Laboratories. The gaseous reactant mixture is introduced into a rotating silica tube heated by a traversing heat source (e.g., oxygen-hydrogen burner). SiO_2 glass particles are formed in the gas phase via homogeneous nucleation and are deposited in the form of a porous mass (soot) ahead of the burner, on the internal wall of the tube. The soot is subsequently vitrified by the traversing burner into a sintered glassy layer. The deposition process continues until the core and cladding layers have formed, and the tube is heated by burner to a higher temperature (e.g., 1800°C) to collapse the tube into a solid preform as shown in Figure 6.20a [70]. The core of the optical fibre has a higher refractive index than

(a) (b)

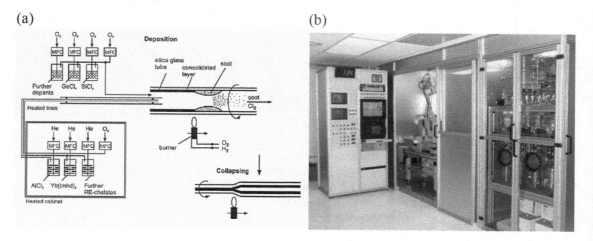

Figure 6.20 **(a)** Schematic of the MCVD process combined with gas phase doping for RE and Al. (From Unger, S. et al., *Laser Phys. Lett.*, 035–103.) **(b)** commercial MCVD system. (From http://www.custom-sys.com/4_Projects_HTML/06%20-%20mcvd.html.)

the surrounding cladding layers. Commercial MCVD glass fibre production equipment (see Figure 6.20b) is highly automated with precise control of the relative movements of the fibre optic preform and burner, and precise regulation of the delivery of several gases simultaneously during the MCVD process.

PMCVD has also been developed to fabricate the optical fibre preform using a traversing high-power microwave cavity to replace the oxy-hydrogen burner; otherwise, the operating principle is similar to MCVD. PMCVD allows the gas plasma to be heated directly without the need to heat up the silica tube itself, hence enabling faster deposition and more accurate grading of the refractive index of the optical fibres at a lower temperature (circa 1000°C) efficiently [65,68]. Furthermore, MCVD can fabricate a large preform capable of producing a longer length of more compact optical fibre (e.g., few hundred km). Such advantages can outweigh the cost of using a more sophisticated and expensive reactor and vacuum system.

The process principles of OVD, which is so called "soot process," has been developed, patented, and used by Corning in the 1970s, whereby the hot glass soot is deposited layer by layer onto a mandrel (graphite or aluminium oxide). Subsequently, the central mandrel is removed. The hollow porous preform is dehydrated and collapsed under sintering in a controlled atmosphere (e.g., He) to form the desired perform [65]. In the AVD process, the preform is fabricated continuously as shown in Figure 6.21 and is the preferred method for industrial production of fibre preform. During VAD, the chemical precursors are fed from the bottom of the reactor into the oxy-hydrogen burner where chemical reactions occur and deposit the glass soot onto a rotating silica rod where the preform is grown in the axial direction. The porous soot preform (~Ø 200 mm × 1,400 mm) is dehydrated and collapsed under sintering in ring heaters [65] to form a sintered perform (~Ø 100 mm × 700 mm about 7,500 fkm ZWPF/Core Preform) which could yield 1,200,000 fkm fibre capacity/machine/year for example [72]. Single crystals can be grown by pulling the rod upward while rotating in the same manner.

6.9 FREE-STANDING SHAPES AND 3D DEPOSITION

Free-standing films with complex shapes can be produced through the detachment of CVD-deposited films from their substrates. These can be thin free-standing foils and membranes or thick free-standing disks and wafers. The most common and widely used free-standing CVD material is based on synthetic diamond, which exploits the relatively small difference in stability between sp2 and sp3; the two allotropes of carbon and grow by the thermal decomposition of a gaseous source of carbon (<5%) with an excess of hydrogen in plasma (temperature ≥2000°C) to form atomic hydrogen and carbon. The plasma can be generated by direct current, hot filament, microwaves, radio frequency and lasers as described in Chapter 1. The formation of diamond depends on having a faster nucleation and growth rate than graphite on a refractory substrate with a low-thermal expansion coefficient, which is stable to carbide formation.

Natural diamond is scarce and has inherently variable properties that limits its applications. Synthetic diamond was first made in the 1950s using high pressure and high temperature diamond anvils. Subsequently, in the 1980s covalent, crystalline diamond was grown using CVD. With the advances in CVD growth and materials processing technology, microwave plasma enhanced CVD (see Chapter 1 for details) has emerged as a commercial synthesis method in the 1990s. High growth rates and the ability to control purity and the doping level has enabled the production of high-quality, free-standing,

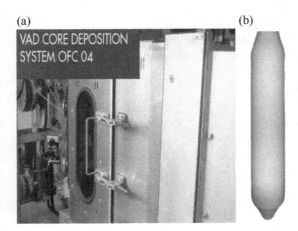

(a)   (b)

VAD CORE DEPOSITION
SYSTEM OFC 04

Figure 6.21 **(a)** An example of a commercial VAD deposition system by Nextrom and **(b)** VAD produced fibre preform. (From Nextrom, *VAD Core Deposition System*, Nextrom.)

polycrystalline diamond with a range of grain sizes (e.g., ultra-nanocrystalline (<10 nm), nanocrystalline (<50 nm) micro crystalline (<500 μm)), and single-crystal CVD diamond with outstanding, consistent properties for practical components in a variety a of high technology and engineering applications. These include optics, acoustics, high power electronics, environmental applications, and sensors that go beyond the tribological applications [73–77], consume around 800 tonnes of CVD diamond annually, about 150 times the amount of natural diamond mined as gemstones [73]. CVD diamond can be classified by its grain size or the properties of a single crystal. The grain size depends on synthesis conditions, substrate and layer thickness. The properties of CVD diamond are summarised in Table 6.12, which shows the typical

Table 6.12 Properties of CVD diamond

Property	Value
Density	$3\text{-}51524 \times 10^2$ kg m^{-3}
Number density	$1\text{-}77 \times 10^{22}$ per cm^3
Lattice type	Lattice constant between 0.356683 ± 0.000001 and 0.356725 ± 0.000003 nm at 298 K
Fracture toughness K_{10}	5 MPa m$^{0.5}$ single crystal 8.5 MPa m$^{0.5}$ polycrystalline
Poisson's ratio	0.1
Young's modulus	1050 GPa
Fracture strength	2.5–3 GPa single crystal surface finish dependent 200–1100 MPa polycrystalline grain size and grade dependent
Fatigue life	>95% strength after >>10^7 cycles to 70% of FS
Fracture plane	[111] and occasionally {110}
Weibull modulus	2.55 single crystal 10 growth surface polycrystalline 20 nucleation surface polycrystalline
Hardness	70–120 GPa single crystal (Plane and direction of indent dependent) $=81 \pm 18$ GPa polycrystalline (Grain orientation dependent)
Friction coefficient (μ)	0.05–0.15 (orientation dependent) In air (requires surface termination)
Vickers hardness[a]	10,000 kg/mm^2
Young's modulus[a]	1050 GPa
Poisson's ratio	0.1
Density	3.515 g/cm^3
Atom density[a]	1.77×10^{23} 1/cm^3
Thermal expansion coefficient	1.0×10^{-6}/K @300K
Sound velocity[a]	17,500 m/s
Friction coefficient	0.1
Specific heat @ 20°C	0.502 J/gK
Debye temperature[a]	1850 ± 10K
Randgap	5.45 eV
Resistivity	10^{11}–10^{16} cm

Source: Element Six, *Element Six Supermaterials*, 2017.
[a] Highest value of all solid materials.

superior physical properties of CVD diamond as well as it is extreme chemical inertness. Such superior properties of diamond derive from its strongly bonded, dense, tightly packed, and rigid structure consisting of tetrahedral covalent bonds between four nearest carbon neighbours, linked in a cubic lattice. The properties of diamond can be optimised and tailored by manipulating the impact of defects and varying the synthesis conditions to grow electrically insulating diamond (Type IIa) where nitrogen is the main impurity [similar to most of the natural diamond (Type I)] or p-type semiconductor boron doped CVD (Type IIb), which is rare in the nature.

The thermal conductivity of CVD diamond is >2000 W/mK, (i.e., five times that of copper). Graphitization of diamond occurs at temperatures above 700°C in an oxygen-containing environment and 1500°C in an inert atmosphere.

6.9.1 Thin free-standing foils and membranes

CVD diamond (optical grade or nanocrystalline) foils and membranes (as free standing or attached to a supporting frame) are manufactured commercially as thin as 50 nm for applications such as ultra-thin X-Ray windows, radiation detectors for heavy ions, luminescence screens that remain in the beam during measurement, stripper foils for the generation of highly ionized particles, and foils for laser acceleration [74].

6.9.2 Thick disks and wafers

Diamond can be synthesized by CVD in the shape of extended disks or wafers with a typical thickness in the range of 0.01–2 mm [74] and can be laser cut to the desired shape. The grain structure and impurity content have a strong effect on the properties of CVD diamond. Finer-grained diamonds have superior mechanical performance, whereas larger grained higher-purity material exhibit higher thermal conductivity and lower optical absorption [73]. The CVD process parameters can be controlled and tailored to produce CVD diamond with different properties.

The synthesised diamond can be of various grades tailored for specific high technology applications. These include "optical grade," "detector grade," "mechanical grade," "thermal

Figure 6.22 CVD Diamond wafers fabricated by microwave plasma CVD: from left to right – boron doped disk , optically transparent grade diamond, mechanical grade, and unpolished disk. (From Diamond Materials, *Diamond foils and membranes*, http://www.diamond-materials.com/EN/products/disks_films_membranes/disks.htm.)

grade" and nano-crystalline boron-doped, and the surface can be as grown or polished as shown in Figure 6.22. Under optimised growth conditions the properties of these disks approach those of perfect diamond single crystals. Free-standing polycrystalline diamond wafers are routinely fabricated into large discs, ≥100 mm [78,79].

Mechanical grade CVD diamond has exceptionally high hardness, very low coefficient of friction and it is widely used as free-standing diamond tool inserts [76] to greatly extend tool life, reducing downtime and operating costs (Figure 6.23). For operation in harsh and extreme operating conditions such as oil and gas drills, CVD diamond tends to be the choice of material as no other material can cope with and withstand such harsh and extreme conditions. Diamond tools fabricated with a {110} orientation are found to be more wear resistant than those fabricated froma {100} plane, possibly due to the ability to prepare a fine surface finish, free of damage which can promote crack initiation, which varies from one diamond surface to another, and the angle the cutting face makes with the hardest diamond wear direction <111> [80]. Furthermore, CVD diamond can also be fabricated to have an ultra-fine edge for extreme precision which is essential for surgical instruments, such as scalpels in ophthalmic and neuro surgery [81].

Optical grade diamond wafers consist of high-purity polycrystalline diamond grown by

(a) (b)

(c) (d)

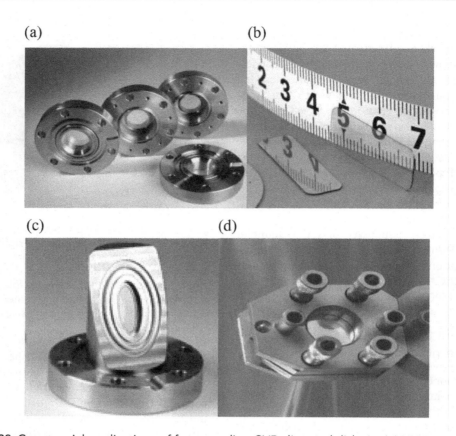

Figure 6.23 Commercial applications of free-standing CVD diamond disks in: **(a)** UHV vacuum window; **(b)** laser windows; **(c)** a diamond Brewster window; **(d)** a diamond liquid cell. (From Sp3 Diamond Technology, What makes a CVD diamond coated cutting tool "best of breed"? www. sp3diamondtech.com/coating-services/#DiaTip; Element Six, Synthetic diamond properties, http://ww.e6.com/wps/wcm/connect/E6_Content_EN/Home/The+power+of+supermaterials/ Synthetic+diamonds+extreme+properties/. http://www.e6.com/wps/wcm/connect/E6_Content_ EN/Home/The+power+of+supermaterials/Synthetic+diamonds+extreme+properties/; Davies, A.R., and Field, J.E., Wear, 256, 153–158, 2004; Godfried, H.P. et al., Advanced high-power lasers, Proceedings of SPIE 3889 553, 2000; Zaitsev, A.M., Optical Properties of Diamond, Springer, Berlin, Germany, 2001; http://www.e6.com/wps/wcm/connect/E6_Content_EN/Home/ Applications/sensors/; http://www.e6.com/wps/wcm/connect/E6_Content_EN/Home/Applications/ Water+and+wastewater+treatment/Diamox/.)

high-power microwave plasma-assisted CVD and can accommodate thermo-mechanical stress and withstand a wide range of operation temperatures in extreme environments. They also have exceptional spectral transparency covering the visible, infrared, terahertz and microwave region which can be used as UHV vacuum windows and for multi-spectral applications (e.g., synchrotron beam lines and space-based spectroscopic systems).

Free-standing CVD diamond can also be used as laser windows (0.3–1.2 mm thick). Its high thermal conductivity, low dn/dT, and excellent infrared transparency mean it can handle much higher power levels without any deleterious effect on the laser beam compared to ZnSe, where the residual absorption of ZnSe results in the formation of a thermal lens which strongly affects the beam quality and the position of the laser focus. Furthermore, the mechanical strength and erosion resistance of free-standing CVD diamond are significantly higher than other infra-red (IR) materials such as sapphire and zinc sulphide because the surface flaws introduced are less important than the intrinsic defects within the bulk.

Such mechanical robustness is essential as IR materials have an important role as protective "windows" to IR optical systems and sensors in high-speed flight against degradation by rain, dust, and sand and ice impacts, where flaws and roughening produced by these impacts can affect both transmission and strength [82].

CVD diamond also has applications in other optical components [83,84]. These include: diamond lenses in terahertz spectroscopy and CO_2 laser surgery; diamond beam splitters for high resolution spectroscopy over a wide spectral range ranging from infrared to terahertz radiation in space bound astronomical spectroscopy and the infrared analysis of the earth's atmosphere; diamond Brewster Windows, free-electron lasers, multi-wavelength IR lasers, or terahertz optical systems and diamond liquid cells for the spectroscopic analysis of liquids. In addition, CVD diamonds have been used for electrochemical sensing to exploit the high levels of sensitivity, selectivity, and responsiveness of corrosion resistance synthetic diamond to transform electroanalysis for industries such as pharmaceuticals, oil and gas, mining, agriculture, and biomedical sciences [85].

CVD free-standing solid boron-doped diamond electrodes have been used for efficient electrochemical oxidation of dissolved contaminants in treating highly contaminated industrial (e.g., textile and pharmaceutical) wastewaters, in systems which are difficult to treat by conventional means. The electrodes have oxidation capacity 2 kg COD h^{-1}, recirculation flow rate up to 50 m^3h^{-1} and can be scalable for effluent streams upwards of <0.5 m^3 [86].

The exceptional thermal conductivity of diamond has been exploited for thermal management applications such as diamond heat spreaders and heat sinks with a typical thickness of 100–500 μm depending on the lateral dimension of the heat source. Such capability outperforms copper, silicon carbide, and aluminium by factors of 3–10, thus helping to address today's single biggest cause of failure in electronics due to heat [87]. Furthermore, diamond surfaces can be: metallised with Au, AuSn, or In layers to promote adhesion and wetting properties; patterned using lithography or shadow masks; electrically isolated between top and bottom metallization and stress relieving slits for stress free mounting. The CVD diamond heat spreaders (see Figure 6.24a) help to enable the next-generation high-power density optoelectronics such as high power RF devices.

The optically transparent, radiation resistant, high thermal conductivity, known photoluminescence defects and low-Z material properties of diamonds are being exploited as free-standing doped diamond fluorescence screens [88] about 20 μm thick for X-ray beam position monitors (see Figure 6.24b) that can remain in the beam unlike the conventional screens which interrupt the X-ray beam and thus require motorised stages to move in and out of the beam. The beam intensity, position, and profile can be monitored readily by measuring

(a)   (b)

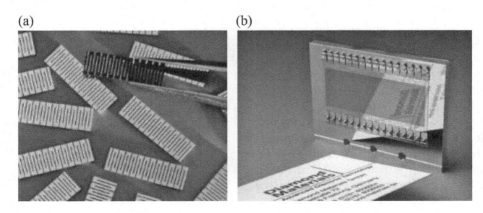

Figure 6.24 Applications of CVD diamond as: **(a)** heat-spreader; **(b)** CVD diamond X-Ray beam monitors. (From Bansal, N.P., *Mater. Sci. Eng.*, A231, 117–127, 1997; Yeh, C., *Handbook of Fiber Optics: Theory and Applications*, Academic Press, San Diego, CA, 1990; Sp3 Diamond Technology, *What makes a CVD diamond coated cutting tool "best of breed"?* www.sp3diamondtech.com/coating-services/#DiaTip.)

the fluorescence of the foil using a simple CCD camera. Polished (surface roughness <10 nm rms) optically grade C diamonds with thickness 20–300 μm (UHV compatible) or 1–50 μm for O-ring sealing are used as white-beam X-ray windows in synchrotron accelerators, where high thermal loads are generated due to the absorption of low-energy X-rays. Furthermore, non-toxic CVD diamond meets safety requirements during operation.

CVD can also be used to create complex geometric shapes with tailorable thicknesses. This has been exploited to produce high stiffness diamond domes with excellent acoustic properties. The upper limit to the frequency at which the dome behaves ideally is called the break-up frequency, and is proportional to the sound propagation velocity, which is itself proportional to $\sqrt{E/\rho}$, where E is Young's modulus and ρ density. Diamond offers performance exceeding that of current materials by a factor of three, and is being using by companies such as Bowers & Wilkins in their flagship 800 Series loudspeaker range.

Electronic grade CVD diamond (single-crystal and polycrystalline plates) has been widely researched [89,90] and exploited in a variety of demanding radiation-detection applications. These include high-energy physics, medical dosimetry in radiotherapy, neutron detection in civil nuclear and homeland security applications, and geophysical prospecting. The superior properties of CVD diamond have made it a robust detector in high-radiation environments. These properties include the high purity, low leakage current (relative to silicon), fast signal response, and radiation tolerance. As reviewed in reference [80], the testing of single crystal diamond as a spectrometer, the FWHM measured for 5.5 MeV α particles was 17 keV, compared to 14 keV using a Si p-i-n diamond, which corresponds to an energy resolution of 0.3% [91,92]. They have also demonstrated temporal resolution of 28 ps (one standard deviation for the correlation between two detector responses) for both polycrystalline and single-crystal diamond. CVD diamond produced by Element Six has been used in the Large Hadron Collider, Compact Muon Solenoid, and ATLAS Beam Condition Monitoring Systems for the discovery of a new particle consistent with the Higgs boson [93]. CVD diamond is currently being explored as a candidate material for quantum computing and secure quantum communication.

The impurities/defects in CVD diamond can act as a solid-state atom trap and be individually manipulated and probed at room temperature. Photons of light emitted from these impurities can be used to read out their quantum information [94–97].

Chae et al. has explored the use of direct current plasma-assisted CVD (DC-PACVD) with a diode configuration (see Figure 6.25) for the fabrication of 8-inch (203.2 mm) free-standing CVD diamond wafers [98]. The deposition system consists of a pair of disk electrodes attached to the cooling block in a vacuum chamber, with the 8-inch Mo disk substrate placed on the Cu anode which also acted as a cooling block. The cathode has negative voltage, whereas both the anode and the chamber wall are grounded. The radial distance from the electrode edge to the chamber side wall, as well as to the top and the bottom plate of the chamber is ~10–50 cm while the inter-electrode distance is typically 5 cm, which is about one order of magnitude smaller than the electrode-wall distance in order to provide a strong electric field in the inter-electrode space. Free standing microcrystalline diamond wafers were grown on a rotating Mo substrate (1200°C–1300°C) using a gas mixture of 5–12 volume % of CH_4 in H_2, at a flow rate of 400 sccm, pressure 100~130 Torr, with a discharge voltage and current of 840–910 V

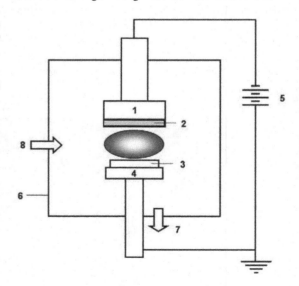

Figure 6.25 Schematic diagram showing the DC-PACVD system (1. cathode holder, 2. cathode, 3. substrate, 4. anode, 5. power supply, 6. reaction chamber, 7. vacuum pump, 8. gas inlet). (From Chae, K.-W. et al., *Diamond Relat. Mater.*, 19, 1168–1171, 2010.)

and 90–110 A, respectively. The typical growth rate under such processing conditions is 7–9 μm/h. After deposition and thermal contraction upon cooling to the room temperature, the free-standing diamond wafer can be easily removed and lifted off the Mo substrate. Dense, void-free and well-facetted polycrystalline diamond wafers can be grown by DC-PACVD as shown in Figure 6.26. The grain shape is the typical cubo-octahedron with some secondary nucleation.

The thickness (circa. 600 μm) and the thermal conductivity (circa. 10 W/cmK) variation over the diamond wafer was within 10% due to the stable and uniform plasma provided by the DC-PACVD diode configuration.

6.9.3 Near net-shape free-standing shaped articles

Near net-shape free-standing shaped mono-lithic parts are manufactured commercially by depositing thick films onto shaped mandrels by CVD and subsequently the removal of mandrels via mechanical or chemical methods to produce free-standing thick-walled (0.1 mm to >2 mm) structures of W, W/Re, BN, SiC, or PyC via the equations in S1 or S2, which are difficult to be manufactured by other techniques [99,100]. The mandrel substrates are designed to produce specific, near net shaped final components, including negative relief features. The mandrel is rotated to obtain uniformly distributed ceramic deposits onto the near net-shapes of the articles. The CVD

reactor and process parameters (e.g., temperature, pressure, and precursors concentrations) can be varied to control the wall thickness, deposition rate and microstructure.

Metal deposition

Metal halide (g) → metal(s) + by-product (g) S1

Ceramic deposition

Metal halide (g) + oxygen/carbon/nitrogen/boron source (g) → ceramic(s) + by-product (g) S2

The advantages of CVD for the fabrication of near net shape free standing shaped articles include:

- Minimal finishing is required to provide the article in its finished shape. This is particularly useful for manufacturing refractory materials that require critical dimensional control and challenging to be shaped by conventional machining methods
- Consumption of less deposit material
- Overall cost reduction and increased production efficiency

The free-standing structures could be of various shapes, including manufactures crucibles, boats, tubes, heating elements, and thermo-wells for a variety of applications, especially in high-temperature environments. For example, as shown in Figure 6.27, refractory structures used at high temperatures, beyond the capability of conventional steels and superalloys and that require greater toughness and

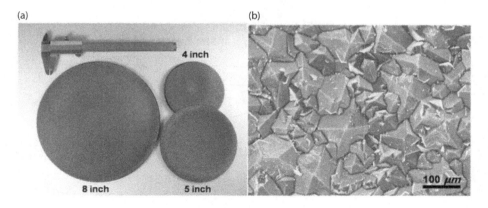

Figure 6.26 (a) Free-standing diamond wafers grown by DC-PACVD method and (b) the SEM image of the dense, void-free and well-facetted polycrystalline diamond wafer. (From Chae, K.-W. et al., *Diam. Relat. Mater.*, 19, 1168–1171, 2010.)

(a)   (b)   (c)

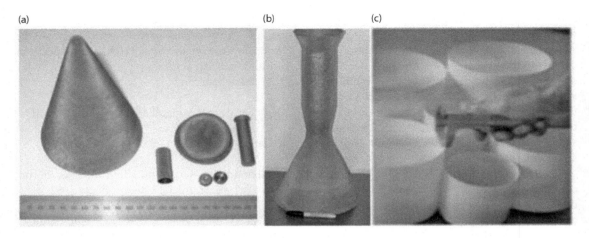

Figure 6.27 CVD fabricated free-standing shaped articles: **(a)**; **(b)** large freestanding solid rhenium thruster (19″ high × 10″ diameter at base). (From http://www.ultramet.com/chemical_vapor_deposition. html.) and **(c)** boron nitride crucible. (From http://www.ultramet.com/chemical_vapor_deposition.html.)

ductility than ceramics (e.g., tungsten, rhenium, tantalum, molybdenum, and niobium); combustion chambers and hot-gas-path components (e.g., valves and tubing); and ceramic crucibles for crystal growth [99,100].

6.9.4 Free-standing micro-objects

Besides the fabrication of free-standing macro objects, research has also explored the possibility of using CVD, especially laser-assisted CVD to demonstrate the feasibility of the fabrication of three-dimensional free standing micro-objects. Figure 6.28 shows a schematic diagram of the experimental set up for the laser assisted CVD for the fabrication of the three-dimensional aluminium grid structure.

The substrate can be rotated in one axis, and moved in all directions via a digital control system. A continuous Ar^+-laser ($\lambda = 514.5$ nm) is

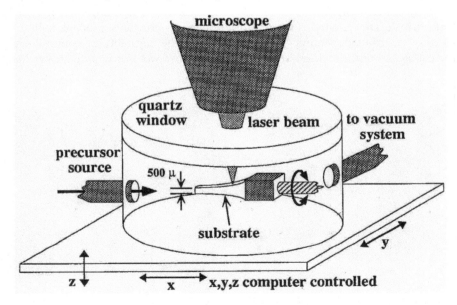

Figure 6.28 Schematic view of the experimental setup for the laser-assisted CVD of free-standing micro objects. (From Lehmann, O., and Stuke, M., *J. Phys. IV Colloq.*, 2, C2-337–C2-342, 1991.)

used as photon source. The laser beam is coupled into the beam path of a microscope objective and passing a quartz window of 2 mm thickness and is focused onto the substrate surface. The adduct precursor aluminum-trihydride-trirnethylamine, $(CH_3)_3N. AlH_3. N(CH_3)_3$ is introduced onto a pre-formed polycarbonate substrate with a focus diameter (1 μm), where chemical dissociation and reaction occur at the vicinity of the laser-beam-heated substrate and deposit a dense grid with a diameter around 10 μm of aluminium. Initially a thin line is created with relatively high scan speed (200–300 μm/s) and low laser output power (20–30 mW) under digital control. Subsequently, the line is amplified with a lower scan speed (ca. 150 μm/s) and higher laser power (up to 60 mW). This allows the aluminium line created by the first step to absorb much of the incoming energy when performing the second step and the high conductivity of aluminium does not result in the undesired local melting of the substrate; however, it would cause a faster line growth. The laser writing of the grid forms the finished 3-D structure as defined by the shape of the 3-D substrate. Insoluble by-products are removed by pulsed UV excimer laser and subsequently the self-supporting 3-D structure is obtained after dissolving the substrate using liquid chloroform [101].

ACKNOWLEDGEMENT

Contribution of David Betteridge to the aluminising and diffusion coatings is acknowledged. Special thanks also to Rodney Wing for the helpful discussion on coating materials for high temperature applications.

REFERENCES

1. Inc, R.P. *CVD Coatings - Chemical Vapor Deposition*
2. Ti Coating, Inc. *Chemical Vapor Deposition Coatings.*
3. Brookes, K.J., *Metal Powder Report*, March–April 2014, pp. 22–27.
4. Walz, D. and C. J. Raub, *Metalloberflache*, 1686. **40**(6):239–241.
5. Fukui, H., Evolutional history of coating technologies for cemented carbide inserts—chemical vapor and physical vapor deposition. *Sei Technical Review*, 2016. **82**.
6. García, J., J. Persson and E. Göthelid, Development of wear resistant coatings for cutting tools. *World PM2016 – SIS 90 Years of Cemented Carbide – Past, present and future Hamburg*, 2016.
7. Ruppi, S., Deposition, microstructure and properties of texture-controlled CVD α-Al_2O_3 coatings. *International Journal of Refractory Metals and Hard Materials*, 2005. **23**(4): 306–316.
8. McDonald, C.F., The role of the ceramic heat exchanger in energy and resource conservation. *Journal of Engineering for Power*, 1980. **102**(2): 303–315.
9. M'Saoubi, R. and S. Ruppi, Wear and thermal behaviour of CVD a-Al_2O_3 and MTCVD Ti(C,N) coatings during machining, *CIRP Annals*, 2009. **58**(1): 57–60.
10. *Walter's New CVD Inserts Offer Gilt-edged Productivity Improvements.* 2009.
11. García, J., J. A. Persson, E. Göthelid, Development of Wear Resistant Coatings for Cutting Tools, 2016. World PM2016-SIS 90 Years of Cemented Carbide-Past, present and future Hamburg, October 13, 2016. Available from: https://www.worldpm2016. com/post-event/presentations/sis-presentations/sis-presentations-hm/49-development-of-wear-resistant-coatings-for-cutting-tools/file.
12. Richter Precision Inc. Coating Technical Data, Available from http://www.richterprecision.com/coating-data/technical-data/.
13. Holzschuh, H., Deposition of Ti–B–N (single and multilayer) and Zr–B–N coatings by chemical vapor deposition techniques on cutting tools. *Thin Solid Films*, 2004. **469**: 92–98.
14. Hu, P. et al., Synthesis, structure, and olefin polymerization behavior of Nickel complexes with carborane [S, C] or [S, S] ligands. *Organometallics*, 2011. **30**(18): 4935–4940.
15. *What makes a CVD diamond coated cutting tool "best of breed"?* Available from: http://www.sp3diamondtech.com/products.php.
16. *Plasma technology for you.* Available from: http://www.hauzertechnocoating.com/en/plasma-coating-explained/dlc-coating/.
17. *PVD, CVD, TD & DCD coatings.* Available from: http://www.richterprecision.com/dlc-coatings.htm.

18. Padture, N.P., M. Gell, and E.H. Jordan, Thermal barrier coatings for gas-turbine engine applications. *Science*, 2002. **296**(5566): 280–284.

19. Yamane, H. and T. Hirai, Yttria stabilized zirconia transparent films prepared by chemical vapor deposition. *Journal of Crystal Growth*, 1989. **94**(4): 880–884.

20. Ferrao, L.P.C., H.K. Bowen, and W.D. Kingery, Chemical vapordeposition of solid-solutions in system zirconia–yttria. *American Ceramic Society Bulletin*, 1973. **52**: 345.

21. Wahl, G. et al., Chemical vapor deposition of TBC: An alternative process for gas turbine components. *Journal of Engineering for Gas Turbines and Power*, 2000. **123**(3): 520–524.

22. Bertrand, G. and R. Mévrel, Zirconia coatings realized by microwave plasma-enhanced chemical vapor deposition.*Thin Solid Films*, 1997. **292**(1): 241–246.

23. Préauchat, B. and S. Drawin, Properties of PECVD-deposited thermal barrier coatings. *Surface and Coatings Technology*, 2001. **142**: 835–842.

24. Vargas Garcia, J.R. and T. Goto, Thermal barrier coatings produced by chemical vapor deposition. *Science and Technology of Advanced Materials*, 2003. **4**(4): 397–402.

25. Kimura, T. and T. Goto, Rapid synthesis of yttria-stabilized zirconia films by laser chemical vapor deposition. *Materials Transactions*, 2003. **44**(3): 421–424.

26. Hidetoshi, M., K. Teiichi, and G. Takashi, Acceleration of deposition rates in a chemical vapor deposition process by laser irradiation. *Japanese Journal of Applied Physics*, 2003. **42**(3B): L316.

27. Goto, T., A review: Structural oxide coatings by laser chemical vapor deposition. *Journal of Wuhan University of Technology-Materials Science Edition*, 2016. **31**(1): 1–5.

28. Demirkiran, A.S., E. Celik, M. Yargan, and E. Avci, Oxidation behavior of functionally gradient coatings including different composition of cermets. *Surface and Coatings Technology*, 2001. **142–144**: 551–556.

29. Choy, K.L., Chemical vapour deposition of coatings. *Progress in Materials Science*, 2003. **48**(2): 57–170.

30. Vyas, J.D. and K.L. Choy, Structural characterisation of thermal barrier coatings deposited using electrostatic spray assisted vapour deposition method. *Materials Science and Engineering: A*, 2000. **277**(1): 206–212.

31. Diffusion Alloys, *Aluminising*. Available from: http://www.diffusion-alloys.com/content/aluminising.

32. Company, S., *Selective chemical vapor deposition (CVD)*. Available from: http://www.sulzer.com/lt/Products-and-Services/Turbomachinery-Services/Repair-Services/Coatings/Selective-Chemical-Vapor-Deposition-SE-CVD.

33. Goward, G.W. and L.W. Cannon, Pack cementation coatings for superalloys: A review of history, theory, and practice. *Journal of Engineering for Gas Turbines and Power*, 1988. **110**(1): 150–154.

34. Xiang, Z.D., J.S. Burnell-Gray, and P.K. Datta, Aluminide coating formation on nickel-base superalloys by pack cementation process. *Journal of Materials Science*, 2001. **36**(23): 5673–5682.

35. Pochet, L.F., P. Howard, and S. Safaie, CVD coatings: From cutting tools to aerospace applications and its future potential. *Surface and Coatings Technology*, 1997. **94**: 70–75.

36. Seelig, R.P. and R. J. Stueber, High-temperature-resistant coatings for superalloys. *High Temperatures-High Pressures*, 1978. **10**:207–213.

37. Alpérine, S. et al., Structure and high temperature performance of various palladium-modified aluminide coatings: A low cost alternative to platinum aluminides. *Surface and Coatings Technology*, 1990. **43**: 347–358.

38. Wing, R.G. and I.R. McGill, Platinum in aluminide diffusion coatings. *Platinum Metals Review*, 1981. **25**: 94.

39. Ionbond UK Ltd, *CVD technology*. Available from: http://www.engineeringarena.co.uk/companies/ionbond-uk-ltd/products/bernex-aluvap-cva-coating-system.

40. Strakov, H. et al., Advanced chemical vapor aluminizing technology: Co-deposition process and doped aluminized coatings. *American Society of Mechanical Engineers*, 2012. **2012**(44717): 201–207.

41. Goward, G.W., Current research on the surface protection of superalloys for gas turbine engines. *Journal of Metals*, 1970. **10**: 31–39.

42. Wierzba, B. et al., Aluminizing of nickel alloys by CVD. The effect of HCl flow. *Chemical Vapor Deposition*, 2014. **20**(1–3): 80–90.

43. Gunda, N., R. Cooke, S.K. Jha, and S.A. Sastri, *Strongly adherent thick coatings of ultra high purity silicon*. Available from: http://www.surmet.com/docs/(2)Article_UHPSi.pdf.

44. Acerde, *Chemical Vapor Deposition (CVD)*. Available from: http://www.acerde.com/process/technology.

45. http://www.kiscoparylene.com/parylene/coating-process/

46. http://vsiparylene.com/parylene

47. http://www.parylene.co.uk/about-parylene

48. http://www.paryleneinc.com

49. http://www.paryleneengineering.com/basics_of_parylene.htm

50. https://scscoatings.com/what-is-parylene/parylene-properties/

51. http://www.scscookson.com/parylene/properties.cfm

52. Lee, C.J., H. Wang, and G.A. Foggiato, U.S. Patent 6,140,456, Precursors for making low dielectric constant materials with improved thermal stability. Issue date: October 31, 2000.

53. Senkevich, J.J. and S.B. Desu, Near-room-temperature thermal chemical vapor deposition of poly(chloro-p-xylylene)/SiO_2 nanocomposites. *Chemistry of Materials*, 1999. **11**(7): 1814–1821. doi:10.1021/cm990042q.

54. Specialty Materials, Inc., *Boron fiber*. Available from: http://www.specmaterials.com/boronfiber.htm.

55. Liu, Y. et al., Yttrium aluminum garnet fibers from metalloorganic precursors. *Journal of the American Ceramic Society*, 1998. **81**(3): 629–645.

56. Layden, G.K., Fracture behaviour of boron filaments. *Journal of Materials Science*, 1973. **8**(11): 1581–1589.

57. Buck, M. and M. Dorf, Boron fibers in composite materials. *Advanced Performance Materials*, 1995. **10**(1–2):13–16.

58. Cristin, F., Naslain, R., Bernard, C., A thermodynamic and experimental approach of sic CVD, Application to the CVD-infiltration of porous carbon composites, In Sedwick, T.O. & Lydin, H. (Eds.), *Proceedings of the 7th International Conference CVD*. 1979: Electrochemical Society, Princeton, NJ, p. 449.

59. Ning, X.J. and P. Pirouz, The microstructure of SCS-6 SiC fibre. *Journal of Materials Research*, 1991. **6**: 2234–2248.

60. Chollon, G., R. Naslain, C. Prentice, R. Shatwell, and P. May, High temperature properties of SiC and diamond CVD-monofilaments. *Journal of the European Ceramic Society*, 2005. **25**: 1929–1942.

61. Flores, O. et al., Ceramic fibers based on SiC and SiCN systems: Current research, development, and commercial status. *Advanced Engineering Materials*, 2014. **16**(6): 621–636.

62. Kanouté, P. et al., Multiscale methods for composites: A review. *Archives of Computational Methods in Engineering*, 2009. **16**(1): 31–75.

63. Bansal, N.P., CVD SiC fiber-reinforced barium aluminosilicate glass—ceramic matrix composites. *Materials Science and Engineering: A*, 1996. **220**(1): 129–139.

64. Bansal, N.P., Mechanical behavior of silicon carbide fiber-reinforced strontium aluminosilicate glass-ceramic composites. *Materials Science and Engineering*, 1997. **A231**: 117–127.

65. Ungar, S., *Fibre Optics: Theory and Applications*. 1990: Wiley, New York.

66. Richardson, D.J., J. Nilsson, and W.A. Clarkson, High power fiber lasers: Current status and future perspectives [Invited]. *Journal of the Optical Society of America B*, 2010. **27**(11): B63–B92.

67. Popp, A., A. Voss, T. Graf, S. Unger, J. Kirchhof, and H. Bartelt, Thin-disk laser-pumping of ytterbium-doped fiber laser. *Laser Physics Letters*, 2011. **8**: 887–894.

68. Pierson, H., *Handbook of Chemical Vapour Deposition*. 1992: Noyes, Park Ridge, NJ.

69. Yeh, C., *Handbook of Fiber Optics: Theory and Applications*. 1990: Academic Press, San Diego, CA.

70. Unger, S., F. Lindner, C. Aichele, M. Leich, A. Schwuchow, J. Kobelke, J. Dellith, K. Schuster and H. Bartel, A highly efficient Yb-doped silica laser fiber prepared by gas phase doping technology. *Laser Physics Letters*: 035–103.

71. http://www.custom-sys.com/4_Projects_HTML/06%20-%20mcvd.html

72. Nextrom, *VAD Core Deposition System*. Nextrom. Available from: http://www.rosendahlnextrom.com/fiber-optics/manufacturing-solutions/telecom-fibers-and-preforms/ofc-04/.

73. Element Six, *Element Six Supermaterials*. 2017. Available from https://www.e6.com.

74. Diamond Materials, *Diamond foils and membranes*. Available from: http://www.diamond-materials.com/EN/products/disks_films_membranes/disks.htm.

75. Balmer, R.S., J.R. Brandon, S.L. Clewes, H.K. Dhillon, J.M. Dodson, I. Friel, P.N. Inglis et al., Chemical vapour deposition synthetic diamond: materials, technology and applications. *Journal of Physics: Condensed Matter*, 2009. **21**: 364221.

76. Sp3 Diamond Technology, *What makes a CVD diamond coated cutting tool "best of breed"?* Available from: www.sp3diamondtech.com/coating-services/#DiaTip.

77. Koizumi, S., C. Nebel, and M. Nesladek, *Physics and Applications of CVD Diamond*. 2008: Hoboken, NJ, John Wiley & Sons.

78. Heidinger, R., G. Dammertz, A. Meier, and M.K. Thumm, CVD diamond windows studied with low- and high-power millimeter waves. *Transactions on Plasma Science*, 2002. **30**: 800–807.

79. Parshin, V.V., G.B.M., S.E. Myasnikova, and A.V. Orlenekov, Dielectric losses in CVD-diamonds in the millimeter-wave range at temperatures 300–900 K. *Radiophysics and Quantum Electronics*, 2004: 974–978.

80. Balmer, R.S. et al., Chemical vapour deposition synthetic diamond: materials, technology and applications. *Journal of Physics Condensed Matter*, 2009. **21**: 364221.

81. Element Six, *Synthetic diamond properties*. Available from: http://www.e6.com/wps/wcm/connect/E6_Content_EN/Home/The+power+of+supermaterials/Synthetic+diamonds+extreme+properties/. http://www.e6.com/wps/wcm/connect/E6_Content_EN/Home/The+power+of+supermaterials/Synthetic+diamonds+extreme+properties/.

82. Davies, A.R. and J.E. Field, The strength of free-standing CVD diamond. *Wear*, 2004. **256**(1): 153–158.

83. Godfried, H.P. et al., Advanced high-power lasers Proc. SPIE3889 553, 2000.

84. Zaitsev, A.M., *Optical Properties of Diamond*. 2001: Springer, Berlin, Germany.

85. http://www.e6.com/wps/wcm/connect/E6_Content_EN/Home/Applications/sensors/

86. http://www.e6.com/wps/wcm/connect/E6_Content_EN/Home/Applications/Water+and+wastewater+treatment/Diamox/

87. http://www.e6.com/wps/wcm/connect/E6_Content_EN/Home/Applications/New+applications

88. Bergonzo, P., D. Tromson, and C. Mer, CVD diamond-based semi-transparent beam-position monitors for synchrotron beamlines: preliminary studies and device developments at CEA/Saclay. *Journal of Synchrotron Radiation*, 2006. **13**: 151–158.

89. Secroun, A. et al., Dislocation imaging for electronics application crystal selection. *Physica Status Solidi*, 2007. **204**: 4298–304.

90. Nesladek, M., A. Bogdan, W. Deferme, N. Tranchant, and P. Bergonzo, Charge transport in high mobility single crystal diamond. *Diamond and Related Materials*, 2008. **17**: 1235–1240.

91. Pomorski, M., E. Berdermann, A. Caragheorgheopol, M. Ciobanu, M. Kis, A. Martemiyanov, C. Nebel, and P. Moritz, Development of single-crystal CVD-diamond detectors for spectroscopy and timing. *Physica Status Solidi*, 2006. **203**: 3152–3160.

92. Berdermann, E., A. Caragheorgheopol, M. Ciobanu, M. Pomorski, A. Pullia, S. Riboldi, M. Traeger and H. Weic, *Diamond and Related Materials*, 2008. **17**: 1159.

93. http://www.e6.com/wps/wcm/connect/e6_content_en/home/applications/electronics/radiation_detector

94. Stoneham, M.A., A.H. Harker, and G.W. Morley, Could one make a diamond-based quantum computer? *Journal of Physics: Condensed Matter*, 2009. **21**(3619): 364222.

95. http://www.e6.com/wps/wcm/connect/E6_Content_EN/Home/Applications/Electronics/QuantumMagnetometry/

96. Jelezko, F. and J. Wrachtrup, *Physics and Applications of CVD Diamond*. 2008: Wiley, New York, Chapter 10.

97. Koizumi, S., C. Nebel, and M. Nesladek, *Physics and Applications of CVD Diamond*. 2008: Wiley, Weinheim, Germany.

98. Chae, K.-W., Y.-J. Baik, J.-K. Park, and W.-S. Lee, The 8-inch free standing CVD diamond wafer fabricated by DC-PACVD. *Diamond & Related Materials*, 2010. **19**: 1168–1171.

99. http://www.cvd.co.uk/ATL%20CVD%20Tungsten.pdf

100. http://www.ultramet.com/chemical_vapor_deposition.html

101. Lehmann, O. and M. Stuke, Generation of three-dimensional free-standing micro-objects by laser chemical processing. *Journal de Physique IV Colloque*, 1991. **2**(C2): C2-337–C2-342.

7

CVD of nanocomposite coatings

YURI ZHUK AND KWANG LEONG CHOY

Nanocomposite coatings consist of two or more immiscible phases, where at least one of the phases is in nm scale. In general, a nanocomposite coating contains nanostructured reinforcement materials/nanofillers in a matrix. There are three types of nanocomposite coatings:

1. 0D-nanocomposites, where the reinforced material is in zero dimension such as nanopowder/nanoparticles
2. 1D-nanocomposites, including reinforcing 1D material such as nanofibres, nanowhiskers, and nanowires
3. 2D-nanocomposites consisting of 2D reinforcing materials with one of the dimensions in nm scale such as nanoplatelets or nanolayered materials

The high surface-area-to-volume ratio of the reinforcing materials leads to significant changes to the properties of the matrix material compared to the equivalent bulk material. The combination of the properties of the nanofillers with the matrix material tends to lead to materials with extraordinary properties. Hence, nanocomposite coatings are attracting increasing interest.

Chemical vapour deposition (CVD) can be used to fabricate nanocomposite coatings via:

1. The *in-situ* synthesis of nanoparticles and the formation of the coating matrix simultaneously
2. The incorporation of nanomaterials (e.g., nanoparticles, nanowires, and nanotubes) into matrix during the CVD of the matrix materials

This chapter reviews in detail the *in-situ* CVD formation of nanocomposite coatings, focusing on the specific example of low-pressure, thermal CVD of Tungsten-Tungsten Carbide Hardide coatings with outstanding toughness, crack, and impact resistance (see Section 7.1). Such established CVD nanocomposite coatings are manufactured commercially for a variety of industrial applications, especially those engineering components operating in abrasive, erosive, and corrosive environments where the coatings can extend the life of the components.

In addition, other variants of CVD have also been explored for the fabrication of nanocomposite coatings as described in Section 7.2. Most of these CVD methods are at the research and development stage.

7.1 THERMAL CVD OF NANOSTRUCTURED TUNGSTEN CARBIDE-BASED NANOCOMPOSITE COATINGS

7.1.1 Background

Section 7.1 highlights the new CVD of tungsten-tungsten carbide nanocomposite coatings (so-called Hardide coatings [1]), their structure, and key properties. This section also gives a few examples of industrial applications of these coatings.

This new family of nano-structured CVD tungsten carbide-tungsten coatings is mainly used to increase the life of critical parts and tools operating in abrasive, erosive, and chemically aggressive environments.

Hardide coatings consist of tungsten carbide nano-particles dispersed in a metal tungsten matrix. This structure gives a combination of high hardness of tungsten carbide with excellent toughness as well as crack and impact resistance of tungsten, which together produce a synergetic effect of outstanding wear and erosion-resistance. From extensive laboratory and field testing, it was found that high hardness combined with enhanced toughness would achieve the optimum protection against both wear and erosion. The coating's hardness inhibits the micro-cutting mechanisms of wear and erosion, while its toughness, ductility, residual compressive stresses, and homogeneous micro-structure prevent fatigue micro-cracking/chipping and platelet mechanisms of erosion.

Hardide coatings are typically 50 microns thick, which is exceptionally thick among hard CVD coatings. They are sufficiently tough to withstand 3000 micro-strain deformations without damage; this deformation will crack or chip most other thick hard coatings. The gas-phase CVD process enables the uniform coating of external and internal surfaces and complex shapes. As a pore-free CVD coating, Hardide coating resists acids and aggressive media, providing barrier against corrosion.

From the coating thickness and surface finish perspectives, the Hardide coating fills the gap between thin-film PVD coatings [2–3] and much thicker, rough, and non-uniform thermal spray coatings. As compared to thin PVD coatings, 50-microns-thick Hardide coating has a higher load-bearing capacity and is much more durable in abrasive and erosive applications such as oil drilling tools withstanding contact pressure in excess of 1.2 GPa without cracking or delamination. Unlike thermal spray coatings, Hardide gas-phase CVD coatings can be applied uniformly to internal surfaces and complex shapes.

Hardide CVD coating is a better-performing replacement for hard chrome plating, which is under pressure from the US OSHA and the EU REACH regulations. Proven applications for Hardide coatings include critical parts of oil drilling tools, aircraft components, pumps, and valves operating in abrasive, erosive, and corrosive environments where the coating can typically triple part life.

7.1.2 Coating deposition method, structure, and composition

7.1.2.1 COATING DEPOSITION METHOD

The Hardide coatings are deposited using thermal CVD from a mixture of reactive gases, which includes tungsten hexafluoride, hydrogen, and a hydrocarbon gas. The Hardide low-temperature and low-pressure CVD method produces a nano-structured material with a unique combination of properties.

The process of parts CVD coating includes the following steps, as illustrated in Figure 7.1:

1. Parts are loaded inside the CVD reactor, which is vacuumed down and heated.
2. Reactive gas mixture is deployed into the reactor.
3. There is a series of reactions taking place on the surface of metal parts which results in the coating crystallization atom-by-atom.
4. The coating can be produced on both internal and external surfaces.
5. The coating process is continuing until the target coating thickness is reached.

Hardide coatings are applied at low temperatures ranging from 480°C up to 600°C; these are significantly lower as compared to more traditional CVD coatings typically produced at a higher temperature, within the range of 800°C to 1000°C [5]. This enables the coating of a wide range of materials, including stainless steel, tool steels stable at 500°C, Ni-, Cu-, and Co-based alloys, and titanium. The typical deposition pressure range is from 0.1 mbar to 100 mbar. The deposited coating has a strong metallurgical adhesion to these substrates with the bond strength typically exceeding 70 MPa (Figures 7.2 and 7.3).

7.1.2.2 NANO-STRUCTURE OF HARDIDE COATINGS

The structure of Hardide coatings was investigated using high-resolution transmission electron microscopy (HRTEM). This analysis of Hardide

(a) (b)

(c) (d)

Figure 7.1 **(a–d)** Various stages of the CVD nanocomposite coating's deposition process.

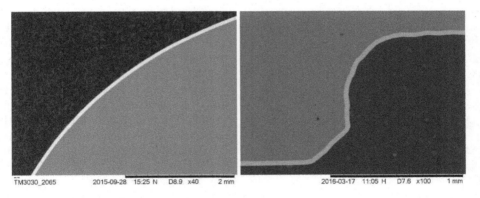

Figure 7.2 Cross-section SEM images of the Hardide coating, showing conformal uniform coating on complex shapes: a spherical ball (left) and a machined part with both internal and external corners (right). Hardide-T type coating thickness is circa 50 microns.

coatings shown that they consist of tungsten carbide nano-particles with the size typically between 1 and 10 nm, dispersed in an alloyed tungsten metal matrix. Figures 7.4 and 7.5 show the HRTEM electron microscopy images of Hardide-T showing the tungsten carbide nano-size inclusions and also the grain boundary structure.

Figure 7.4 shows that the tungsten carbide precipitate and the metal tungsten matrix have their crystalline planes aligned with each other, forming a coherent precipitate structure. This structure is produced as a result of a complex system of several physical processes (adsorption/desorption, condensation), and a series of chemical transformations taking place on the surface/gas phase boundary. These physical and chemical processes result in the coating crystallization atom-by-atom, building a dense pore-free

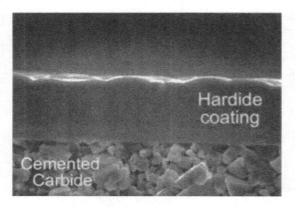

Figure 7.3 Cross-section SEM images of the Hardide coating, showing dense pore-free coating structure.

Figure 7.4 HREM micrograph of precipitate in Hardide-T as-deposited on copper substrate. The inter-atomic distances (1.49 and 1.76 Å) directly taken from the precipitate region are matched best to the lattice constants of W_2C tungsten carbide (110 plane distance is 1.49 Å and 102 plane is 1.74 Å). (Micrograph produced at the Oxford University Department of Materials).

structure with the tungsten and tungsten carbide constituents bonded together at the atomic level. TEM analysis also shows that the Hardide coatings are free from inter-granular inclusions, impurities, porosity, and other defects which

can weaken the material's mechanical properties and make it brittle. Alloying elements further enhance its properties. This structure and composition help produce an advanced hard material with outstanding toughness and crack and impact resistance.

7.1.2.3 TYPES OF HARDIDE COATINGS AND THEIR COMPOSITION

There are five types of CVD of Hardide coating, as detailed in Table 7.1.

All of these coatings consist of a tungsten carbide/tungsten nano-structured composite produced by CVD. Unlike thermal spray-deposited tungsten carbide coatings, Hardide does not use a cobalt or nickel metal matrix binder. The main differences between the Hardide-T, A, M, H, and W-type coatings is the content of tungsten carbides, their phase composition, and structure. The hardest type of the coating, Hardide-H, consists of highly pure binder-free tungsten carbides which are extremely hard although relatively brittle. The lowest hardness coating, Hardide-W, is a metal tungsten with some alloying elements enhancing its properties. The multi-layer coating, Hardide-M, includes layers of various compositions having different hardness and toughness values. By varying the ratio between the thickness and properties of each individual layer of Hardide-M coating, one can adjust the overall coating characteristics to meet specific application requirements. Similar control of the coating hardness/toughness was also achieved at the nano-structure level when the Hardide-T and Hardide-A coatings were developed. Hardide-A coating has lower residual stresses as compared to Hardide-T coating. These coatings, which are the most widely used in industrial applications, consist of tungsten carbide nano-particles dispersed in an alloyed metal tungsten matrix. This structure gives it a unique combination of properties: high hardness is combined with excellent toughness, impact and crack-resistance. Hardide-T low-stress coating technology produces a hard coating with the thickness of 50 microns or more—uniquely thick among CVD hard coatings.

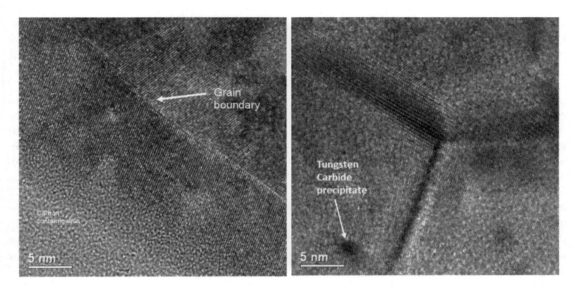

Figure 7.5 (left) and (right) High-resolution TEM micrographs of tungsten metal matrix grain boundaries in two different Hardide-T coating samples, also showing a Tungsten carbide nano-size precipitate. Only slight contrast changes are seen along the grain boundaries indicating that there are no significant precipitates. This feature appears similar in all grain boundaries seen during the study of several Hardide-T coating samples. (Micrographs produced at Oxford University Department of Materials).

Table 7.1 Types of Hardide coatings, their key characteristics, and applications

Type	Hardness	Toughness	Thickness	Applications
Hardide-T (Tough)	1100–1600 Hv	Excellent	Typically 50 μm	Oil tools, pumps, valves, actuators
Hardide-A	800–1200 Hv	Excellent	Typically 50–100 μm	Hard chrome replacement, primarily for aerospace applications
Hardide-M (Multi-layer)	1200–2000 Hv	Good	Typically 50 μm	Abrasion/erosion resistance
Hardide-H (Hard)	2000–3500 Hv	Satisfactory	5–12 μm	Self-sharpening blades
Hardide-W	400 Hv	Excellent	Typically 50–100 μm	X-ray anodes

7.1.3 Benefits of the coating nanostructure

7.1.3.1 COMBINATION OF HIGH HARDNESS WITH ENHANCED TOUGHNESS AND RESISTANCE TO IMPACT AND DEFORMATIONS

Toughness and resistance to impact and deformations are properties of significant practical importance especially for applications involving shock loads and impact. Brittleness and poor impact resistance are among the main drawbacks of traditional WC/Co hard metals. The thermal spray coatings such as high-velocity oxygen fuel (HVOF) WC/Co are known to crack and spall under high-load and high-cyclic-fatigue conditions [6]. These drawbacks restrict the use of cemented carbides and spray-deposited coatings on tools and wear parts operating in conditions where shock loads, substrate deformations, and impact may cause fracture and catastrophic failure. The Hardide coating can provide a solution to these problems, thanks to its enhanced toughness.

Accurate measurement of fracture toughness of coatings is complicated by the substrate influence and also by the residual stresses in the coating/substrate system. Two methods were used in an attempt to measure the fracture toughness of Hardide-T coating: repeat nano-impacts and also indentations using cube corner diamond indenters.

MicroMaterial Ltd. NanoTest Vantage NTX Instrument was used to make a series of 100 nano-impacts into the same spot of the coated samples using a sintered alumina indenter with a spherical tip. Figure 7.6 shows the nano-indenter in contact with a Hardide-T sample (left) and also the SEM image of the coating area subjected to the series of 100 impacts (right). The Hardide-T coating sample shows only a small indentation left in the impact area, but no cracks were formed and in general this test failed to fracture the coating. In fact, the coating dynamic hardness, dynamic depth, and coefficient of restitution during the 100 nano-impacts showed no signs of the coating brittle behaviour at all.

Four different tip geometries were used in order to determine the mechanical behaviour of the Hardide-coated systems under distinct stress fields: Berkovich, Vickers, Spherical, and Diamond cube corner indenters. The traditional Berkovich and Vickers indenters failed to initiate cracks into Hardide coatings; therefore, a sharper diamond cube corner indenter was used repeatedly in an attempt to introduce cracks into Hardide-T coating. The length of these cracks could be used to estimate fracture toughness of the coating material. Figure 7.7 shows SEM images of the indentations made with the four types of indenters: all of them failed to produce measurable cracks extending from the corners of the indentation.

As a result, both nano-impact and various indentations tests failed to produce cracks in the Hardide-T coating and thus could not measure its fracture toughness. Meanwhile, using some comparative fracture toughness values for other materials showing similar behaviour an estimate was made that the fracture toughness of Hardide-T coating is higher than $9 \text{ MPa} \times \text{m}^{1/2}$. This value represents the lower limit of the fracture toughness of bulk hard materials which do not show brittle behaviour and do not form cracks under the indentation conditions similar to those described above.

Other laboratory and field tests of Hardide coatings also produced evidence of their excellent fracture toughness, resistance to impact and shock loads. In laboratory testing, Hardide coatings could withstand deformations of 3000 microstrain without producing any defects such as micro-cracks. In NACE corrosion tests, coated samples that were pre-stressed to this deformation strain were immersed in aggressive solutions for 30 days and the coatings still protected the substrate from the corrosive attack—proving the deformation produced no cracks even at microscopic level.

Figure 7.8 below illustrates the Hardide coating's ability to survive impact and significant substrate deformations without spalling or cracking: photo on the left—a crater after a shock impact on a

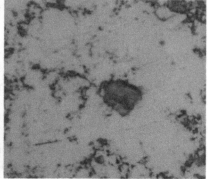

Figure 7.6 Sintered alumina nano-indenter in contact with a Hardide-T sample (left) and also the SEM image of the coating area subjected to the series of 100 impacts (right).

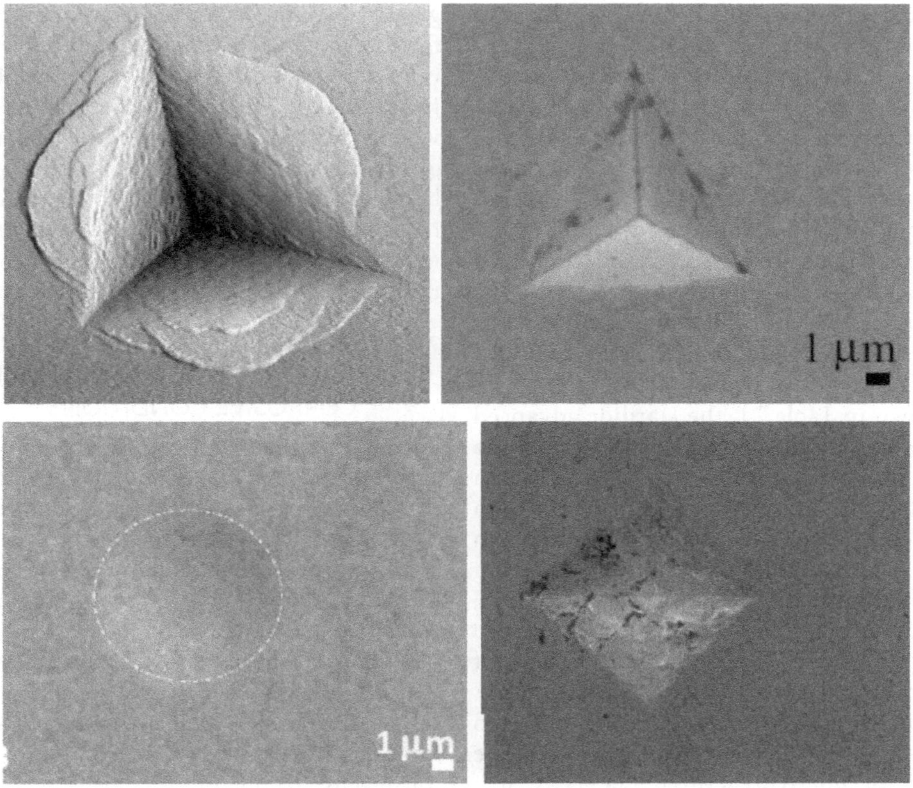

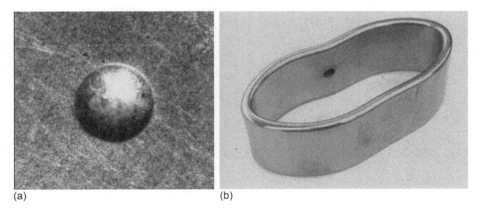

Figure 7.7 SEM images of the indentations made with four types of indenters: a diamond cube corner indenter (top left), Berkovich indenter (top right), spherical indenter (bottom left), and Vickers indenter (bottom right). None of the indenters produced cracks in the Hardide-T coating samples. (Micrographs produced by the University of Southampton and the Universitat Politechnica de Catalunia.)

(a)   (b)

Figure 7.8 Illustrates the Hardide coating's ability to survive impact and significant substrate deformations without spalling or cracking. (a) a crater after a shock impact on a Hardide-T-coated sample, (b) a coated steel ring crushed in a vice.

Hardide-T-coated sample, on the right—a coated steel ring crushed in a vice.

During operation in erosive or abrasive environments, brittle hard material could suffer from micro-cracking and fatigue erosion leading to premature failure. For these reasons, toughness and ductility are very important for practical applications of hard materials. In reality, hardness and toughness are often contradicting characteristics. Most traditional materials can be either hard (but brittle—like glass or cemented carbide), or tough and ductile (but soft—like copper).

As shown in Table 7.1, the Hardide advanced CVD coating technology enables control of the hardness and toughness of the coating, which can be tailored to meet specific application requirements. This control was achieved through extensive experimentation, mainly involving the following factors: coatings composition, nano-structure, intrinsic stresses, and also presence of alloying elements which can affect mechanical properties even in minute amounts. Nano-structured materials are expected to combine properties which are not compatible in macro- or micro-structured materials, like hardness and toughness. Hardide-T is a good example as it has high hardness (1100–1600 Hv) while demonstrating unique toughness, crack, and impact resistance. Hardide-T coated onto a component manufactured out of a tough alloy

gives an unprecedented combination of surface wear resistance and the ability to survive impacts and shock loads—which are illustrated further below. The combination of high hardness with enhanced toughness protect against various mechanisms of wear and erosion. As shown below, the Hardide coating's wear and erosion resistance are superior even to some of the tested materials which have higher hardness but poor toughness.

7.1.3.2 COATING RETAINING SURFACE FINISH AFTER OPERATION IN ABRASIVE AND CORROSIVE CONDITIONS

The as-deposited Hardide coating reproduces the substrate surface finish, adding some roughness due to its nodular morphology. A uniform coating thickness also facilitates the finishing of Hardide-coated parts: in most applications, a good finish can be achieved by polishing, without the need for expensive and complicated post-coating grinding operations. Using standard polishing techniques and commercially available polishing materials the coating can be polished to 0.2–0.3 microns Ra. Simple geometry coated parts can be superfinished to 0.02 microns Ra finish.

Due to their nano-structure, Hardide coatings retain their surface finish even in abrasive or corrosive environments. This is illustrated in Figure 7.9 showing SEM images of the worn surfaces of two types of tungsten carbide coatings after the ASTM

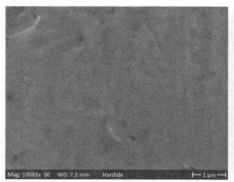

Figure 7.9 SEM images of the worn surfaces after the ASTM G65 abrasion test, procedure E: (left) the Hardide-T coating sample, and (right) the WC-6%Co carbide grade; both materials have the same hardness of about 14.5 GPa. Sample on the right shows selective wear of soft metal binder, making its surface rough and abrasive. (From Konyashin, I. et al., *Int. J. Refract. Met H.*, 49, 203–211, 2015.)

G65 abrasion test, procedure E. Image on the left is the nano-structure Hardide-T coating sample, and the sample on the right is the WC-6%Co cemented carbide grade, which has average tungsten carbide grain size ~1 micron. Both coatings have the same hardness of about 14.5 GPa [4]. The micro-structured coating sample on the right shows selective wear of soft cobalt metal binder, leaving hard and sharp grains of tungsten carbide phase protruding. This loss of the binder makes the coating surface rough and abrasive with exposed sharp edges of tungsten carbide grains. In contrast to this, there is no selective wear or leaching of the softer phase from the nano-structured Hardide coating.

Nano-structure and absence of porosity in Hardide coatings also help prevent its surface roughening in corrosion or chemically aggressive media. Figure 7.10 shows an example of such surface degradation of HVOF WC/CoCr coating surface after exposure to seawater-corrosive media caused CoCr binder leaching, creating pits. Rougher surfaces as shown on Figures 7.9 (Right) and 7.10 can become extremely abrasive for counter-body surfaces, for example if the coated shaft is rotating in a sliding bearing or against an elastomeric seal. The hard grains with sharp edges visible on these images become in fact grinding tools accelerating wear of the bearing or seal material many-fold.

This ability of the Hardide nano-structured CVD coatings to retain good surface finish in abrasive or corrosive environments makes them non-abrasive for the counter-body, such as a bearings or elastomeric seals.

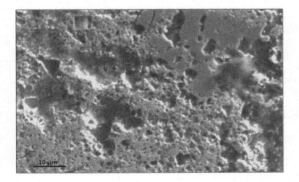

Figure 7.10 HVOF WC/CoCr coating surface after exposure to seawater—corrosive media caused CoCr binder leaching, creating pits.

7.1.3.3 ABSENCE OF POROSITY AND CORROSION-PROTECTIVE PROPERTIES OF HARDIDE COATING

Traditional hard coatings such as flame-sprayed or hard chrome have micro-pores and micro-cracks which open when the substrate deforms under load and allow the solution to attack the substrate. This sort of porosity is almost unavoidable when the coatings are formed by micrometre-scale particles, for example thermal spray powders or ceramic grains. It is not possible to pack such particles with 100% density; there would always be some gaps and residual porosity, which could be of either closed or open type. These pores are sometimes sealed using epoxy or other organic sealants with low viscosity which would penetrate into the open pores and then polymerise to seal them. These sealed coatings have several disadvantages. Firstly, the use of organic sealant limits the temperatures to which the coatings can be exposed as many sealants would degrade or oxidise above 200°C. The second problem is that the sealants can only seal pores which are open to the surface and when the coating gradually wears as the coated parts are used, the deeper concealed pores-which are not sealed and thus are prone to corrosion-would open. In contrast, the Hardide coating is practically pore-free as applied and does not require additional sealing in most applications.

The CVD coating deposition mechanism is fundamentally different: the coating is crystallised atom-by-atom on the part surface. Together with the coating nanostructure this enables the production of pore-free layers. The Hardide coating is free from through-porosity starting from the coating thickness of less than 1 micron. The coating is crystallised from the gas-phase atom-by-atom; the highly mobile reaction products fill micro-pores and defects in the coating as it grows. The coating porosity was measured in a metallographic cross-section of a coated sample, using Gwyddion v2.19 Image analysis software to measure the area of dark pores as percentage of the total coating section area. Figure 7.11 shows the results for two coatings: HVOF spray WC/Co sample porosity was measured at 2.55%, while the Hardide-T coating sample showed 0% porosity. Another porosity analysis method measured the difference between theoretical and actual material density. The Hardide coating porosity was estimated at less than 0.04% by volume.

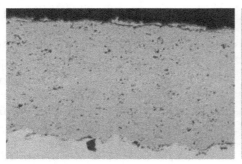

Figure 7.11 Porosity measurements in coatings metallurgical cross sections: left—HVOF WC/Co has 2.55% porosity (dark spots), right—Hardide coating has 0% porosity.

Tungsten and tungsten carbide have high chemical resistance, especially to acids (including H_2S) [7]. Unlike sprayed WC/Co tungsten carbide coatings, Hardide does not use cobalt, which can be affected by acids; this is especially important for processing sour oil. As a result, the pore-free Hardide coating is resistant to aggressive chemicals and can isolate and protect critical parts and tools from corrosion and chemical attack. These protective properties were demonstrated in the Hardide testing in corrosive and chemically aggressive conditions, described below.

Figure 7.12 shows the results of comparative neutral salt spray corrosion tests to ASTM B117-07, where mild steel plates were coated with three coatings: hard chrome, HVOF WC/Co, and Hardide. Hard Chrome samples were badly corroded and removed from test after just 288 hours exposure. HVOF-coated samples showed heavy rust stains and the coating blistered due to the intensive corrosion of the steel plate beneath. The Hardide coating samples showed only light staining. Unlike various soft anti-corrosion coatings, Hardide coatings offer the additional benefit of enhanced wear and erosion resistance. Hardide coatings can be used at temperatures up to 400°C-where organic coatings and sealants cannot.

Hardide-T was tested by Bodycote Materials Testing for resistance to H_2S in the NACE sulphide stress cracking test in a solution of 5% NaCl,

Figure 7.12 Samples of three different coatings after salt spray corrosion tests: left—HVOF after 480 hours; centre—Hard Chrome after 288 hours; right—Hardide coating after 480 hours.

0.5% acetic acid, saturated with H_2S [8]. Samples were tested in three-point bent conditions with coating elongation up to 3000 micro-strain. During the 30-day test, the uncoated control sample cracked across the full 20 mm width and suffered from extensive micro-cracking and pitting while the same substrate coated with Hardide-T showed no micro or macro-cracking or degradation after the same test. This confirmed the non-porous structure of Hardide-T, as under 3000 micro-strain deformation any existing micro-cracks or defects would open up channels for the aggressive fluid to attack the substrate.

The zero porosity of Hardide-T is important for applications with gas valves where any porosity can result in gas diffusion through the coating layer leading to a potentially explosive mixture.

7.1.3.4 ENHANCED WEAR AND EROSION RESISTANCE OF HARDIDE COATINGS

7.1.3.4.1 Wear and abrasion resistance

Wear and abrasion resistance of Hardide coatings have been extensively tested, both in the laboratory and in industrial environments. Figure 7.13 presents the results of abrasion resistance tests performed in accordance with the ASTM G65 standard—Procedures A and B [9]. The results showed that the Hardide coating wear rate is 40 times lower than abrasion-resistant steel AR-500, 12 times lower than

hard chrome, and four times lower than thermally sprayed WC coatings.

The Hardide coating technology allows a wide range of flexibility in the control of the coating hardness, e.g., Hardide-A coating can range from 800 Hv to 1200 Hv and Hardide-T can range from 1100 Hv to 1600 Hv. In order to investigate the effect of the coating hardness on its wear resistance, a series of coating samples were tested using modified ASTM G65 method [10] with wet silica sand abrasive under the following conditions:

Test Rig	Modified ASTM G65
Test Duration	21 minutes in 3-minute increments
Load	130 N
Wheel Speed	75 rpm
Abrasive	David Ball PLC 60/85 mesh fuse sand, batch PML/1009/09-330
Counterface	Shore durometer A 60 neoprene
Condition	Wet (medium; pH 6.3 mains water)

The results are presented in Figure 7.14 as the volumetric wear rates for coatings with different hardness. For comparison, a control uncoated Inconel 718 sample was tested in the same conditions and showed a wear rate of 116.68×10^{-6} cm^3 s^{-1}.

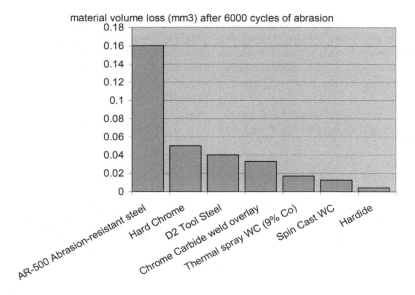

material volume loss (mm3) after 6000 cycles of abrasion

Figure 7.13 Results of ASTM G65 tests of Hardide coating abrasion resistance as compared to the results for other hard materials.

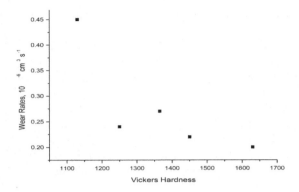

Figure 7.14 Results of modified ASTM G65 wear test: Variation of Hardide coating's wear rate with hardness. The control uncoated Inconel 718 sample shown the wear rate of 116.68×10^{-6} cm^3 s^{-1} in the same test.

As can be expected, the coating wear rate decreases as its hardness increases. The most significant improvement in wear resistance was observed when the coating hardness increased from 1130 Hv to 1250 Hv and the wear rate reduced by almost 50%. This step change can be explained by the coating hardness exceeding that of silica sand. Quartz has hardness up to 1100 Hv, so the coating hardness exceeding the hardness of the abrasive material suppresses the micro-cutting abrasive wear mechanism. A further increase of the coating hardness from 1250 Hv to 1630 Hv gave only a moderate reduction in the wear rate by some 20%–25%. When the coating hardness is significantly higher than the abrasive hardness, other wear mechanisms become dominant such as fatigue micro-cracking of the coating surface after repeated deformations. Toughness and fatigue-resistance of the coating become more important to resist this wear mechanism.

As compared to an uncoated Inconel 725 sample, the Hardide-T coating showed a wear rate reduced

by a factor of ×260 or even by ×585 against wet-sand abrasive. This illustrates the difference that can be achieved by the use of the hard coating: 260–580 microns wear of an Inconel part could result in significant leaks in a hydraulic system, slack, or loss of performance of mechanisms, while wear of just 1 micron of the same part with Hardide coating would not have any noticeable effect in most applications.

In addition to hardness and fracture toughness, the coating residual stresses can also influence its wear-resistance. Hardide coatings have a low thermal-expansion coefficient (TEC): 4.3×10^{-6} per °C which is significantly lower than the TEC of common substrate materials like 316 austenitic stainless steel which has a TEC 16×10^{-6} per °C. When the coating is cooling down from its application temperature of 500°C, this mismatch between the coating and the substrate thermal expansion results in the coating being in a compressive stress which is beneficial for its wear-resistance and fatigue properties.

7.1.3.4.2 Erosion Resistance

Erosion resistance tests were performed in accordance with ASTM G76-95; aluminium oxide (particle size 50 μm) was used as the erosive material and the gas jet velocity was 70 m/sec. Table 7.2 and Figure 7.15 below present the Hardide coating results and comparative results for other hard materials at various angles of impact-90°, 60°, 45°, and 30°. The Hardide coating's erosion rate was 0.017–0.019 mm^3/g, which was significantly better than the erosion rate of the tested types of cemented carbide, white iron, hard chrome, and chrome carbide weld overlay. Hardide coating resists erosion three times better than steel and more than two times better than WC cladding.

Table 7.2 Erosion rates of hard materials tested to ASTM G76-95: Alumina in gas jet at 70 m/sec

Angle of target	Hardide	Chrome carbide weld overlay	White iron	AR-500 abrasion-resistant steel	WC cladding	Hard chrome
		Erosion rate, mm^3/g*1000				
30	17					
45	19	71	76	53	36	25
60	18	66	64	48	41	26
90	18	60	40	40	50	30

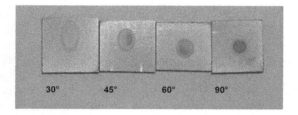

Figure 7.15 Hardide-coated samples after ASTM G76-95 erosion tests with alumina particles at 70 m/sec under different angles.

It is interesting that the Hardide coating erosion rate was almost constant at all the tested angles from 30° to 90° and variation was within the +/−5% of the mean value while with other hard-facing materials the variation was more than +/−15%. Harder materials like WC cladding have higher erosion rates at 90° angle when each particle impact releases more kinetic energy. This erosion at a high impact angle involves brittle micro-cracking of the hard material, producing Hertzian ring cracks and cone cracks leading to material loss via chipping. For some materials, it also involves the platelet erosion mechanism. Less hard and more ductile materials like White Iron show an opposite trend: a higher erosion rate at 30° due to the micro-cutting erosion mechanism: hard particles moving along the surface are 'ploughing' and removing material by a micro-cutting action. Hardide coatings combining ultra-hardness with toughness and ductility are capable of resisting all these erosion mechanisms equally well. The coating's ultra-hardness inhibits the micro-cutting mechanisms of wear and erosion, while its toughness, ductility, residual compressive stresses, and homogeneous micro-structure prevent fatigue micro-cracking/chipping and platelet mechanisms of erosion.

The Hardide coating also significantly outperformed various hard materials in a sand/water erosion test conducted by Southampton University (RIfI) using the following test conditions:

- Erodent: angular sand 300 μm average grain size in water
- Slurry concentration of ~2.1% w/w
- Jet velocity: 24 m/s
- Impingement angle: 90 degrees
- Total test time: 120 minutes

The following erosion rates were measured:

Hardide-T Erosion Rate: HVOF coating	1.38 mg per hour
85WC-10Co-4Cr:	5–15 mg per hour
Stainless steel	171 mg per hour

Hardide-T erosion rate was measured to be 7× lower as compared to the average value of HVOF 85WC-10Co-4Cr and 125× lower than stainless steel.

In both wear and erosion tests, Hardide outperformed HVOF spray WC/Co coatings of similar hardness: Wear rate was reduced by a factor of 4× and sand erosion rate was lowered by a factor of 7×. The possible explanation for this enhanced performance is the synergetic effect of high hardness with enhanced toughness and fatigue-resistance of the nano-structured Hardide coatings.

7.1.4 Examples of Hardide nanocomposite coating applications

Experience of practical use of the Hardide coatings is at least as important as the laboratory testing. Several examples of Hardide applications are described in this section.

7.1.4.1 COATING COMPLEX SHAPES AND INTERNAL SURFACES

The CVD of Hardide coatings are particularly suitable for coating complex-shaped parts with precise dimensions, or where internal surfaces need protection against wear and erosion.

For many coating technologies, internal surfaces or items of a complex shape are very difficult or even impossible to coat due to the line-of-sight nature of the coating processes. For example, spray coatings (including HVOF, Plasma Spray, D-gun) can be applied to external surfaces which are easily accessible for the spray gun so that the spray nozzle can be kept at a distance to prevent overheating of the part being coated. When spray-coating complex-shaped items, it is difficult to avoid building a thicker coating layer on the more exposed edges while applying a thinner layer in the 'shadowed' areas. This can distort the part shape. Similar limitations exist for PVD coatings, where planetary rotation of the parts being coated helps achieve coating uniformity on the external surfaces but the coatings cannot be deposited uniformly inside deep holes. Electrolytic processes such as hard chrome plating can coat inside but often build a thicker layer on the edges where the current density is higher, thus creating the 'dog-bone' shape. As a result, complex-shaped items such as a pump impeller, a hydraulic cylinder ID, or an extrusion die could not be coated by these traditional methods.

Hardide coatings are applied from gas media at a low pressure by CVD technology. The coating is crystallised atom-by-atom from the gas phase on every hot surface in contact with the reactive gas mixture. When this mixture is pumped through a hydraulic cylinder bore, the coating grows uniformly inside the part. To illustrate this important ability, Figure 7.16 shows some examples of complex-shaped items coated with Hardide. Figure 7.17 shows a magnified section of M5 thread with 50-micron-thick Hardide coating. The coating is not only uniform on the edges and between the threads, it also accurately follows imperfections in the steel surface. This ability to coat uniformly internal, as well as external, surfaces and complex shapes opens many new applications where previously hard coating was impossible.

7.1.4.2 COATING SEVERE-SERVICE BALL VALVES AND THEIR PERFORMANCE

Ball valves similar to those shown in Figure 7.18 can suffer from abrasion by sand or stone chippings present in the fluids or from erosion by accelerating flow when the valve is being closed/opened. Hardide coatings make the valve parts scratch-proof and able

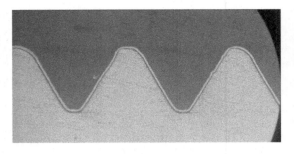

Figure 7.17 A cross-section SEM of Hardide coating on an M5 thread shows ability to coat complex shape items uniformly; even slight imperfections in the substrate are accurately conformed.

Figure 7.18 Ball valves coated with Hardide.

to resist abrasion and erosion. This prevents valve leaks and significantly increases service life.

Such Hardide coatings are used by LG Ball Valves/EnerMech Ltd, a UK producer of ball valves since 2003. Most of the Hardide-coated

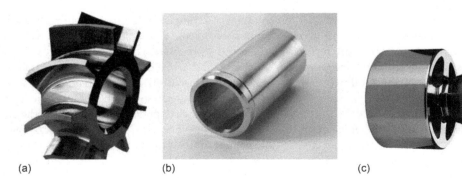

(a)   (b)   (c)

Figure 7.16 Examples of complex shape parts coated with Hardide coating: **(a)** an impeller **(b)**, a pump cylinder ID coated (centre), **(c)** a down-hole tool flow diverter (right).

LG Ball Valves' valves are used in topside applications in the oil and gas industry and they are in service in the UK, Norway, and South Africa as well as in high-pressure oil refinery applications. The Hardide-coated valves have been *in service* for between one and two years with no failures reported.

In an instant coffee manufacturing application, hard-chrome-plated ball valves suffered from intensive abrasion and erosion and had to be replaced every few days. Since being Hardide-coated, they have been in continuous service for over 18 months.

Hardide-coated LG Ball Valves' valves are also used successfully in speciality chemicals manufacturing where chemical resistance is required. In these cases, the coated valves have been in service for more than six months while previously the valves were failing every few days or weeks. Hardide-coated valves are also in use in cryogenic equipment controlling liquid Helium at a temperature of $-196°C$ and pressure of 200 bar, an application which is very abrasive for valves.

After two years of working in co-operation and impressive slurry test results, Hardide has been approved for use on a new line of ball and seats by Flowserve—one of the leading providers of flow-control products. The Hardide coating enabled Flowserve to offer 316 stainless steel as the base metal for use in severe service applications that require metal-to-metal seating, including abrasive and slurry applications. In the Flowserve slurry tests Hardide coated 316 balls and seats remained operational after more than 70,000 cycles in slurry, whereas Stellite would fail in 29,000 cycles.

7.1.4.3 APPLICATIONS WITH OIL DRILLING AND DOWNHOLE TOOLS

Hardide coatings are used successfully in several advanced down-hole tools including:

- Mud-driven hydraulic parts for directional drilling tools
- High-loading bearing pins
- Grippers for down-hole tractors

In each of these applications the Hardide-coated parts are operating in a highly abrasive and erosive drilling mud environment. In some cases, the mechanical abrasion is combined with chemical attack by acidic fluids and H_2S.

Figure 7.19 Hardide-coated Inconel pin/bush assembly survived intense repeated hammer impacts without fracture or flaking despite significant deformations of the substrate (×5).

When the down-hole tools are moved or operated they often experience impact or shock loads and can be significantly deformed. In these conditions brittle materials could fail catastrophically. If these actions cause fracture or chipping of the hard materials, this may lead to a catastrophic equipment failure. One example is shown in Figure 7.19. This is a pin/bush assembly of an oil down-hole tool which must resist abrasion by sand and stone chippings and also survive impact and shock loads. After hammer-testing, the coated parts retain their integrity and can continue operating under harsh conditions.

Hardide coatings have extended the life of critical parts for these tools and reduced the downtime costs. In some applications, Hardide-T appeared to be the only suitable solution as traditional hard materials were too brittle and difficult to machine due to the complex part geometry, while other coating technologies were not able to reach the important hidden-surface areas.

7.1.4.4 APPLICATIONS IN PUMPS

Hardide-T coating is used on the inside cylinder and outside piston of a positive displacement pump handling abrasive viscous fluids at a pressure of up to 2800 psi. In this application, the main coating advantages were the ability to coat internal surfaces, enhanced wear-resistance, and also reduced wear of packing counter-surfaces. The coating has tripled the pump life.

7.1.4.5 HARDIDE AS A HARD CHROME REPLACEMENT

Hardide is an attractive replacement for hard chrome, which is to be phased-out due to environmental and health and safety considerations. Hard-chrome plating is widely used as a wear-resistant and anti-galling coating with some degree of corrosion protection, but the hexavalent chrome salt solutions used in the coating production, and the process effluents, are known carcinogens which represent major health, safety, and environmental problems [6]. Restrictive pollution-control legislation, such as EU REACH and US OSHA, apply further pressure on the plating companies, which increases the cost and reduces the future availability of hard-chrome plating. In response to this, some of the large users of chrome plating such as aircraft manufacturers launched programs to identify new and more environmentally friendly replacement technologies. HVOF thermal-spray coatings are often selected as a suitable replacement [11] but they could not be applied to internal surfaces, have a very rough finish as applied, and require expensive and complicated grinding which is not possible on complex-shaped parts.

In order to meet industry demand for an alternative solution, Hardide has developed a new type of coating, Hardide-A, specifically for aerospace applications. This coating has hardness similar to hard chrome (800–1200 Hv) and can be applied with the same thickness as hard chrome (typically between 50 and 100 microns). This makes it easier for hard chrome users to adapt drawings and specifications and reduces the cost of a switch to Hardide-A. As Hardide-A is free from the micro-cracks typical of hard chrome, it has much better corrosion resistance—see Figure 7.9—and also demonstrates excellent fatigue properties. Airbus is one of the prospective customers conducting qualification testing of the Hardide-A coating as a replacement for hard chrome. Applied by CVD technology, Hardide-A is particularly suitable for the coating of internal surfaces and complex shapes which are difficult to coat by other coating technologies.

Most of the Hardide coatings' applications are structural components where coating hardness and toughness provide excellent resistance to wear and erosion, combined with corrosion-protection. Biomedical applications have been deliberately avoided as these tend to require extremely long and complicated testing. The Hardide coatings' combination of properties makes it a promising material for some functional applications where Hardide coating has a function different from wear- and corrosion-resistance, i.e., as a catalyst. Meanwhile, these functional applications have not yet progressed to the commercial stage and scale and mostly remain at the R&D stage.

7.1.5 Summary

Nano-structured Hardide coatings offer a combination of protective properties including wear and erosion resistance, protection against aggressive chemicals and corrosion, as well as toughness, impact, and crack resistance. The coatings' structure and composition give them a combination of ultra-hardness with excellent toughness and ductility. From extensive laboratory and field testing it has been demonstrated that this combination of hardness with toughness achieves excellent protection against both wear and erosion. The coating hardness inhibits the micro-cutting mechanisms of wear and erosion, while its toughness, ductility, residual compressive stresses, and homogeneous micro-structure prevent fatigue micro-cracking/chipping and platelet mechanisms of erosion. In some tests, Hardide coatings outperformed even harder brittle materials.

The ability to coat internal surfaces and complex shapes opens new potential applications for hard coatings on critical parts. Being pore-free, the coating protects the substrate from attacks by acids. These properties are realised in various applications of Hardide for downhole tools, pumps and valves operating in oil and gas facilities, food manufacturing, refineries, cryogenic equipment and power generation. Typically, the coating triples the operational life of critical parts in abrasive conditions. The use of Hardide enables the advanced design of engineering systems operating in abrasive and corrosive environments and under shock loads.

7.2 PECVD AND LPCVD OF HARD/SUPERHARD TERNARY AND QUATERNARY NANOCOMPOSITE COATINGS

Plasma-enhanced CVD (PECVD) has been used to deposit ternary and quaternary hard nanocomposite coatings such as Ti–Si–C–H [12] and TiN/SiN$_{1.3}$ coatings [13]. For the deposition of nanocomposite of Ti–Si–C–H coatings, TiCl$_4$, CH$_4$, H$_2$,

Ar precursor gases were used at flow rates of 25, 16, 100, and 50 sccm, respectively. Si contents were achieved by introducing SiH_4 as an additional working gas at low rates of 0, 0.7, 1.5, 3.0, and 7.0 sccm for respective samples. Electromagnetic excitation at a radio frequency of 13.56 MHz was used to generate reactive plasma during deposition. The hardness of the nanocomposite coating deposited at 400°C was between 22 and 35 GPa depending on the hydrogen content. The typical microstructure of the coating consists of face-centred cubic titanium carbide nanocrystallites (nc-TiC) surrounded by an amorphous matrix of hydrogenated carbon (a-C:H) and in the presence Si from the silane gas, the amorphous DLC-type phase was found to be alloyed with correspondingly different amounts of silicon (a-C:Si:H) [12].

Nanocomposite $TiN/SiN_{1.3}$ hard coatings were deposited using the deposition conditions summarised in Table 7.3 [13]. For an optimum concentration of Si (5–10 at.%), nanocomposite coatings with high values of microhardness (25 and 45 GPa) and of Young's modulus (270 and 350 GPa) were achieved for the films deposited at 300°C and 500°C, respectively. These mechanical properties are superior to TiN (H;22 GPa, E;200 GPa) and $SiN_{1.3}$(H;18 GPa, E;160 GPa) coating materials.

CVD of TiCxNy coatings has been widely used for cutting tool applications as reviewed in Chapter 1. Further improvements have been achieved by adding Si to form the quaternary TiSiCN nanocomposite coatings to improve hardness and oxidation resistance using PECVD [14–17] and physical vapour deposition (PVD) methods [18,19]. These deposition methods allow coating deposition to occur at a lower temperature. In addition, the quaternary TiSiCN nanocomposite coatings also decreased the friction coefficient for sliding against steel and alumina. Nanocomposite coatings deposited by PECVD and PVD exhibited similar hardnesses and properties [17,20].

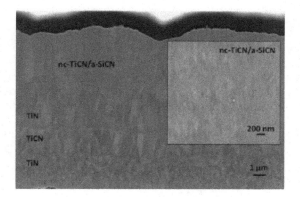

Figure 7.20 SEM image of a TiSiCN top coating with a silicon content of 4.14 at.% on a TiN/TiCN/TiN interlayer system (deposition at 850°C, $SiCl_4$/$TiCl_4 = 8.2$). The inset shows a higher magnification of the TiSiCN top coating. (From Endler, I. et al., *Surf. Coat. Technol.*, 215, 133–140, 2013.)

In order to reduce the cost of coating production by avoiding the use of plasma and high vacuum, Endler et al. [20] explored the use of low-pressure thermal CVD (LPCVD) to deposit quaternary TiSiCN nanocomposite coatings onto hard metal substrates at a moderate temperature within the range of 800°C–900°C and 6 kPa to enhance coating hardness, wear, and oxidation resistance. TiSiCN nanocomposite coatings were deposited from a mixture of the precursors $TiCl_4$, $SiCl_4$, and acetonitrile with flow rates of 2.5–13.9 sccm, 20.4 sccm, and 7.9 sccm, respectively. Figure 7.20 shows the high-resolution SEM of 5 μm thick TiSiCN coatings with silicon contents between 2 at.% and 22 at.%. The coatings are nanocomposites consisting of TiCxNy nanocrystals and an amorphous phase of carbon-rich SiCxNy. TiSiCN with 2.2 at.%–11.7 at.% silicon was stable up to 900°C and the coatings with a silicon content of 11.7 at.% have high hardness of up to 4100 HV as compared to TiSiN with a hardness of 3700 HV and stability up to 700°C [20].

Table 7.3 Deposition conditions for the nc-$TiN/SiN_{1.3}$ films

Process	Time [min]	Substrate temperature [°C]	Pressure [mTorr]	Bias voltage [V]	Gas flow [sccm]				
					Ar	N_2	H_2	$TiCl_4$	SiH_4
Pretreatment (1)	10	300, 500	40	−600	40	—	—	—	—
Pretreatment (2)	30	300, 500	300	−600	—	40	—	—	—
Deposition of TiN/$SiN_{1.3}$	60	300, 500	200	−600	40	15	100	9	0–3

Source: Jedrzejowski, P. et al., *Thin Solid Films*, 426, 150–159, 2003.

Superhard nanocomposite coatings consisting of nanocrystalline (nc) phases, such as nc-TiN/a-BN and nc-TiN/a-BN/a-TiB$_2$ with a hardness of 40–50 GPa, have been deposited by plasma-induced CVD with a high-frequency discharge inside a silica glass reactor by Veprek et al. [21–23]. The reactor was inserted into an electrical oven with controlled temperature. The substrate holder was used as one electrode, whereas the other electrode was a grounded nickel sheet attached tightly around the silica tube. The processing parameters and the substrates used have been summarised in Table 7.4 [21]. The microstructures of the coatings with the variation of B contents are shown in Figure 7.21.

Table 7.4 The deposition parameters used for the deposition of nc-TiN/a-BN/a-TiB$_2$ coatings by plasma CVD

Deposition parameters	
Frequency	13.56 MHz
Flow rate of H$_2$ (purity 99.999%)	50 sccm
Flow rate of N$_2$ (purity 99.999%)	5 sccm
Flow rate of TiCl$_4$ (purity 99%)	1.7–1.9 sccm
Flow rate of BCl$_3$ (purity 99.9%)	0–1.8 sccm
Pressure	3 mbar
HF power	100–110 W
Deposition temperature	550°C–600°C
Coating thickness	5–10 Am
Deposition rate	0.65–1.0 nm/s
Substrate	Stainless steel, silicon (111), α-Fe, Mo

Source: Karvankova, P. et al., Surf. Coat. Technol., 200, 2978–2989, 2006.

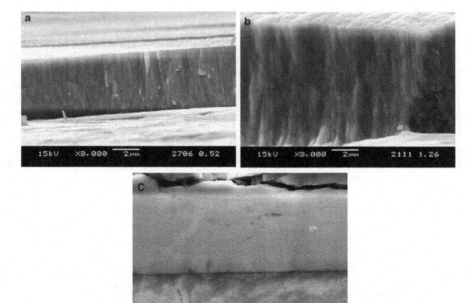

Figure 7.21 The development of the morphology of the nc-TiN/a-BN/a-TiB$_2$ coatings with increasing content of boron in the coatings: **(a)** [B] = 0.9 at.%, coverage = 0.52, H = 34.4 GPa, crystallite size 30.8 nm; **(b)** [B] = 2 at.%, coverage = 1.26, H = 47 GPa, crystallite size = 34.7 nm; **(c)** [B] = 5.25 at.%, coverage = 0.87, H = 48.4 GPa, crystallite size = 8.6 nm. (From Karvankova, P. et al., Surf. Coat. Technol., 200, 2978–2989, 2006.)

A mechanism of formation was proposed due to the spinodal nature of the phase segregation in the coating system during the film growth [22,23].

The maximum hardness of the nc-TiN/a-BN and nc-TiN/a-BN/a-TiB$_2$ coatings was obtained at the percolation threshold when there was about a continuous monolayer of thin a-BN between the TiN nanocrystals. The chlorine content in the coating would influence the hardness and coating adhesion. The chlorine content in the nc-TiN/a-BN/a-TiB$_2$ coatings prepared using the conditions in Table 7.1 was in the range of 0.5–1.75 at.%. A chlorine content above 2% leads to poor coating adhesion and a lower hardness. The maximum boron content in the coatings prepared was circa 15 at.%. Such nanocomposite coatings have reasonably high thermal stability and oxidation resistance, although both are lower than the corresponding values of the nc-TiN/a-Si$_3$N$_4$ and nc-TiN/a-Si$_3$N$_4$/a-TiSi$_2$ coatings.

7.3 A HYBRID PVD AND CVD OF HARD NANOCOMPOSITE FILMS

7.3.1 Single-layer nanocomposite film

A hybrid CVD/PVD system (Figure 7.22) has been used for the synthesis of nanocomposite Ti–Si–N

thin films whereby silicon was incorporated into crystalline TiN by bubbling a silicon-based liquid precursor (e.g., tetramethylsilane, TMS) at a flow rate between 0 and 60 sccm into a cathodic arc process at a substrate temperature of 350°C. The reaction deposition of TiN is produced from the Ti arc cathode in the presence of nitrogen (flow rate of 140 sccm) at a deposition pressure of 0.8 Pa. The arc source was operated at a DC arc current of 145 A and a bias voltage of −150 V was applied to the substrate (100) conducting silicon wafers (resistivity 0.05 Ω cm) [24,25]. The TiN crystalline size decreased from 33 nm to 4 nm with the increase of TMS. The hardness of the films was found to be strongly dependent on the Si content as shown in Figure 7.23. Figure 7.24 shows a high-resolution TEM image of a Ti–Si–N film and indicates that the nanocomposite coating consists of TiN crystallites and amorphous Si$_3$N$_4$. The presence of the amorphous Si$_3$N$_4$ phase in the film has resulted in high hardness of 41 GPa with ~5% Si content. The mechanism could be due to any dislocation movement during high stress being trapped at the grain boundaries by the amorphous layer surrounding the nanocrystallites resulting in the increase in the material strength.

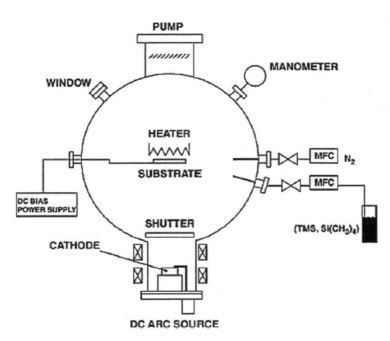

Figure 7.22 Schematic diagram of the hybrid cathodic arc and CVD system. (From Bendavid, A. et al., *Surf. Coat. Techol.*, 201, 4139–4144, 2006.)

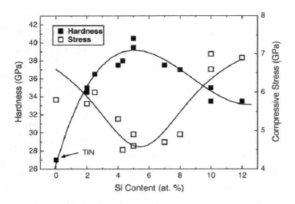

Figure 7.23 Dependence of the measured hardness and compressive stress of the Ti–Si–N films on the silicon content in the films. (From Bendavid, A. et al., *Surf. Coat. Techol.*, 201, 4139–4144, 2006.)

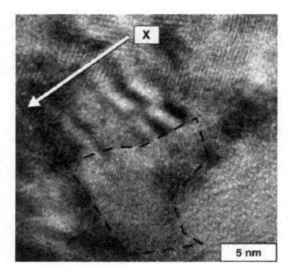

Figure 7.24 High resolution TEM image of Ti–Si–N film with–5 at.% silicon content. (From Bendavid, A. et al., *Surf. Coat. Techol.*, 201, 4139–4144, 2006.)

7.3.2 Multilayered nanocomposite coatings

Multilayered nanocomposite TiSiAlN coatings were deposited on M2 steel by a hybrid physical/chemical vapour deposition process for potential high-speed cutting applications [26], where resistance to high-temperature oxidation is crucial. TiSiAlN coatings were deposited by the reactive cathodic arc evaporation of a sintered 40/60 Ti/Al target in N_2/Ar

atmosphere (PVD). A current between 70 and 90 A was applied to the targets and cathode voltage was between −18 and −22 V and the bias voltage between −80 and −150 V. The substrate temperature was fixed at 450°C. SiH_4 was used as precursor for Si. The flow ratio of SiH_4/N_2 was varied between 0.2 and 0.25 for a total pressure of 0.7 to 1.3 Pa [27]. SiH_4 flow was alternated during the deposition in order to deposit successive alternate layers of TiAlN and TiAlSiN, as shown in Figure 7.25a. Such a hybrid process allowed the control of the silicon enrichment along the coating thickness for the targeted applications. The average silicon enrichment for the Si-rich layer was 5.9 at.% and 1.4 at.% for Si-poor ones (respectively 4.1 and 0.8) for the multilayer. Both films were synthesised by applying a serrated silane partial pressure during deposition, leading to a multilayered structure with a 700 nm period. The stratified structure of both TiSiN and TiSiAlN coatings is shown in the profile analysis (Figure 7.25b) using the glow discharge optical emission spectroscopy (GDOES) chemical analyses. The low silicon content consisted of dominant columnar features, whereas the Si-rich layers exhibited a glassy-like structure compatible with a nanocomposite structure. The multilayer structure and the nanometric size of the grains in layers containing a high Si content were observed by cross-section TEM. The complex nanoscale distribution of the phases in the multilayer nanocomposite coatings were stable at 800°C and preserved the best of the characteristics of TiN and SiNx. Such nanocomposite coatings exhibited a high durability in terms of mechanical and oxidation behaviours. During thermal cycling SiNx would tend to show a strong tendency to cracking, which would favour continuous oxidation. However, in this multilayer nanocomposite coatings, the silicon-containing coatings exhibited a high resistance due to the better thermomechanical behaviour of TiN. The coatings also were highly resistant to thermal cycling; this seems to be due to the synergy between the shield effect of the SiNx network, which would limit the oxidation process and the intrinsic 'deformability' of TiN layers that could withstand the volume change of the substrate due to temperature variations. Thus, the multilayer nanocomposite TiSiN/TiSiAlN coatings enhanced the best properties of their two single phases, i.e., mechanical properties of TiN and the intrinsic refractory properties of SiNx. An addition of aluminium led to the formation of the outer

(a)   (b)

Figure 7.25 **(a)** Cross-section TEM micrograph of a FIB-cut TiSiAlN coated sample. Results of Si content (Si/(Ti+Al)) in at.%) analyses by EDS measurements are mentioned directly for each layer. **(b)** Correlation between the chemical nature of the film and the H_2 partial pressure. Left-axis: GDOES profile chemical analysis of the TiSiN coated steel, right-axis: H_2 partial pressure (by-product of silane decomposition) measured during the deposition process versus the coating thickness. (From Mège-Revil, A. et al., *Thin Solid Films*, 518, 5932–5937, 2010.)

refractory alumina layer, which contributed to the improvement of the oxidation resistance [26].

7.4 LASER-ASSISTED CVD OF OXIDE-BASED NANOCOMPOSITE FILMS

Other variants of CVD such as laser-assisted CVD have also been used to fabricate nanocomposite films of Al_2O_3–ZrO_2 using Al(acac)$_3$ (acac: acetylacetonate) and Zr(dpm)$_4$ (dpm: dipivaloylmethanate) precursors, which were heated at 433 and 501 K, respectively, and transported by Ar carrier gas into the reaction chamber. The Al to Zr molar ratio in the source vapour was 30 mol% Zr. O_2 was introduced into the chamber through a double-tube nozzle. The mass flow rates of Ar and O_2 were maintained at 0.084×10^{-6} and 0.17×10^{-6} m^3 s^{-1}, respectively. The chemical reaction was initiated using an InGaAlAs diode laser beam (wavelength 808 nm, continuous-wave mode, output power 95–170 W), and alpha-Al_2O_3–ZrO_2 and gamma Al_2O_3–ZrO_2 nanocomposite films were produced at 1207 and 1000 K respectively, at a deposition time of 0.6 ks and pressure 100 Pa [28]. The details of the laser CVD apparatus are described in reference [29]. The Al_2O_3–ZrO_2 nanocomposite consists of 10-nm-width of t-ZrO_2 nanodendrites in an Al_2O_3 columnar matrix as shown in Figure 7.26 [28]. The gamma-Al_2O_3–t-ZrO_2 nanocomposite films exhibited a higher nanoindentation hardness (28.0 GPa)

with a lower heat effusivity of 4788 J s$^{-1/2}$ m^{-2} K^{-1}), as compared to alpha- Al_2O_3 film with a lower hardness of 26.8 GPa and a lower higher effusivity (8284 J s$^{-1/2}$ m^{-2} K^{-1}) [28].

7.5 SEQUENTIAL CVD DEPOSITION OF OXIDE-BASED NANOCOMPOSITE FILMS

ZnO-TiO_2 nanocomposites were synthesised by sequential CVD deposition. ZnO nanoplatelets (host) were initially grown on Si(100) and Al_2O_3 substrates, followed by the dispersion of TiO_2 nanoparticles (guest) leading to the formation of ZnO-TiO_2 nanocomposite deposits with an average thickness of 140 nm, whose characteristics were directly affected by the host matrix porosity and the guest amount and dispersion, tailored by varying the TiO_2 deposition time. The chemical precursors for Ti and Zn sources were Ti(O/Pr)$_2$ (dpm)$_2$ and Zn(hfa)$_2 \bullet$TMEDA (O/Pr: iso-propoxy; dpm: 2,2,6,6-tetramethyl-3,5-heptanedionate; hfa:1, 1,1,5,5,5-hexafluoro-2,4-pentanedionate; and TMEDA: N, N,N',N' tetramethyl ethylenediamine). CVD was performed at relatively low temperatures (350°C–400°C) in nitrogen plus wet-oxygen atmospheres, without further heat treatment [30].

For gas sensing, the performance of the nanocomposites depends on their composition and morphology. The gas-sensing characteristics of ZnO-TiO_2 were evaluated for the detection

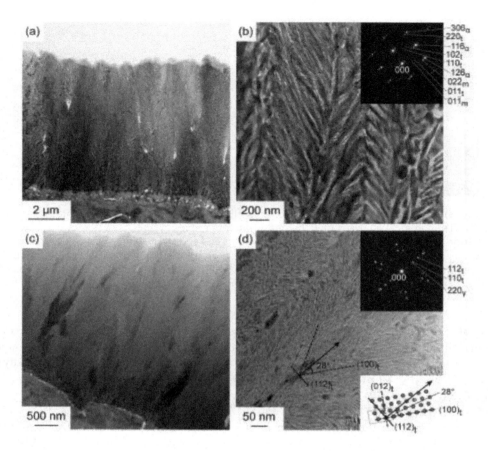

Figure 7.26 TEM bright-field images (a, c) and SAED patterns (b, d) of alpha-Al$_2$O$_3$–ZrO$_2$ **(a, b)** and gamma-Al$_2$O$_3$–t-ZrO$_2$ **(c, d)** nanocomposite films. (From Ito, A. et al., *J. Eur. Ceram. Soc.*, 34, 155–159, 2014.)

of volatile organic compounds (CH$_3$COCH$_3$, CH$_3$CH$_2$OH, and CO) and their performances were better than the pristine ZnO systems, and could be explored for sensing devices for environmental purposes and food-control monitoring [30].

7.6 AACVD OF OXIDE-BASED NANOCOMPOSITE FILMS AND COATINGS

7.6.1 AACVD of Au in transition metal oxides nanocomposites

Metal/metal oxides nanocomposite films such as Au in transition metal oxides have been deposited using AACVD. This can be achieved using a preformed gold (mean diameter of 10 nm) colloid in toluene precursor and co-deposited with [W(OPh)$_6$] at 450°C to form an Au-WO$_3$ nanocomposite coating [31]. Alternatively, nanocomposite coatings can be produced using a simpler one-step process [32], where gold particles are formed *in-situ* during matrix deposition. Au particles were incorporated into transition metal oxide thin films (e.g., TiO$_2$, WO$_3$, and MoO$_3$) with Au:M (M = Ti, W, Mo) atomic ratios ranging from 0.1 to 1.9 and deposited using a single-step process from a solution of [HAuCl$_4$] and a second precursor of [Ti(OiPr)$_4$], [W(CO)$_6$], or [Mo(CO)$_6$], and acetone, methanol, and toluene as solvents onto silica glass at 200°C to 400°C. The gold nanocomposite material exhibited surface plasmon resonance absorption peaks and metal-like reflectivity which is dependent on the concentration of gold particles within the film and the refractive index of the metal oxide matrix.

7.6.2 AACVD of Pt-SnO$_2$ nanocomposite coatings

Pt nanoparticles (3–5 nm) dispersed on polycrystalline SnO$_2$ (8–25 nm) thin films were fabricated by co-deposition of Pt and SnO$_2$ from ultrahigh frequency spraying of a solution using the aerosol-assisted CVD apparatus as shown in Figure 7.27. The precursor solution was prepared by dissolving dibutyltin diacetate and 0%–10% platinum acetylacetonate in acetylacetone (0.1 M concentration). The precursor aerosol (droplet size 2–4 μm) was produced from an ultrasonic beam (850 kHz). The aerosol sol was conveyed by purified air closed to the heated substrates where both the precursor mixture decomposes and pyrolysis occurs leading to film formation at a deposition temperature of 480°C–550°C [33].

The microstructure of the nanocomposite films was characterised using TEM and HRTEM as shown in Figure 7.28. The experimental conditions produced well-crystallised and quasi-spherical Pt particles without anomalous morphology due to multi-twinning.

The SnO$_2$ grain size decreases as the Pt concentration increases (0%–11%). Co-deposition with a low Pt concentration limits the grain size of metallic particles. Such microstructure is attractive for catalysis and gas sensor applications [33]. Recently, AACVD has also been used for the growth of gas-sensitive SnO$_2$ nanorods (NRs) via a non-catalysed vapour solid mechanism onto silicon micromachined platforms, together with *in-situ* functionalisation of the NRs with Au nanoparticles via co-deposition with a gold precursor. Such nanocomposite structure exhibits enhanced sensing properties towards H$_2$ compared to those based on non-functionalised NRs [34].

7.6.3 AACVD of IF-WS$_2$/ Cr$_2$O$_3$ nanocomposite coatings

AACVD has been used to deposit nanocomposite coatings by incorporating inorganic fullerene-like tungsten disulphide (IF-WS$_2$) nanoparticles into a Cr$_2$O$_3$ ceramic matrix to reduce friction and improving wear resistance of Cr$_2$O$_3$.

The IF-WS$_2$ nanoparticles (90–220 nm) were preformed and made by reacting tungsten trioxide spherical particles with H$_2$S in a fluidized bed reactor [35,36]. Figure 7.29 shows a schema tic diagram of the synthesis of IF-WS$_2$ where the conversion of the oxide into WS$_2$ occurs from the outside of the particle towards its centre and creates the hollow onion-like fullerene structure.

The nanocomposite coatings were deposited by mixing the preformed inorganic fullerene-like tungsten disulphide (IF-WS$_2$) nanoparticles (0.23 g/L) with 0.05M chromium nitrate nanohydrate in

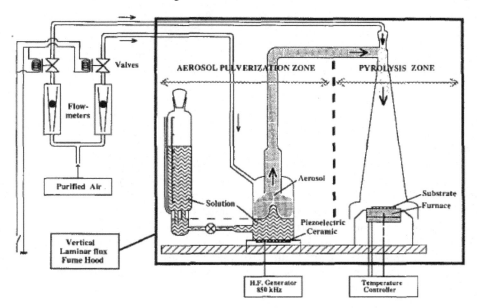

Figure 7.27 Aerosol-assisted CVD for thin-film deposition. (From Labeau, M. et al., J. *Solid State Chem.*, 102, 434–439, 1993.)

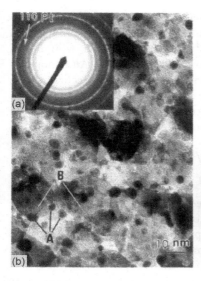
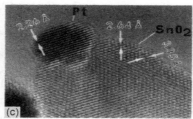

Figure 7.28 Microstructural analysis of Pt(11%)/SnO$_2$ film obtained at 540°C: **(a)** diffraction pattern and **(b)** bright field image of (A. quasi-spherical Pt particles and B, irregular SnO$_2$ grains) of the transmission electron microscopy (TEM) study; **(c)** HRTEM of the nanocomposite film showing the interaction between Pt particle and SnO$_2$ grain induces a distortion of SnO$_2$ grain; **(d)** no multi-twinning appears in the Pt particle structure from the HRTEM micrograph. (From Labeau, M. et al., J. *Solid State Chem.*, 102, 434–439, 1993.)

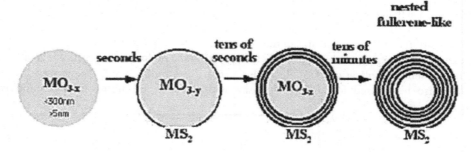

Figure 7.29 Schematic diagram of the synthesis IFLM (M represents tungsten). (From Hou, X. et al., *ECS Trans.* 25, 1135–1141, 2009.)

alcoholic solvent to obtain a uniform suspension via an ultrasonic bath. The precursor mixture was atomised to form aerosol droplets using an ultrasonic generator, at a frequency of 1.7 MHz, with nitrogen as a carrier gas [38,39] and led to the formation of the desired IF-WS$_2$/Cr$_2$O$_3$ nanocomposite coatings at deposition temperatures of 280°C–300°C. Post-annealing of the nanocomposites was carried out in Ar at 500°C for 1hr to obtain the desired crystalline Cr$_2$O$_3$ and microstructure in the coatings. Figure 7.30 shows the schematic illustration of the deposition of nanocomposite coating using AACVD method.

The cross-section SEM images of the Cr$_2$O$_3$ coating with and without IF-WS$_2$ are shown in Figure 7.31. The incorporation of 6wt% IF-WS$_2$ increased the

surface roughness of the coating, and the water contact angle also increased significantly from 8° to 105° (see the insets in Figure 7.31). Thus, the incorporation of IF-WS$_2$ nanoparticles could be used as an effective approach to enhance the hydrophobic and anti-sticking properties of Cr$_2$O$_3$ coatings. The TEM study (see Figure 7.32) confirmed the incorporation of IF-WS$_2$ nanoparticles into the Cr$_2$O$_3$ matrix and that the hollow onion-like IF-WS$_2$ structure is preserved during deposition and annealing. However, agglomeration of nanoparticles in the coating is observed.

Ball-on-test results shown in Figure 7.33 demonstrate that the incorporation of IF-WS$_2$ nanoparticles into the Cr$_2$O$_3$ coating has significantly reduced

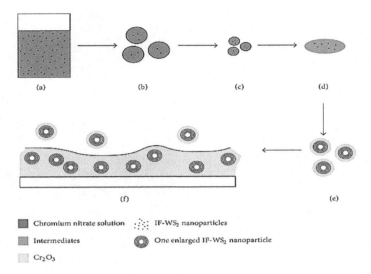

Figure 7.30 A schematic diagram of the deposition of nanocomposite coating using AACVD method. **(a)** Formation of precursor dispersion consisting of chromium nitrate and IF-WS$_2$ nanoparticles, **(b)** atomization of precursor dispersion, **(c)** evaporation of solvent, **(d)** decomposition and chemical reactions of precursor intermediates in gas phase, **(e)** nucleation and growth on IF nanoparticle surface, and **(f)** deposition of IF-WS$_2$/Cr$_2$O$_3$ on the substrate surface. (From Hou, X. et al., *J. Nanomater.*, Article ID 219039, 2013.)

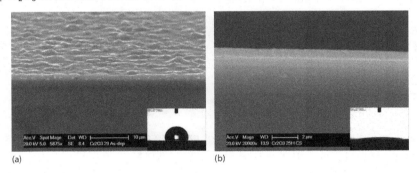

Figure 7.31 The cross-section SEM of the as Cr$_2$O$_3$ coating without **(a)** and with **(b)** IF-WS$_2$. The insets show the water contact angle measurement of the coatings.

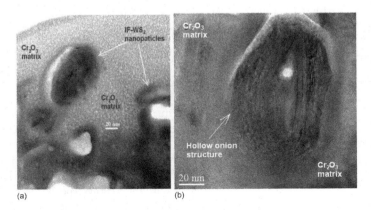

Figure 7.32 **(a)** TEM of the nanocomposite coatings; **(b)** Hollow IF-WS$_2$ nanoparticle in Cr$_2$O$_3$. (From Hou, X. et al., *ECS Trans.*, 25, 1135–1141, 2009.)

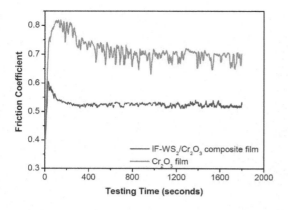

Figure 7.33 Coefficient of friction of the deposited coatings under ball-on-flat test. (From X. Hou, K.-L. Choy, *Thin Solid Films*, 516, 8620–8624, 2008.)

the coefficient of friction by ~25% compared with the pure Cr_2O_3 coating, without the incorporation of nanoparticles. The results demonstrate that AACVD is a promising method for the incorporation of nanoparticles into ceramic matrices to form nanocomposite coatings.

7.7 LOW-PRESSURE CVD OF SILICON/GRAPHITE NANOCOMPOSITE ELECTRODE

Low-pressure CVD has been used to fabricate silicon/graphite composite electrodes in order to combine the high capacity of silicon and the good cyclability of graphite for Li-ion batteries. The silicon was deposited on the graphite (SFG6 grade, mean particle size 6 μm) powder that was spread on a tilted substrate holder using silane diluted in 90% Ar and Ar/H_2, which were also mixed in the reactor as shown in Figure 7.34. The precursor

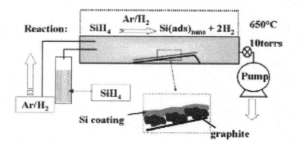

Figure 7.34 Schematic view of the CVD equipment. (From Alias, M. et al., J. Power Sources, 174, 900–904, 2007.)

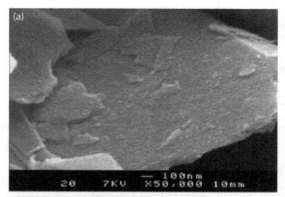

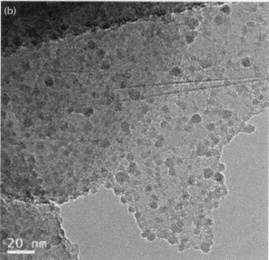

Figure 7.35 Microscopy images of SFG6 4/4 (4 × 3/4 h) sample: **(a)** SEM image (top); and **(b)** TEM (bottom) detail of the sample with small spherical silicon particles on a surface of a graphite sheet. (From Alias, M. et al., J. Power Sources, 174, 900–904, 2007.)

was thermally decomposed at 650°C and silicon nanoparticles were deposited and covered the surface of the graphite flakes for 45 min. The Si-coated graphite powder was mixed and reintroduced into the reactor. The CVD process was repeated several times, removing the sample between each deposition to obtain a nanocomposite layer.

Figure 7.35a and b show the SEM and TEM, respective images of the SFG6 4/4 nanocomposite electrode sample obtained four depositions of 45 min each, which shows a porous homogeneous layer made of circa 5-nm spherical-shaped silicon particles. The cyclic voltammetry indicated that silicon participates in the lithiation and delithiation reactions of

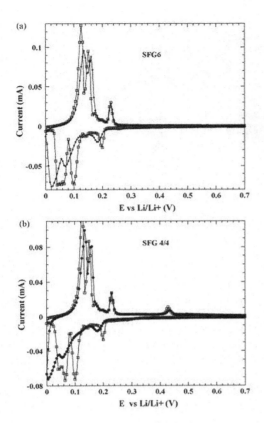

Figure 7.36 First cycle (round) and second cycle (square) cyclic voltammograms of battery using: **(a)** graphite SFG6 electrode and **(b)** Si-coated graphite composite (SFG6 4/4) electrode.

the composite electrodes. The use of 3.6 wt% of Si led to 27% gain in capacity, confirming that composite silicon/graphite electrodes can combine the advantages of both components, consistent with the theoretical calculation. A further increase in the silicon content (10.7 wt%) led to an initial capacity of 780 mAh g⁻¹ as shown in Figure 7.36; however, it affected the cycling ability of the composite negative electrode. The cycling behaviour of the composite electrode seems to be dependent on the silicon content and on the microstructure of the deposit. Low silicon content allows an increase of capacity with improved cyclability. Thus, the cyclability of the nanocomposite still needs to be optimised and improved.

7.8 CVD OF POLYMER-SILICA AEROGEL-BASED NANOCOMPOSITES

CVD together with polymerisation of methyl cyanoacrylate were introduced on silica aerogels (e.g., 650 mg) at approximately 7 mg/h at ambient temperature and pressure to create a silica-polycyanoacrylate nanocomposite as shown in Figure 7.37 [42]. Dry nitrogen gas (0.1 L/min) carrying methyl cyanoacrylate was delivered to silica aerogels and the polymerisation of methyl cyanoacrylate was initiated by the absorbed water. The resulting nanocomposites were stronger (>30-fold)

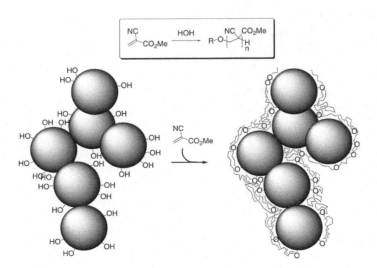

Figure 7.37 CVD and polymerization of methyl cyanoacrylate on the surface of silica aerogels to generate a polycyanoacrylate silica nanocomposite aerogel. (From Boday, D.J. et al., *Chem. Mater.*, 20, 2845–2847, 2008.)

Figure 7.38 SEM images of uncoated aerogel (left), with particle diameters of ~10 nm and of polycyanoacrylate-coated aerogel (right) with particle diameters ~35 nm. (From Boday, D.J. et al., *Chem. Mater.*, 20, 2845–2847, 2008.)

than untreated aerogels, with rupture strength of unmodified and polycyanoacrylate coatings on the silica aerogel nanocomposites of 0.539N and 17.6N, respectively, while the density only increased (3×) to 0.235 g/cm^3. The aerogels changed from transparent to opaque white as a result of the coarsening of silica nanoparticles comprising the aerogels as shown in Figure 7.38. The untreated aerogel immediately adsorbs liquid water and breaks into many pieces, whereas the CVD modified aerogels float intact on water without wetting or cracking as shown in Figure 7.39. The nanocomposites were inherently hydrophobic (water contact angle 129°),

Figure 7.39 Hydrophobic polycyanoacrylate-silica nanocomposite aerogels float in water while unmodified silica aerogels adsorb water wet, burst into fragments, and sink to the bottom of the container. (From Boday, D.J. et al., *Chem. Mater.*, 20, 2845–2847, 2008.)

making them more likely to withstand degradation and weathering [42,43].

The use of amine-modified silica aerogels produces even stronger and lower-density nanocomposite aerogels compared to the nanocomposites from the unmodified silica aerogels according to the scheme in Figure 7.40. The amine groups act as initiators for the cyanoacrylate polymerisations, resulting in higher molecular weight cyanoacrylate macromolecules than those observed with unmodified silica and that were covalently attached to the silica surface. The densities of nanocomposites of amine-modified silica aerogels were up to 0.220 g/cm^3 and up to 31 times stronger (flexural strength) than the precursor aerogel and about 2.3 times stronger than an unmodified silica aerogel of the same density [44].

7.9 CONCLUSIONS

CVD and its variants provide flexible and versatile methods for the deposition of nanocomposite films and coatings via: (i) *in-situ* synthesis of nanoparticles and the formation of matrix coating simultaneously or (ii) the incorporation of nanomaterials (e.g. nanoparticles, nanowires, and nanotubes) into the matrix during CVD of the matrix materials.

Thermal CVD has been demonstrated successfully for the deposition of dense, pore-free, and hard nano-structured CVD tungsten carbide/tungsten nanocomposite coatings (i.e., Hardide coatings) to greatly increase the life of critical parts and tools operating in abrasive, erosive, and chemically aggressive environments. Furthermore, the inherent non-line-of-sight CVD process enables the Hardide coating to be coated on internal surfaces as

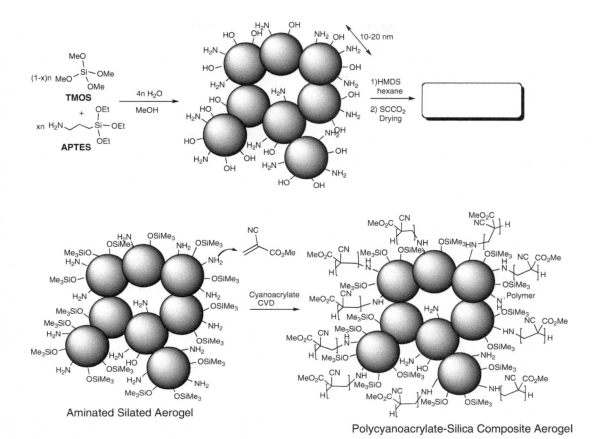

Figure 7.40 Process for preparing the aminated silica aerogel/poly (methyl cyanoacrylate) composites. (From Boday, D.J. et al., *Appl. Mater. Interfaces*, 1, 1264–1369, 2009.)

well as complex shapes, which opens new potential applications for hard coatings on critical parts.

Other variants such as plasma-enhanced CVD and laser-assisted CVD have also demonstrated the capability of producing hard and super-hard ternary and quaternary non-oxide-based nanocomposite coatings. However, these techniques tend to be a little more expensive and may limit the wider industrial applications. Hybrid CVD/PVD techniques have also been developed for the deposition of single-layer and multilayer hard nanocomposite coatings for tooling and tribological applications.

CVD and laser CVD have been explored for the deposition of oxide-based nanocomposite coatings. Various configurations of aerosol-assisted CVD (AACVD) have also been developed for the deposition of oxide-based nanocomposite coatings. Although AACVD-based methods are low-cost and promising techniques, issues with the agglomeration of the nanomaterials/nanofilters affect the reproducibility and properties of the nanocomposite coatings. Thus, there is a lack of adoption of AACVD by industry until such technical challenges are resolved. In addition, CVD has also been explored for the deposition of polymer-based nanocomposites.

ACKNOWLEDGEMENT

KLC thanks Dr. Matthew Jackson and Dr. Zhenyu Zhang for proofreading and formatting this chapter.

REFERENCES

1. T. Shelley, Super-Hard Coating goes deep inside, Eureka, November 1999, p. 21.
2. http://www.richterprecision.com/ coating-data/technical-data/.
3. https://www.ionbond.com/technology/.

4. I. Konyashin, B. Ries, D. Hlawatschek, Y. Zhuk, A. Mazilkin, B. Straumal, F. Dorn, and D. Park, Wear-resistance and hardness: Are they directly related for nanostructured hard materials? *Int. J. Refract. Met H.*, 49 (2015) 203–211.

5. ASM International publication. Surface Hardening of Steels: Understanding the Basics (#06952G). 2002 ASM International. Materials Park, OH.

6. An Updated Thintri MARKET STUDY: 2009: Chrome Plating Alternatives: Thermal Spray, Electroless Plating, and Others, from http://www.thintri.com/chrome-plating-report.htm.

7. E. Lassner, W.-D. Schubert, *Tungsten: Properties, Chemistry, Technology of the Element, Alloys and Chemical Compounds*, Kluwer Academic/Plenum Publishers, New York, 1999.

8. Bodycote Materials Testing test report: 30 DAY SULPHIDE STRESS CRACKING (SSC) TEST TO NACE TM0177-2005/ASTM G39–Method B (1 bar H2S).

9. ASTM G65-94, Standard test for measuring abrasion using the dry sand/rubber wheel apparatus, 1996 Annual Book of ASTM Standards, Volume 03.02, ASTM.

10. M.G. Gee, A.J. Gant and W.P. Byrne, AnNPL Rotating Wheel Abrasion Test, NPL Measurement Note, CMMT(MN) November 30, 1998.

11. Hard Chrome Plating Alternatives–Thermal Spray, from http://www.hazmat-alternatives.com/DoD_Programs_Altsums-HCPA-TS.php.

12. H.W. Strauss, R.R. Chromik, S. Hassani, J.E. Klemberg-Sapieha, *In-situ* tribology of nanocomposite Ti–Si–C–H coatings prepared by PE-CVD. *Wear* 272 (2011) 133–148.

13. P. Jedrzejowski, J.E. Klemberg-Sapieha, L. Martinu, Relationship between the mechanical properties and the microstructure of nanocomposite TiN/SiN1.3 coatings prepared by low temperature plasma enhanced chemical vapor deposition. *Thin Solid Films* 426 (2003) 150–159.

14. L. Shizhi, S. Yulong, P. Hongrui, Ti-Si-N films prepared by plasma-enhanced chemical vapor deposition. *Plasma Chem. Plasma Process.* 12 (1992) 287.

15. I. Endler, E. Wolf, A. Leonhardt, V. Richter, Preparation, characterization and wear behaviour of PACVD cermets. *Surf. Coat. Technol.* 72 (1995) 37.

16. P. Jedrzejowski, J.E. Klemberg-Sapieha, L. Martinu, Quaternary hard nanocomposite TiCxNy/SiCN coatings prepared by plasma enhanced chemical vapor deposition. *Thin Solid Films* 466 (2004) 189–196.

17. S. Veprek, M.G.J. Veprek-Heijman, P. Karvankova, J. Prochazka, Different approaches to superhard coatings and nanocomposites. *Thin Solid Films* 476 (2005) 1.

18. C.T. Guo, D. Lee, P.C. Chen, Deposition of TiSiN coatings by arc ion plating process. *Appl. Surf. Sci.* 254 (2008) 3130.

19. Y.H. Cheng, T. Browne, B. Heckerman, E.I. Meletis, Mechanical and tribological properties of nanocomposite TiSiN coatings. *Surf. Coat. Technol.* 204 (2010) 2123.

20. I. Endler, M. Höhn, J. Schmidt, S. Scholz, M. Herrmann, M. Knaut, Ternary and quarternary TiSiN and TiSiCN nanocomposite coatings obtained by Chemical Vapor Deposition, *Surf. Coat. Technol.* 215 (2013) 133–140.

21. P. Karvankova, M.G.J. Veprek-Heijman, D. Azinovic, S. Veprek, Properties of superhard nc-TiN/a-BN and nc-TiN/a-BN/a-TiB2 nanocomposite coatings prepared by plasma induced chemical vapor deposition. *Surf. Coat. Technol.* 200 (2006) 2978–2989.

22. S. Veprek, S. Reiprich, A concept for the design of novel superhard coatings. *Thin Solid Films* 268 (1995) 64.

23. S. Veprek, The search for novel, superhard materials. *J. Vac. Sci. Technol., A, Vac. Surf. Films* 17 (5) (1999) 2401.

24. A. Bendavid, P.J. Martin, E.W. Preston, J. Cairney, Z.H. Xie, M. Hoffman, Deposition of nanocomposite thin films by a hybrid cathodic arc and chemical vapour technique. *Surf. Coat. Techol.* 201 (2006) 4139–4144.

25. A. Bendavid, P.J. Martin, T.J. Kinder, E.W. Preston, The deposition of NbN and NbC thin films by filtered vacuum cathodic arc deposition. *Surf. Coat. Technol.* 163 (2003) 347.

26. A. Mège-Revil, P. Steyer, S. Cardinal, G. Thollet, C. Esnouf, P. Jacquot, B. Stauder, Correlation between thermal fatigue and

thermomechanical properties during the oxidation of multilayered TiSiN nanocomposite coatings synthesized by a hybrid physical/chemical vapour deposition process. *Thin Solid Films* 518 (2010) 5932–5937.

27. D. Pilloud, J.F. Pierson, P. Steyer, A. Mege, B. Stauder, P. Jacquot, Use of silane for the deposition of hard and oxidation resistant Ti–Si–N coatings by a hybrid cathodic arc and chemical vapour process. *Mater. Lett.* 61(2007) 250.

28. A. Ito, Y. You, T. Ichikawa, K. Tsuda, T. Goto, Preparation of Al_2O_3–ZrO_2 nanocomposite films by laser chemical vapour deposition. *J. Eur. Ceram. Soc.* 34 (2014) 155–159.

29. A. Ito, H. Kadokura, T. Kimura, T. Goto. Texture and orientation characteristics of alpha-Al_2O_3 films prepared by laser chemical vapor deposition using Nd:YAG laser. *J. Alloys Compd.* 489 (2010) 469–474.

30. D. Barreca, E. Comini, A.P. Ferrucci, A. Gasparotto, C. Maccato, C. Maragno, G. Sberveglieri, and E. Tondello. First example of ZnO-TiO_2 nanocomposites by chemical vapor deposition: Structure, morphology, composition, and gas sensing performances, *Chem. Mater.* 19 (2007) 5642–5649.

31. R.G. Palgrave and I.P. Parkin, Aerosol assisted chemical vapor deposition using nanoparticle precursors: A route to nanocomposite Thin Films, *J. Am. Chem. Soc.* 128 (2006) 1587–1597.

32. R.G. Palgrave and I.P. Parkin, Aerosol assisted chemical vapor deposition of gold and nanocomposite thin films from hydrogen tetrachloroaurate(III), *Chem. Mater.* 19 (2007) 4639–4647.

33. M. Labeau, B. Gautheron, F. Cellier, M. Vallet-regi. E. Garcia and J.M. Gonzalez Calbet, Pt nanoparticles dispersed on SnO_2 thin films. A microstructural study. *J. Solid State Chem.* 102 (1993) 434–439.

34. S. Vallejos, S. Selina, F. Annanouch, I. Gràcia, E. Llobet and C. Blackman, Aerosol assisted chemical vapour deposition of gas sensitive SnO_2 and Au-functionalised SnO_2 nanorods via a non-catalysed vapour solid (VS) mechanism. *Sci. Rep.*, volume 6, Article number: 28464 (2016) doi:10.1038/srep28464.

35. R. Tenne, L. Margulis, M. Genut, and G. Hodes, Polyhedral and cylindrical structures of tungsten disulphide. *Nature* 360 (1992) 444.

36. L. Margulis, G. Salitra, R. Tenne, and M. Talianker, Nested fullerene-like structures. *Nature* 365 (1993) 113.

37. X. Hou, K.-L. Choy, V. Serín, and N. Fleischer, Processing and characterization of nanocomposite coatings using Aerosol-assisted Chemical Vapor Deposition (AACVD) Method. *ECS Trans.* 25 (8) (2009) 1135–1141.

38. X.H. Hou, J. Williams, K.L. Choy, Processing and structural characterization of porous reforming catalytic films. *Thin Solid Films* 495 (2006) 262.

39. X. Hou, K.-L. Choy, Synthesis of Cr_2O_3-based nanocomposite coatings with incorporation of inorganic fullerene-like nanoparticles. *Thin Solid Films* 516 (2008) 8620–8624.

40. X. Hou, K.-L. Choy, N. Brun, and V. Serín, Nanocomposite coatings codeposited with nanoparticles using aerosol-assisted chemical vapour deposition. *J. Nanomater.* (2013), Article ID 219039. doi:10.1155/2013/219039.

41. M. Alias, O. Crosnier, I. Sandu, G. Jestin, A. Papadimopoulos, F. Le Cras, D.M. Schleich, T. Brousse, Silicon/graphite nanocomposite electrodes prepared by low pressure chemical vapor deposition. *J. Power Sources* 174 (2007) 900–904.

42. D.J. Boday, K.A. DeFriend, K.V. Wilson, Jr., D. Coder, and D. A. Loy, Formation of polycyanoacrylate-silica nanocomposites by chemical vapor deposition of cyanoacrylates on aerogels, *Chem. Mater.* 20 (9) (2008) 2845–2847.

43. P.J. Mankidy, R. Rajagopalan, H.C. Foley. Facile catalytic growth of cyanoacrylate nanofibers. *Chem. Commun.* 2006, 1139.

44. D.J. Boday, R.J. Stover, B. Muriithi, M.W. Keller, J.T. Wertz, K.A. DeFriend Obrey and D.A. Loy, Low-density nanocomposites by chemical vapor deposition and polymerization of cyanoacrylates on aminated silica aerogels. *Appl. Mater. Interfaces* 1 (7) (2009) 1264–1369.

Chemical vapour infiltration of composites and their applications

MARIA-BEATRICE COLTELLI AND ANDREA LAZZERI

Because materials are an integral part of technological development, the study of Materials offers great opportunities for advancements. Many approaches promote such development, spanning the range from very fundamental understanding of the chemical composition and bonding in a material to the implementation of viable processing routes to make high-quality materials efficiently and at acceptable cost. Within this realm is a fascinating world that includes the study of the structure of materials and the connections between materials processing and structure on one hand and between the structure of materials and their properties on the other.

Samuel M. Allen, Edwin L. Thomas
The Structure of Materials

8.1 INTRODUCTION

Ceramic matrix composites (CMCs) represent advanced materials combining the chemical and thermal resistance of ceramics with the mechanical strength of ceramic reinforcements, thus with an improved fracture toughness with respect to ordinary ceramics, which possess a low fracture toughness. Different routes can be followed to produce them (Figure 8.1), starting from a gas, a liquid, or a solid precursor (in the latter case suspended in a slurry). In general, the sintering method, starting from solid powder precursor, results in shrinkage stresses in the material, making it more brittle than those produced by starting from liquid or gas precursors. In all the cases, for the preparation of continuous fibre-reinforced composites, a fibrous preform is treated with precursors, leading to the deposition of the matrix material in the pores of the preform and resulting in a corresponding increase in density of the material.

For both the gas and liquid routes, chemical reactions occur. Hence, precursor chemicals must be employed, reacting in the preform pores to produce a stable ceramic compound[1]. The starting reagents are thus chemically reactive or decompose at high temperatures (about 1000 K) and, thanks to a redox reaction, form *in-situ* the inert ceramic matrix. Usually this is accompanied by the evolution of—often hazardous—gaseous by-products. As an example, in the case of polymer impregnation and pyrolysis, the degradation of the polymer at high temperatures results in the emission of gaseous chemicals produced by the pyrolysis. The structure of them is a function of the structure of the employed liquid polymeric resin. In the case of phenolic resins[2], the main products are phenol, cresols, and trimethylphenol, but the formation of

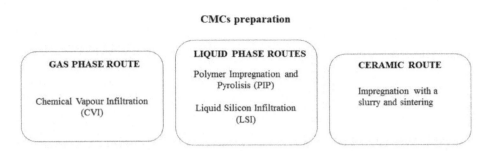

CMCs preparation

GAS PHASE ROUTE

Chemical Vapour Infiltration
(CVI)

LIQUID PHASE ROUTES

Polymer Impregnation and
Pyrolisis (PIP)

Liquid Silicon Infiltration
(LSI)

CERAMIC ROUTE

Impregnation with a
slurry and sintering

Figure 8.1 Classification of methods for preparing CMCs.

condensed ring-based molecules is also possible.[3] Usually, mixtures of different gases are obtained in gas or liquid-infiltration methods. For the ceramic route the matrix is formed by a finely dispersed powder in a liquid slurry which is deposited among the fibres, followed by sintering at high temperature, as in traditional ceramic processing, to obtain the final composite. In this case, the process is not reactive, as the ceramic particles are simply sintered together. However, it is necessary to consider that the slurry liquid medium is evaporated or thermally decomposed during the process. The process for CMC preparation is thus complex and the processing parameters, as well as the input and output reagents and product flows, must be well controlled to achieve the desired final density and homogeneity for the composite. The selection of a suitable preparation method can be fundamental also for controlling the level of adhesion between the matrix and the preform fibres, with reactive processes usually granting a better adhesion; thus, it is the most used and investigated preparation route for CMCs.

The present chapter is mainly focused on chemical vapour infiltration (CVI). This process was first developed in 1962 by Bickerdike[3] for increasing the density of porous carbon. Thereafter, Jenkin[4] patented the process for infiltrating porous alumina preforms in 1964 with chromium carbide. In the early 1970s at Karlsruhe University, the infiltration of SiC was first investigated by Prof. Fitzer's group.[5] Prof. Naslain, in parallel, at Bordeaux University, was developing an isothermal isobaric CVI (I-CVI) with the aim of preparing carbon fibre (CF)-reinforced silicon carbide composites to overcome the low resistance to oxidation of C/C composites.[6] Caputo and Lackey[7] of the

Oak Ridge Laboratory used a vacuum method to obtain preforms of SiC fibres and infiltrated by chemical vapour deposition (CVD) matrices of Si_3N_4 and SiC. The commercialization of the I-CVI for CMCs was first carried out at Du Pont (USA) and at Societé Europeènne de Propulsion (SEP). Currently, the leading companies for these technologies in Europe are CVT (Germany) and Archer Technicoat (UK), and in the USA the TevTech and FirstNano companies. In 1984, the forced CVI technique (F-CVI) was also described by Prof. Lackey's group.[8] Currently the rapid CVI (r-CVI) technique patented by CVT is one of the most advantageous methods for industrial applications, whereas the possibility of making the process faster and the CMCs more homogeneous is explored by developing CVI equipment using microwave for heating.

8.2 DEFINITIONS

The CVI is considered an extension of the CVD technology. In this latter technique, used for special coatings and electronic applications since 1960s, the coating is deposited from gaseous precursors at high temperatures on a completely dense solid surface. On the contrary, in CVI the deposition takes place within **porous preforms** usually made of fibres. The gaseous precursor thus penetrates in the preform pores and undergoes chemical reaction, thus depositing in the pores. In this way, the matrix material grows into the fibrous porous structure (preform) in a continuous layer-by-layer way, thus forming the composite matrix. Besmann et al.[8] evidenced that the CVI technique has the advantage of avoiding the damage of reinforcement filaments in CMCs production thanks to the layer-by-layer growth of the matrix.

8.2.1 Fibres for CMCs

Fibres for CMCs should withstand the relatively high temperatures required to produce ceramics, without significant damage[9]. For this reason, not all the fibres can be used. The fibres that can be used in general in composites can be classified in organic or inorganic.[10] Among the organic fibres polymer-based and carbon fibres can be produced (Figure 8.2). Only the latter can be used in CMCs. In fact, the polymeric ones cannot resist above 500°C. Among the inorganic fibres we can distinguish the metallic fibres from the non-metallic ones. The former are not suitable for producing CMCs because they usually melt below 1000°C. Non-metallic ones can be glass fibres or ceramic fibres. Glass fibres are not suitable for CMCs because their glass transition and softening point is below 800°C. Ceramic fibres, consisting of oxide or non-oxide fibres, are both suitable for CMCs. Hence, overall, the three types of fibres that can be used in CMCs are carbon fibres, oxide ceramic fibres, or non-oxide ceramic fibres.

The physical properties of fibres are determined mainly by three structural parameters: bond type, crystallinity, and molecular orientation. Carbon fibres are characterized by the existence of covalent bonds (200–600 kJ/mol) on the graphitic plans and Van der Waals bonds (1 kJ/mol) in between the different planes. Inorganic fibres are characterized by the presence of ionic bonds (800–15,000 kJ/mol).[10]

The selection of the fibres is an important key point to obtain a high-performance material. Several types of Al_2O_3, C and SiC fibres are currently available on the market.

Commercial carbon fibres are available with a tensile strength up to 7060 MPa and a modulus up to 588 GPa. The commercial oxide fibres consist of Al_2O_3 or Al_2O_3/SiO_2 and show a tensile strength up to 3100 MPa and a Modulus up to 380 GPa. The commercial non-oxide fibres consist mainly of SiC and show tensile strength up to 5865 MPa and Modulus up to 420 GPa.[10]

For the preparation of inorganic fibres, indirect or direct preparation methods can be followed. An indirect production method is CVD process consisting of the gas-phase deposition of inorganics on carrier fibres, like for example the deposition of SiC on carbon fibres or tungsten wires. Another indirect method is relic process exploiting absorbing fibres (usually cellulosic) that are impregnated with a salt solution or sols. The organic part is burned off and the salts or sols are converted in ceramics.[10]

A direct fibre production is the spinning doping based on molecularly dispersed precursors, in which soluble salts are used in the spinning dopes and then they are converted in ceramic by calcination.

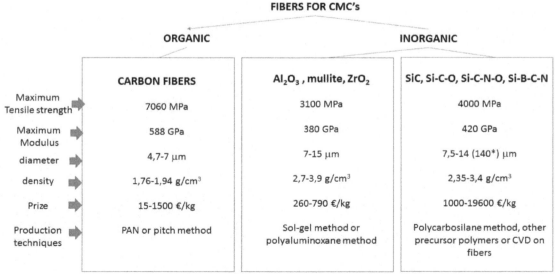

FIBERS FOR CMC's

	ORGANIC	INORGANIC	
	CARBON FIBERS	Al_2O_3, mullite, ZrO_2	SiC, Si-C-O, Si-C-N-O, Si-B-C-N
Maximum Tensile strength	7060 MPa	3100 MPa	4000 MPa
Maximum Modulus	588 GPa	380 GPa	420 GPa
diameter	4,7-7 μm	7-15 μm	7,5-14 (140*) μm
density	1,76-1,94 g/cm³	2,7-3,9 g/cm³	2,35-3,4 g/cm³
Prize	15-1500 €/kg	260-790 €/kg	1000-19600 €/kg
Production techniques	PAN or pitch method	Sol-gel method or polyaluminoxane method	Polycarbosilane method, other precursor polymers or CVD on fibers

* Maximum diameter value for SiC with a core fiber (C or W)

Figure 8.2 Classification and properties of fibres used for reinforcing CMCs.

Usually, organic polymer like poly(ethylene oxide), poly(vinyl alcohol), or poly(vinyl pyrrolidone) are also added in order to modulate the rheological behaviour to allow the spinning process.

In another direct process, spinning dopes are based on colloidally dispersed precursors (sols) and the same polymers are used. Coarse ceramic particles can also be used to replace salts or sol-based spinning dopes. The last method consists of the use of an inorganic polymer that is successively pyrolised to remove the organic components. An important example consists of the use of silanes.

Oxide fibres are commercially available, based on alumina or mixture of alumina and silica, and show high values for tensile strength and modulus; they are stable to oxidation at high temperature, but they are prone to creep under load at 1100°C and for this reason they cannot be used in CMCs to be used at higher temperature. Moreover, the maintenance of oxide fibres at high temperature causes an increase of grain dimensions, resulting in embrittlement.

Commercially available non-oxide fibres, mainly based on SiC or Si-C-(N)-O, have tensile strength and modulus often higher than oxide ones and do not have problems of creep at high temperature thanks to their amorphous structure (Figure 8.4b). Their disadvantage is the low resistance to oxidation and the higher is the content of oxygen and the lower is this resistance.

CFs, if in a non-oxidizing atmosphere, can be used up to 2800°C (at 3730°C it sublimates). In carbon fibres, a graphitic structure is present, but it is arranged more randomly than in pure graphite. The morphologic structure is not planar but rather undulated, called turbostratic. If carbon fibres are obtained from polyacrylonitrile (PAN) as precursor, the structure is turbostratic, but if the carbon fibres are obtained by pitch the structure is almost graphitic (Figure 8.3b).

As the Van Der Waals interactions between the different stacks are weaker than the bonds inside the stacks it is important to orient the graphitic plans along the fibres' main axes to enhance their resistance in this direction. This is achieved by keeping the fibres in tension at high temperature. Carbon fibres can be produced starting by PAN, by pitch, or by regenerated cellulose.

In the PAN process, the precursor PAN fibre is converted into carbon fibre through the following sequence of steps.[10] First, thermosetting is conducted in an oxidizing atmosphere of 200°C–300°C, and the shrinkage of the fibres is avoided so that a structure of six-member heteroaromatic rings is developed in the fibres. In the next step, carbonization occurs much faster than the earlier step and is carried out in a nitrogen atmosphere of 1000°C–1700°C in which water vapour, ammonia, and hydrogen cyanide evolve as gaseous by-products. During this step, the crosslinked fibre molecules rearrange to form

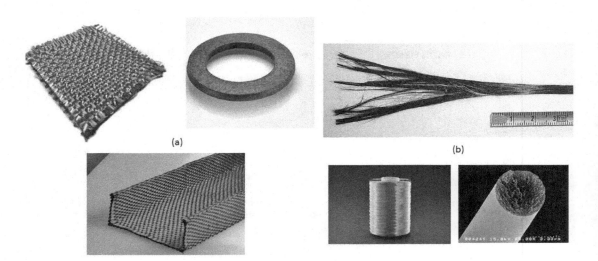

(a)

(b)

Figure 8.3 Examples of carbon preforms **(a)** and fibres **(b)**.

carbon networks of high rigidity. The resulting product is known as "unsized carbon fibre". Due to the high cost of the petroleum-based precursors and the associated processing, CFs remain a specialty product for aerospace, sporting goods, high-end automotive, and specialist industrial applications.

Lignin has a clear cost advantage over textile-grade PAN as precursor for low-cost CF production since its cost is not linked with oil price, as in the case of PAN. With more biomass refineries in the near future, the lignin by-product from the conversion of cellulosic feedstock may represent an abundant, valuable resource for CFs production. The lignin-based CFs have the potential to be less energy intensive, as they are estimated to require 5% less primary and emit 22% less CO_2-equivalent gas emissions than the conventional PAN-based textile-grade acrylic fibres.[11] Lignin-based carbon fibres (LCFs) have been produced by pyrolysis of lignin fibres most commonly spun using the melt spinning technique.[12,13] The currently preferred method involves the preparation of a suitable lignin blend that is processed into fibres by extruding filaments from a melt or swollen solvent gel, under inert atmosphere. The final morphology of the carbon materials depends not only on the chemistry of the precursor but also on the processing methodology. Lignin-phenol-formaldehyde resin (LPF)-derived submicron carbon fibres[14] were successfully prepared by using LPF, which was synthesized by using lignin as partial substitution of phenol in phenol-formaldehyde resin as precursor. The carbon fibres were obtained through electrospinning the LPF/polyvinylpyrrolidone(PVP)/ dimethylformamide(DMF) solution followed by successive stabilization with BN preceramic polymer and carbonization processes.

8.2.2 Preforms for CMCs

As yet stated, CMCs are obtained by CVI technology infiltrating gases in a solid preform. The production of preforms is not uniformly established on an industrial point of view both in the field of SiC preforms and C preforms. Much research is still ongoing on new methods and materials. In the latter context, the tendency of using bio-based (especially from agricultural or industrial wastes) materials for producing C preforms was widely considered in most recent researches in agreement with the general increased interest in circular economy principles.

SiC preforms (Figure 8.4a) can be prepared by sintering SiC powder of an appropriate average particle size mixed with a multicomponent binder.

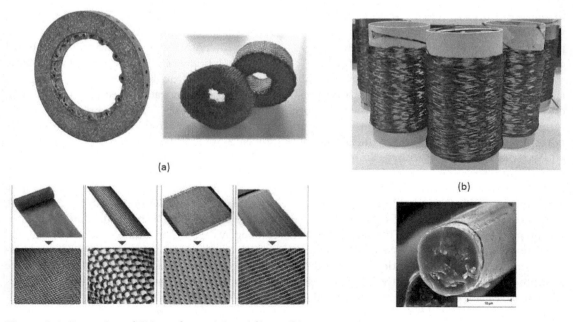

(a)

(b)

Figure 8.4 Examples of SiC preforms **(a)** and fibres **(b)**.

The prepared part is then placed in a vacuum furnace to carbonize the binder. The binder chemistry must support the achievement of accurate shapes and acceptable surface roughness, a resistant preform and maintenance of the part shape during the first furnace infiltration. Ozer et al.[15] prepared SiC preforms with a similar approach by using SiC powders, polyvinyl alcohol (PVA), and colloidal silica.

Langguth et al.[16] prepared different SiC preforms by pyrolysis of polysilanes and polycarbosilanes.

Methylhydroxypolysiloxanes have been proven to be suitable ceramic precursor binders for the manufacture of porous SiC particle preforms with defined ceramic volume fractions by warm pressing and subsequent pyrolysis treatment. Two techniques can be used for introducing the organic binder precursor into the ceramic powder: rotation evaporation and spray drying. These techniques are assessed with respect to yield and distribution homogeneity of binder in the SiC preform by Thunemann et al.[17] In particular, it is shown that preforms made from spray-dried powder are better than those made from rotation evaporated powders. This is evidenced by the fact that when using spray drying, higher preform strengths are obtained and considerably lower binder concentrations are needed. Polymer volume fractions as low as 2,5 vol.% introduced by spray drying, proved to confer sufficient strength to the warm-pressed and pyrolysed preforms to withstand the forces imposed during successive pressurized liquid-metal infiltration by squeeze casting.

Porous silicon carbide preforms were also fabricated by Roy et al.[18] using polymer wax (polyethylene) powder of different dimensions as pore formers. Wax content was varied to introduce open porosities up to 64 vol%. The amount of porosity and its structure increases with the amount of total wax used and process conditions.

Selective laser sintering (SLS)[19] coupled with post-process infiltration, also known as indirect SLS, provides a rapid, cost-effective method for the preparation of preforms. The advantages of SLS are well known: net-shape capability with good dimensional stability, little or no need for post-process machining, relatively inexpensive raw materials, and extensive geometrical freedom. Preforms generated by SLS are bound by a binder and have a continuous porosity network.

Preforms are frequently produced starting from SiC fabrics. The development of radiation-resistant SiC composites (for applications in the nuclear sector) is based on the use of highly crystalline SiC fibres. Recently, highly crystalline SiC fibres have been developed including Hi-Nicalon™ Type-S20 (Nippon Carbon Co., Ltd), Sylramic™21 (Dow Corning Co.), and Tyranno™ SA22 (Ube Industries, Ltd).[20]

Preforms of boron nitride/silicon carbide (BN/SiC) coated Sylramic SiC fibres have been used as a starting assembly for the fabrication of SiC/SiC composites.[21] The preforms are fabricated by stacking 2-D woven mats of SiC fibres in a fixture and then infiltrating the fixture with BN and SiC coatings by CVI. A variety of processes are used for further densification including CVI, polymer infiltration and pyrolysis, and melt infiltration (MI). For MI, the fibre preforms are infiltrated with SiC particle slurry and then with molten silicon metal to fabricate SiC/SiC composites. By using CVI the damage to the fibres and interface coating during preform fabrication is minimum when compared with other consolidation processes that involve pressure. Complex-shaped preforms can be easily manufactured by CVI process.

Fu et al.[22] studied the effect of Si-content on the microstructure and mechanical properties of Si–SiC composites fabricated by 3D-printing. For the fabrication of the preforms, printing powder was prepared from mixtures of Si powder, α-SiC, and dextrin powder $(C_6H_{10}O_5)_n$ with n = 10–200. The dextrin powder, which served as a binder when coming in contact with the injected water-based printer solution, was kept at a constant volume fraction in all powder blends and 3D-printing was carried out using a water-based printing solution, containing glycerin.

Zhu et al.[23] prepared dense SiC by infiltrating molten Si at 1550°C into porous preforms composed of pyrolised rice husks (RH) and extra SiC powder in different ratios.

Sometimes preforms are obtained by combining several layers of carbon or SiC fabrics. In general, fabric of carbon, also 3D needled,[24] can be obtained starting by carbon fibres. In general, the traditional carbon-based preforms are formed by mould pressing of powder or fibres. However, the size and shape of the preforms are limited by the mould configuration. Moreover, the density and porosity at different part of the carbon preform is not uniform (Figure 8.3a).

The main source of the carbon preforms generally includes graphite powder, soot, carbon black, carbon fibre, etc. However, some binders such as resin and pitch are necessary during the green-body-forming process using above carbon sources. Mesocarbon microbeads (MCMB)[25] have been reported to be a superior precursor for high-performance carbon/graphite material because it has good self-sintering behaviour, high yield of carbon, and homogeneous shrinkage. MCMB derived from liquid pitch or organic compounds consists of aromatic oligomers where aromatic layers stack approximately parallel to each other in the same direction. Therefore, no additive is necessary in the forming and sintering processing when MCMB is used as carbon source.

SiC ceramic tubes[26] were prepared using preforms made from cotton fabric/phenolic resin composite to obtain SiC matrix composites reinforced with CF (C_f/SiC). Coir fibre, an agricultural waste material of immense economic importance in Indian subcontinent, was used as a precursor to biomorphic SiC ceramics. Fibreboards made of coir fibres were converted to carbon preforms[27] by controlled thermal processing.

Three-dimensional needled carbon fibre felt, with 80 vol.% unidirectional continuous fibre and 20 vol.% chopped fibre was used as reinforcement and the volume fraction of carbon fibre was 27%. C/C preforms with modulated densities were fabricated by CVI with propylene as the precursor gas.[28] Hence, CVI technique is useful for refining the preparation of preforms through preliminary infiltrations.

Biomorphic SiC components with homogeneous properties were manufactured[29] from sawdust from pine wood using a novel method to produce preforms, without addition of any extra binder, but with enough mechanical strength to be carbonized up to 1400°C without deformation. Sawdust preforms were fabricated by uniaxial hot pressing.

The tendency of using bio-based (especially from agricultural or industrial wastes) materials for producing C preforms was widely considered in most recent researches in agreement with the general interest in promoting circular economy principles. The design of the preforms (Figure 8.5) and their preparation is part of the preparation of the composites and the procedures adopted for preforms preparation are often quite long and expensive. Hence, there is much current research on alternative methods for the preparation of reinforcements and preforms for CMCs.

Several combinations of properties can be achieved by properly modulating the composition and shape of matrix and reinforcements. In particular, the percentage by weight and the shape of the reinforcing fibres is much important. As an example, the properties of a C_f/SiC composite are different if the CFs are dispersed reinforcing elements or if they consist of a fabric or a porous continuous preform. The properties of the final composite, in terms of mechanical properties, thermal resistance, and chemical resistance, can be predicted by taking into account the properties of the materials constituting the composite.

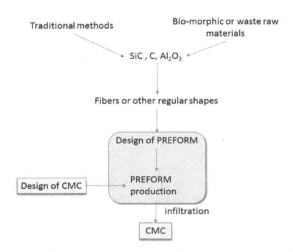

Figure 8.5 Scheme evidencing the importance of the design and manufacturing of preforms in the production of CMCs.

Table 8.1 Main reactions used in CVD and CVI to produce CMCs matrices

Matrix of CMC	Main reactions employed in CVD or CVI
C	$CH_{4(g)} \rightarrow C_{(s)} + 2H_{2(g)}$
SiC	$CH_3SiCl_{3(g)} \xrightarrow{H_2} SiC_{(s)} + 3HCl_{(g)}$
B$_4$C	$CH_{4(g)} + 4BCl_{3(g)} + 4H_{2(g)} \rightarrow B_4C_{(s)} + 12HCl_{(g)}$
B	$BCl_{3(g)} + \frac{3}{2}H_{2(g)} \rightarrow B_{(s)} + 3HCl_{(g)}$
Al$_2$O$_3$	$2AlCl_{3(g)} + 3CO_{2(g)} + 3H_{2(g)} \rightarrow Al_2O_{3(s)} + 6HCl_{(g)} + 3CO_{(g)}$
ZrO$_2$	$ZrCl_{4(g)} + 2CO_{2(g)} + 2H_{2(g)} \rightarrow ZrO_{2(s)} + 4HCl_{(g)} + 2CO_{(g)}$
TiC	$CH_{4(g)} + TiCl_{4(g)} \xrightarrow{H_2} TiC_{(s)} + 4HCl_{(g)}$
WC	$2W + 2CH_{4(g)} \rightarrow 2WC + 4H_{2(g)}$
TiN	$2TiCl_{4\,(g)} + N_{2(g)} + H_{2(g)} \rightarrow 2TiN + 8HCl$
BN	$BCl_{3(g)} + NH_{3(g)} \rightarrow BN_{(s)} + 3HCl_{(g)}$
	$BF_{3(g)} + NH_{3(g)} \rightarrow BN_{(s)} + 3HF_{(g)}$
Si$_3$N$_4$	$3SiCl_{4(g)} + 4NH_{3(g)} \rightarrow Si_3N_{4(s)} + 12HCl_{(g)}$
TiSi$_2$	$TiCl_{4(g)} + 2SiCl_{4(g)} + 6H_{2(g)} \rightarrow TiSi_{2(s)} + 12HCl_{(g)}$
TiB$_2$	$TiCl_{4(g)} + 2BCl_{3(g)} + 5H_{2(g)} \rightarrow TiB_{2(s)} + 10HCl_{(g)}$

8.2.3 Matrices for CMCs

The matrices of CMCs can be oxide or non-oxide based. A list of the most common matrices is reported in Table 8.1. The oxide-based ones consist of alumina (Al$_2$O$_3$) or zirconia (ZrO$_2$). Non-oxide-based matrices consist of carbon (C), silicon carbide (SiC), boron carbide (B$_4$C), boron (B), titanium carbide (TiC), tungsten carbide (WC), titanium nitride (TiN), boron nitride (BN), silicon nitride (Si$_3$N$_4$), titanium silicide (TiSi$_2$), and titanium boride (TiB$_2$).

8.3 VARIANTS AND CLASSIFICATION OF CVI TECHNIQUES

CVI methods are classified by considering several aspects, such as the heating method (radiative or inductive), the uniformity of temperature in the preform, the use of plasma microwave or radiofrequencies, the type of reactor, the pressure regime (atmospheric/low pressure and constant/variable), and the immersion of the preform in a precursor liquid.

The *isothermal/isobaric CVI (I–CVI)* (Figure 8.6) is the oldest approach (in use since the 1960s) for the densification of refractory composites. Nowadays, this process is used for the densification of C$_f$/C composites and SiC$_f$/SiC composites, both employed in aerospace applications. The matrices more commonly produced by isothermal, isobaric CVI including SiC, C, BN, Al$_2$O$_3$, and B$_4$C. In a I-CVI reactor the fibrous preform is set in the isothermal zone of a hot -wall deposition chamber. The gaseous precursor flows inside the preforms by diffusion and out the preforms by convection. Deposition is performed at relatively low temperatures (decreasing reaction rates, long deposition times), and under reduced pressures (atmospheric or sub-atmospheric) to promote the diffusion of the gaseous species in the pores. In these conditions, it is possible to avoid an early sealing of the pore entrance by the deposit. Therefore, the deposition yield and deposition rate are low. Keeping the preform porosity open until the end of the densification process is the key point of the CVI process. If the outside temperature of the preforms is higher than the interior one, the deposition will be preferential at the exterior of the preform and not advance into the interior of the sample.

In the *thermal-gradient CVI*, a thermal gradient through the thickness of the porous preform enables the gaseous precursors to enter the preform through the cold side and move through the preform by chemical diffusion, decomposing near the hot side to deposit the solid matrix, allowing a reduction of the porosity near the hot side. Hence, the highest temperature region moves toward the

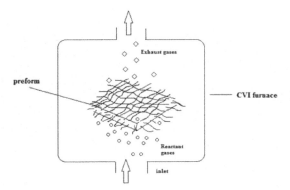

Figure 8.6 Schematic representation of I-CVI process.

cold wall. Only large individual parts with non-complex shapes can be densified via thermal-gradient methods. As the diffusion mechanism governs all the reaction, the infiltration times in thermal-gradient CVI are usually long, on the order of several days.

In *Forced Flow CVI (F-CVI)*, the precursor reagents flow through the preform and along the pore under high pressure ($P_1 = 100–200\,kPa$), while the by-products (and the unreacted species) are removed at a lower pressure, P_2. The F-CVI process exists in isothermal or in the thermal-gradient version. In the former a forced flow of reagents passes through a preform where the decomposition of the gases takes place. The process goes on until some portion of the preform reaches a sufficiently high density. At this point, the process can be considered terminated. Usually, the regions of the preforms, where the density is higher, are those nearer the preform surface (where the reagent concentration at its maximum value) since the reagent gases enter there first. In the thermal-gradient F-CVI, an inverse thermal gradient can be applied along the pore and this allows the decomposition of the solid matrix near the hot side, thus reducing the porosity near the hot side. The main advantage of the F-CVI is the significant reduction in the overall infiltration time, in comparison to the processes described before, but it is difficult to apply to complex shapes.

In *pulsed-flow CVI (P-CVI)* the furnace and the fibrous preform pore network are first evacuated by pumping, and then the reactants are injected very quickly into the reaction chamber and inside the preforms, by increasing impulsively the pressure. The solid deposition takes place during a predetermined time (residence time). In this process,

the pore network of the preform is periodically fed with fresh reactants when the reaction products are cyclically removed. If the composition of the precursor is changed during the different injection steps, P-CVI allows the formation of multi-layered ceramic deposits. The industrialization of the P-CVI is at a lower maturity level than the processes described earlier and is generally considered not to have much advantage.

The *rapid CVI* (r-CVI) (Figure 8.7a) was developed and patented by the German CVT company.[30] In this process, a porous structure is subjected in the reaction zone to the flow of a gas containing at least one gaseous precursor, wherein the partial pressure of the precursor and the dwell time of the gas are set at a given temperature. In such a manner, the deposition reaction of the precursor occurs in the porous structure in the partial pressure range of the saturation adsorption. Moreover, the reaction of the precursor is limited in each stage of the infiltration in such a manner that during the flow through the reaction zone no more than 50% of the precursor is deposited as a solid phase in the porous structure. The exposure of the porous structure to the flow occurs in a stack of superimposed layers through ring-shaped, vertical circumferential gaps as well as through transverse gaps which are open towards the circumferential gaps.

The *microwave-heated CVI (MW-CVI) (Figure 8.7b)* method exploits microwave radiation for the furnace heating. This kind of heating technique is potentially attractive because of its potential for generating a controllable inverse temperature profile during the heating of a ceramic fibre preform. By employing this technique, the centre of the preform can be hotter than the surface,

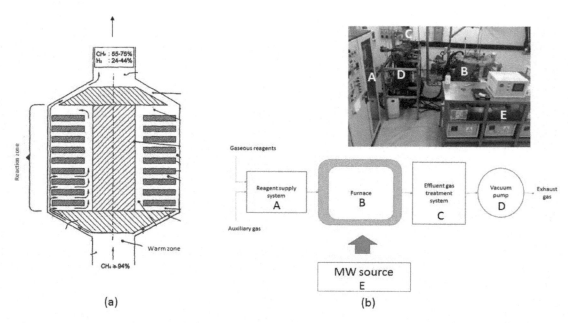

Figure 8.7 Scheme of the rapid CVI process **(a)** (From Hegermann, R. and Goetz, P., Method for the chemical vapor infiltration of refractive substances, WO2013104685, 2013.) and of MW-CVI process **(b)** (From "High-frequency ELectro-Magnetic technologies for advanced processing of ceramic matrix composites and graphite expansion" (HELM), FP7-NMP.2011.4.0-1, GA n°. 280464, website: http://www.helm-project.eu/.)

Table 8.2 Classification of CVI techniques

CVI method	Pressure	Temperature profile	Heating type
Isothermal/isobaric CVI (I-CVI)	Reduced and constant	Homogeneous	Radiative
Thermal-gradient CVI	Reduced and constant	Higher on one side	Radiative
Forced-flow CVI (F-CVI)	Higher pressure for reagents and lower for by-products removal	Isothermal or higher on one side	Radiative
Pulsed-flow CVI (P-CVI)	Reduced and impulsively increased with reagents introduction	Homogeneous	Radiative
Microwave-heated CVI (MW-CVI)	Reduced and constant	Higher in the centre	Microwave

potentially by up to several hundred degrees centigrade. This allows the MW-CVI process to initiate at the centre and then the reaction front to move toward the surface as the dielectric characteristics of the preform change. Hence, premature pore closure due to crusting is completely avoided. Thus, thicker and centrosymmetric preforms can be reasonably infiltrated thanks to this inverse temperature profile. Moreover, the densification can occur an order of magnitude faster than with I-CVI.

The industrialization of the MW-CVI is at a lower maturity level than the processes described earlier, as research projects about its potentiality are still in course.

The different methods for production of CMCs by CVI are reported in Table 8.2.

Up to now, the isothermal CVI is the most used in industrial applications, probably because it is more versatile. Anyway, the long infiltration times and the high energy consumption are pushing industries to consider more rapid processes. Currently, r-CVI and MW-CVI are trying to solve this issue, but they are relatively new and thus not yet much applied in industry for this reason.

8.4 CVI PROCESSING STEPS, DEPOSITION MECHANISMS, AND CHEMISTRY

The precursor gases react and decompose, and an inert solid (the composite matrix) is formed within the pore network of the heated substrate. The general reaction occurring in a CVI process can be written as:

$$\alpha A_{(g)} + \beta B_{(g)} \rightarrow \gamma C_{(s)} + \delta D_{(g)}$$

where A and B represent the reagent gases, D is the product gas, and C is the solid ceramic matrix. The stoichiometry of the reaction, not coincident but related to its kinetic mechanism, much affects the process. Hence, the most successful reactions are those where β is zero, that is those starting from a unique reactant gas. Furthermore, chemical reactions take place between the gaseous species that flow by diffusion within the pores.

In addition, for the application of the CVI technology to a given porous substrate it is necessary that the pores are interconnected and the substrate is thermally and chemically stable under the CVI conditions. In fact, the temperature ranges between 900°C and 1200°C and the process is often carried out at reduced pressure. The principal scope of CVI is to increase the density of the porous preform by a factor ranging from 2 to 10, depending on the properties required for the final application. The CMCs that take more advantages from the CVI technology are silicon carbide (SiC) ceramic reinforced with SiC fibres (SiC_f/SiC composites) and carbon (C) ceramic reinforced with C fibres (C_f/C composites). SiC_f/SiC composites and C_f/C composites produced by CVI present excellent physical and thermal properties.

In the isothermal CVI (I-CVI) process, schematized in Figure 8.6, the temperature of the preform is maintained constant inside a hot-wall furnace. As the composite is generated by the slow deposition of the matrix in the preform pores, several days are required for infiltrating a preform containing 40–50 volume % of fibres. The process requires a combination of low reagent concentrations, low temperatures, and low pressures, to avoid inhomogeneity or the occlusion of the outer preform surfaces. Thus, the time necessary for obtaining the desired density is long and the process is quite expensive. The wall thickness of preforms is limited to a few millimetres, and thicker preforms require usually two or more infiltration steps. In the latter case, the infiltrated preforms are treated with abrasive machines to remove the surficial crust formed during the previous infiltration to open the preform porosity, and then the infiltration is repeated.

The chemistry of CVI infiltration can be considered like the chemistry of the chemical vapour deposition (CVD) process, employed for producing ceramic surface coatings, extensively described by Kumpta et al. in the book "Chemical Aspects of Electronic Ceramics Processing." Table 8.1 lists the main reactions necessary for producing different ceramic matrices in CVI and CVD process.

For the preparation of C, the most versatile deposition process used for C_f/C composites is the thermal decomposition of a hydrocarbon gas starting from a porous preform typically made from polyacrylonitrile (PAN). The nature and phase of the precursor and the way to deliver energy represent the two basic parameters affecting the different types of allotropic forms of carbon. The precursors are usually alkanes (such as methane or propane), but sometimes non-saturated hydrocarbons (such as acetylene or propylene) or aromatic compounds (benzene and derivatives) can be used as they decompose at a lower temperature than the alkanes.

Two classes of reactions, homogeneous (in the gas phase) or heterogeneous (on the surface), can be considered.

In the first type of reactions, radical processes are favoured. So, free radicals are formed which are recombined immediately. In this step, non-saturated species such as allenes, propyne, and butadiene are formed. After the formation of the latter, the aliphatic C3 and C4 compounds undergo cyclization processes, while at longer deposition times, aromatics and polyaromatics are produced.[31]

Heterogeneous reactions affect the rate of formation of pyrocarbons, as they are produced by the nucleation and the growth processes on a given substrate. The presence of a catalyst (for example, a transition metal or a gaseous reagent such as H_2,

O_2, or Cl_2) is a predominant factor that influences the kinetics of this surface reaction.

For SiC preparation, experiments[32] have demonstrated that methyltrichlorosilane (MTS) is an ideal precursor for producing high-quality beta-silicon carbide (β-SiC) by using the CVD technique.

The main reasons are:

• The Si: C ratio in MTS is the same as the Si: C ratio of the SiC deposit.
• The by-product HCl derived using MTS as a precursor might contribute to suppressing the deposition of pure Si crystal during the growth of SiC films or crystals.
• The H_2 carrier gas may favour the reduction of the Si–Cl bonds on the growing surface of SiC, thus resulting in an increase in deposition rate.

The reaction reported in Table 8.1, for the deposition/infiltration of SiC, is faster in the presence of H_2 behaving as a catalyst. Consequently, the mechanism reported for the reaction consists of two steps:

1. Homogeneous decomposition of MTS in the presence of hydrogen

$$CH_3SiCl_3 + H_2 \rightarrow CH_4 +:SiCl_2 + HCl$$

2. Heterogeneous reaction for SiC deposition

$$CH_4 +:SiCl_2 \rightarrow SiC + H_2 + 2HCl$$

This mechanism can explain the catalytic activity of H_2. However, on the other hand, further kinetic studies showed that the kinetic equation of the deposition reaction was of the type:

$$v = k_0 [CH_3SiCl_3] \exp\left(-\frac{E_a}{RT}\right)$$

Hence, it was observed that the reaction was of the first order with respect to MTS and an apparent activation energy of 120 kJ/mol was determined. Consequently, the slow step of the reaction is reasonably the molecular decomposition of MTS.

$$CH_3SiCl_3 \rightarrow \cdot CH_3 + \cdot SiCl_3$$

But this reaction step has an activation energy of 320 kJ/mol, not in agreement with the determined value.

However, specific kinetic studies about MTS decomposition showed the formation of three products (CH_4, HCl, and $SiCl_4$) when hydrogen or helium were used as carrier gases. $SiCl_3\cdot$, CH_3, and Cl· radicals were also detected. It was also observed that the decomposition is faster in hydrogen than in helium. This observation was only tentatively explained by considering the collision partner M:

$$CH_3SiCl_3 + M \rightarrow \cdot CH_3 + SiCl_3 + M$$

Hydrogen is more effective than helium as collision partner as it can take part to the radical mechanism. Although the study of the mechanism is still controversial, the most important reactions are believed to be the following[31]:

1. First-order MTS decomposition

$$CH_3SiCl_3 \rightarrow CH_3 + SiCl_3$$

2. Bimolecular H-atom abstraction from MTS by Cl:

$$CH_3SiCl_3 + Cl \rightarrow CH_2SiCl_3 + HCl$$

The atomic chlorine can be generated by:

$$SiCl_3 \rightarrow :SiCl_2 + Cl$$

The mechanism for the formation of the final product was not fully understood. Interestingly, also the 1,2 elimination of HCl was kept into account, despite the fact that the formation of $CH_2 = SiCl_2$, which is the product of this reaction, was not observed experimentally.

However, the formation of an electron-rich species with an increased electronic density on carbon, capable of rapidly reacting with electron-poor species like $SiCl_2$, can better justify the formation of the C-Si linkages in the final structure. Reasonably, the formation of the final SiC structure is mediated by the formation of chlorinated carbosilanes structure with alternating Si and C atoms.

The halogenated silane precursors are the most common but they have the disadvantage of producing acidic by-products (e.g., HCl).

Trimethylsilane (TMS) was also studied as precursor of amorphous SiC by CVD process.[33] The deposition rate showed a maximum trend as a function of TMS flow and decreased by increasing the temperature. Thus, a higher deposition rate (above 500 nm/minute) is achieved at low temperatures. All the produced films showed a compressive stress in the range of 200 to 5000 MPa and showed no temperature-stress hysteresis effects. On the whole, the efficiency of the silane-based reactions is not considered as high as the process obtained from MTS and this latter is currently the most employed to produce CMCs.

Hence, carbosilanes containing a 1:1 Si:C ratio have been studied[34] and compared as single-source low-pressure CVD (LPCVD) precursors. All of the precursors examined produced polycrystalline, stoichiometric, ß-SiC at Si(100) substrate temperatures from 800°C to 1000°C and a reactor pressure of 0,67 Pa. Interestingly several differences were revealed between linear and cyclic precursors at the highest temperature studied (1080°C). In fact, at T above 1080°C, the cyclic precursors growth rate/thickness continues to increase linearly while the films become C-rich, whereas the growth rate of the linear precursors decreases, the film stoichiometry remains 1:1, and several interesting morphological features are produced.

Titanium nitride is an extremely hard ceramic material, often used as a coating on titanium alloys. The synthesis of TiN can be achieved by different processes where the source of nitrogen is often N_2, but also ammonia or amines can be employed. Moreover, usually the Ti source consists of $TiCl_4$, but also Ti complexes, such as

tetrakis(dimethylamido)titanium, were successfully employed.[33]

Tungsten carbide (WC) is a very resistant and durable material, especially with respect to fatigue and abrasion. The synthesis of WC can be achieved by starting from pure W, but also from WO_3.[34] By considering the first approach, kinetic studies showed the formation of W_2C as intermediate compound, suggesting that it is possible to select the composition in terms of the two carbides by properly selecting the conditions of the process.

TiN and WC are often synthesized industrially by CVD but neither is industrially used in CVI. In general, the reactions that can be used in CVD are potentially suitable for CVI. However, up to now only some of them are industrially exploited. Among them those giving C, SiC or borides are the most common.

Oxide CMCs can be prepared by CVI by exploiting the reaction reported in Table 8.1 related to the production of Al_2O_3 and ZrO_2. These reactions are generally more complex than the ones for obtaining SiC or C, as three reactive gases must be fed in the reactor: $AlCl_3$ (or $ZrCl_4$), H_2, and CO_2.

The challenge for oxide/oxide composites is to provide an alternative to SiC/SiC composites with the aim of decreasing the cost of manufacture and improving the thermal stability in air at high temperature. Oxide/oxide composites based on weak matrix concept without interphase and consisting of a porous alumina matrix reinforced by fibres of the same nature, are under investigation at Onera, a USA company.[35] In this process, the infiltration occurs integrated with the sintering of alumina slurry. By using submicron alumina powders and specific additives necessary for shaping and for control of the sintering shrinkage, the first composites were obtained by developing a method of infiltration into fibrous reinforcements.

In CVI process, the deposition of the matrix material on the preform fibres changes the morphology of the obtained composite as successive layers of matrix material are deposited on the fibres, thus leading to less porous materials, as evident in Figure 8.8. In this case, SiC was deposited on SiC fibres.[32]

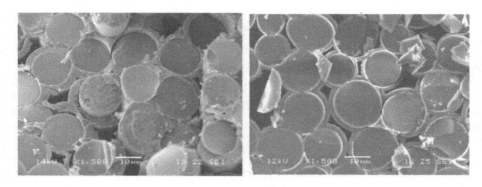

Figure 8.8 Morphology of two infiltrated SiC$_f$/SiC samples.

8.5 CVI EQUIPMENT

In general, a CVI apparatus consists of a reagent supply, the furnace where the preform is placed, and the effluent gas treatment system. The equipment for the CVI process is schematically shown in Figure 8.9.

A gas supply system feeds the furnace and it includes mass flow controllers, automatic pressure control systems, and rotameter flow indicators.

Standard furnaces used for CVI are used up to about 1200°C. The furnace vessels are manufactured with stainless steel, including the water jacket, furnace, and all flanges. Graphite hot zones are used on the CVI vacuum furnaces and also graphite resistance heating elements and low-density fibrous graphite insulation. Temperature monitoring along the process is carried out with a thermocouple or with an infrared pyrometer. Interestingly, the sample holder must allow the permeation of the preform by the reactant gases. Hence, the configuration of the process must be suitably designed keeping into account the geometrical constrains of the specific system. In fact, the latter aspect much impacts the effective achievement of a homogeneous composite.

Gaseous precursors for CVI processes must be volatile but at the same time stable for both safety and processing reasons. In general, precursor reagents provide only a single compound to the infiltrated preform, while by-products are volatilized during the process. However, in some cases, precursors may provide more than one compound, allowing a simplified delivery system with a reduction of the number of reactants required by the infiltration. As shown in Table 8.1, the most-used CVI precursor gases are CH_4, CH_3SiCl_3, BCl_3, $TiCl_4$, NH_3, $SiCl_4$, or $AlCl_3$. Commercial CVI equipment is produced by Archer technicoat (ATL, UK), CVT, CVD Equipment Corporation, TevTech, FirstNano, etc. They consist of a furnace, a control

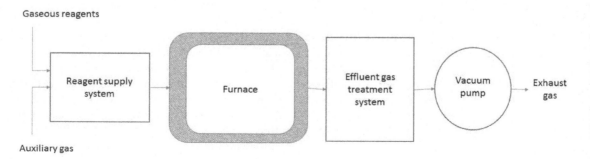

Figure 8.9 Scheme of the CVI equipment.

cabinet where the temperatures and gas flows are regulated, a system for pumping the gases, and a scrubber for treating the gaseous effluents coming from the furnace (Figure 8.6a). The reactor is contained within a water-cooled stainless steel vacuum vessel. The internal graphite working chamber is heated by a single-zone graphite resistance heater and consists of an insulated reaction chamber in a vacuum chamber. Several types of graphite sample holders are available to support samples within the working zone. The insulation is made of CFs. A pyrometer provides excellent temperature control. The gas supply system uses mass-flow controllers to deliver permanent gases into the reactor. The reactor is evacuated by a chemically resistant rotary pump. The effluent gas passes in a wet-scrubber unit which neutralises the acid by-products of the CVI process. A cold trap can also be fitted before the vacuum pump if this is required.

The control system is housed in a single cabinet with all the controls on its front panel. In Figure 8.10 a commercial laboratory-scale CVI plant picture is reported. This plant has an overall dimension of 125 mm diameter × 250 mm height. Industrial CVI plants commercially available have dimensions of 1300 mm diameter × 1400 mm height, for example (Figure 8.10).

8.6 PROPERTIES OF CVI-PRODUCED COMPOSITES

Monolithic ceramic materials show high modulus and strength at elevated temperatures, but their use as structural components is severely limited because of their brittleness.

However, fibre-reinforced ceramic-matrix composites (CMCs), obtained by incorporating fibres in ceramic matrices, not only exploit their attractive high-temperature strength but also reduce the propensity for catastrophic failure.

Composites made with a discontinuous whisker-like reinforcement were studied and prepared by the conventional techniques for monolithic ceramics, such as slip casting or injection moulding followed by full sinterization up to the maximum density in oven at high temperature. The processes can be also combined with hot isostatic pressing.

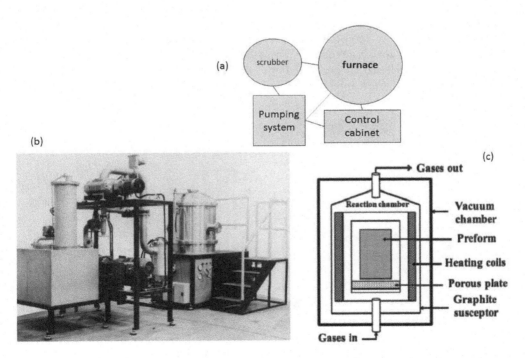

Figure 8.10 Commercial CVI equipment: **(a)** scheme; **(b)** picture; **(c)** furnace scheme.

CMC materials based on SiC with a discontinuous SiC reinforcement containing up to 75% of reinforcing phase show excellent properties of resistance to abrasion in both dry, slurry, and sliding conditions and also resistance to thermal shock and chemical attack; thus, they are used in the mining sector and in many anti-abrasion applications.[36] CMCs with a discontinuous reinforcement show a higher toughness than monolithic ceramics, as evidenced by Boulanger et al.[37] investigating composites consisting of an alumina matrix and SiC whiskers as reinforcement. The impact toughness was reported to increase by increasing the content of whiskers up to a plateau and only a slight decrease as a function of temperature was noticed up to 1500°C. Crack deflection and whisker pull-out are considered as the major energy-dissipating mechanisms in fracture.[38] At 1250°C, however, whisker pullout appeared to be much more extensive than crack deflection.[37] As reported more recently also by Parlier et al.,[35] it is the result of achieving during processing a fibre-matrix bonding sufficiently weak to allow debonding and consequently the relaxation of the stress field at the crack tip and the bridging of matrix cracks by the fibres (Figure 8.11).[35]

When the reinforcement is continuous the toughness is further improved and, currently, CMCs with a continuous reinforcement are the most common. In this kind of composite, the use of a proper preform or fabric is required. The fibres of the preform or fabric can be pretreated by a CVI coating to achieve a weak interphase. The thickness of this coating is lower than 0.5 microns and is deposited on each fibre of the preform with the aim of favouring the debonding during the crack propagation in the brittle matrix, thus improving the impact resistance. The CMCs with a continuous reinforcement consist of two classes: oxide matrix CMCs and non-oxide matrix CMCs. The latter class have been studied for 30 years, but commercial products such as brakes parts for aerospace shuttles have been available for 15–20 years, and those having a SiC matrix are the most important. In fact, their properties (Table 8.3) show very high values of elastic modulus and tensile strength, especially in composites prepared by CVI. On the other hand, the composites prepared by PIP or LSI show generally lower properties, with the exception of porosity, that is minimized by using the LSI techniques. The CVI method is, however, showing the best reproducibility. The low pressure required in the I-CVI allows a more uniform distribution of density, but the main disadvantage is the slower deposition rate and longer deposition times than the other processes.

The mechanical properties, the stability at high temperatures (about 800°C–1800°C), and the chemical resistance makes SiC matrix composites suitable for applications such as hot-gas recirculating fans, heat exchangers, gas-fired radiant burner screens, gas-fired radiant burner tubes, gas turbine engine combustion liners and tip shrouds, hot-liquid filtration, canned motor pump containment shells, brakes, structural components in aerospace industry, in nuclear reactors, and in heat engines.

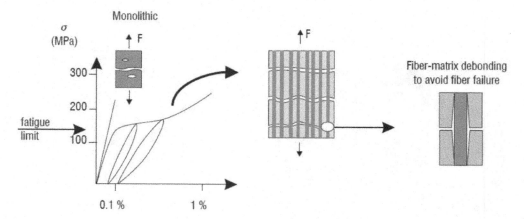

Figure 8.11 Damage tolerant behaviour of CMCs compared to monolithic ceramic. (From M. Parlier, M. et al., *J. Aerosp. Lab*, 3, AL03-09, 1–12, 2011.)

Table 8.3 Properties of SiC matrix composites prepared by different methods

Property	Unit	CVI-SiC$_f$/SiC	CVI-C$_f$/SiC	PIP-C$_f$/SiC	LSI-C$_f$/SiC
Density	g/cm²	2.5	2.1	1.8	1.7–1.8
Fibre content	Vol%	40	45	46	42–47
Elastic modulus	GPa	170–230	90–100	65	60–80
Tensile strength	MPa	150–200	350	250	240–270
Strain at break	%	0.3–0.5	0.9	0.5	0.8–1.1
Porosity	%	10	10	10	2–5

Source: Naslain, R. et al., J. Phys. Colloq., 50, C5-191–C5-207, 1989.

In the preparation of CMCs, slurries of compounds are applied to fibres or preform. Since the matrix is not fully densified in the composites, further sintering may occur, depending on the conditions of temperature and duration in service. The unavoidable consequence of the shrinkage associated with matrix sintering is a decrease in the mechanical properties of the composites, mainly due to degradation of load transfer between fibres and matrix (matrix microcracking, debonding between fibre and matrix). The preparation of CMCs by CVI allows for overcoming this issue.

8.7 NUMERICAL MODELLING OF THE CVI PROCESS

Applying a finite element method to CVI, Kehzi et al.[39] developed a program to simulate the isothermal CVI process, in particular the fabrication of C$_f$/C composites for predicting process parameters as density and porosity. The basic principles of Finite Element Modelling (FEM) and mass transport are at the basis of the model. Then finite element equations and the elemental stiffness matrices were derived. The distribution of density and porosity was thus predicted by the model in agreement with experimental results.

Danielsson et al.,[40] simulated a silicon carbide CVD process, including inductive heating and fluid dynamics. For consolidating the model, experiments were carried out by using SiH$_4$ and C$_3$H$_8$ as precursor gases. The simulations were conducted in a horizontal hot-wall CVD reactor, but the results are more general and can be applied to any reactor configuration since no adjustable parameters were used to fit experimental data. It was shown that including etching and parasitic growth on all reactor walls exposed to gas flow much improved the accuracy of the simulations.

A two-dimensional model was developed by Wei et al.[41] for isothermal CVI process in order to investigate physicochemical phenomena and infiltration-induced structural changes in C$_f$/SiC composites. The mathematical model was implemented to simulate the densification behaviour of the C/SiC component of a small-scale thruster liner for rocket engine.

The results showed that infiltration efficiency is high at first and then decreases dramatically, which is in agreement with the corresponding experimental results. The dependence of densification behaviour on infiltration temperature was also investigated and the calculation results showed that the densification rate increases while density uniformity of overall composites decreases significantly at elevated temperature.

Vignoles et al.[42] developed a computational tool for the modelling of CVI of carbon/carbon composites, which is based on 3D images acquired by X-ray Computerized Micro-Tomography with a very fine resolution, such that the fibres are clearly distinguishable from each other. Preliminary image processing was necessary in order to perform segmentation between void and solid phases. Then, morphological and transport properties are computed in the images. Random walkers are used for simulating gas transport in rarefied regimes and in continuum. The study helped in defining the concept of infiltrability.

More recently Deck et al.[43] (Figure 8.12) developed a computer model for F-CVI applied to SiC$_f$/SiC that simulates the transport of the SiC precursors, the deposition of SiC matrix on the fibres surface, and the effect of by-products on the process. The critical parameters in the process were temperature and reactant concentrations. Thanks to this model it was possible to identify infiltration conditions maximizing composite density while

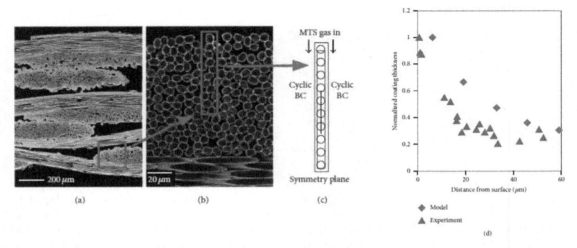

Figure 8.12 **(a)** Cross-section of stacked fabric layers in SiCf/SiC, **(b)** cross-section of fibre bundle, **(c)** model geometry showing MTS introduction and boundary conditions, and **(d)** comparison between simulated and experimentally observed SiC deposition gradients into the fibre perform. (From Deck, C.P. et al., *Sci. Technol. Nucl. Ins.*, ID 127676, 10 pages, 2013.)

minimizing the fabrication time. Very recently, Tang et al.[43] developed a model including global reactions and 1074 elementary reactions (involving 285 species in both the gas phase and the solid surface) to simulate the deposition of pyrocarbon on carbon fibres in CVI using propane. The global reactions consisted of the direct dehydrogenation reactions from the hydrocarbon species in the gas phase. The elementary reactions included 66 surface species and 250 elementary steps. Isobaric experiments having a duration between 0.5 and 4 seconds and performed at different temperatures (in the range 1173–1323 K) were performed to compare their results with the simulated trials. The simulated and experimental results were in good agreement. On the basis of this study it was concluded that the pyrocarbon precursors are mainly small unsaturated species (such as acetylene and ethylene) and methyl radicals, and the deposition kinetics could be quantitatively described considering the deposition of these peculiar species.

Although not many papers can be found in literature about modelling of CVI, it is evident that the modelling activity has supported the development of this technique since the 1990s. The modelling,

performed considering different modelling scales, was important not only for prediction of properties (process optimization) but also for better understanding the kinetics of the process as a function of chemical and physics parameters, taking into account the peculiar reactions occurring in the specific CVI plants.

8.8 CORRELATIONS BETWEEN PROCESS, MORPHOLOGY, AND FINAL APPLICATION

As shown in Table 8.3, the mechanical properties observed for SiC matrix composites are better when they were produced by CVI. In fact, the advantages of CVI versus other CMC fabrication methods, such as hot pressing or liquid infiltration, consist of the minimization of the mechanical damage of the fibres due to the much lower pressure and temperature used in CVI, with respect to those in other fabrication methods, and also in the higher purity of the matrix produced by CVI. In fact, by hot-pressing sintering, for example, the use of additives can result in the presence of impurities or defects in the produced part.

For composites produced by CVI, the residual porosity is about 10%–15% by volume. In a measure depending on the materials, processing, and application the residual porosity can affect the performance of the final product. Residual porosity may be open or closed, and interconnected or not. A higher pressure or temperature can accelerate the deposition but favours the occlusion of preform pores. In fact, the material deposited during CVI often closes the open porosity on the outer surface before the desired final densification is achieved. Therefore, the composite parts must be removed from the furnace, and an intermediate step of machining to open the closed pores is required. Then, the composites have to be reheated before the CVI process continues up to the required final densification degree. In C_f/C composites, the number of intermediate machining steps mainly depends on the thickness of the component to be densified via CVI. For example, for the preparation of C_f/C components with a thickness of 40 mm, the number of machining steps can be up to three.

The morphology of the composite is usually controlled thanks to the correct design of the preform, its coating treatment to favour debonding during failure, and the control of all the parameters of the process, such as temperature and pressure, which strongly influence the crystalline structure of the deposited matrix. In the case of SiC the deposition of β-crystals with a face-centred cubic (FCC) crystalline structure is obtained.[44]

WORKED EXAMPLE 8.1

In a CVI plant, a deposition yield of 2 atom moles/day is obtained in a SiC_f preform having a weight of 98 g and 100 mm × 100 mm × 20 mm dimensions. Considering that only a porosity of 12 vol% is admitted in the SiC_f/X composite, what is the minimum estimated time and final density for the process for (a) a SiC_f/SiC composite; (b) a SiC_f/C composite (molar mass of C = 12 g/mole; molar mass of Si = 28 g/mole; density of SiC is 3.21 g/cm³; density of C is 2.27 g/cm³).

(a) As the porosity will be 12% vol, the final density of the SiC_f/SiC composite will be:
final density of the SiC_f/SiC composite = 3.21 (1-0,12) = 2.82 g/cm³
The volume of the SiC_f fibres in the preform is 98 g/ (3.21 g/cm³) = 30.5 cm³
In the SiC_f/SiC composite, considering 12 vol% of porosity, the volume to be filled by SiC is:
Volume to be filled = Volume of preform – volume of porosity – volume of the SiC fibres =
(10 · 10 · 2) – (10 · 10 · 2)· 0.12 – 30.5 = 145.5 cm³
The mass of SiC to be deposited in the preform is ρ_{SiC} x Volume to be filled
= 3.21 g/cm³ · 145.5 cm³ = 467.0 g
The atomic molar mass of SiC can be calculated as the average value between Si and C:
atomic molar mass of SiC = (12 + 28)/2 = 20 g/mole.
Hence, the moles of deposited atoms will be: 467.0 g/(20 g/moles) = 23.35 atom moles.
As the deposition yield is 2 atom moles/day, the necessary days can be calculated as:
23.35/2 = 11.7 days
(b) The volume to be filled is the same as in the (a) part.
The mass of C to be deposited = ρC Volume to be filled = 2.27 g/cm³ · 145.5 cm³ = 330.3 g
The moles are 330.3/(12 g/mole) = 27.5 moles
As the deposition yield is 2 atom moles/day, the necessary days can be calculated as:
27.5/2 = 13.8 days
The final density will be:
Final density of SiC_f/C composite: (30.5 cm³ · 3.21 + 145.5 cm³·2.27)/200 cm³ = 2.14 g/cm³

Progresses have been made concerning property tailoring of CMCs through matrix modification, interphase design, and reinforcement decoration.[45] The matrix modification can be made by the use of proper active fillers able of strengthening the matrix and modulating its porosity.

Regarding interphase design, it was found that with the deposition of PyC/SiC interphase on carbon fibres, the composites show typical non-brittle fracture behaviour.[46] The positive effect of PyC/SiC interphase on mechanical properties of composites can be due to the existence of PyC/SiC interphase that can act as diffusion barrier during the later stage of composite fabrication process, which can effectively prevent the diffusion of silicon toward carbon fibre. Moreover, the weakly bonded interphases can arrest and deflect the matrix cracks, thus protecting the fibres from early failure through notch effect. Concerning the decoration of reinforcement, the example of decoration of carbon fibres with carbon nanotubes in SiC matrix composites can be cited. In this case, the decoration significantly strengthens the matrix of the materials, and accordingly fibre bundle pull-out is observed in them. In conclusion, for obtaining high-performance CMCs, special attention should be paid to the composition and microstructure optimization of the material, together with the development of special fabrication processes.

8.9 APPLICATIONS OF CVI

Nowadays, CMCs are used for numerous high-performance applications (Figure 8.13). These need light, heat-resistant, and extremely robust materials.[47] An important factor that has hindered the progress of CMCs is the high temperatures usually employed for production. The production and processing are long and time consuming. Hence, manufacturing process must be made more economical, productive, and efficient. CVI has high potentiality as it provides CMCs with very good properties.

Another important issue is the differences in coefficients of thermal expansion between the matrix and the reinforcement. This leads to thermal stresses on cooling from the processing temperature. Anyway, whereas the thermal stresses can generally be relieved in metal matrix composites by plastic deformation of the matrix, this is not possible for CMCs and cracking of the matrix can occur. The nature of the cracking depends on the

whether the reinforcement shrinks more or less than the matrix on cooling, as it determines the character (tensile or compressive) of the local thermal stresses.

CMCs are used and could be used more in many high-temperature processes.[48] In fact, they show very high thermal shock and creep resistance, which enables design with large mechanical and thermal hot-gas valves for the control of the gas flow in gas-fired high-temperature furnaces. Compared to metallic valves, the service life of the CMC components is much longer, thus balancing their higher purchasing costs.[47] Similar advantages are obtained when CMC components are used as batch carriers in metal hardening. These C/C-grids have small heat capacity, thus reducing energy consumption and allowing fast heating and cooling cycles. Other applications of CMCs in high-temperature processes are flame tubes, heat exchangers, protective tiles, and various high-temperature holders.[47–49]

CMCs can also be used in extreme environments like gas turbines and heat exchangers in advanced thermal system, a topic extensively reviewed by Sommers et al.[49] Thanks to the use of these materials the operating temperatures of gas turbines have been increased, thus improving energy efficiency.

The lightweight and high-temperature properties of CMCs are ideal for new designs of the turbines as they require blades with very high rotational speeds. First products will be soon on the market.[50] Other CMC applications in aerospace[51] are body flaps, shrouds, and thermal protection systems. The anisotropic thermal expansion of C/SiC can be used to design components with zero thermal expansion in one or two directions. These components are in precision optics, in satellite communication or microelectronics, or for calibration of dimensional control tools. Another use of non-oxide CMCs is in fusion energy systems (the first wall and the blanket of a fusion reactor) for generation of nuclear energy.

CVI process is used especially for producing C_f/C composites or C_f/SiC composites. By CVI carbon or SiC is deposited on the fibres in the interior of the preform. The high wear resistance and the favourable friction properties of CMCs allow for their application in brakes, clutch-plates, and sliding contact bearings. CVI technology is used

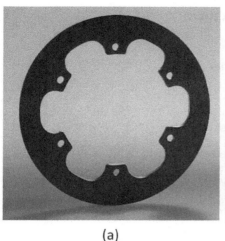
(a)

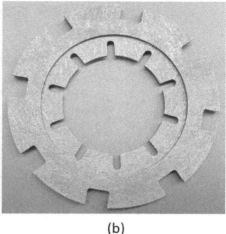
(b)

Figure 8.13 Examples of CMC-based products: **(a)** brake disc; **(b)** nozzle for rocket.

in brake discs in Formula 1 racing cars, as well as in the aerospace industry. When Formula1 racing cars are braked at full speed, this places an extreme load on the brakes and the parts subject to wear, in particular the ceramic brake discs.[52] With the aid of CVI infiltration it is possible to make brake discs extremely dense and efficient (Figure 8.7a) and brakes based on CMCs were used in Porsche at the beginning of 2000.

Through further treatments such as siliconisation, temperature-resistant and oxidation-resistant materials for rocket nozzles are being developed for the market (Figure 8.13b).

As an attempt to summarize the different applications of CMCs (Figure 8.14), we can consider: the applications aimed at improving energy production, including nuclear plants and engines; and the applications for the vehicles, including aerospace and high-speed cars. Other applications in mechanical engineering are also possible such as cutting tools, dies, and abrasives. The uses of CMCs in the defence industry are also important. Interestingly, CMCs are used in antiballistic plates thanks to their optimum combination of low density, high stiffness, and high impact strength. Recently, He et al.[53] investigated the ballistic performance and damage characteristics of quasi three-dimensional needle-punched C_f/SiC composites prepared by CVI by penetration experiments. High porosity and many micro-thermal stress cracks may directly lead to the lower ballistic performance. The damage characteristics of C_f/SiC

composites include matrix cracking, fibre bundle cracking, interfacial debonding, fibre fracture, and fibre bundle pull-out, playing important roles in energy absorption.

Very recently an application of CVI in the field of nano-structured materials for catalysis was also reported in literature.[54] In particular a catalyst was prepared by CVI as a Shape-Selective Hydrogenation Catalyst infiltrating a palladium complex (allyl-Pd(Cp)) in a metal-organic frameworks (MOF) based on Zr complexes. The resulting Pd-MOFs (consisting of nanostructured materials based of Palladium nanoparticles in the MOF structure) are shape-selective catalysts, as shown for the hydrogenation of carbonyl compounds with different steric hindrance. The CVI method is thus suitable for the preparation of CMCs parts, but also peculiar chemical reactions especially necessary for the development of nano-structured systems will be potentially possible in the near future.

8.10 ADVANTAGES AND DRAWBACKS OF CVI

The CVI technique is the most used techniques for producing non-oxide CMCs. The advantages of the CVI technique consisted of:

- The higher reproducibility than other techniques

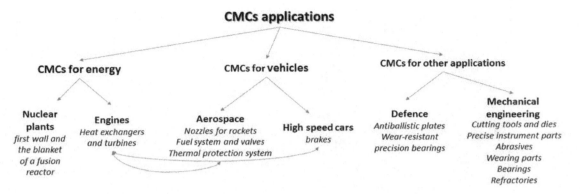

Figure 8.14 Scheme of CMCs applications.

- The possibility to densify simultaneously a large number of complex preforms in huge furnaces
- The minimization of mechanical damage to fibres during processing
- The higher purity of the produced matrix
- The low pressure allowing a more uniform distribution of density, a decrease in the flow rates of the inlet precursor, and a reduction in the formation of deleterious by-products (tar and soot, for example, for C matrix)

The drawbacks of the CVI technique consisted of:

- A porosity of 10–15 vol% is normally obtained. The LSI technique allows obtaining lower values (2–5 vol%).
- Necessity of doing iterative cycles because of pores' occlusion in thick composites
- Long deposition times, generally longer than 10 days
- High capital costs of the plant

8.11 RECENT INVESTIGATIONS ON CVI PROCESS

In the last decade, with the aim of making the CVI process faster, some new processes were investigated. Regarding the CVI equipment, the necessity of decreasing the time of infiltration[55] led HERAKLES aerospace company to design a plant to maximize the CVI furnace loading and also the efficiency of the process by introducing

in the CVI furnaces many different preforms to be infiltrated.

In the case of very light carbon preforms, the CVI isothermal infiltration was thoroughly investigated. With the aim of improving the mechanical properties of low-density carbon open-cell foams of two different pore sizes, these foams were infiltrated with pyrocarbon by CVD at reduced pressure and using pure propane as precursor.[56] Starting from very porous vitreous carbon foams, the CVI has allowed preparing pyrocarbon foams with well-balanced relative densities, ranging from 0.03 to 0.15 g/cm^3. Compressive crushing tests were carried out to investigate the influence of the pore size, the pyrocarbon type, and the relative density on the mechanical properties of the pyrocarbon-infiltrated foams. The results showed that they retain their non-brittle and dissipating behaviour up to relative densities of 0.15 g/cm^3. The stiffness, crushing strength, and dissipated energy increase significantly by increasing the relative density.

Tungsten can be a good candidate for the plasma-facing components of a future fusion reactor, but its use is strongly restricted by its brittleness. An innovative concept to overcome this problem is tungsten fibre-reinforced tungsten composite (W_f/W) prepared by CVI.[57] An *in-situ* fracture experiment was performed in a scanning electron microscope for close observation of the propagating crack. Stable crack propagation accompanied with rising load-bearing capacity was observed.

An innovative CVI process, called reactive CVI, was also studied by Ledain et al.[58] CMC samples were prepared by using a hybrid process in which the ceramic powder route was combined with reactive CVI, a new gas-phase method. The aim of this process was avoiding the occlusion of porosity resulting in high residual porosity in infiltrated samples. In fact, in this technique, the carbide growth occurs from the conversion of a carbon-bearing powder and slows down with increase in carbide thickness due to solid-state diffusion limitation. This self-limitation was demonstrated to allow a self-regulation of the growth between the interior and the surface of the sample and thus to ensure a better homogeneity. The chosen carbide was TiC obtained by reacting H_2 and $TiCl_4$ with a carbon powder introduced by slurry impregnation in felts. The rate of infiltration was demonstrated to be constant and the maximum density (0.15 g/cm^3) was achieved in about 17 h.

A microwave-heated methyl trichlorosilane-based CVI technique was used by Binner et al.[59] to form SiC$_f$/SiC composites from SiC fibre preforms preimpregnated with SiC powder using two different fabrication techniques (vacuum bagging or electrophoretic impregnation). Infiltration rates were increased thanks to the loading with powder. Moreover, preferential infiltration occurred in regions where the SiC powder was most concentrated. The consequence was density gradients in the final composites. Nevertheless, average densities as high as 75% of the theoretical value could be achieved in a 10 h process whereas an isothermal CVI process can last weeks. Hence, a decrease in infiltration time was obtained thanks to MW heating granting a reverse heating gradient in the furnace. Therefore, the MW-CVI process is, overall, up to now the most promising technique.

8.12 PERSPECTIVES AND CHALLENGES FOR CVI

Fibre-reinforced CMCs and densification of such composites by a variety of infiltration routes is an open research area. The research activities are continually driven by the superior physical properties of composites, compared to their monolithic counterparts. Moreover, it is necessary to find a good compromise between the material quality and the processing conditions, such as reduction in time, decrease in energy consumption, and adequate plant expenses.

New applications of CMC components produced by CVI are under study or development.

Examples are valve components, with mechanical resistance at high temperature and pressure shocks, heat shields for racing cars, and casings for rocket engines. The improvement of ceramic fibre thermal and chemical stability in order to perfectly match the material requirements for applications at high temperatures, such as heat exchangers and gas turbines, is also of interest.

The high thermal stability allows applications in materials for nuclear reactors, and in particular Riccardi et al.[60] investigated these materials for applications in fusion reactors.

The isothermal, isobaric CVI, the oldest CVI process, is widely used in industrial applications. The main drawback is the very long processing time due to the diffusion mechanism associated with low overall precursor efficiency (a few percent).

New methods to develop rapid infiltration techniques have been explored. In particular, the thermal-gradient CVI, associated with an *in-situ* vaporized liquid precursor (rapid CVI) for C$_f$/C composites,[30] and microwave-assisted processes (MW-CVI), for SiC$_f$/SiC[61] composites are becoming available. Even if these processes exhibit less flexibility with respect to the well-established I-CVI, they may be more appropriate for specific matrix compositions and/or specific application fields.

The MW-CVI process has recently received much attention for its potential to produce a controllable temperature profile inside a preform with a temperature gradient from the centre of the preform to its periphery. Therefore, the deposition of the ceramic matrix proceeds from the inside to the outside, avoiding the problems connected with the occlusion of the outer pores of the preform. This kind of process was investigated in the framework of the European Project "High-frequency Electro-Magnetic technologies for advanced processing of CMCs and graphite expansion" (HELM).[61] A pilot MW-CVI plant was designed and built during the project activities with the support of modelling activities and promising results about the infiltration rate were achieved, although more research should be made to optimize the system.

By MW-CVI process, SiC matrices could be obtained and preforms of very complex geometry could be successfully infiltrated, at operating temperatures between 900°C and 1200°C with a potential reduction of deposition times from weeks to days.

Together with the advantages of microwave heating described above, some technical difficulties are associated with the possible insurgence of thermal instabilities (*thermal runaways*).

This can be a major problem in materials, like ceramics, characterized by low thermal conductivity resulting in a nonhomogeneous distribution of temperature with some points much hotter than the surrounding material (hot spots). As, for many ceramic materials, dielectric properties increase exponentially with temperature, the hot spots can reach extremely high temperatures even leading to local fusion of the material.

These inhomogeneities in local temperature can even give rise to high internal stresses in the material due to the differential thermal expansion, with formation of cracks and even leading to the fracture of the sample during heating. Therefore, in order to make the most of the microwave heating in this type of application, the prediction of the temperature profile, the electromagnetic field, and the prevention of thermal instabilities are of fundamental importance.

Another issue in the coupling of microwave technology and conventional-based CVI is the possibility of plasma formation because of simultaneous presence of gases at reduced pressure and irradiation by low-frequency electromagnetic waves. The plasma, consisting of charged particles moving in the reactor, changes the gases' reactivity and generates local magnetic fields, making the outcome of the process unpredictable and allowing also local arcing. Currently, the fine tuning of gas pressure in the CVI-MW furnace seems the best method to avoid plasma formation.

8.13 SUMMARY

CMCs represent advanced materials combining the chemical and thermal resistance of ceramics with the mechanical strength of ceramic reinforcements, thus with an improved fracture toughness with respect to ordinary monolithic ceramics, which possess a low fracture toughness. Different routes can be followed to produce them. The CVI consists of the deposition of the ceramic matrix in the pores of a preform or fabric obtained by ceramic-based fibres. The reaction of deposition for the different matrices are heterogeneous gasphase reactions because a solid product (C, SiC, TiN, WC, etc.) is formed. These reactions are often

catalysed by proper gaseous species. Whereas the general reactions are known, the kinetic of the reactions are often not completely understood.

The equipment necessary for a traditional CVI process consists of a reagent supply system, a furnace, an effluent gas treatment system, and a vacuum pump. The heat is transferred from the furnace wall to the sample area by convection. Hence, some thermal inhomogeneity can affect the process. For this reason, a different version of the CVI equipment was investigated, such as the gradient CVI. Another drawback of the conventional CVI is the long infiltration time. The attempt of developing the F-CVI or microwave (MW)-heated CVI was made for overcoming CVI drawbacks. The latter attempt, introducing also a reversed temperature profile thanks to the use of MWs, represents the most promising innovation in CVI and was one of the topics recently investigated in the framework of the HELM project.

The applications of these composites are in the aerospace field, in the heat exchangers and turbines sectors, in the brakes for airplanes or sportive cars, and in the nuclear sector.

ACKNOWLEDGEMENTS

Dr. Calvin Prentice of Archer Technicoat (ATL) company (High Wycombe, UK) is thanked for helpful discussion about CVI. European Commission is thanked for having supported the HELM project (GA n°280464).

ADDITIONAL READINGS

Arkles, B., J. J. Sullivan, Chemical aspects of electronic ceramics processing, edited by P. N. Kumta, A. F. Hepp, D. B. Beach, *Materials Research Society*, Symposium Proceedings Volume 495, Warrendale, PA, 1988.

Leclercq, G., M. Kamal, J. M. Giraudon, P. Devassine, L. Feigenbaum, L. Leclercq, A. Frennet et al. Study of the preparation of bulk powder tungsten carbides by temperature programmed reaction with CH_4/H_2 mixtures, *Journal of Catalysis* 158, 142–169, 1996.

Raether, F. Ceramic matrix composites–An alternative for challenging construction tasks, *Ceramic Applications*, 1(1), 45–49, 2013.

Roman, Y. G. Forced flow chemical vapour infiltration, Eindhoven: Technische Universiteit Eindhoven, 1994. doi:10.6100/IR418123.

Zhang, L., S. Dong, H. Zhou, Y. Kan, F. Zhou, Z. Wang, 3D C$_f$/ZrC-SiC composites fabricated with ZrC nanoparticles and ZrSi$_2$ alloy, in *Ceramics International* 40(8 Part A), 11795–11801, 2014.

REFERENCES

1. A. Lazzeri, CVI processing of ceramic matrix composites, in *Ceramics and Composites Processing Methods*, 1st ed., edited by Narottam, P. B. and Boccaccini, A. R., The American Ceramic Society, John Wiley & Sons, New York, 2012.

2. L. Zhang, H. Zhou, Y. Kan, F. Zhou, Z. Wang, 3D Cf/ZrC-SiC composites fabricated with ZrC nanoparticles and ZrSi$_2$ alloy, *Ceramics International*, 40(8 Part A), 11795–11801, 2014.

3. R. L. Bickerdike, A. R. G. Brown, G. Hughes, H. Ranson, The deposition of pyrolytic carbon in the pores of bonded and unbonded carbon powders, in *Proc. Fifth Cod. Carbon*, Vol. I, edited by Mrosowski, S. Studebaker, M. C. and Walker, P. L., Pergamon Press, New York, 575–583, 1962.

4. W. C. Jenkin, Method of depositing metals and metallic compounds throughout the pores of a porous body, U.S. Patent 3,160,517, December 8, 1964.

5. E. Fitzer, Dynamische instabilitäten bei heterogenen gas/feststoffreactionen, *Chemie Ingenieur Technik*, 41(5–6), 331–339, 1968.

6. R. Naslain, F. Langlais, R. Fedou, The CVI processing of ceramic matrix composites, *Journal de Physique Colloques*, 50(C5), C5-191–C5-207, 1989.

7. A. J. Caputo, W. J. Lackey, Fabrication of fiber-reinforced ceramic composites by chemical vapor infiltration, ORNL/TM-9235, 1984.

8. T. M. Besmann, B. W. Sheldon, R. A. Lowden, D. P. Stinton, Vapor-phase fabrication and properties of continuous-filament ceramic composites, *Science*, 253, 1104–1110, 1991.

9. W. J. Lackey, T. L. Starr, Fabrication of fiber-reinforced ceramic composites by chemical vapor infiltration: Processing, structure and properties, in *Fiber Reinforced Ceramic Composites*, edited by K. S. Mazdiyasni, Noyes Publications, Park Ridge, NY, 397–450, 1990.

10. B. Clauss, Fibers for ceramic matrix composites, in *Ceramic Matrix Composites*. edited by W. Krenkel Copyright © 2008 WILEY-VCH Verlag GmbH & Co. KGaA, Weinheim, Germany, 2008.

11. S. Das, Life cycle assessment of carbon fiber-reinforced polymer composites, *International Journal of Life Cycle Assessment*, 16, 268–282, 2011.

12. K. Sudo, K. A. Shimizu, New carbon-fiber from lignin, *Journal of Applied Polymer Science*, 44, 127–134, 1992.

13. S. Kubo, Y. Uraki, Y. Sano, Preparation of carbon fibers from softwood lignin by atmospheric acetic acid pulping, *Carbon*, 36, 1119–1124, 1998.

14. Z. Guo, Z. Liu, L. Ye, K. Ge, T. Zhao, The production of lignin- phenol-formaldehyde resin derived carbon fibers stabilized by BN preceramic polymer. *Materials Letters*, 142, 49–51, 2015.

15. Ozer, A. Kizasoz, K. A. Guler, A. Karaaslan, Aluminum matrix composite fabrication by infiltration of SiC preforms using squeeze casting technique, *8th International PhD Foundry Conference*, 7–8 June 2011, Brno University of Technology Czech Foundrymen Society – CFS. https://www.academia.edu/7711923/Aluminum_matrix_composite_fabricat%C4%B1on_by_infiltration_of_SiC_preforms_using_squeeze_casting_technique.

16. K. Langguth, S. Bockle, E. Moller, Polysilane-derived porous SiC preforms for the preparation of SiC-glass composites, *Journal of Materials Science*, 30, 5973–5978, 1995.

17. M. Thünemann, O. Beffort, S. Kleiner, U. Vogt, Advanced Engineering Materials, cfi/Ber. DKG 82 10 D, 1–5, 2005.

18. S. Roy, K. G. Schell, E. C. Bucharsky, P. Hettich, S. Dietrich, K. A. Weidenmann, A. Wanner, M. J. Hoffmann, Processing and elastic property characterization of porous SiC preform for interpenetrating metal/ceramic composites, *Journal of the American Ceramic Society*, 95(10), 3078–3083, 2012, doi:10.1111/j.1551-2916.2012.05347.x.

19. B. Stevinson, D. L. Bourell, J. J. Beaman, Dimensional stability during post-processing of selective laser sintered ceramic preforms, *Rapid Prototyping Journal*, 14(3), 149–154, 2006.

20. T. Hinoki, E. Lara-Curzio, L. L. Snead, Mechanical properties of high purity sic fiber-reinforced cvi-sic matrix composites, http://web.ornl.gov/~webworks/cppr/y2001/pres/116946.pdf.

21. R. T. Bhatt, Y. L. Chen, Heat Treatment Effects on Microstructure of SiC Fiber Preforms, NASA/TM—1999-209292, 1999.

22. Z. Fu, L. Schlier, N. Travitzky, P. Greil, Three-dimensional printing of SiSiC lattice truss structures, *Materials Science and Engineering A*, 560, 851–856, 2013.

23. D. Zhu, M. Gao, H. Pan, Y. Pan, Y. Liu, S. Li, H. Ge, N. Fang, Fabrication and mechanical properties of SiCw(p)/SiC-Si composites by liquid Si infiltration using pyrolysed rice husks and SiC powders as precursors, *BioResources*, 9(2), 2572–2583, 2014.

24. Y. Cai, S. Fan, X. Yin, L. Zhang, L. Cheng, Y. Wang, Microstructures and mechanical properties of three-dimensional ceramic filler modified carbon/carbon composites, *Ceramics International*, 40, 399–408, 2014.

25. Y. Zhao, H. Xia, R. Tang, Z. Shi, J. Yang, J. Wang, A low cost preparation of C/SiC composites by infiltrating molten Si into gelcasted pure porous carbon preform, *Ceramics International*, 41, 6478–6487, 2015.

26. G. Amirthan, M. Balasubramanian, Helium gas permeability of SiC tubes produced using cotton fabric, *Ceramics International*, 41, 3589–3594, 2015.

27. A. Maity, D. Kalita, N. Kayal, T. Goswami, O. Chakrabarti, P. G. Rao, Synthesis of biomorphic SiC ceramics from coir fibreboard preform, *Ceramics International*, 38, 6873–6881, 2012.

28. Z. Li, P. Xiao, X. Xiong, B-Y. Huang, Preparation and tribological properties of C fibre reinforced C/SiC dual matrix composites fabrication by liquid silicon infiltration, *Solid State Sciences*, 16, 6–12, 2013.

29. N. R. Calderon, M. Martinez-Escandell, J. Narciso, F. Rodriguez-Reinoso, Manufacture of biomorphic SiC components with homogeneous properties from sawdust by reactive infiltration with liquid silicon, *Journal of the American Ceramic Society*, 93(4), 1003–1009, 2010, doi:10.1111/j.1551-2916.2009.03572.x.

30. R. Hegermann, P. Goetz, Method for the chemical vapor infiltration of refractive substances, WO2013104685, 2013.

31. P. Delhaes, Review—Chemical vapor deposition and infiltration processes of carbon materials, *Carbon*, 40(5), 641–65, 2002.

32. B. Cioni, A. Lazzeri, Modeling and development of a microwave heated pilot plant for the production of SiC-based ceramic matrix composites, *International Journal of Chemical Reactor Engineering*, 6, A53, 2008.

33. P. N. Kumta, A. F. Hepp, D. B. Beach, B. Arkles, J. J. Sullivan, Chemical aspects of electronic ceramics processing, *Editors, Materials Research Society, Symposium Proceedings*, Volume 495, Warrendale, PA, 1988.

34. G. Leclercq, M. Kamal, J. M. Giraudon, P. Devassine, L. Feigenbaum, L. Leclercq, A. Frennet, J. M. Bastin, A. Lofberg, S. Decker, M. Dufour, Study of the preparation of bulk powder tungsten carbides by temperature programmed reaction with CH_4/H_2 mixtures, *Journal of Catalysis*, 158, 142–169, 1996.

35. M. Parlier, M.-H. Ritti, A. Jankowiak, Potential and perspectives for oxide/oxide composites, *Journal Aerospace Lab*, 3, AL03-09, 1–12, 2011.

36. W. Krenkel, Carbon fiber reinforced CMC for high-performance structures, *International Journal of Applied Ceramic Technology*, 1(2), 188–200, 2004.

37. C. L. Boulanger, A. P. Majidi and T.-W. Chou, Fracture mechanisms in sic-whisker reinforced alumina, in *High Temperature/High Performance Composites*, Cambridge University Press, April 5–7, 1988, Reno, NV, edited by F. D. Lemkey, S.G Fishman, A. G. Evans and J. R. Strife.

38. A. P. Majidi, T.-W. Chou, Elevated Temperature Studies of Continuous and Discontinuous Fiber Reinforced Ceramic Matrix Composites, The American Society of Mechanical Engineers, Presented at the Gas Turbine and Aeroengine Congress and Exposition—June 4–8, 1989—Toronto, Canada, 89-GT-124, 1989.

39. L. Kehzi, L. Hejun, J. Kaiyu, H. Xianghui, Numerical simulation of isothermal chemical vapor infiltration process in fabrication of

carbon-carbon composites by finite element method, *Science in China, Series E*, 43(1), 77–85, 2000.

40. O. Danielsson, A. Henry, E. Janzen, Growth rate predictions of chemical vapor deposited silicon carbide epitaxial layers, *Journal of Crystal Growth*, 243, 170–184, 2002.

41. X. Wei, L. F. Cheng, L. T. Zhang, Y. D. Xu, A two-dimensional model for densification behaviour of C/SiC composites in isothermal chemical vapour infiltration, *Modelling and Simulation in Materials Science and Engineering*, 14, 891–904, 2006, doi:10.1088/0965-0393/14/6/001.

42. G. L. Vignoles, C. Germain, O. Coindreau, C. Mulat, W. Rosa, Fibre-scale modeling of C/C processing by chemical vapour infiltration using Xray CMT images and random walkers, *ECS Transactions*, 25(8), 1275–1284, 2009, doi:10.1149/1.3207733.

43. C. P. Deck, H. E. Khalifa, B. Sammuli, C. A. Back, Modeling forced flow chemical vapour infiltration fabrication of SiC/SiC composites for advanced nuclear reactions, *Science and Technology of Nuclear Installations*, ID 127676, 10 p., 2013, doi:10.1155/2013/127676.

44. Z. P. Tang, W. Xu, A. J. Li, Z. W. Zhang, R. C. Bai, J. S, Wang, Modeling of carbon deposition from propane in chemical vapor infiltration, *New Carbon Materials*, 31(1), 78–86, 2016.

45. S. Dong, Z. Wang, H. Zhou, Y. M. Kan, X. Zhang, Y. Ding, L. Gao, B. Wu, J. Hu, Research progress in SiC-Based Ceramic Matrix Composites, *Journal of the Korean Ceramic Society*, 49, 4, 295–300, 2012.

46. Y.-Z. Zhu, Z. R. Huang, S. M. Dong, M. Yuan, D. L. Jiang, Correlation of PyC/SiC interphase to the mechanical properties of 3D HTA C/SiC composites fabricated by polymer infiltration and pyrolysis, *New Carbon Materials*, 22(4), 327–331, 2007.

47. M. Rosso, Ceramic and metal matrix composites: Routes and properties, *Journal of Materials Processing Technology*, 175, 364–375, 2006.

48. M. Belmonte, Advanced ceramic materials for high temperature applications, *Advanced Engineering Materials*, 8, 692–703, 2006.

49. A. Sommers, Q. Wang, X. Han, C. T'Joen, Y. Park, A. Jacobi, Ceramics and ceramic matrix composites for heat exchangers in advanced thermal systems: A review, *Applied Thermal Engineering*, 30, 1277–1291, 2010.

50. K. L. Luthra, Emerging Applications and Challenges in using Ceramics at General Electric, Ceramic Leadership Summit 2011 August 2, 2011, http://ceramics.org/wp-content/uploads/2011/08/gs1-emerging-apps-ge-luthra.pdf.

51. J. A. DiCarlo, Advances in SiC/SiC Composites for Aero-Propulsion, NASA/TM—2013-217889, 1–22.

52. W. Krenkel, *Ceramic Matrix Composites: Fiber Reinforced Ceramics and Their Applications*, John Wiley & Sons, Weinheim, Germany, 2008, 93–95.

53. X. He, X. Cheng, Q. Wang, P. Wang, Ballistic performance and damage characteristics of chemical vapor infiltration quasi 3D-Cf/SiC composites, *Journal of Wuhan University of Technology-Mater. Sci. Ed.*, 31(1), 118–122, 2016. doi:10.1007%2Fs11595-016-1340-9.

54. I. Luz, C. Rösler, K. Epp, F. X. Llabrés i Xamena, R. A. Fischer, Pd@UiO-66-Type MOFs prepared by chemical vapor infiltration as shape-selective hydrogenation catalysts, *European Journal Inorganic Chemistry*, 23, 3904–3912, 2015.

55. S. Bertrand, F. Lamouroux, S. Goujard, C. Descamps, Chemical Vapour Infiltration Apparatus Having A High Loading Capacity, WO2014013168, 2014.

56. G. Chollon, S. Delettrez, F. Langlais, Chemical vapour infiltration and mechanical properties of carbon open-cell foams, *CARBON*, 66, 18–30, 2014.

57. J. Riesch, T. Höschen, C. Linsmeier, S. Wurster, J.-H. You, Enhanced toughness and stable crack propagation in a novel tungsten fibre-reinforced tungsten composite produced by chemical vapour infiltration, *Physica Scripta*, T159, 2014, 014031 (7pp) doi:10.1088/0031-8949/2014/T159/014031.

58. O. Ledain, S. Jacques, L. Maillé, Consolidation and conversion of carbon powders into TiC by reactive chemical vapour infiltration, *Journal of the European Ceramic Society*, 36, 481–488, 2016.

59. J. Binner, B. Vaidhyanathan, D. Jaglin, Microwave heated chemical vapour infiltration of SiC powder impregnated SiC fibre preforms, *Advances in Applied Ceramics*, 112, 4, 235–241, 2013, doi:10.1179/17436761 12Y.000000007159.

60. B. Riccardi, L. Giancarli, A. Hasegawa, Y. Katoh, A. Kohyamae, R. H. Jones, L. L. Snead, Issues and advances in SiC_f/SiC composites development for fusion reactors, *Journal of Nuclear Materials*, 329–333, 2004, 56–65.

61. "High-frequency ELectro-Magnetic technologies for advanced processing of ceramic matrix composites and graphite expansion" (HELM), FP7-NMP.2011.4.0-1, GA n°. 280464, website: http://www.helm-project.eu/.

Index